INELASTICITY OF MATERIALS
An Engineering Approach and a Practical Guide

Series on Advances in Mathematics for Applied Sciences – Vol. 80

INELASTICITY OF MATERIALS
An Engineering Approach and a Practical Guide

Arun R Srinivasa
Texas A&M University, USA

Sivakumar M Srinivasan
Indian Institute of Technology, Madras, India

NEW JERSEY • LONDON • SINGAPORE • BEIJING • SHANGHAI • HONG KONG • TAIPEI • CHENNAI

Published by

World Scientific Publishing Co. Pte. Ltd.
5 Toh Tuck Link, Singapore 596224
USA office: 27 Warren Street, Suite 401-402, Hackensack, NJ 07601
UK office: 57 Shelton Street, Covent Garden, London WC2H 9HE

Library of Congress Cataloging-in-Publication Data
Srinivasa, Arun R. (Arun Ramaswamy)
  Inelasticity of materials : an engineering approach and a practical guide / by Arun R Srinivasa & Srinivasan M. Sivakumar.
    p. cm. -- (Series on advances in mathematics for applied sciences ; v. 80)
  Includes bibliographical references and index.
  ISBN-13: 978-981-283-749-3 (hardcover : alk. paper)
  ISBN-10: 981-283-749-3 (hardcover : alk. paper)
  1. Hardness--Mathematical models. 2. Brittleness--Mathematical models. 3. Plasticity--Mathematical models. 4. Materials--Testing--Computer simulation. 5. Strains and stresses.
I. Sivakumar, Srinivasan M. II. Title.
  TA418.42.S75 2009
  620.1'126--dc22
                                              2009008937

British Library Cataloguing-in-Publication Data
A catalogue record for this book is available from the British Library.

Copyright © 2009 by World Scientific Publishing Co. Pte. Ltd.

All rights reserved. This book, or parts thereof, may not be reproduced in any form or by any means, electronic or mechanical, including photocopying, recording or any information storage and retrieval system now known or to be invented, without written permission from the Publisher.

For photocopying of material in this volume, please pay a copying fee through the Copyright Clearance Center, Inc., 222 Rosewood Drive, Danvers, MA 01923, USA. In this case permission to photocopy is not required from the publisher.

Printed in Singapore.

Sri Ramajayam.

This book is dedicated:

to my father Prathivadibhayankaram Srinivasa Ramaswamy Iyengar, my mother Saroja, my wife Prabha, my children Vishnu and Divya, and to my Ph.D. advisor Prof. P.M. Naghdi .

–Arun Srinivasa

to my dad, Panchanathan Srinivasan as he turns 80 and, my mom, Pattammal to my family, Leela, Satchit and Aniketh, and to my friend Dr. Dilip Veeraraghavan.

–Sivakumar Srinivasan

We also dedicate this book to our friend and mentor Prof. K.R. Rajagopal.

Foreword

It is a pleasure, as well as a responsibility, to write a foreword[1] to a book co-authored by two colleagues, when one has worked very closely with one of them for over a decade and interacted with the other for over two decades, lest one loses objectivity while penning it. As good luck would have it, the authors have made my task all the easier by writing a book that approaches the subject matter in an inimitable manner. The book is written as though one would deliver the material in class; the informality with which the subject matter is introduced is a welcome change to the sterile and formal pedagogy that one finds in most texts. Also, the emphasis of the book is towards educating the student by educing that which is inherent rather than that of teaching. The idea of presenting the material in a question and answer format makes the material very readable and easy to understand, the motivation behind many of the concepts and procedures are then well motivated and rendered transparent. Ernest Rutherford is supposed to have said that if one truly understands the material, one should be able to explain it to a barmaid. While the authors are not the sorts that hold conversations with bar maids, they seem to have taken Rutherford's comments to heart. The book is accessible to anyone that has taken an undergraduate course in mechanics of materials.

A very interesting feature of the book is its intended audience. One could cull material to offer a regular senior level undergraduate course, honors level undergraduate course, a graduate course in plasticity theory, and it also offers insights into the foundations of the field that will be both new and appealing to the experts in the field. The healthy balance that has been achieved between the theoretical and numerical and computational material, between worked out examples and home work problems, is commendable. The student is led through the complex maze of plasticity is small steps so that he can gain a proper understanding of the underpinnings of the subject. Whether it is a lumped parameter approach to one dimensional problems in inelasticity, small stain inelasticity or inelasticity associated with finite deformation, the authors are able to provide a clear and rationale thermodynamic basis. The authors appeal to the maximization of the rate of entropy production,

---

[1] According to Fowler [Fowler (1926)], the word foreword is to be preferred to preface as the former is of Anglo-Saxon origin while the latter is of Latin origin.

an approach that has proved very fruitful in describing the response of a plethora of bodies that produce entropy in disparate manner. The book is peppered with exercises with hints as to how to solve them. The exercises have been carefully selected so that the student will obtain useful, interesting and illuminating insights into subtle and interesting issues concerning inelasticity.

The initial material concerning one-dimensional response appealing to springs and dashpots provides the basis for the more abstract mathematical setting for the three dimensional response of inelastic bodies within the context of a sound thermo-mechanical framework. The initial lumped parameter model of a spring accounts for how the material stores energy and the dashpot how dissipates energy. The lumped parameter modeling places the main ideas concerning inelasticity within the grasp of a senior undergraduate student or a beginning graduate student. This early treatment of inelasticity is followed by a detailed and careful discussion of small strain plasticity. The reader is provided a reasonably thorough account of the historical development of small deformation plasticity. The authors start with a discussion of rigid-perfectly plastic bodies and then provide a reasonably exhaustive treatment of the behavior of elastic-plastic bodies. Plane strain and plane stress problems are discussed in great detail and the student is made aware of how the classical Airy's stress function can be put to use gainfully to study plane stress and plane strain problems for inelastic bodies. Numerical solutions are discussed which can be used to solve problems in more complex geometries. The final three chapters in the book are devoted to advanced topics in plasticity; a general introduction to inelasticity involving large deformations, a succinct treatment of crystal plasticity and three case studies devoted to shot peening, equal channel angle extrusion, and aging tissues. The final chapter devoted to case studies takes the reader to the very boundaries of the subject, showing them how to tackle complicated problems that have important technological ramifications. A person that has mastered the contents of the book will have no difficulty in tackling important open problem in inelasticity.

I found the book to be most instructive and unhesitatingly recommend it to my colleagues to use it as a text book for a senior undergraduate level or beginning graduate level course by restricting the course to the first two parts of the book, or to an advanced graduate course by mainly covering the final part of the book, or as a source of reference to one carrying out research in the field of inelasticity.

-K. R. Rajagopal
Distinguished Professor, Regents Professor,
Forsyth Chair in Mechanical Engineering,
Professor of Mathematics,
Professor of Biomedical Engineering,
Professor of Civil Engineering,
Professor of Chemical Engineering,
Texas A&M University
College Station, Texas, 77843, USA.

About the book...

The two of us have known each other since our days as undergraduate students at IIT, Madras and staying in the same dorm or Hostel—Godavari. We have collaborated with each other over many years and have shared a passion for teaching. This book is an outgrowth of teaching courses on plasticity at Texas A&M University, at the University of Pittsburgh and at the Indian Institute of Technology, Madras with both of us being quite dissatisfied with what we have taught. Both of us found plasticity to be fascinating in so many ways and have used it as a doorway to other theories, but we found that students were curiously unmoved by the subject. Not surprisingly we blamed it on the students and the material—how could anything be wrong with the way we teach.

But we did notice a curious thing—students were always fascinated by in class demonstrations and the application of plasticity to engineering tasks. Then, one summer, we decided to do a class lecture on thermodynamics and its applications—the students' interest was intense and they peppered us with questions. They had taken thermodynamics as undergraduates but had never really understood its place in the world—but they had had enough glimpses of its wide reach to really pique their curiosity; and here was a speaker talking to them about the thermodynamics of hurricanes and thunderstorms!

In talking things over, it dawned on us that our presentation style was at fault—we spent way too much time lecturing and not enough time challenging students and helping them discover the beauty of the subject. The difference is like that between a museum lecture on Egyptian artifacts where each item is brought before us and its relevance explained and actually visiting Egypt and seeing the artifacts in their original setting.

This book is a result of our intense discussions on how to present this matter to the students. After many false starts, we have finally managed to arrive at a consensus on how this book was to be written. We have written it in an informal style (like a conversation) with jokes thrown in[2]. We also agreed that it has to be a textbook and not a monograph and that it had to be engaging. We do not want it

---

[2] so don't be shocked by an occasional smiley face ":-))" in the text

to meet the fate of the book which was "so interesting that once you put it down, it was hard to pick up again"—a fate that is all too common to monographs.

Thus it is written more like a workbook. The text is frequently interrupted by exercises, and homework, just like what you would do in a class. We frequently make use of web research to help students discover some of the remarkable applications of plasticity. YouTube^{TM} is a great resource to help students see videos and animations and have served as an invaluable complement to the material in the book. Most of the chapters start with an engineering scenario where you would have to make a decision, and you have to use your knowledge of inelasticity to help you make the decision.

One of the major new features in the book is the extensive use of programming exercises to help students model realistic material behavior. We make extensive use of MATLAB'S ODE suite to solve many homogeneous deformation problems as well as other numerical techniques. In our opinion, this has been one of the features that has to be sorely lacking in many plasticity texts (since analytical solutions are all but impossible other than for a few special and simple cases).

Thus, the book has homework exercises and lots of programming exercises. In our classes, we have found that these programming exercises help students anchor the concepts firmly in their mind since it provides us a way to give instant feedback to the students.

We have also provided a number of case studies that can be used by students and instructors to see actual applications of the theory in real situations. Also, based on student feedback we have taken pains to ensure that there is not a heavy dependence of chapters on each other. The notable exceptions are Chapter 2 and Chapter 6—they form the foundations of the book.

Organization of the book

This book is written in a modular fashion, which provides adequate flexibility for adaptation in classes that cater to different audiences such as senior-level students, graduate students, research scholars, and practicing engineers. A graphical look at the set of chapters and their organization is given in Fig. 0.1.

Acquiring the capacity to model inelastic behavior and to choose the right model in a commercial analysis software has become a pressing need for practicing engineers with the advent of a host of new materials ranging from shape memory alloys to bio-materials to multiphase alloys. Even with the traditional materials, there is a continued emphasis on optimizing and extending their full range of capability in the applications. This book builds upon the existing knowledge of elasticity and thermodynamics, and allows the reader to gain confidence in extending ones skills in understanding and analyzing problems in inelasticity. By reading this book and working through the assigned exercises, you, the reader will gain a level of comfort and competence in developing and using inelasticity models.

This book is written in three parts:

Part I, the first part, is primarily focused on lumped parameter models and simple structural elements such as trusses and beams. In *Chapter 1*, the subject is introduced through discussing the appropriate sample scenarios in which inelastic modeling and analysis plays an important role. Further, in this chapter, formulation of the elastoplastic equations, together with the different spring and dashpot elements are introduced from a purely mechanistic approach directly exploiting an analogy. These analogies are meant to replicate, in a purely mechanical manner, the stress-strain response of many types of inelastic materials.

Chapter 2 on lumped parameter systems forms the most important chapter of this part which sets the tone of the book. The lumped parameter models considered as "zero" dimensional models use primarily springs, dashpots and masses, and the appropriate set of variables to define the state of the material including temperature to derive response equations. Both mechanistic and energetic approaches are used to demonstrate how governing equations for the inelastic behavior can be obtained. The maximum rate of dissipation hypothesis is introduced here.

The reader is seamlessly taken to a scenario of solving one-dimensional problems in the next chapter (*Chapter 3*) where a one-dimensional structure is solved for quantities appropriate to inelasticity. This chapter discusses how constitutive equations can be formulated for materials with hardening and how one can use the existing knowledge of the "accumulated" plastic strain calculated from the model for predicting the fatigue life of a one-dimensional component. Examples have been introduced along with MATLAB codes to provide the reader with a chance to try out examples.

More often than not, the typical structural element that an engineer encounters or considers in bearing mechanical loads are the beam elements. *Chapter 4* is dedicated to this structural action (beam action) for this reason. The reader is taken through a complete exercise of either modeling an existing beam or designing a beam for dimensions for inelastic deformations. An algorithmic approach to solving these one dimensional boundary value problems is given along with the appropriate MATLAB codes.

Two more solved real life examples have been introduced in *Chapter 5*. One is that of a crane girder made of a truss system. The other is that of a seismic damper frame that uses shape memory alloy cables for damping. The idealization of the systems, the analysis and modeling exercises are carried out and solved for appropriate quantities needed for design and understanding.

These essentially form the chapters in Part I dealing mainly with lumped systems and simple structural elements.

Part II of this book focuses on small deformation multi-dimensional inelasticity. Sufficient material is included on how to numerically implement an inelastic model and solve either using a simple stress function type of approach or using commercial software. Simple examples are included. There is also an extensive discussion of thermodynamics in the context of small deformations.

The simplest and most commonly used among the plasticity models are the J2 plasticity models. With a view on dealing with this model more thoroughly, *Chapter 6* provides a complete idea of using J2 plasticity model. In this chapter, the reader is taken through the history of the development of plasticity models that have led to the present prominence and popularity of the J2 plasticity models. The chapter explains, to start with, how to arrive at the governing set of equations for a rigid plastic model in this class of J2 plasticity models. Then, the model is relaxed to introduce elasticity into the model. This makes it suitable for structural applications. To make it more realistic hardening should be added to the model. A nice definition of hardening for the three dimensions is introduced subsequently before presenting the process of modeling for the same as extension from a one-dimensional model. The chapter ends with examples of different loading histories after explaining the procedure obtaining the tangent moduli of the material.

Now that the reader is ready for solving boundary value problems involving multi-dimensional inelasticity, *Chapter 7* deals with formulating and solving simple boundary value problems, especially, the two-dimensional problems. It is shown in the chapter how to derive an analogical set of equations of inelasticity that look similar to the biharmonic operators that are popular in isotropic small deformation elasticity. Airy stress functions are introduced, shown how they can be obtained for a set of cases before attempting to solve the equations.

J2 plasticity is not application in many applications and one has to resort to other yield surfaces and conditions. *Chapter 8* explains how the concepts learnt so far can be extended to other yield conditions and behaviors such as pressure dependency, anisotropic moduli and anisotropic yield.

Thus far, there remain many questions unanswered in a curious reader's mind in relation to the thermodynamic principles adopted and their basis. An entire chapter (*Chapter 9*) is devoted to explaining the thermodynamic principles involved in modeling these materials. This chapter is fundamentally different from the others in the book since it is more philosophical in nature and presents a point of view as to how one could unify the constitutive relations for different dissipative materials using a common thermodynamical framework. The approach presented in this chapter is more explanatory than procedural.

In general, in spite of such simplifications, the resulting boundary value problem may be complex for a simple analytical approach to work in solving it. Over simplification may lead to unreliable results. The direct numerical simulation comes to the rescue. Simulation of a plastic deformation process in a realistic engineering application such as components undergoing collapse and fatigue or a metal forming process, gives insight into the mechanics that occurs in the appropriate process and helps make design decisions. In *Chapter 10*, the authors walk you through a process of numerically solving the governing equations of a IBVP for the simulation of a general small deformation application. Simple examples are solved to elucidate the effectiveness of this solution. A meshless modeling exercise is introduced to give wider possibilities for simulation.

Part III: More advanced situations such as finite deformation inelasticity, thermodynamical ideas and crystal plasticity are dealt with in Part III. Some advanced case studies are also included.

In order to model finite elastoplastic deformation, either one has to be very familiar with the notation and terminology. *Chapter 11* provides a quick refresher on the concepts in continuum mechanics apart from an overview of the notations and terminologies used. The chapter is introduced essentially for completeness and is not a substitute of a continuum mechanics course. Specifically, the treatment is from the perspective of understanding the concepts introduced in later chapters on finite deformations focusing only on basic foundations for such discussions.

Chapter 12 discusses on finite deformation inelasticity for isotropic materials. The central idea is to find a way to adapt an existing finite strain elasticity to formulate an inelastic model. A novel idea is introduced in which one need not break his or her head on what objective rates but directly use the finite deformation elastic model and the thermodynamic restrictions in order come up a constitutive law. Several questions and issues that may arise are answered with clarity. The relationships with other models are also discussed to make it clear to the reader the connections that exist between the proposed model and the models existing in the literature.

We have addressed modeling an isotropic material for inelastic finite deformations but, in general, especially in metal forming applications, one may need to deal with anisotropic materials. *Chapter 13* deals with polycrystal plasticity where the orientations of the individual grains and their slip mechanisms play a role in the anisotropy of the material. An anisotropic plasticity model is introduced here borrowing concepts of crystal plasticity such as resolved shear stress on slip systems and of power law dashpots. Polycrystalline aggregates can be approximately simulated as a collection of single crystals (all with the same velocity gradient (the so-called Taylor assumption) or all with the same stress (the so-called Sach's assumption). There have been attempts to improve on these two assumptions for various special conditions. Finally, we end the discussions in *Chapter 14* with simple case studies of shot peening process, equal channel angular extrusion process, and ageing of skin tissues. The use of commercial packages in solving these problems are explained.

How to use this book

"You cannot build muscles by watching an exercise video"— these words[3] are the guiding motto of this book! We strongly believe that you get the most out of this book by actually doing all the exercises and not by simply looking at the answers. Don't take our word for anything in the book. Do it yourself and you will understand it in your own way. We can only give you an explanation, no one but yourself can give you an understanding. So the way the book is written, it is essential for you to get a notebook and interrupt your reading and doing the exercises that appear in the middle of the narrative.

If you are trying to learn plasticity by yourself by following this book, then you have to first assess your capabilities. Are you familiar with elasticity? or only strength of materials? If it is the former then the material up to Chapter 10 does not require additional mechanics knowledge. On the other hand if your knowledge is strength of materials then you can do the first six chapters and then attempt the remaining after taking a class on elasticity. If you want to do the numerical examples, learn MATLAB! you will thank us for it[4].

If you are more familiar with continuum mechanics, then after Chapter 2, skip to Chapter 10 and above.

Suggested course outlines

As mentioned in the previous section, the book is intended for different audiences ranging from the senior-level students, the graduate students, the research scholars, to practicing engineers.

Depending on which part of the book is covered, different pre-requisites may be needed. For example, for the first part of the book (part I) that deals with introductory concepts and structural elements, an elementary knowledge of structural / strength of materials and introductory thermodynamics may be necessary. Part II that deals the small deformation inelasticity of continua, requires that the student has undergone an advanced structural mechanics or an elasticity course apart from introductory thermodynamics. Part III, as is clear, required the student to have undergone a continuum mechanics course and is familiar with earlier parts of the book.

We suggest the following that would allow the instructor to pick the topics according to the focus areas of the students.

---

[3]The quote is to be credited to Dr. Bil Schneider, Designer of the TransHab inflatable space module and friend and colleague of one of the authors.

[4]While we absolutely love MATLAB, we have no affiliation with it–nor do we have stock in MATHWORKS (what a pity!) since it is a privately held company. But trust us when we tell you that for prototyping some numerical methods, MATLAB is hard to beat. Both of us have written programs in FORTRAN, C, C++, LabView, etc., but MATLAB is great for the tasks that we throw at it and its learning curve is really flat (if you are comfortable with Matrices).

Part	Chapters	No. of weeks	Level	Pre-requisites
I	1-5	8	Senior	(a) Strength of materials (b) Intro. to Thermodynamics
II	6-10	10	Grads Res. scholars	(a) Elasticity / Adv. Strength of Mat. (b) Intro. to Thermodynamics
III	11-14	8	Senior Grads Res. Scholars	(a) Elasticity (b) Continuum thermodynamics

For example, if it is a semester course (14 weeks) for just the senior level and first year graduate students, one can omit Part III of the book and only deal with small deformations (Parts I and II). Some portions of the book need not be covered. For example, one can omit Chapters 8 and 9, without affecting the flow of the book i.e. the 14 weeks' portion could be Chapters 1-7 and 10.

For a two-semester course on inelasticity, or if it is an advanced one-semester course for students with a continuum background, we suggest after a refresher on parts of Chapter 2 (sections 2.5-2.8) and Chapter 6 (J2 plasticity), Chapters 8 and 9 could be introduced in the first part of the course. The second part of the course could be devoted to the finite deformations part of the book (Part III).

Our addresses and websites are given below: Our websites will connect you to the book website which has MATLAB codes, solutions to exercises and additional material that could be used for teaching and learning purposes.

Arun R. Srinivasa,
Department of Mechanical Engineering,
Texas A & M University,
College Station, TX 77843, USA.
http://www2.mengr.tamu.edu/FacultyProfiles/facultyinformation.asp?LastName=asrinivasa

and

Sivakumar M. Srinivasan,
Department of Applied Mechanics,
Indian Institute of Technology Madras
Chennai 600036, INDIA.
http://apm.iitm.ac.in/smlab/mss

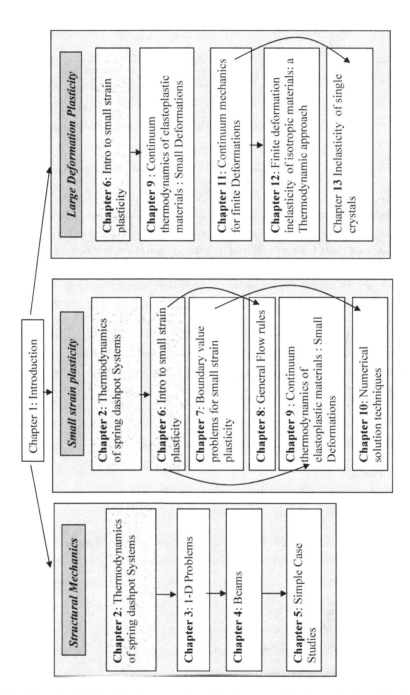

Fig. 0.1 A flow chart illustrating the grouping of the chapters. The dependencies of the chapters are shown by the arrows.

Acknowledgments

I thank my parents who have sacrificed so much, for my sake, and my loving wife who enabled me to take a four-month sabbatical to write this book while she looked after our children all by herself. I also thank my children Vishnu and Divya who showed me how to be joyful and restored my sense of perspective.

I would like to thank my advisor Prof. P. M. Naghdi, who believed in me and introduced me to the fascinating world of thermodynamics and inelasticity in his own inimitable way and my mentor Prof. K. R. Rajagopal who vastly broadened my vision well beyond mechanics. I cherish my interactions with them, and their words and actions continue to guide me.

I would like to acknowledge the insightful conversations that I had with my good friend Dr. Jeff Froyd from the center for teaching excellence at Texas A&M University who educated me on so many aspects of effective teaching. They were invaluable to me in figuring out how to present the material. I would also like to acknowledge Prof. M. S. Ananth, Director, IIT Madras, for allowing me to stay in the sylvan surroundings of IIT while we wrote this book. The environs of IIT inspire me like no other place I know. My undergraduate days there and my sabbatical stay there are definitely some of the best days of my life.

–Arun R. Srinivasa.

First of all, I should thank Arun for inviting me to coauthor this book. It was a great learning experience all the way and I relished every moment of our conversations which many times went beyond just the technical matter.

I started learning the ABC of plasticity when I was with Prof.G.Z.Voyiadjis, my Ph.D. guide. I thank him for providing that crucial start of this continuing journey.

I wish to thank Prof. K. Ramamurthy, chairman, CCE, my department chair, Prof. K. Ramesh and our Director Prof. M.S. Ananth for facilitating this through the Golden Jubilee Book Writing Scheme. Without this, it would have been a painful struggle to complete this task. They were also proactive in effecting Arun's visit to IITM that saw a step on the gas in this effort.

I remember the ready support my dad gave when I wanted to pursue higher studies in a situation that demanded my going for a job. It was my brother, Raman who provided the foundation stone to the technical me today that participated in writing this book. Watching my wife, Leela do the juggling between the work place and home deftly was more than just a performance to watch. The time I had with my children during bedtime gave me the much needed relaxation and joy. I thank my family for all this.

Special thanks to Baskaran Bhuvaraghan and Arun Chandran who helped in solving some of the problems and case studies inserted in this book and Mr.Sreekumar for help with svnhost.

The constant poking from my B.Tech classmates Aditya Gurajada and Shekar Ramakrishnan kept my desire burning, culminating in the completion of the book.

Thanks to them. I would like to thank my friend and mentor Prof. K. R. Rajagopal who taught me so many things well beyond just mechanics. The guidance and motivation from him kept me going.

–Sivakumar M. Srinivasan.

We would also like to acknowledge our conversations with Dr. Chandrasekaran, who was a consultant at IIT Madras. His enthusiasm for work, humility and deep knowledge served as an example to all of us who were in the Active materials lab at IIT Madras.

We acknowledge the tireless work of Pritha Gosh, Shriram Srinivasan, Satish Karra, Shreyas Balachandran, Guruprasad, Srikrishna Doraiswamy, H.S. Srivatsan, A. Jayavel, K. Jayabal, K.Srivatsan, S.Balaji, Arun Chandran, R.S. Priyadarsini, and R. Suresh Kumar—the students who proofread the material, checked the equations, and pointed out places where our explanations were murky and needed clarification. Without their eagerness to learn and their help, this book would not have been possible. Thanks to Dr. Mangala Sunder and Mr. Britto Jones for help with diagrams. The help and support from France Prakash and Dr. A. Arockiarajan deserve a special mention.

Lastly, we might have inadvertently missed out, in the above, some of those who have directly or indirectly helped us with this effort. Our sincere thanks to them all.

–Arun R. Srinivasa
& Sivakumar M. Srinivasan

Contents

Foreword vii

About the book... ix

Acknowledgments xvii

INTRODUCTION TO INELASTICITY
Simple Structural Elements   1

1. Introduction to Inelasticity   3
   - 1.1   What is inelasticity?   5
   - 1.2   Why should one study inelasticity?   5
   - 1.3   A sample scenario   6
   - 1.4   Observation, experimentation and modeling of material response   8
   - 1.5   Experimental observations and model development   9
   - 1.6   Plastic deformations and the friction block analogy   13
   - 1.7   Creep and time dependent behavior   19
   - 1.8   Summary and Conclusions   26

2. Thermodynamics of Inelastic Materials: A lumped parameter approach   27
   - 2.1   Notion of a lumped parameter model   31
     - 2.1.1   The role of lumped parameter models   31
     - 2.1.2   Elements of lumped parameter systems: Extending the results of Chapter 1   32
     - 2.1.3   Equations of motion   33
   - 2.2   General considerations of springs and dashpots   36
     - 2.2.1   Linear and nonlinear springs   36
     - 2.2.2   Viscous dashpots   39
     - 2.2.3   Frictional dashpots   40
     - 2.2.4   Springs and dashpots in 2 dimensions   43

	2.3	Energy formulation of the equations of motion	47
	2.4	Mechanical power theorem	48
		2.4.1 Frequently asked questions and clarifications	52
	2.5	Thermal considerations: The Helmholtz and Gibbs potentials . . .	65
		2.5.1 Extension of the power theorem	69
		2.5.2 Finding the Helmholtz and Gibbs potentials for a system .	76
		2.5.3 Determination of the equations of motion and the temperature of the system	78
	2.6	Solving the thermo-mechanical evolution equations for an inertialess system	85
	2.7	Lumped parameter models: Gibbs potential approach	86
	2.8	Projects and exercises	88
3.	Inelastic Response of Truss Elements		89
	3.1	An example problem	91
	3.2	Inelastic bars under axial loading	93
		3.2.1 Task 1: Make a preliminary list of the variables of interest	93
		3.2.2 Task 2: Make simplifying assumptions about the response	93
		3.2.3 Task 3: Modeling of the response of the truss	95
	3.3	Prescribed displacements	110
	3.4	The Bauschinger effect	112
		3.4.1 Modeling the Bauschinger effect	114
	3.5	An example cyclic loading problem	116
		3.5.1 Stress cycling	116
		3.5.2 Strain cycling	117
	3.6	General loading conditions	119
	3.7	Summary	122
4.	Elastoplastic Beams: An Introduction to a Boundary Value Problem		125
	4.1	An example task	127
	4.2	Modeling of a thermoelastoplastic beam	127
	4.3	The elastoplastic beam	130
		4.3.1 Task 1: Make a preliminary list of variables of interest . .	130
		4.3.2 Task 2: Make simplifying assumptions on the response . .	131
		4.3.3 Task 3: Modeling the response of the beam	132
		4.3.4 Task 4: Formulate the solution strategy	139
		4.3.5 Problem 1: No axial confinement	140
		4.3.6 Problem 2: Beam with axial confinement	144
		4.3.7 Statically indeterminate problems / general beam problems	145
	4.4	Summary	149
	4.5	Projects	149

5. Simple Problems ... 151
　5.1　Case Study I: Rehabilitation of a crane girder ... 153
　5.2　Case Study II: Passive Damping of a frame structure using super-elastic shape memory alloys (SMA) bracings ... 158
　　5.2.1　Modeling the material response ... 160
　　5.2.2　Modeling the SMA braced frame ... 164
　　5.2.3　Dynamic analysis of the SMA braced system ... 170
　　5.2.4　Simulation studies ... 171

INELASTICITY OF CONTINUA
Small Deformations　　　　　　　　　　　　　　　　　　　177

6. Introduction to Small Deformation Plasticity ... 179
　6.1　An example task ... 181
　6.2　The J2 rigid-plasticity model ... 182
　　6.2.1　3-D kinematics ... 182
　　6.2.2　3-D equations of motion ... 183
　　6.2.3　Constitutive laws: The J2 rigid plasticity equations ... 183
　　6.2.4　Standard form for rigid plastic (Kuhn-Tucker form) ... 187
　　6.2.5　Example problems ... 188
　6.3　The J2 elasto-plasticity model ... 190
　　6.3.1　The value of ϕ and the "tangent modulus" ... 194
　　6.3.2　Non-dimensionalization ... 200
　6.4　Hardening and the plastic arc length ... 201
　　6.4.1　Definition of hardening, softening and perfectly plastic behavior ... 201
　6.5　Finding the response of the material ... 206
　　6.5.1　Finding the tangent modulus ... 207
　6.6　Summary ... 209
　6.7　Projects ... 211
　6.8　Homework ... 211

7. The Boundary Value Problem for J2 Elastoplasticity ... 213
　7.1　The governing equations ... 215
　　7.1.1　3-D case ... 215
　　7.1.2　Compatibility equations ... 217
　7.2　Plane problems ... 218
　　7.2.1　Plane strain ... 218
　　7.2.2　Plane stress ... 221
　7.3　The stress function and the equations of compatibility ... 226
　7.4　Boundary conditions for the stress function ... 227

7.5	Numerical solution	231
	7.5.1 Modifying the MATLAB PDE toolbox to solve elastoplasticity problems	232
7.6	Concluding remarks and summary	233
7.7	Projects	233

8. Examples of Other Yield Surfaces: Associative and Non-associative Plasticity — 237

8.1	The general, small strain elastoplastic model	239
	8.1.1 General characteristics of yield surfaces	239
8.2	Examples of general yield surfaces	242
	8.2.1 Pressure dependent yielding and the strength-differential or S-D effect	242
	8.2.2 Three-invariant or isotropic yield functions	242
	8.2.3 Anisotropic yield functions	247
8.3	Examples of plastic potentials and non-associative flow rules	249
8.4	Changes in the size, location and shape of yield surfaces and methods to quantify them	252
	8.4.1 Isotropic hardening	256
	8.4.2 The Bauschinger effect, the back stress $\boldsymbol{\alpha}$ and certain predictable characteristics of cyclic loading	257
	8.4.3 Quadratic yield surface plasticity with kinematic hardening	258
	8.4.4 Homogeneous motions and governing differential equations	261
8.5	Summary	262
8.6	Projects	263
8.7	Exercises	264

9. Thermodynamics of Elasto-plastic Materials: The Central Role of Dissipation — 265

9.1	Generalization of J2 plasticity: the rationale	267
9.2	Foundations of the thermodynamics of continua	270
	9.2.1 Governing balance laws for small deformation	270
	9.2.2 The macroscopic state variables	272
9.3	The entropy and the equation of state	276
	9.3.1 Equations of state	277
	9.3.2 Heating and working	279
9.4	Equivalent forms of the equation of state: Legendre transformations	281
	9.4.1 The Helmholtz potential	282
	9.4.2 Further transformations: The Gibbs potential	285
9.5	The "heat" or entropy equation and dissipative processes	286
	9.5.1 Results in terms of the Helmholtz potential	288

		9.5.2	Generalization of the state variables and obtaining small-strain, continuum versions of spring-dashpot models	290
	9.6	Constitutive Laws for $\dot{\varepsilon}_p$ and the satisfaction of the second law		294
		9.6.1	Viscoelastic and viscoplastic models	295
		9.6.2	Rate independent models without yield criteria	296
	9.7	The maximum rate of dissipation criterion		297
		9.7.1	Graphical understanding of the MRDH	304
	9.8	Rate-independent plasticity: How to get the yield function and the flow rule by using MRDH		305
		9.8.1	The yield function	306
	9.9	The Bauschinger effect and history dependence		309
		9.9.1	A simple model for the Bauschinger effect	312
	9.10	Summary		316
	9.11	Homework		318

10. Numerical Solutions of Boundary Value Problems   319

10.1	Background		321
10.2	The modeling exercise		322
10.3	The mesh-free method		326
	10.3.1	Moving least squares approximation	327
	10.3.2	Weak form of the balance law	334
	10.3.3	Generating and solving the nonlinear algebraic equations .	337
10.4	Integration of the plastic flow equations		339
	10.4.1	Non-dimensionalization and its importance	339
	10.4.2	The convex cutting plane algorithm	342
	10.4.3	Homogeneous deformation examples	345
10.5	Numerical examples of boundary value problems		346
	10.5.1	Rod with varying cross-section	346
	10.5.2	Plate with a hole subjected to tension	347

INELASTICITY OF CONTINUA
Finite Deformations   353

11. Summary of Continuum Thermodynamics   355

11.1	Overview of kinematics for finite deformation		357
	11.1.1	Temporal and spatial gradients of the motion	359
	11.1.2	Local motion: Deformation of line, area and volume elements	360
	11.1.3	Sequential versus simultaneous action	364
	11.1.4	Stretch and rotation	366
11.2	Strain measures		373
11.3	Dynamics and thermodynamics of motion		377

11.3 Dynamics and thermodynamics of motion 377
    11.3.1 Thermodynamics or how to avoid creating perpetual motion machines 380
    11.3.2 Constitutive equations for thermoelasticity 383
    11.3.3 Isotropic elastic materials: Large strain 386
11.4 Invariant and objective tensors 389
    11.4.1 When is a tensor objective? 393
    11.4.2 Other rotating frames 396
    11.4.3 Why do we not need to worry about objective rates? . . . 396

12. Finite Deformation Plasticity   399

12.1 Introduction to finite deformation inelasticity 401
    12.1.1 The objective of a finite inelasticity constitutive model . . 403
12.2 Classification of different macroscopic inelasticity models 405
12.3 Recapitulation of small deformation inelasticity results 407
12.4 How to develop a minimalist model for isotropic inelastic materials subject to finite deformation 408
    12.4.1 The modeling procedure 410
12.5 Dissipative behavior of inelastic materials 417
    12.5.1 Isotropic hardening in finite plasticity 430
12.6 Loading/unloading criteria and the value of ϕ; The advantages of strain space yield functions 432
    12.6.1 Results in terms of eigenvalues 437
12.7 Numerical implementation of the plastic flow equations using the Convex Cutting Plane Algorithm 439
12.8 Summary 442
12.9 Moving natural states, aka multiplicative decompositions 442
12.10 Geometrical significance of $\mathbf{D}_p$ 448
12.11 The model can be extended to viscoplasticity and other dissipative responses 449
12.12 Homework projects and exercises 453

13. Inelasticity of Single Crystals   455

13.1 Why is the study of the plasticity of single crystals important? . . 457
13.2 Crystals and lattice vectors 459
    13.2.1 A word regarding our approach 459
    13.2.2 Introduction to crystal lattices 460
13.3 Lattice deformation and crystallographic slip 462
13.4 Deformation and slip of single crystals 466
13.5 The Helmholtz potential and equation of state for a crystal 470
    13.5.1 Thermomechanical equation of state 472
13.6 The dissipation function and constitutive equations for slip 472

		13.6.1	Independence of slip systems and other equations for the rate of slip 475

13.6.1 Independence of slip systems and other equations for the rate of slip 475
13.6.2 Explicit expressions for the resolved shear stress in terms of the lattice vectors 476
13.6.3 The evolution equation for $\mathbf{a}_i$ 477
13.7 Hardening 483
13.7.1 Phenomenological model of hardening: Single slip 485
13.7.2 Hardening with multiple slip 486
13.8 Concluding remarks 492
13.9 Exercises and projects 493

14. Advanced Case Studies 495

14.1 Case Study I: Shot Peening - A Process for Creating Wear Resistant Surfaces 497
14.1.1 Background and problem statement 497
14.1.2 Simplifications and assumptions on modeling the system . 498
14.1.3 Sample results of the analysis 500
14.2 Case Study II: Equal Channel Angular Extrusion - A Materials Processing Route 501
14.2.1 Background and problem statement 501
14.2.2 Need analysis and specification of the system 503
14.2.3 Embodiment of the model 505
14.2.4 Concluding remarks 508
14.3 Case Study III: Modeling of an Aging Face - An Application in Biomechanics 509
14.3.1 Background and problem statement 509
14.3.2 Need analysis and specification of the system 510
14.3.3 Embodiment of the model 511
14.3.4 Concluding remarks 513

Bibliography 515

Index 531

PART 1
INTRODUCTION TO INELASTICITY
Simple Structural Elements

Chapter 1

Introduction to Inelasticity

Learning Objectives

By learning this material, you should be able to:

(1) identify the situations and conditions in a given application that lead to inelastic response and the related engineering decision or judgment to be made;

(2) make simplifying assumptions in order to extract information from the mechanical behavior of the material, useful for making the appropriate engineering decisions;

(3) identify some of the characteristic tests to be performed on the material;

(4) extract useful information out of the tests and relate to the application in question;

(5) model the behavior by combining basic concepts of characteristic mechanical response of materials and by using analogies.

This is a book about the modeling of inelastic behavior. A model is NOT like a detailed photograph of an event —it is meant to be a cartoon or a caricature, exaggerating some features and ignoring others altogether. Models do NOT capture reality, only a grotesque distortion of it. We use these models to gain qualitative insight and to make decisions—it is foolish to expect perfect fidelity with every aspect of data. One can get that only at the expense of loss of predictability and insight. A good modeler (like a good cartoonist) has an instinctive eye for which features of reality to exaggerate and which to throw away. We hope that this book (especially the case studies) will help you gain an understanding of this process.

> **CHAPTER SYNOPSIS**
>
> While, in general, mechanical structures are built to take the appropriate design loads, it is desirable to make sure that these structures don't go through excessive irreversible deformations. How does one go about determining the need for studying inelasticity in such applications? How can we answer simple design questions by creating simple models and analyzing for stresses and strains? In this chapter, we walk you through how one would go about deciding on the tests to be conducted, how observations made are turned into working hypotheses that lead to development of systematic experiments which also eventually lead to prescribing characteristic testing protocols. The analysis of the application uses the model developed to answer some of the design questions such as how safe is the structure and what is the estimated life span of the structure.
>
> Simple mechanistic models for inelastic materials can be created by judiciously combining the three key mechanical elements: springs, viscous dashpots and frictional dashpots. This needs a basic understanding of the characteristic behavior of each of these elements.

Chapter roadmap

It is important to understand the need for studying inelasticity of materials from the perspective of practical applications. One can go through sample scenario provided in section 1.3 to understand the typical steps involved in a prediction process and their importance briefed in section 1.4. Section 1.5 takes you through a process of making pertinent observations useful for the modeling exercise for a typical elasto-plastic deformation observed in the experiments.

To understand how to model plastic deformations in a simple way, an analogy of a friction block is used and the corresponding phenomena occurring in the friction block and the plastic deformation process are tabulated in section 1.6. There are other phenomena that are time-dependent in the form of creep and relaxation that may have to be modeled. If one is interested in understanding how one builds a model using viscous dashpot analogies and what characteristic behavior we generally anticipate because of the time dependent deformations, the student can go through section 1.7.

Those focusing on purely elasto-plasticity can skip section 1.7 and go to the next chapter that formalizes and extends the type of modeling introduced here apart from giving different perspectives of understanding the inelastic behavior.

1.1 What is inelasticity?

Inelasticity is the study of materials that are not elastic. This could mean such a wide range of things that "anything" could be studied. We actually mean something more specific than such a broad meaning. By "inelasticity" we mean the study of *persistent shapes and their changes due to external stimuli in materials*[1].

Such materials are "hysteretic"[2] and so we very quickly enter into the realm of internal friction in materials! This is at the heart of inelasticity. This book seeks to study various manifestations of persistence of shapes and hysteresis in a variety of materials from a thermodynamical point of view. This is one of the defining characteristics of this book and sets it apart from many other books devoted to specific kinds of hysteretic effects.

In the process of learning the material in this book, you will see that many of the well accepted notions of what it means to be a solid or a fluid will gradually disappear. Rather we enter a sort of "gray" area where materials show solid-like or fluid-like behavior. At the two ends of the spectrum are the ideal rigid solid (where no changes in shape can occur) and an inviscid fluid (where no shape is persistent). We will find materials that have intermediate properties between these two extremes.

1.2 Why should one study inelasticity?

Most structures are built to "withstand" the elements, i.e. the persistence of their configuration is essential to their functioning. We do not want cars and buildings to be "flexible". So, they are designed to not change their shapes under normal operating conditions. However, no structure can be built to withstand all external stimuli. A car may crash, a structure may be subject to a hurricane or an earthquake or a blast from an explosive, a nuclear power plant may have a "core melt down". In these circumstances, designers have to find out exactly how a structure is likely to fail and will it fail in a manner that protects the safety of society (this is called "fail-safe design")—enter inelasticity.

Thus, ensuring the survival of a structure or a critical component under rare and unexpected situations requires the use of inelasticity; for example, we use inelasticity to study the structural integrity of the passenger compartment in a car when it crashes. Similarly, another situation wherein inelastic analysis is required, occurs when there are unanticipated changes to the load on a structure—a new larger engine is to be mounted on a bracket, or a new storey is to be added to a building

---

[1] The Meriam-Webster dictionary (available online) defines the word persistent to mean "existing for a long or longer than usual time or continuously", "retained beyond the usual period", "continuing without change in function or structure," etc.

[2] Again, Meriam-Webster to the rescue: "hysteretic", means "a retardation of an effect when the forces acting upon a body are changed (as if from viscosity or internal friction)".

etc. Approval of these modifications requires extensive modeling of inelasticity. The study of the causes of failure of a component (called "failure analysis") in order to prevent similar failures also requires extensive knowledge of inelasticity.

A different field in which inelasticity is heavily used, is the optimization of materials processing operations. You have to change the shape and/or microstructure of a new kind of material and you need to find the best way to do this without physical trials (which are very costly)—this requires knowledge of large deformation inelasticity.

Finally, materials like shape memory alloys, smart polymers etc. are naturally hysteretic and simulating their response requires knowledge of inelasticity.

> ⊙ **1.1 (Exercise:). Examples of the need for inelasticity**
> *Find different specific engineering scenarios requiring the use of inelasticity (either from your own experience, talking to friends or colleagues or searching the web for examples) and state the reasons why knowledge of inelasticity is needed.*

1.3 A sample scenario

Let us consider a situation in engineering where modeling the material for an 'other than elastic' response becomes important.

First, consider a cable stayed bridge that is being built to take certain types of loads for its function. The functional requirement of the cable stayed bridge is, in general, to be able to provide for carrying a road on which vehicles can ply safely and the road is maintained, to the extent possible, a flat and even surface. The flat and even road provides for a psychological feeling of safety for those driving the vehicles. This condition poses a challenge of avoiding excessive deflections of the bridge.

In order to design such a bridge the actual load on the bridge has to be estimated. The load is of four kinds. We will list them in the order of predictability and frequency of occurrence:

- The first type of load is the dead load - the load that acts on the structure all the time. Loads such as self-weight of the structure, the road and its components belong to this category. This is completely predictable and practically constant through the life of the bridge.
- The second type of load is the live load - the load related to the movable objects, in this case - primarily the vehicular loads. These are much more variable but are still entirely predictable from studies of traffic flow patterns and other similar bridges. These also occur throughout the life of the bridge.
- The normal prevailing wind conditions also contribute to the design load. These loads vary in magnitude and only a statistical idea of the variation and

limits can be obtained about these loads to arrive at the design load. Generally, a design load is prescribed based on the probability of maximum load occurrence. In general, the bridge is designed in such a way that the material used in constructing the cable stayed bridge takes the load well within its functional capabilities. This is called "safe life design".

We will take, for example, a component of the bridge, namely, its main cable, that is expected to take a design working load of $P = P_{design}$, say, that includes a critical combination of dead load, live load and wind load. This means that the material should be capable of taking the stress due to the load P without collapsing or undergoing excessive deformations. While designing for such a load that has a high probability of occurrence in the building, the engineer would wish to make sure, in general, that each time the live load is applied and removed, the structure and therefore the material, bounces back to its original configuration to retain the functional geometry of the structure. Hence, the engineer would design the column such that the load P creates a stress that is well within the regime of the material response that avoids permanent or excessive deformations. This way, he or she makes sure that the bridge is safe against failure (*safe-fail or safe-life*) for such loads. Here *failure* means failure to provide the core functionality of the bridge. This aspect of design makes heavy use of elastic response and requires minimal knowledge of inelasticity.

- On the other hand, there is a small probability of the occurrence of strong wind conditions during a storm or strong motions during an earthquake of high magnitude, though brief in period compared to its design life (expected safe life). Typically, during the life of a cable stayed bridge (of say a hundred years), such a high magnitude loading may occur, once or twice, may be. It will not be economically viable to think of a design for safety against failure for such loads. Instead, the thinking would be to design such that there is enough warning before the failure occurs so that precautionary measures can be taken to reduce the loss of human life and property. In other words, the designer makes sure that the failure is a safe one giving enough warning - usually called the *fail-safe* design. This part of the design requires extensive knowledge of inelasticity and may involve even large deformations.

Therefore, it is not only important to make sure that the material that constitutes the structural components respond reversibly, but also to make sure a sudden collapse does not occur. The structure may not return completely to its original geometry but may assume a slightly different geometry under such a situation. It is also important to assess whether it is safe for its functional use after such a high magnitude loading has occurred. In all such cases discussed above, characterizing the irreversible response of the material that constitutes the main cable, modeling it appropriately and analyzing for the structural behavior under such critical loads becomes important.

Now, let us examine the kind of loads we see in this kind of example. They are usually a superposition of the normally occurring dead and live loads together with the sudden bursts of cyclic forces due to strong wind or earthquake loads, i.e. *cyclic loading with a bias*. The material may or may not be sensitive to the *frequency* of such a cyclic loading, a possibility which needs to be investigated. This load occurs over a very small period of time and the number of cycles is small over its entire design life.

1.4 Observation, experimentation and modeling of material response

The process of designing against failure is a long and involved process and involves a whole community of researchers much of whose work is completely hidden from practicing engineers. This process broadly involves five major steps:

(1) **Observation**:
    Initially, testing the material for loads that mimic the application generally helps in realizing a simple working hypotheses (or "model") of the material response. In this phase, we are not sure what the response would be and precisely what aspects of the test needs to be controlled. We merely "observe" the phenomena that occur and construct a working hypothesis[3].

(2) **Hypothesis testing with controlled experiments**:
    Once a working hypothesis is constructed, we now have an idea as to what parameters are important and what is not and now we can carry out controlled experiments to see if the hypothesis is true [4]. By means of different controlled experiments, it is possible to test the applicability of the working hypothesis given sufficient evidence in the controlled test results. This procedure is often an iterative one since the working hypothesis has to be tuned to make sure it is sufficiently realistic and practical to model the response of the material under the given set of loading conditions. Now we have a reasonable model and its parameters may be measured by anyone provided the tests to be performed are properly described.

(3) **Development of testing protocols**:
    Thus, once this modeling exercise is done, a set of material characterization tests are prescribed to obtain the relevant data. This is what is found in the descriptions in the ASTM manuals for testing different materials. The data should be interpreted and parameters appearing in the model are estimated.

(4) **Application of the model to a practical situation**:
    After the testing is done, it is followed by the analysis for stresses, strains and

---

[3] Imagine Newton observing the falling of an apple (which is an uncontrolled experiment) and hypothesizing that the bodies are attracted to each other by some force).

[4] Imagine Galileo running experiments with a evacuated tube with a feather and a penny to test his hypothesis that all objects fall at the same rate in the absence of air.

displacements in the structural components or the entire structure. Here is the role of strength of materials, elasticity, finite element methods etc.
(5) **Decisions based on calculations**:
Why do we make these calculations? Most of you are used to problems in mechanics that read "find the force in ... ". Why do we need this information? Because a design or safety or financial decision is to be based on this calculation. This is at the heart of all our tasks.

The first task in this process is to list the kind of action that the member goes through in order to bear the design load. In the cable stayed bridge example, the cable primarily takes the load by stretching and hence is primarily subjected to axial tensile loads. Thus our working hypothesis is that the cable is a straight bar subjected to axial forces. Given this situation of loading on the component, namely, the main cable, a suitable test needs to be carried out in order to find the mechanical characteristics of the cable. From the application point of view, therefore, we are interested in obtaining the following information on the material that the cable is made of (apart from its initial elastic behavior):
(a) The stress beyond which the material of the cable ceases to act reversibly.
(b) The kind and order of deformations that the material of the cable may undergo under different magnitudes of the load.
(c) The behavior of the material under a cycle or cycles of loading.
(d) The behavior of the material under different frequencies or rates of strong loading, and
(e) other situations that may relate to the secondary effects in the material but significant enough to affect the design life. Such effects take place when there are excessive deflections in tall structures, thin-walled members, etc.

The nature of stress to be applied for testing, suitable for this application, could be simply an axial (tensile) load test. The typical geometry could be a circular specimen mimicking the geometry of the actual structural component, more specifically, a wire testing is ideal. It is important to separate out or remove, as far as possible, the geometry effects when finding the material behavior.

⊙ **1.2 (Exercise:). Testing Protocols**
Find an ASTM standard testing protocol for testing cable wires and describe it in a short paragraph.

1.5 Experimental observations and model development

As discussed above, the simplest of tests - the tension test can be conducted on a mild steel wire specimen (since the main cable of the bridge is made of mild steel), say. The task, in this exercise, is to come up with some of the working hypotheses

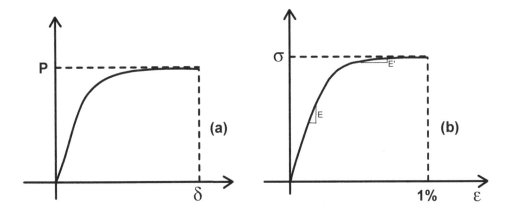

Fig. 1.1 (a) Load vs. Elongation for the mild steel wire specimen tested (b) Engineering stress vs. engineering strain.

that can closely model the different characteristics of material under this kind of loading.

The test concerns measuring the load, P, needed to elongate a wire with an initial diameter, $D = 0.5mm$ and of initial length, $L = 200mm$, by a certain given extension, δ. The test is initially conducted at a very slow speed. This is usually called the *quasi-static* loading condition. The testing is done so that the range of elongation is within 1% of the total initial length of the wire. This range is taken to reflect the fact that there may be a noticeable change seen in the structure if the elongation goes beyond this range and this would violate the actual working conditions. Figure 1.1a shows the load vs. elongation plot for the test conducted. If we assume that practically the same stress and strain occur at all points of the wire during this test, it is possible to convert the load vs elongation plot of the wire shown in Fig. 1.1a into an approximate stress vs. strain of the material (see Fig. 1.1b). This is done so that the geometry dependence is removed to a large extent, showing the representative mechanical behavior of the material under the quasi-static monotonic load applied. The stress is obtained by normalizing with respect to the initial area of cross-section of the wire, $\sigma = P/A$ with $A = \pi D^2/4$, and the strain is obtained by normalizing the elongation with respect to the initial length of the wire, $\varepsilon = \delta/L$.

The following simplifying assumptions are made in extracting the results from the above test:
- An initial approximation that the change in area is negligibly small,
- The initial length of the wire is high compared to the extension it undergoes,
- Variation in stress and elongation across the cross section is negligible.

Let us examine the stress-strain curve obtained by monotonically elongating the wire till 1% as shown in Figs. 1.1b and 1.2. Two salient features can be observed. (a) There is an almost linear response seen till a point or a zone, A and thereafter,

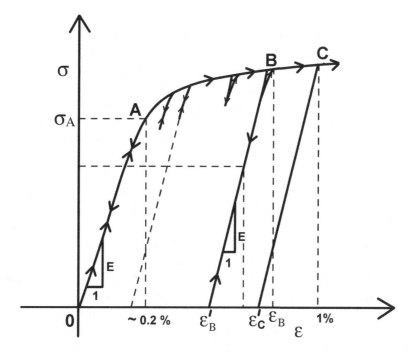

Fig. 1.2 Response under loading-unloading at various stages. This response is used to find out zones of reversibility and irreversibility in response.

(b) the response curve is nonlinear with the slope of the curve gradually decreasing from an initial value of E. The curve approaches a limit as the strain increases to about 1%.

If we seek to answer the first question that arose from the application point of view, i.e. about the question of mechanical reversibility, it is important to conduct the test in such a way that there is unloading involved. In Fig. 1.2, the results of such an exercise, carried out at different stages of the response, are shown. One can notice that beyond the point A, the reversibility is lost, i.e., the loading curve is different from the unloading curve. Also, for this material, this point or zone almost matches with the point of deviation from linearity. Therefore, the first hypothesis that can be proposed is that "a material which is loaded up to the point A will retrace its path upon the gradual removal of the load". Now a more controlled experiment up to the point A reveals that the strain, ε, changes proportionally and reversibly with the stress, σ. Such a behavior, as one can already recognize, can be termed *elastic* behavior and is represented by a "spring". Therefore, we refine our hypothesis for an application in which the material experiences very small strains (roughly less than 0.2% as in Fig. 1.2), that the response is almost linear and reversible, and thus, can be approximated by

$$\varepsilon < 0.2\% \Rightarrow \sigma = E\varepsilon, \tag{1.1}$$

where E is the constant of proportionality which is also the slope of the curve OA

in Fig. 1.2.

Let us examine this model more closely. Notice that the total stress is directly related to the total strain and therefore, irrespective of how the strain is reached, whether monotonic loading or unloading loading, etc., the model predicts the stress to be dependent only on the strain. This automatically encompasses reversibility since, an increase in stress, followed by an equal magnitude of reduction in stress will leave the strain unaltered at the end of this process. Therefore, this model is applicable for a reversible process of application of stress or strain. Let us call this model element as *spring element*. We will revisit this notion after discussing models related to the subsequent deformations that occurs in the material.

Next, examining what happens when the material is loaded beyond the point A, we find that there is a deviation from the linear behavior noticed. Also, upon unloading from the point B, it is clear that the unloading curve is different from the loading curve indicating that it has undergone an irreversible deformation. Complete unloading as shown in Fig. 1.2 introduces a remnant strain in the material. When the material is again loaded almost up to point B, we find that the material responds almost linearly and retraces the unloaded stress-strain curve, indicating an elastic response. We therefore hypothesize that, if the reference is shifted to ε'_B, then the response can be modeled simply by the above described simple linear elastic model. Further controlled experimentation reveals that the stress and strain can be now related simply by,

$$\sigma = E(\varepsilon - \varepsilon_{B'}). \tag{1.2}$$

Note that we have an additional piece of information. While the reference point (i.e. the x intercept) has changed, *the slope of the stress-strain line* did not change. Alternatively, for an infinitesimal increment in strain, $d\varepsilon$, there is a corresponding proportional infinitesimal increase in stress, i.e.

$$d\sigma = E d\varepsilon. \tag{1.3}$$

Integrating stress from 0 to σ and strain from $\varepsilon_{B'}$ to ε, we get the relationship in equation (1.2). In the above, we find that there is no time parameter used indicating that the response is modeled as a time-independent response. Therefore, without loss of generality, we can rewrite (1.3) as,

$$\dot\sigma = E\dot\varepsilon, \tag{1.4}$$

as long as the x-intercept does not change. Note that the time derivative is a dummy derivative since the equation is a homogeneous differential equation of order one in time.

Instead of unloading from B, if we proceed to deform the material more beyond B to, say a point C as shown again in Fig. 1.2, and then unload at point C, we see that the behavior is similar. Upon unloading and reloading at point C, we notice that there is a near linear elastic behavior noticed till the response curve reaches the point C again so that one can model this elastic behavior by a shift in reference to ε'_C. Again, the $\sigma - \varepsilon$ response shows the same straight line but with a new x intercept. This suggests that we make a bold extension to our hypothesis, namely that

$$\sigma = E(\varepsilon - \varepsilon^p) \quad (1.5)$$

where ε^p is a changing x-intercept. We now notice that we need to investigate how ε^p changes with σ.

Thus, one can realize that each time an additional irreversible deformation takes place, there is an accompanying shift in the reference or in other words, an additional residual strain is observed to occur. Since this shift happens only upon reaching the stress pertaining to B or to C, the shift is realized only after a threshold stress is reached. To make this task simple and meaningful, let us remove the linear elastic part from the response curve in Fig. 1.2 by plotting σ versus $\varepsilon - \sigma/E$ and re-plot it to obtain a response as shown in Fig. 1.3. Note the intricate interplay between hypothesis and measurement: each set of measurements gives us new or refined hypothesis which in turn suggests new experiments.

1.6 Plastic deformations and the friction block analogy

As one can see now, the unloading-reloading curves are vertical curves and the x-axis represents the residual or the remnant strain, ε_p, say. We now make a further extension of our hypothesis and state that "the change in the remnant strain with stress is analogous to the change in the position of a block sliding on a rough surface with friction". This is a major and rather stunning advance in our understanding and we need to investigate this very closely.

To elucidate the connection between this behavior and the friction-block model, consider a mass m connected at point **A** to an inextensible string **AB** (see Fig. 1.4) and sliding on a horizontal rough surface with coefficient of friction μ (we shall assume that the coefficient of static and dynamic friction are the same). We pull the end **B** of the string *gradually* by a force F. This force is directly exerted on the block by the string as a function of time [5]. We quickly see that if we pull on the block suddenly, there will be a "jerky" motion of the block. So the analogy is not perfect unless we do the experiment very slowly and ignore inertial effects.

The following qualitative features of the motion of the block and the force exerted by the string can be verified[6] :
(1) If the magnitude of the applied force F is below $\mu m g$ then the block A will not begin to move.

---
[5] You can try out this experiment yourself to see the effects.
[6] The reader is urged to do this experiment by moving a block very slowly and convince themselves of the results. A load cell or a spring balance can be attached to the string to determine the force with which the string is pulled.

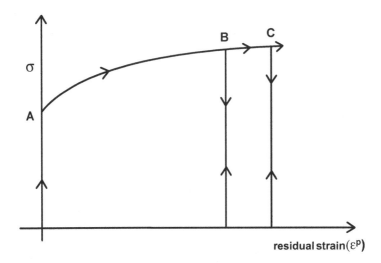

Fig. 1.3 Response with remnant strain - the unloading-reloading curves are vertical.

(2) The magnitude of the applied force cannot exceed $\mu mg = \mu N$ where g is the magnitude of the acceleration due to gravity.
(3) The block **A**, when it moves, always moves in the direction of the force of the string F.
(4) The magnitude of the velocity of **A** is non-negative.
(5) As long as the block continues moving, the applied force remains at μN. When the block is moving, if the force drops below μN, the block will stop moving.

If the force F with which the string is pulled increases monotonically with time, we can use the five points listed above to conclude that

- The block does not move until F becomes equal to μN.
- Once the force reaches μN, then there is a possibility of the block moving. It depends upon how we move the end **B**. For example, if we stop the point **B** once $F = \mu N$, then it should be clear to you that block **A** will not move. If, on the other hand, F continues to exist effecting the movement u_B, then the block **A** will move in the direction of the force F.
- Finally the motion of the block **A** will be such that the magnitude of the force is always μN. The resulting force displacement diagram is shown in Fig. 1.5. Comparing this behavior with what is shown as response in Fig. 1.3, in which there was unloading and reloading responses shown, can you see that there is an analogy possible? We are going to exploit this analogy to develop the basic structure of the constitutive equations that are related to this behavior as given in the following table.

Friction block	Material behavior
Displacement of **A** : u_A Force on **A**: F	Plastic strain: ε^p Applied stress: σ
Conditions on the motion of A	Conditions on the plastic strain ε^p
1. If the magnitude of the applied force F is below μN then the block **A** will not begin to move.	1. If the magnitude of the applied stress σ is below σ_y then $\dot{\varepsilon}^p = 0$.
2. *Yield Condition:* The magnitude of the applied force cannot exceed μmg where g is the magnitude of the acceleration due to gravity (ignoring inertia). $f(F) = (\|F\| - \mu N) \leq 0$	2. The magnitude of the applied stress cannot exceed σ_y $f(\sigma) = (\|\sigma\| - \sigma_y) \leq 0$
3. *Flow Rule:* The block **A**, if it moves, always moves in the direction of the force of the spring F. $\dot{u}_A = \lambda sgn(F)$.	3. The plastic strain rate is always in the direction of the applied stress σ. $\dot{\varepsilon}^p = \lambda sgn(\sigma)$.
4. The magnitude of the velocity of **A** is non-negative. $\lambda \geq 0$.	4. The magnitude of the plastic strain rate is non-negative. $\lambda \geq 0$.
5. *Loading Condition:* If applied force is less than μN then the block will not move, if the block moves, then the applied force must be equal to μN. $\lambda f(F) = 0$.	5. If applied stress is less than σ_y then $\dot{\varepsilon}^p = 0$, if $\dot{\varepsilon}^p \neq 0$, then the applied stress must be equal to σ_y. $\lambda f(\sigma) = 0$.
6. *Consistency Condition:* As long as the block continues moving, the applied force remains at μN. $\lambda \dot{f}(F) = 0$.	6. As long as $\dot{\varepsilon}^p \neq 0$, the applied stress remains at σ_y. $\lambda \dot{f}(\sigma) = 0$.

In the above table, the symbol *sgn* stands for the signum function and λ is the magnitude of the motion. Also we observe that condition (5) is actually two statements—either $\lambda = 0$ or $f(F) = 0$—packaged into one function for convenience. Notice that (6) above seems a redundant, and indeed it is actually something that we can do away with; however, it is an essential equation from which we will calculate the value of λ. It becomes meaningful in a multi-dimensional situation.

Now, let us examine the response of this model with respect to varying loads. If the stress reaches σ_y and exceeds slightly, the strain builds up. If the element is strained to an extent, say, ε_1 and unloaded, the strain stays at ε_1. Upon further

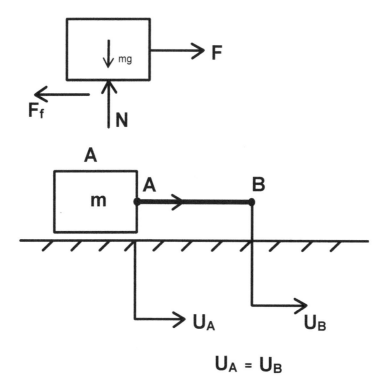

Fig. 1.4 A block of mass m connected to an inextensible string on a frictional horizontal surface. The end B is gradually pulled by a force F.

loading, no straining occurs till one reaches the stress limit σ_y beyond which again strain builds up. On unloading, the built up strain remains. Thus, the residual or permanent strain build up that occurs in the actual material can be modeled using this element. The response when plotted along with the actual response as presented in Fig. 1.6, we see that the model closely captures the behavior exhibited by the material. It should be noted that once σ_y is reached, there is no further stress necessary to increase the strain or, in other words, there is a flow causing strain as long as the stress is kept at σ_y. Such an idealization is called the *perfectly plastic model*.

> ⊙ **1.3 (Exercise:). Response under stress cycling**
> *Combine a spring element and a friction block element, as shown in Fig. 1.7, to describe the model response under such stress cycling as was done in the experiment described above. Plot the response and make appropriate observations based on the results obtained.*

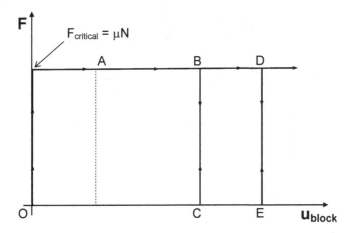

Fig. 1.5 Force-displacement response for the frictional block model.

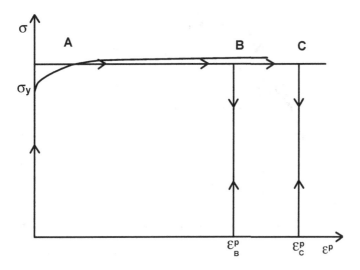

Fig. 1.6 Stress-strain plot for the friction element and comparison with the actual stress-remnant strain response.

⊙ 1.4 (Exercise:). Hardening Behavior

Using appropriate combinations of the spring and the friction block elements, it is possible to simulate behavior such as strain hardening. Show that, a sloping curve as shown in Fig. 1.8 is realizable using a combination of a spring attached in series to a set of a friction element and a spring element connected in parallel.

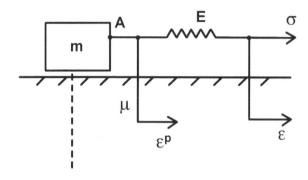

Fig. 1.7 Spring-friction block model and the corresponding response of the model.

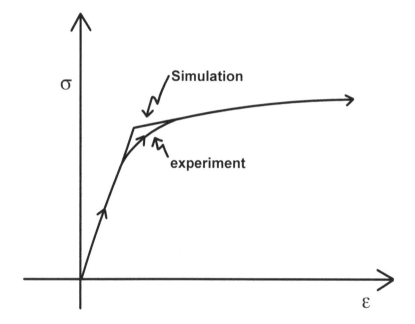

Fig. 1.8 Stress-strain elasto-plastic response with a sloping plastic curve.

Having completed the task of modeling of the elastic-plastic response, the behavior of the cable under the given loads can be simply solved by inserting the model into the set of governing equations that model the entire cable and attempt to solve the resulting set of equations. We will return to this task in later chapters where we describe the solution of such boundary value problems.

It should also be understood that the model should reflect the requirement from the application. For example, if the intention is to find the collapse load, it is enough to try a simple elastic-perfectly plastic idealization or model. However, if it is important in the application to find the strain at a given stressed condition for a material that hardens considerably, one should model the material as an elastic-plastically hardening model. If the task is to find the accumulation of the plastic strain critical to the prediction of fatigue in the material, it is worth modeling the nonlinear hardening near the initial stages of yielding and the changes in the yield stress with plastic strain accumulation better. A detailed treatment of hardening models are presented in later chapters.

1.7 Creep and time dependent behavior

Now, turning our attention to the other considerations related to the cable, another important phenomenon to be modeled is the deformation that may occur over time due to the sustained nature of some of the loads such as the dead loads and mean live loads in the structure. These deformations may take place over long periods of time. A typical example is the sagging of a clothesline over a period of months or years. This effect is seen especially predominantly in polymeric materials with which the clotheslines typically are made of. This phenomenon is called the *creep* in the material.

Since these deformations are due to sustained loading, the simplest test could be to observe the change in displacements over time in wires that are loaded with different constant forces. A key point to observe is that: *It is vital to plot forces, displacements and other data versus time and NOT versus each other if we want to develop models for this behavior. Plotting traditional stress versus strain curves, while useful for some purposes, turns out to be extremely restrictive for modeling purposes.*

Let us investigate this behavior by applying a constant load on a wire. Figure 1.9 shows the response in terms of strains, $\varepsilon(t)$ (normalized with respect to the initial length of the wire) with respect to time (creep curves) and which is called "engineering strain" for different normalized loads σ_{const}, (normalized with respect to the initial area of cross-section) which are referred to as "engineering stresses".

There are a few points that can be noted from the response shown in Fig. 1.9:

(1) The start point of each of the responses on the strain axis is different for different constant stresses for which creep test has been carried out. This is followed by a gradually increasing strain with time. We hypothesize therefore that the response to a constant suddenly applied load can be broken up into a "instantaneous" strain and a time evolving strain. By plotting the different constant stresses with respect to the corresponding instantaneous strains one finds that the curve is nearly straight with a slope $1/E_2$, say. So we refine our hypothesis to state that the total strain is $\varepsilon = \sigma/E_2 + \varepsilon^p(t)$. This is similar to the case of plasticity, but there ε^p did not change unless stress changed.

(2) The strain versus time curve reaches an asymptotic slope which is different for different constant stresses with progress of time. Plotting the asymptotic slope (which is the variation of strain with respect to time, or, strain rate) with

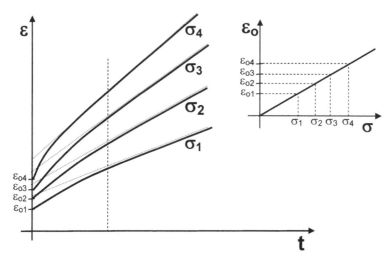

Fig. 1.9 Creep curves for constant load. Note that the cross-sectional area will keep changing. In this plot, the engineering strain is plotted over time under a constant engineering stress.

respect to the constant stress applied, one finds that the plot is nearly linear with a slope η_2. In other words, the material seems to have "viscosity" like honey or other fluids of that kind.

(3) The curves vary nonlinearly between the instantaneous strain and the asymptotic slope with a gradual decrease in slope.

At this stage you may wonder: is this material a solid? since it keeps deforming continuously under an applied load, we could call it a "fluid" but it also has elasticity! While it is possible to classify these materials as solids or fluids by different criteria, for the purposes of our initial investigation, we don't really care, as long as we can meaningfully and consistently model the response over the time scale of interest.

Before we proceed further, let us now introduce an element that resembles the phenomenon of linear viscous resistance, generally called a linear *dashpot* element. For this element, the relationship between the rate of strain ($\dot{\varepsilon}$) and the resistance σ can be written as,

$$\sigma = \eta \dot{\varepsilon}, \tag{1.6}$$

where η is the viscosity coefficient which, for our simple model, is generally a constant. It is simple to see that for a constant stress applied, the strain increases linearly with time for a pure dashpot. The rate of increase of strain, $\dot{\varepsilon}$, in the element is such that the rate increases with the increase in the applied constant stress σ_{const} as shown in Fig. 1.10.

As you can see, for a given stress, the strain in a spring immediately reaches a value equal to σ/E_2 while, in a dashpot, for an applied stress, the value of the strain is initially zero but indefinitely increases upon sustained constant stress. Therefore,

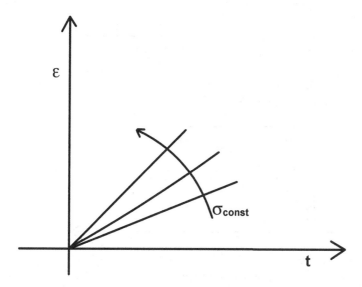

Fig. 1.10 Response curves for a viscous dashpot - note that the slope (strain rate) is higher for higher constant stresses.

our first hypothesis is that "if a spring and a dashpot are connected in parallel as shown in Fig. 1.11, initially dashpot resists the stress applied more than the spring but as time progresses, the dashpot gives way and the spring gradually takes over". The response, $\varepsilon(t)$ can be derived to be:

$$\varepsilon(t) = \sigma_{const}/E_2 \left(1 - e^{-\frac{E_2}{\eta_2}t}\right) \tag{1.7}$$

The response of this combination is shown in Fig. 1.11. There are some features we can observe from this configuration:
(1) the slope of the initial response is equal to the stress divided by the viscosity of the dashpot, σ_{const}/η_2
(2) The strain asymptotically reaches a strain equal to σ_{const}/E_2 which would have been the strain if dashpot were absent
(3) There is an exponential decay of increase in strain with respect to time.

Alas! this model while it has many features that mimic creep, does not allow the material to creep at all. This model does a good job of representing the early stages of response but does not do a good job with the long time behavior. We need to extend out model further. Our next exercise tries a different model.

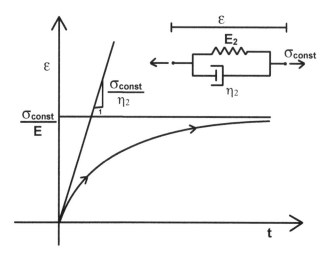

Fig. 1.11 A configuration with a spring and a dashpot connected in parallel. The response is also shown. Observe that the initial slope is related to the dashpot characteristics and the asymptotic slope is related to the spring characteristics.

⊙ 1.5 (Exercise:). Visco-elastic Response

Connect a spring element of elastic constant E_1 with a dashpot element with η_1 in series as shown in Fig. 1.12. Derive and show that the response ($\varepsilon(t)$) of this combination / configuration of elements for a constant stress applied σ_{const}, $t \geq 0$ at the ends is as shown in Fig. 1.12. List the characteristics of the response.

As you can see from the figure, this new model has features that mimic the long term behavior but not the early stages of response. So as you might suspect, with a stroke of pure genius :-))) we connect the two models together as you will see in the next exercise:

⊙ 1.6 (Exercise:). A Creep Example

Show that the response shown in Fig. 1.9 can be approximately modeled by a set of spring, E_2 and dashpot, η_2 in parallel and a set of spring, E_1 and dashpot, η_1 in series as shown in Fig. 1.13.

This model is quite reasonable for many creeping materials. Of course there is no need to stop with just two springs and dashpots. There are a number of books that are devoted to increasingly complex response functions and the use of many springs and dashpots in various arrangements.

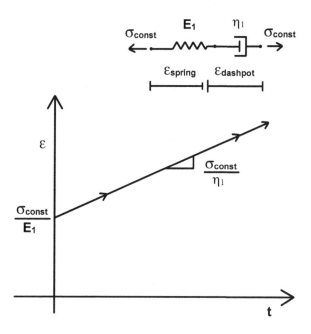

Fig. 1.12 Spring and dashpot elements in series. The strain response over time is shown under a constant stress σ_{const}.

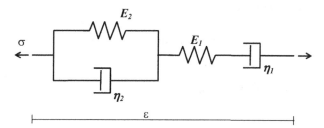

Fig. 1.13 A set of series and parallel connections of springs and dashpots to closely model the actual behavior of the material.

⊙ **1.7 (Exercise:). A Relaxation Example**
Another test that is often carried out is the application of a particular strain to the material and measure the stress response. This is often attributed to the relaxation in the material. This is especially important in applications in which there is a strain controlled situation existent.
Find such a response for the configuration - a spring E_1 and a dashpot η_1 in series under a constant strain ε_{const} $t \geq 0$. List out the characteristic features of this response.

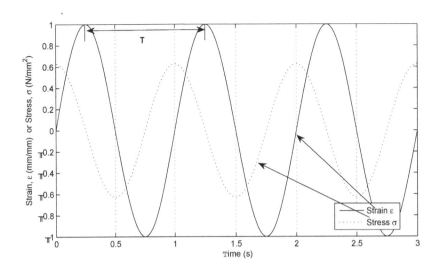

Fig. 1.14 The stress response with time for a dashpot element under an applied strain that varies sinusoidally with time.

Frequency Response

Sometimes, a typical experiment that is conducted is a strain or stress that is cycled over a mean stress or mean strain with a specified frequency. The corresponding stress or strain response is obtained.

Assume for example, that a dashpot with η_1 is subjected to a strain that is varying with time about a zero mean strain and an amplitude of ε_0, as,

$$\varepsilon(t) = \varepsilon_0 \sin\omega t \quad (1.8)$$

The response of the stress can be found out to be simply as,

$$\sigma(t) = \mu\dot{\varepsilon} = \mu\omega\varepsilon_0\cos\omega t \quad (1.9)$$

Figure 1.14 shows the variation of the stress response, σ, and the applied strain, ε with time. In order to understand the behavior better, one can plot the stress vs. strain. Figure 1.15 shows responses for different frequencies. This clearly shows that a dashpot stress-strain response is sensitive to the frequencies of the strain applied. Another point to note is that the stress-strain response is irreversible indicating an inelastic time dependent dissipative response.

Often, in applications, this kind of behavior helps in damping, i.e. dissipation of energy. The common practice of quantifying the damping is the energy dissipation per cycle of straining or loading. One can find that the damping or the energy dissipation is different for different frequencies. In contrast, if the response is computed for the friction block element to the same time varying strain with a frequency, the response is seen to be independent of the frequency of the strain applied (Fig. 1.15).

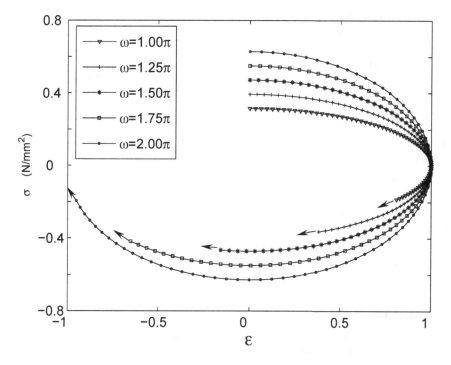

Fig. 1.15 Stress-strain response under a sinusoidally varying applied strain for the dashpot element and the frictional element.

A Multidimensional Case

The considerations so far are devoted to a simple one-dimensional cases. An example that has a different requirement is sheet metal forming. Here, formability is an important consideration. Imagine you are working for a large can-making company (such as ALCOA). Cans are made by a process called deep drawing where a sheet of metal is spread over a hole and a punch slowly stretches the sheet to form a cup[7]. These and similar processes involve stretching a sheet in two dimensions. This will cause the sheet to thin out[8]. The question for you is "find out how much you can stretch the sheet before it will tear". Your observation leads to the conclusion that the material essentially "bulges out". A typical test needed here is a bulge test. In this test, a sheet with a grid pattern drawn on it is subjected to a uniform pressure and changes in the grid pattern due to bulging under pressure are measured. They are a measure of the planar stretches that have taken place. The principal planar stretches plotted with respect to the pressure provides input to the formability diagrams needed for an engineer to get an idea of how far the

---

[7] Rather than us drawing a figure of deep drawing, just go to *YouTube* (www.youtube.com) and search under deep drawing, you will see some really great videos of machines and animations of the process. If a picture is better than a thousand words, a video is better than a thousand pictures.

[8] Any one who has rolled dough has the experience of this thinning out and tearing.

sheet can be stretched in different directions. In these exercises, the sheet may be anisotropic - with different properties in different planar directions, initially. In this case, simple spring dashpot models will be woefully inadequate and you need to know much more about continuum mechanics to do this sort of model. This is partly the purpose of the third part of the book.

1.8 Summary and Conclusions

- Inelasticity is essential to the design of structures to fail safely in unexpected catastrophic conditions. It is also useful in the forming industry in making more efficient die designs. Typical applications include
  (1) collapse load determination
  (2) excessive deformations
  (3) fatigue
  (4) creep and relaxation
  (5) formability in metal forming operations
- The five principal steps in the prediction process are:
  (1) observation and hypothesis construction,
  (2) controlled Experimentation for hypothesis testing,
  (3) development of testing protocols for measuring model parameters,
  (4) simulation of real situations using the model, and,
  (5) using the results to make decisions.

Of these the first three are usually not evident to practicing engineers but are essential to modern technology.

- A simple analogy for an inelastic material can be obtained by combining springs, dashpots and friction elements in various ways.
- The analogy does not work for all cases (especially for multidimensional loading) and knowledge of continuum mechanics is essential for these latter stages.
- There are various categories of application or design focus related to inelasticity that are possible.

Chapter 2

Thermodynamics of Inelastic Materials: A lumped parameter approach

Learning Objectives

After learning the material in this chapter, you should be able to

(1) describe the elements of a lumped parameter system;

(2) describe the steps involved in the modeling of such systems and obtaining the corresponding differential equations;

(3) describe the kinetic equations for a dry friction damper and contrast with that for a viscous damper, identifying their similarities and differences;

(4) write down the kinetic equations for dampers in 2 dimensions;

(5) write a MATLAB program to solve a simple spring damper system ignoring inertial effects;

(6) describe the physical meaning of the Helmholtz and Gibbs potentials and their relationship with the internal energy of a lumped mass system;

(7) obtain the mechanical and thermal evolution equations for a given a lumped parameter system, from the knowledge of either its Helmholtz potential or Gibbs potential;

(8) write a program to simulate the response of such a thermo-mechanical system using a built-in ordinary differential equation (ODE) Solver in MATLAB;

CHAPTER SYNOPSIS

If you consider a system of springs, dashpots and masses, which are interconnected and subject to an external load, it is possible to write the governing equations that describe the evolution of its thermomechanical state in the form of a set of ordinary differential equations of the form

$$\mathbb{A}(\mathbf{q}, t)\dot{\mathbf{q}} = \mathbf{f}(\mathbf{q}, t),$$

where the vector $\mathbf{q}$ is a column vector composed of the mechanical variables and the temperature of the system. This chapter describes how to arrive at this set of equations.

If we assign a variable to signify the extension of each spring and dashpot, you will typically end up with more variables than the "degrees of freedom" of the system. We can express the connectivity of the system by means of constraint equations. The equations of motion can then be written down from considerations of the power dissipated and the constraint forces from Gauss' theory of least constraint (see [Gauss (1829); O'Reilly and Srinivasa (2001)]).

For a large class of materials, the equations of motions *for an inertialess dissipative system* can also be derived from the maximum rate of dissipation hypothesis; the procedure will enforce any additional constraints on the systems automatically.

The central result described in detail in this chapter is the following:

Consider an inertialess lumped parameter system that is subject to known external forces. Let the degrees of freedom in the system be q_i, $i = 1, 2, ..., n$ and the external forces be $Q_i^{ext}(t)$ (many of them may be zero) so that the external power supply is $\sum_i^n Q_i^{ext} \dot{q}_i$. Let the kinematical variables be subject to m equality constraints of the form

$$f_\alpha(q_i, T, \dot{q}_i) = 0, \quad (\alpha = 1, \ldots, m) \tag{2.1}$$

where T is the temperature. Let the isothermal work potential (or Helmholtz free energy) be $\psi(q_i, T)$ and the dissipation potential (or the rate of dissipation) be ζ. Then the following results hold:

(1) The evolution of the state of the system satisfies the mechanical dissipation equation which states that "The difference between the mechanical power supplied and the rate of increase of the isothermal work potential (keeping temperature fixed) is the rate of mechanical dissipation which is non-negative", i.e.

$$\sum_i^n Q_i^{ext} \dot{q}_i - \frac{\partial \psi}{\partial q_i} \dot{q}_i = \zeta \geq 0 \tag{2.2}$$

(2) Any evolution equation for $\dot{q}_i$ consistent with the requirement that ζ be non-negative, and consistent with the constraints is acceptable from the perspective of the second law of thermodynamics. So, in principle you can assume any evolution equation you like for the mechanical variables, substitute it into (2.2) and see if ζ is greater than zero. If it is, you have satisfied the second law of thermodynamics. This does not mean that these evolution equations are correct. If you want the evolution equations to be consistent with Newton's laws for the mechanical variables, then a slightly different approach is warranted as described in the following items (3a) and (3b).

(3a) The Newton's laws of motion for a non-inertial system (together with the Gauss's theory of least constraint) are given by[1]

$$Q_i^{ext} - \frac{\partial \psi}{\partial q_i} = Q_i^d + \sum_\alpha \mu_\alpha \frac{\partial f_\alpha}{\partial \dot{q}_i}, \qquad (2.3)$$

where Q_i^d are the generalized dissipative forces or generalized driving forces[2] and the variables μ_α are called "Lagrange multipliers" and are obtained by satisfying the constraint equations (2.1). This form of incorporating nonlinear constraints is called "Gauss' theory of least constraint" [Gauss (1829)].

(3b) Typically the generalized forces are assumed to be functions of q_i, T and $\dot{q}_i$. Any constitutive equation for Q_i^d can be chosen, provided that it satisfies the non-negativity of the rate of dissipation, i.e. as long as it satisfies

$$\sum_i \left(Q_i^d + \sum_\alpha \mu_\alpha \frac{\partial f_\alpha}{\partial \dot{q}_i} \right) \dot{q}_i = \zeta \geq 0. \qquad (2.4)$$

(4) A convenient way (and by no means the only way) to satisfy (2.4) is to assume a constitutive equation for the rate of dissipation $\zeta = \zeta(q_i, T; \dot{q}_i)$ and then require that

The state of the material evolves in such a way as to maximize ζ subject to all the constraints on the system.

This assumption allows us to *simultaneously (a) obtain equations of motion (b) incorporate constraint forces and (c) obtain constitutive equations for the dissipative forces that automatically satisfy the non-negativity of the rate of dissipation.*
The resulting equations are

$$Q_i^{ext} - \frac{\partial \psi}{\partial q_i} - \mu \frac{\partial \zeta}{\partial \dot{q}_i} - \sum_\alpha \mu_\alpha \frac{\partial f_\alpha}{\partial \dot{q}_i} = 0, \qquad (2.5)$$

where μ and μ_α are obtained by satisfying (2.4) and (2.1). Specifically, we can show that

$$\mu = \frac{\zeta - \sum_\alpha \mu_\alpha \sum_i \dot{q}_i \partial f_\alpha / \partial \dot{q}_i}{\sum_i \dot{q}_i \partial \zeta / \partial \dot{q}_i}. \qquad (2.6)$$

---

[1] If the system has inertia, the right-hand side of (2.3) is not zero but is equal to the rate of change of momentum.
[2] The Generalized forces may not be "forces" at all in any conventional sense, but their products with the corresponding generalized velocities are related to the power supplied.

(5) The "heat equation" for the determination of the temperature is given by

$$-T\frac{d}{dt}\left(\frac{\partial \psi}{\partial T}\right) = \zeta - h(T - T_\infty) + r, \tag{2.7}$$

where h is the heat transfer coefficient and T_∞ is the temperature of the surroundings and r is any other non-mechanical source of energy (such as electrical heating).

Chapter roadmap

The rest of this chapter explains how to obtain the above results and shows several examples of the application of this set of equations to various lumped parameter systems. This is a rather long and involved chapter and has two major parts— one devoted to pure mechanics and one devoted to thermodynamics of lumped parameter systems. We first introduce the notion of the fundamental mechanical elements — springs, viscous and frictional dashpots — and obtain equations that represent their response. We then show that we can obtain the equations of motion by conventional means (i.e. applying Newton's Laws) or by using the power theorem to motivate the form of the equations[3] or, as advocated in this book, by using the notion of maximization of the rate of dissipation. Next we show how these ideas can work for two-dimensional systems also where applying Newton's laws become truly cumbersome.

The next half of the chapter extends these notions to thermomechanics introducing many of the ideas of thermodynamics (such as entropy and heat capacity) and applying them to lumped parameter systems. We finally end with a full thermodynamical example and then show how an alternative method using Gibbs potentials can also be used.

On first reading, one may skip the thermodynamics aspects and go on straight to Chapter 3. But we have found that students are fascinated by the thermodynamics of springs and dashpots since a typical mechanical, civil or aerospace engineering student is more familiar with strength of materials and Castigliano's two theorems[4] rather than ideas of Helmholtz and Gibbs potentials. The thermodynamics part is written as an extension of Castigliano's theorems rather than from ideal gases as is done in a conventional text on thermodynamics. So it might be well worth the effort to read this chapter.

---

[3]This is a sort of "poor man's Lagrangian mechanics" since it only works for inertialess systems.
[4]Carlo Augusto Castigliano was an Italian mathematician and physicist working for the Northern Italian Railways. Working for the railways, would be the same as working for Google in terms of its importance and reach, since the railways did to atoms what Google does to bits. (Think about that for a while :-))

PART A
LUMPED PARAMETER MODELS: PURELY MECHANICAL PROBLEMS

2.1 Notion of a lumped parameter model

2.1.1 *The role of lumped parameter models*

You are working for a "smart materials product group" or a company that develops active vibration dampers or sound deadening systems or are working with the control systems group in a product team. You may be exploring some simple models for a new component that can be quickly implemented on a computer. The task in front of your team is to help design the component and build an efficient controller that will enable the component to work as expected. You do not have too much time[5] and your specific job is to come up with some "plant model" (this is the control systems terminology for constitutive equation[6] + equations of motion) so that different control strategies could be tried.

If you start this process by talking about continuum mechanics and the governing partial differential equations of elasticity, your control systems group will roll their eyes and tell you, not unkindly, that most control algorithms have been developed only for discretized systems where the state of the system is represented by only a small, finite set of "state variables" whose evolution with time is typically represented by ordinary differential equations [7]. So your central challenge can be stated as:

How do you construct a "plant model" for your system?

There are at least three common ways to do this:
(1) set up the PDEs using continuum mechanics and then use a finite element method (or other equivalent numerical discretization schemes) to get sets of ordinary differential equations,
(2) introduce a finite set of generalized coordinates to represent the macroscopic configuration of the model and use ideas of Lagrangian mechanics to set up the equations for a finite degree of freedom system, or,
(3) use combinations of springs, dashpots etc, and develop a "lumped parameter model" which is based on a spring dashpot mass analogy that will "mimic" the response of the component and then set up its equations.

---

[5] Isn't this always true?

[6] A constitutive equation is mathematical description of how a particular element—say a spring or a dashpot—responds to an external stimulus. Finding is really the foundation of thermodynamics and much of modeling.

[7] preferably linear so that they can take Laplace transforms and start talking about Bode' plots, poles and zeros etc. They may go beyond that. But, for practical purposes: No Partial Differential Equations (PDEs).

At this stage of the game, going for a PDE and then discretizing it seems like an overkill. It is very hard to do and even harder to implement. However the benefits are the possibility[8] of very high accuracy and predictive capability.

The most versatile and powerful direct tool for creating finite DOF models is the Lagrangian mechanics approach. But this is not always necessary and is quite hard to explain[9].

In this chapter, what we will do is to develop a sort of a "hybrid approach": we will show how to set up a "lumped parameter model" using springs and dashpots which allows for a certain visual approach but at the same time show how the resulting equations can be generalized beyond just pure mechanics and beyond springs and dashpots. In other words, we will incorporate elements of Lagrangian mechanics (though not a full version of it) into a spring-mass-dashpot based approach.

2.1.2 Elements of lumped parameter systems: Extending the results of Chapter 1

The formulation of the elastoplastic equations, together with the different spring and dashpot elements, given in the previous chapter, is a purely mechanistic approach directly exploiting an analogy. These analogies are meant to replicate, in a purely mechanical manner, the stress-strain response of many types of inelastic materials.

As we saw in Chapter 1, materials that are inelastic show the following characteristics:
(1) Yield: A threshold load or deflection below which the material behaves elastically (like a spring) and beyond which it behaves like a system with friction.

(2) Stress relaxation: If the material is stretched beyond its elastic limit and held fixed, the stress slowly decreases or "relaxes" with time.
(3) Creep: If the material is subject to a fixed load beyond its threshold, the strain will increase gradually with time, i.e., the material creeps.
(4) Retardation: If the load on a material is suddenly removed, then it tends to gradually recover its original shape.

All of these phenomena can be reproduced by suitable combinations of springs, masses and dashpots. These three elements represent different conceptual aspects of the response of a material. For example, masses represent inertial aspects, i.e.

---

[8]We say "possibility" because the accuracy depends upon the continuum model and the reliability of the experiments to determine coefficients as well as the discretization procedure. Finite Element Methods, contrary to popular belief among students who are mesmerized by its color pictures, are not "magic eight balls" that can give you answers to any question you might have. Experts in the area will tell you that there is a lot of experience and technical knowledge that you need to bring to bear to get a decent continuum model and its finite element discretization.

[9]We urge the student to take a class on this beautiful and deep approach to mechanics—you will appreciate its elegance and will gain much insight.

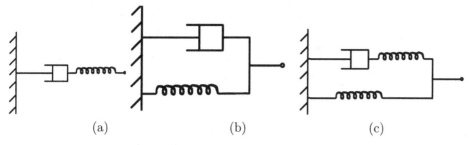

Fig. 2.1 Three commonly used models for inelastic behavior (a) the Maxwell model (b) the Kelvin-Voigt model and (c) The standard solid model. Of these the Maxwell model is that of a "fluid-like" material which does not return to its original configuration upon removal of load (i.e. it has no retardation effect at all). The other two are solid-like models.

the tendency of the object to continue moving after the forces are removed, springs represent the fact that work can be recovered from the body once it is deformed and dashpots represent the irrecoverable loss of work and the consequent "internal heating" of the material.

The three most common combinations of such elements shown in Fig. 2.1 are
(a) the Maxwell model composed of a spring and a dashpot in series,
(b) the Kelvin-Voigt model which is a spring and a dashpot in parallel, and
(c) a standard solid which is a Maxwell element in parallel with a spring.

2.1.3 *Equations of motion*

> ⊙ **2.1 (Exercise:).**
> *Using Newton's laws, obtain the governing equations of motion for each of the systems shown above. (Hint: First identify and label the connections between the different "elements". Then apply Newton's laws to each connection.)*

> ⊙ **2.2 (Exercise:).**
> *By considering the way in which you obtained the differential equations for the above models, write out a step by step instruction on how to get governing equations for any combinations of springs dashpots and masses.*

We will now formalize the procedure for writing the equations of motion of a lumped parameter system. You should compare your method with the list given here and see in what way it is similar or different:
(1) Label each joint where springs and dashpots are connected (and where masses are lumped) as 1, 2, 3, ... etc.
(2) Assign a coordinate to each joint as $x_1, x_2, \ldots$ etc.
(3) Draw the free body diagram (FBD) of each joint and apply Newton's laws to

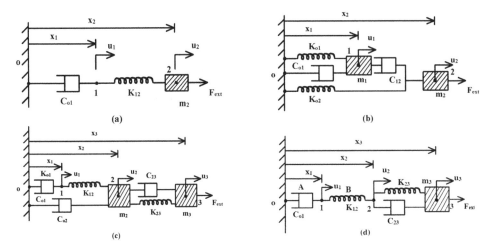

Fig. 2.2 Some spring mass models to illustrate lumped mass modeling of beams. In (a) and (b) the nodes are numbered and the positions are marked as x_i and the displacements as u_i. While (b) and (c) represent "solid-like" behavior with the elements returning to their respective equilibrium positions upon removal of the load, models (a) and (d) represent "fluid-like" behavior since they will not return to its original configuration due to the absence of a spring connection to the ground.

each joint.
(4) Substitute the constitutive equations for each spring and each dashpot (making sure that the signs for the forces are right).

Although it is common to use a linear spring and a linear dashpot, it is not essential to do so, and we will see several models where the dashpot does not have viscous damping but is of a dry friction type and the spring may be a stiffening or softening spring.

⊙ **2.3 (Exercise:). Governing equations**
Apply Newton's laws to each junction of the spring-mass dashpot system shown in Fig. 2.2a and show that the governing equations for this system are (assuming that all the elements are "pulling")

$$\text{Kinematical equations:} \quad \frac{du_1}{dt} = v_1, \quad \frac{du_2}{dt} = v_2 \quad (2.8)$$

$$\text{Kinetic equations:} \quad F^{ext} - F_{12} = m_2 \frac{dv_2}{dt}, \quad F_{12} - F_{01} = 0 \quad (2.9)$$

$$\text{Constitutive equations:} \quad F_{12} = K_{12}(u_2 - u_1), \quad F_{01} = C_{01} v_1. \quad (2.10)$$

Of the above set of equations, the first is trivial and represent just the relationship between the displacement and velocity.

The second equation is a statement of Newton's second law applied at each joint. Since there is a mass at only one of the joints, we see only one acceleration term. One of the ways to write the equations of motion when there is no inertia is to simply recognize that the force at any joint must sum to zero. This will further simplify, if we assume that every spring is in tension and so it "pulls" on the joints and every dashpot is extending so that it also "pulls" on the joint and none of the elements "push" on a joint, it is relatively straightforward to write the equations of motion. It is possible to formalize this, but we will resist this, trusting to your common sense to figure out the forces. Also we use the notation F_{ij} to represent the forces due to the elements connecting joint i to joint j. You can use any notation for this that you like as long as you are consistent and draw the free body diagrams (FBDs) right.

Newton's laws can also be interpreted with a flow analogy: imagine a current flowing through the network from the ground. The force is like the current and the mass is like a capacitor (for accumulating charge). If you study Fig. 2.2a carefully and imagine that it is like a pipe for fluid (which represents the current) to flow from one end to another, the same fluid or force (or current) flows *through* the dashpot (resistor) and the spring (inductor) and flows into the capacitor. A different amount of force (current) flows out the other end, the difference being accumulated in the capacitor. In this formulation, Newton's law (2.9), is the mechanical counterpart of the capacitor equations. Similarly, springs become inductors (with inductance being the spring stiffness) and dashpots become resistors (with resistance being the damping coefficient). This kind of analogy works very well in one dimension but is particularly difficult for forces in two and three dimensional problems, nevertheless the analogy reveals that there is a hidden similarity between mechanical and electrical effects. While a full exploitation of this similarity can be accomplished by means of Lagrangian mechanics, we will be able to utilize parts of the latter.

The last group of equations, i.e. (2.10) are called "constitutive relations".

CONSTITUTIVE RELATIONS

Constitutive relations are empirical relationships that represent the response of the different elements such as springs and dashpots to external stimuli. Such relationships are also called "plant models" in the control systems literature. The bulk of this book is devoted to the development and implementation of a consistent and common framework for finding suitable constitutive relations for different kinds of inelastic elements.

Notice that the constitutive equations are easily written if we assume that each element is "pulling" at its joint.

⊙ 2.4 (Exercise:). More lumped parameter models (LPMs)

For the models shown in Figs. 2.2(b,c,d), (a) locate the nodes and number them (with zero for the ground) (b) Properly label the forces and displacements and obtain the kinematical, kinetic and constitutive relations equations for each. Assume that the springs and dashpots are not necessarily linear but are general functions.

Hint: Using the circuit current flow analogy for the forces will considerably simplify the derivation of the equations. The labeling has already been done for Fig. 2.2(b) to assist you here. Note that there are two joints labeled 1 and 2. Ans: (assuming that all the elements are "pulling"): Fig. 2.2(b):

$$\text{kinematical equations: } \frac{du_1}{dt} = v_1, \quad \frac{du_2}{dt} = v_2, \quad v_{12} = v_2 - v_1 \quad (2.11)$$

$$\text{kinetic equations: } F^{ext} - F^d_{12} - F^s_{02} = m_2 \frac{dv_2}{dt} \quad (2.12)$$

$$F^d_{12} - F^s_{01} - F^d_{01} = m_1 \frac{dv_1}{dt} \quad (2.13)$$

$$\text{Constitutive equations: } F^s_{01} = f(u_1), \quad F^s_{02} = g(u_2),$$
$$F^d_{12} = h(v_2 - v_1), \quad F^d_{01} = p(v_1). \quad (2.14)$$

In the above solution, we have used the notation F^s for spring forces and F^d for dashpot forces. In the above set of equations, we note that the functions f, g, h and p have to be specified. Of course, the matter is simple if they are linear, in which case, we have

$$f(u_1) = K_{01} u_1, \quad g(u_2) = K_{02} u_2, \quad h(v_2 - v_1) = C_{12}(v_2 - v_1), \quad p(v_1) = C_{01} v_1, \quad (2.15)$$

but a general nonlinear spring and dashpot has to satisfy some additional criteria if they are to be acceptable. It is to this aspect that we turn next.

2.2 General considerations of springs and dashpots

2.2.1 Linear and nonlinear springs

We are not going to belabor the point that springs are "conservative", i.e., that the work that we put in to deform the spring can be recovered and that the conservative forces can be written as gradients of a potential. So, we begin by listing several common types of springs and their corresponding potentials (see Fig. 2.3(i)):

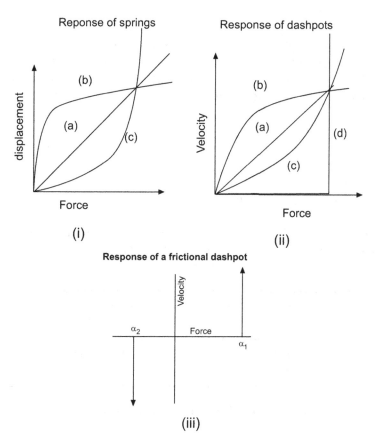

Fig. 2.3 Response of conservative components (springs) and dissipative components (dashpots). Note that the graphs have been drawn with the forces along the x axis. Thus, for the spring response (a) is a linear spring (b) is a stiffening spring and (c) is a softening spring. For the dashpot, (a) is a linear dashpot (b) is a thickening dashpot,(c) is a thixotropic dashpot and (d) is a frictional dashpot. Note that, for the frictional dashpot seen in (iii), there is a threshold force below which no motion is possible whereas as soon as the force reaches a threshold, any velocity in the direction of the force is possible. The actual velocity will be determined only by the motion of the other components.

Types of springs:

$$\text{General power-law spring: } F_{spr} = ku^\alpha \Rightarrow u = (F_{spr}/k)^\beta, \beta = 1/\alpha. \quad (2.16)$$

$$\text{Associated strain energy: } \psi = \frac{ku^{\alpha+1}}{\alpha+1}. \quad (2.17)$$

$$\alpha = 1 \Rightarrow \text{linear spring: } F_{spr} = ku \Rightarrow u = F_{spr}/k \quad (2.18)$$

$$\alpha = 1/n < 1 \Rightarrow \text{softening spring: } F_{spr} = ku^{1/n} \Rightarrow u = (F_{spr}/k)^n \quad (2.19)$$

$$\alpha = n > 1 \Rightarrow, \text{stiffening spring: } F_{spr} = ku^n \Rightarrow u = (F_{spr}/k)^{1/n}. \quad (2.20)$$

Typically n is a positive odd integer.

⊙ **2.5 (Exercise:). Modeling spring stiffnesses**
(1) Why is n required to be odd? (2) Consider a spring whose response is given by $F = k_0 x + k_1 x^3$. Is this a stiffening or softening spring? why? What is the strain energy associated with the spring? How about $F = k_0 x + k_1 x^{1/3}$?

In general, we note that if a spring is connected to two joints i and j, its potential function is of the form $\psi(u_i - u_j)$. Then the magnitude of the force of the spring is $\|F\| = \|\partial \psi / \partial (u_i - u_j)\|$ and its direction at joint i is the same as that of $u_j - u_i$, while at the joint j it is along $u_i - u_j$[10].

For a general system, we begin by stating a result that follows from a statement of Castigliano's first theorem which can be paraphrased as[11]:

CASTIGLIANO'S FIRST THEOREM

For any system with generalized coordinates q_i, if we write the strain energy ψ as a function of q_i then the partial derivative of ψ with respect to q_i will yield the corresponding generalized force.

Thus, for any general spring, the force of the spring is related to its deflection by

$$F_{\text{spring}} = \frac{d(\text{strain energy})}{d(\text{deflection of spring})} \qquad (2.21)$$

If the system consists of multiple springs, then we can define a strain energy for the whole system as the sum of the strain energy of all the spring components and we can write a general equation of the form

$$F_{i^{th}\ \text{spring}} = \frac{d(\text{strain energy of system})}{d(\text{deflection of } i^{th}\ \text{spring})}. \qquad (2.22)$$

Notice that a spring is a "repository" of mechanical work so that at times, the power supplied to the spring can be negative—mechanical power flow out of the spring. Thus, a spring can be both a source as well as a sink of mechanical power.

---

[10] It turns out that this complication regarding directions is only a nuisance in one dimension. In two or more dimensions, the complication is completely removed as we shall see shortly.

[11] The following includes our first official use of the word "generalized coordinates". These are coordinates used to represent the configuration of a system. Unlike cartesian coordinates, these can be angles, lengths, or any physical quantity we choose to use to indicate the relative arrangement of the elements or constituent parts of a system.

2.2.2 Viscous dashpots

A typical dashpot exhibits a *velocity dependent force*, i.e., the force on a dashpot is a function of the relative velocity of its ends. *The equation relating the force on a dashpot to the relative velocity of its ends is called a kinetic equation.* This is the counterpart of the force displacement relationship for a spring (which is called a state equation). Just as different springs had different force displacement relationships, different types of dashpots have different types of force velocity relationships. Figure 2.3(ii) shows a comparison between force-velocity relationships for different dashpots. Note that there is one exceptional case for the dashpot namely curve (d). This is a rather special situation and will be considered separately.

Types of dashpots:

$$\text{General power-law dashpot: } F = \mu v^\alpha \Rightarrow v = (F/\mu)^\beta, \beta = 1/\alpha. \quad (2.23)$$
$$\alpha = 1, \text{ linear dashpot: } F = \mu v \Rightarrow v = F/\mu. \quad (2.24)$$
$$\alpha = 1/n < 1, \text{ thixotropic dashpot: } F = \mu v^{1/n} \Rightarrow v = (F/\mu)^n. \quad (2.25)$$
$$\alpha = n > 1, \text{ stiffening dashpot: } F = \mu v^n \Rightarrow v = (F/\mu)^{1/n}. \quad (2.26)$$

Typically n is a positive odd integer. For the dashpot, all we can say is that *the power supplied to any purely dissipative system cannot be negative*, i.e., a dashpot can *never* be a source of mechanical power, it is always a sink.

⊙ **2.6 (Exercise:). Commonly occurring dissipative materials**
Find out what the word "thixotropy" means and give examples of thixotropic materials that you commonly encounter. Food items are very good examples. Experiment and find out if Honey is thixotropic. How about Ketchup or Paint? Blood?
Make a list of 5 common items such as pizza dough, bread dough etc. and explore whether they have "springiness". What kind of viscous behavior do they have.

DASHPOTS DISSIPATE MECHANICAL ENERGY

One aspect of the second law of thermodynamics can be taken to mean that, *you cannot extract mechanical power through friction.* Thus, if the pulling force exerted *by* the dashpot on one of its two ends is F^d and the rate of increase of its length is v then the constitutive equation for a dashpot must always satisfy $F^d v \geq 0$.

Thus, we choose equations for dashpots to satisfy this criterion, then we can guarantee that such a material will not become a perpetual motion machine. For example, if we assume that the dashpot in Fig. 2.2(a) is linear, so that

$$F = C_{01} \dot{u}_1 \quad (2.27)$$

then the power supplied to the dashpot is $F\dot{u}_2 = C_{01}\dot{u}_2^2 \geq 0$ so long as C_{01} is non-negative.

In general, just as springs oppose relative extension, dashpots oppose relative motion. In other words, a spring "pulls" at its joints whenever it is elongated, whereas a dashpot "pulls" at its joints whenever the joints are moving apart. Finally, we wish to emphasize that it is not always necessary to write an expression for the force in terms of the velocity for a dashpot. In many cases, it is more convenient and more natural to write the velocity in terms of the force, or that some function of the force and velocity is zero. But in all these cases, it is important to satisfy the condition that the product of the force and velocity be non-negative. The next exercise shows an example where it is common to write the velocity as a function of the force.

⊙ 2.7 (Exercise:). Creep Functions
For elastoplastic materials undergoing creep deformation, a common kind of a dashpot that reflects some of the properties is one that is of the form $v = AF|F|^n e^{-Q/kT}$, with $n = 3$ for dislocation creep and $n = 1$ for superplastic flow, and Q is an "activation parameter". Does this satisfy the second law of thermodynamics in the sense that the power supplied to the dashpot must always be non-negative? Plot the velocity versus force for $n = 1, 3$ and assuming that $A = 1, Q = (300k)J$ and for $T = 300K$ and $500K$. (Look up the value for Boltzmann's constant k.) Such expressions are common for thermally activated phenomena.

2.2.3 Frictional dashpots

An exceptional and important special case of a dashpot is a dashpot with dry friction. This is the lumped parameter equivalent of perfectly plastic materials and deserves special consideration because of its rather special nature and the difficulty in grasping the essence of it.

As a means of motivation, we will first consider dry friction as a limiting case of thixotropy. To see this, we first note that the constitutive equation for a power law dashpot is of the form

$$F = \mu v^\alpha \Rightarrow v = (F/\mu)^{1/\alpha} \qquad (2.28)$$

The above form, which deals only with the magnitude and not the direction of the force, hides a key element of this constitutive formulation. We recall that for any dashpot, we need to satisfy the condition that the power supplied to it is always non-negative. In order to do this easily, it is much better to write the constitutive equation for viscous forces in terms of "magnitude" and "direction" as follows:

- For any dashpot in 1 dimension, *the second law of thermodynamics is satisfied if the force always opposes the velocity*, i.e.

$$F = -\chi v \Rightarrow v = -\phi F, \phi \geq 0. \quad (2.29)$$

This first step, which is true for ANY dashpot in 1 dimension, is a mathematical way of saying that "the velocity is in the opposite direction of the force".
- Since the viscous dashpot always consumes mechanical power, i.e. $Fv = -\phi F^2 \leq 0$ we see that $\phi \geq 0$.
- Next we can prescribe any functional form for the parameter ϕ (or equivalently χ) so long as the function is always positive, i.e.

$$\phi = \phi(F). \quad (2.30)$$

This two-step process can be applied to any kind of dashpot. If we apply it to a **power law dashpot**, we will get

$$F = -\chi v, \quad \chi = \kappa(\|v\|)^\alpha \quad \text{or} \quad v = -\phi F, \quad \phi = \frac{1}{\kappa}(\|F/\kappa\|)^{\frac{1-\alpha}{\alpha}}. \quad (2.31)$$

Now what happens when α tends to 0? Note that, by putting $\alpha = -1$ in (2.31), we get

$$F = -\kappa \frac{v}{\|v\|} = -\kappa sgn(v) \quad (2.32)$$

where *sgn* stands for the *signum* function[12]. In other words, when v is not zero, the force is constant and is equal to μ and opposes the sliding. If v is zero, the force is indeterminate. How does this square with your usual description of friction? We hope you recall that "friction opposes relative motion and adjusts its value according to the other external forces until it reaches its maximum value of μ times the normal force". This is what we have here: the "indeterminacy" when v is zero is a way of saying that its value depends upon the other forces. Thus (2.32) is exactly what we want for dry friction.

In view of the importance of understanding frictional phenomena, we will further elaborate on the constitutive equation for dry friction in the following way:

- The sliding is always opposed to the force applied : $v = -\phi F$.
- A frictional dashpot always consumes mechanical power, it cannot produce it, i.e. if F^f is the force produced by the dashpot on other bodies, then the power supplied by the dashpot $F^f v = -\phi(F^f)^2 \leq 0$. Since $(F^f)^2$ is positive, we have $\phi \geq 0$. Thus, the above two conditions are the same as for a viscous dashpot. The next two conditions are different and are the key to the constitutive equations for friction.
- The magnitude of the frictional force cannot exceed a critical value: $f(F) = \|F\| - \kappa \leq 0$.

---

[12] $sgn(x) = 1$ if $x > 0$ and -1 if $x < 0$. Become familiar with this function; it plays a key role in all friction type behavior.

- Sliding can only occur if the frictional force is at its maximum. If it is less, no sliding occurs: $\phi = 0$ if $f(F) < 0$.

- The friction constitutive equations can be summarized in a very compact form and are referred to as the "Kuhn-Tucker conditions or inequalities", and read

$$v = -\phi F^f, \quad \phi \geq 0, \quad f(F^f) \leq 0, \quad \phi f(F^f) = 0. \tag{2.33}$$

The above equations form an *implicit set of equations that have to be solved to obtain* ϕ. This is in direct contrast to viscous dashpots where ϕ is given explicitly in terms of the friction force. This is the reason why solving dry friction problems is harder than viscous drag problems and requires trial and error methods.

At this stage, we are not interested in solving the equations of motion, but, just setting them up. When we are ready for solution, we will see how to tackle these problems. Equation (2.33) is vital to the understanding of frictional dashpots and needs to be memorized.[13]

To help you with the memorization we will restate it in a succinct form

---

CONSTITUTIVE EQUATIONS FOR DASHPOTS

For any dashpot in one dimension,
(a) The sliding velocity is opposed to the force exerted *by* the dashpot ($v = -\phi F$).
(b) It can only consume mechanical power ($\phi \geq 0$).
(c) For viscous friction, the magnitude of the velocity can be specified as an explicit non-negative function of the frictional force ($\phi = \phi(F) \geq 0$).
For dry friction dashpot, we cannot get an explicit expression for ϕ instead we have
(a) the frictional force cannot exceed a critical value

$$f(F^f) = \|F^f\| - \kappa \leq 0, \quad \text{and,} \tag{2.34}$$

(b) sliding occurs only if the frictional force is at the critical value, else, there is no sliding, i.e.

$$\phi f(F^f) = 0. \tag{2.35}$$

---

[13] We are of the opinion that you can't look up everything: you should be able to memorize some expressions along with a verbal description of them. Just memorizing the expressions is a mistake. The verbal description is vital.

⊙ **2.8 (Exercise:). Nonlinear spring dashpot models**
Consider the spring dashpot model shown in Fig. 2.2(d). Assume that (1) the mass m_3 is zero (no inertia), (2) the springs are both power law stiffening springs (say, $F = kx^3$) for each spring, K_{12} and K_{23} where k is a constant and x is the extension, (3) the dashpots, C_{01} and C_{23}, are shear thinning dashpots (i.e. $v = -\phi F_{fric}, \phi = \mu \|F_{fric}\|^3$).
Obtain the equations of motion for the system.

(Hint: Don't substitute for the dashpot forces in terms of the velocity, rather obtain expressions for the dashpot forces in terms of the spring and external forces)

Ans:
$$\dot{x}_3 - \dot{x}_2 = -\mu \|F(t) - k(x_3 - x_2)^3\|^3 (F(t) - k(x_3 - x_2)^3),$$
$$k(x_2 - x_1)^3 = F(t), \qquad (2.36)$$
$$\dot{x}_1 = -\mu \|F(t)\|^3 F(t).$$

What happens if both the dashpots become frictional dashpots with critical force μ?

Ans:
$$\dot{x}_3 - \dot{x}_2 = \phi_1 \left\{ F(t) - k(x_3 - x_2)^3 \right\},$$
$$k(x_2 - x_1)^3 = F(t),$$
$$\dot{x}_1 = \phi_2 \left\{ F(t) \right\}^3. \qquad (2.37)$$
$$\phi_i \geq 0, \quad f_1 = \|F(t)\| - k(x_3 - x_2)^3 - \mu \leq 0,$$
$$f_2 = \|F(t)\| - \mu \leq 0, \quad \phi_1 f_1 = \phi_2 f_2 = 0.$$

Note that the final form for the dashpots (whether with viscous or dry friction) is a set of ordinary differential equations together with some algebraic equations. For frictional systems, we also have to satisfy the Kuhn-Tucker inequalities. We will show how to solve these equations later. At this stage all we want to emphasize is the following:

For inertialess systems, Newton's laws together with the constitutive equations result in a combination of differential equations, algebraic equations and possibly some inequalities.

2.2.4 Springs and dashpots in 2 dimensions

Imagine that you have a refrigerator of mass m that you have to move on a greasy floor. You are using a small elastic chord of stiffness k to "slowly" move the refrigerator around. This is a 2-dimensional problem. How would you set up the equations of motion for this problem?[14] Figure 2.4 shows a "plan view" of the problem. As usual, we number the joints (in this case there are 2) and we introduce as kinematical variables, the position vectors $\mathbf{q}_1$ and $\mathbf{q}_2$ of the joints[15]. Let us list the forces:

- External force $\mathbf{F}$ applied by you to one end of the elastic cable.
- Force exerted by the spring at joint B (which is the point where you are pulling): $\mathbf{F}_{spr} = -k(\mathbf{q}_2 - \mathbf{q}_1)$. Notice the negative sign. This can also be written as $\mathbf{F}_{spr} = -\partial \psi / \partial \mathbf{q}_1$ where $\psi = 1/2 k \left\| \mathbf{q}_1 - \mathbf{q}_2 \right\|^2$.
- Force exerted by the spring at joint A (which is the point of connection of the refrigerator): $\mathbf{F}_{spr} = k(\mathbf{q}_1 - \mathbf{q}_2) = -\partial \psi / \partial \mathbf{q}_2$.
- Force exerted on joint A (the refrigerator) by the ground: Now we are somewhat in a quandary—there are normal and frictional forces on the refrigerator. However, have no fear, we can use our common sense to figure it out:
  (a) We don't really care about the normal force. The viscous friction force exerted by the ground on the refrigerator must be in a direction that opposes sliding of the refrigerator: $\mathbf{F}_d = -\chi \dot{\mathbf{q}}_1$ and
  (b) the magnitude of the force is proportional to the speed of sliding (if the grease on the floor is linear viscous) : $\chi = \mu = const$. Again, we note that we first figured out the direction and then the magnitude of the friction forces. This is a trick that is very useful.

Now it is a relatively straightforward matter to obtain the equations of motion by using Newton's Laws at joint A and joint B to give

$$\text{joint A: } \sum \mathbf{F} = 0 \Rightarrow k(\mathbf{q}_2 - \mathbf{q}_1) - \mu \dot{\mathbf{q}}_1 = 0,$$
$$\text{joint B: } \sum \mathbf{F} = 0 \Rightarrow \mathbf{F} - k(\mathbf{q}_1 - \mathbf{q}_2) = 0. \tag{2.38}$$

---

[14] 2-dimensional problems are more useful for understanding the constitutive equations for dashpots than 1 dimensional problems and we have introduced this problem to help you gain a greater understanding of the way in which constitutive equations are written.

[15] From now on, we will start using the standard mechanics notation of "q" for coordinates rather than x.

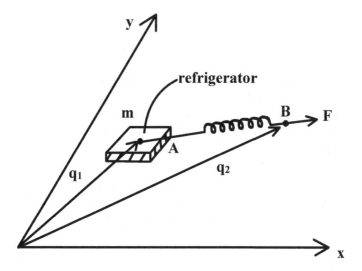

Fig. 2.4 The two-dimensional model for a refrigerator moved around slowly over a greasy floor using a small elastic chord AB. The points A and B are represented using $\mathbf{q}_1$ and $\mathbf{q}_2$ respectively.

⊙ **2.9 (Exercise:). Moving a refrigerator**
What would the equations of motion be if the refrigerator was sliding on grease which has a power law viscosity of the form $\|\mathbf{F}\| = \mu \|\mathbf{v}\|^n$?
What if it were dry friction with a critical force of κ?
Ans:

$$\text{joint } A: \sum \mathbf{F} = 0 \Rightarrow k(\mathbf{q}_1 - \mathbf{q}_2) - \chi \dot{\mathbf{q}}_1 = 0. \tag{2.39}$$

For power law viscosity, $\chi = \mu \|\mathbf{v}\|^{n-1}$, for friction we have to get the Kuhn Tucker inequalities again:

$$\chi \geq 0, f = \|k(\mathbf{q}_1 - \mathbf{q}_2)\| - \chi \leq 0, \chi f = 0. \tag{2.40}$$

Thus our two-step process of assigning direction first and then magnitude that we introduced for the dissipative forces works in 2 dimensions also. However, there is a new wrinkle. Until now, we have used the word "friction force opposes the velocity" to mean that "friction force is exactly opposite to the direction of the velocity". However, consider the case of floor where there are parallel lines scoured on the floor. For this case it should be obvious to you that moving perpendicular to the scour marks may be more difficult than moving parallel to them: the force of friction is not exactly opposite to the velocity.

Thus, in 2 dimensions, we can have a richer structure: The friction force may only form an obtuse angle with the velocity, i.e. we can have a situation where all we need is that $\mathbf{F}\cdot\mathbf{v} \leq 0$. In this case, we can state the kinetic law for any viscous drag in the following form

- The direction of the velocity is determined by the friction force: $\mathbf{v} = -\phi\mathbf{M}(\mathbf{F})$. where $\mathbf{M}$ is some function of $\mathbf{F}$[16].
- Friction cannot generate mechanical power but only consume it. Hence the direction of the velocity must make an obtuse angle with the force, i.e. $\mathbf{F}\cdot\mathbf{v} \leq 0 \Rightarrow \mathbf{M}\cdot\mathbf{F} \leq 0$ and $\phi \geq 0$.
- The value of ϕ is to be specified either by a constitutive equation in terms of the force (for viscous drag) or by the Kuhn-Tucker inequalities (for dry friction). In other words, we must ensure that (1) sliding does not occur until the friction force reaches some critical condition and (2) once sliding occurs, the friction force continues to satisfy the critical condition, i.e.

$$\phi \geq 0, f(\mathbf{F}) \leq 0, \phi f = 0 \qquad (2.41)$$

Notice that we don't require $f(\mathbf{F})$ to be just the norm or magnitude of $\mathbf{F}$ but can be any reasonable function.

As you can see, while in one dimension viscous and dry friction seem very different, they are actually very similar in 2 dimensions or higher, differing only in the way that the value of ϕ is specified[17].

Finally, we note that force can be interpreted as "momentum flow", just as we can interpret "heat" as energy flow. Therefore, if the force is in the same direction as that of the velocity then the momentum increases: Momentum is flowing in. In the same way, if force is in the OPPOSITE direction of the velocity then the momentum decreases, so the momentum flows OUT.

The notation F_{ij} is used for the force "flowing" through the system from the node j into the node i. Thus we can interpret F_{ij} as force flowing INTO the node i from j. F_i^{ext} pointing in the direction of the positive velocity at node i is treated as force flowing OUT. The change in momentum at a node is given by the force flowing out of the node sans the net forces flowing into the node. Furthermore, we will use the symbol F_{ij}^s for forces through components from which work can be recovered, and F_{ij}^d for forces through components from which mechanical work cannot be recovered.

---

[16] Although it is conventional to think that the force is a function of the velocity, it is much more convenient for solution of ODEs (especially for dry friction) to write the velocity as a function of the force. You should get used this switching back and forth between the two ways of writing it.

[17] You might have noticed that we keep repeating the verbal description of how friction acts, repeatedly before we write the corresponding expression. This has been done deliberately so that you see that *every symbolic expression is a short form for some assumption or declarative statement*. It has been our experience that too often students can write down and manipulate symbolic expressions but struggle to articulate the meaning of the steps. We hope that by our repeating the statements, the verbal description of a symbolic expression will become a subconscious act on the part of the students and researchers who read this book.

2.3 Energy formulation of the equations of motion

The approach using springs and dashpots, while invaluable for gaining insight, is not easily extendable to the thermo-mechanical responses. Furthermore, it may be necessary to create systems with various components connected together in different ways to mimic the response of a realistic material (with creep, relaxation, hardening and other complex interactions between them), especially ones where temperature is involved. For example, it is not a simple matter to account for thermal expansion and coupling of thermal and mechanical response for many polymeric materials using a simple mechanical analogy.

Fortunately, we can approach the same task in a different way—one that has considerable generality and also one that is commonly used in strength of materials for dealing with complex systems—energy based approaches and specifically techniques that can be viewed as generalizations of Castigliano's first and second theorems.

The procedure associated with the energy formulation is the following

---

ENERGY FORMULATION OF THE EQUATIONS OF MOTION

For any system, we assign the mechanical state of the system through the displacements of the various components and the thermal state of the system through the temperature;
Then if we specify the Helmholtz potential (which is the generalization of the strain energy) as a function of the states, we can find all the conservative forces on the system as well as the entropy and the heat capacity of the system in a rather straightforward manner through differentiation.

Later in this chapter, with the Gibbs formulation, we will see that the forces in the body can be used rather than the displacements, in order to specify the state of the body.

Our aim in considering this approach is to develop a systematic way to obtain *state evolution equations*, (i.e. equations of the form $\dot{s} = f(s, t)$ where s is a list of state variables of the system) in such a way that the fundamental laws of physics can be met.

The main point is that *for spring dashpot systems without inertia, we do not have to write Newton's Laws and draw free body diagrams. We can find the equations of motion by simply writing the power theorem (which we will discuss shortly) in terms of a minimal set of variables and then reading off the coefficients of the velocity components.*

To understand how this can be done, consider again the simple mass, spring and damper in series as shown in Fig. 2.2(a). The thermodynamic state of the system is represented by

(1) the displacements of junctions u_1 and u_2 where the elements are connected to each other,
(2) the velocities of the junctions v_1 and v_2 which are just the time derivatives of the displacements,
(3) the forces $\mathbf{F}_{ij}$, that act through each element and,
(4) the temperature T of the system.

The forces are classified as internal ($\mathbf{F}_{ij}$) and external forces ($\mathbf{F}_i^{ext}$) on the system.

Essentially, the central task is the following:

Given the external forces on the system as well as the temperature of the surroundings, $\longrightarrow$ *find the evolution equations for the internal state of the system (displacements, internal forces and temperature) as a function of time.*

For the spring-mass-damper system being considered (Fig. 2.2(a)), the variables are $\{u_1, u_2, v_1, v_2, \mathbf{F}_{01}, \mathbf{F}_{12}, T\}$— a total of 7 variables. We, thus, need seven independent algebraic or differential equations to determine the subsequent evolution of the system.

The state description of the system is not yet complete, since we do not have enough equations for the unknowns yet. A quick count of unknowns and equations (in (2.8) and (2.9)) reveals that for the system (Fig. 2.2a) there are 7 unknowns $\{u_1, u_2, v_1, v_2, F_{01}^d, F_{12}^c, T\}$ and only four equations (the two kinematical and two kinetic equations). We need three more equations—one for the temperature and two related to the response of the spring and the dashpot.

2.4 Mechanical power theorem

In order to obtain a thermodynamic approach to the specification of the evolution equations, let us, for the moment consider that the temperature is fixed, i.e. the process is *isothermal* and that the process occurs so slowly that inertia can be neglected; *the process is quasi static*. We are going to approach the model from the point of view of mechanical power: The applied external force on the system supplies mechanical power to the system. We track the mechanical power as it "flows" through the system and we will gain further insight into the mechanisms of storage and dissipation. For any system, it is intuitively evident that

> **THE POWER THEOREM**
>
> For an isothermal process, in the absence of inertial effects, the external mechanical power supplied to the body is equal to the mechanical power dissipated by the dashpots and that supplied to the springs. Furthermore the power supplied to the springs is equal to the rate of increase of the potential energy of the springs which is a function of the relative displacements of the ends of the springs.

We are rather unhappy about the use of the word "potential energy" since the word "energy" is highly misused and abused. We will in the future refer to it as simply "work potential" rather than potential energy and try to reserve the word "energy" to mean internal energy of the system and not the various other "energy-like" quantities.

In order to use the above theorem, we need to be able to calculate the power supplied:

- **Power of external forces**: The power supplied by the external forces is the inner product of the force and the velocity of the point of application of the force, i.e.

$$P_{ext} = \mathbf{F}_{ext} \bullet \mathbf{v}. \tag{2.42}$$

- **Power supplied to the springs**: *The power supplied to the springs is simply the sum of the rates of change of the work (or energy) potential of all the springs.*

$$P_{springs} = \frac{d\psi}{dt}. \tag{2.43}$$

- **Power dissipated to the dashpots**: Thus, the only terms left are those that correspond to the power dissipated by the dashpots which is always non-negative.

By considering the power theorem and grouping terms in a specific way, we will show that there is a way to write down the force on each dashpot *without the need to consider Newton's Laws explicitly*.

We will do this in a rather roundabout way. Bear with us and you will (eventually) see the advantage of this procedure. We will illustrate this procedure by considering the spring dashpot system in Fig. 2.2(d). We are reproducing the figure here for your convenience

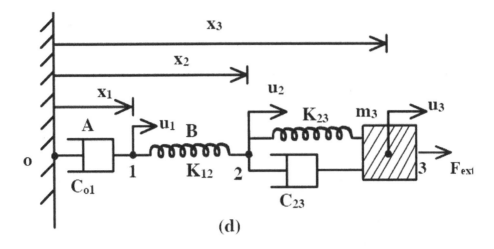

(d)

(1) **DOFs of the joints:** We will choose as many kinematical variables as is convenient to describe the system. Specifically, we will assume that x_i are the displacements of the joints. For the system in Fig. 2.2(d), there are three joints whose coordinates are $x_i, (i = 1, 2, 3)$ from left to right.

(2) **DOFs of the springs:** We will introduce variables x_i^e (the superscript "e" is for "elastic") to represent the relative displacements of the ends of each spring. For the system in Fig. 2.2(d), there are two springs, K_{12} and K_{23} whose relative displacements are assumed to be $x_i^e, (i = 1, 2)$ respectively. At this time, we don't care if these are related to the coordinates / displacements x_i of the joints or not.

(3) **DOFs of the dashpots:** We will introduce variables x_i^p (the superscript "p" is for "plastic") to represent the relative displacements of the ends of the dashpots C_{01} and C_{23}, respectively.

The set of degrees of freedom will be referred to as the q_i, $i = 1, 2, ..., 7$ and the set of external and dissipative forces will be referred to as Q_i, $i = 1, 2, ..., 7$. For this system, we have, $q_i = x_i$, $i = 1, 2, 3$, $q_4 = x_1^e$, $q_5 = x_2^e$, $q_6 = x_1^p$, $q_7 = x_2^p$. Q_i are the generalized forces pertaining to the degrees of freedom, q_i.[18]

(4) Now, write down the work potential for the system in terms of the spring DOFs. For the system in Fig. 2.2(d), there are two springs and assuming that they are linear, the work potential for the system is $\psi(x_1^e, x_2^e) = \frac{1}{2}K_{12}(x_1^e)^2 + \frac{1}{2}K_{23}(x_2^e)^2$.

(5) Next write down the power theorem for this system in terms of the respective DOFs introduced: For the system in Fig. 2.2(d), the external power supplied is[19] $F^{ext}\dot{x}_3$. The power supplied to the springs is $\dot{\psi}$ and the power dissipated by the dashpot is $F_1^d \dot{x}_1^p + F_2^d \dot{x}_2^p$, where F_j^d are the forces exerted by the joints, on the dashpots, j.

---

[18]This is to be consistent with the notation of Lagrangian mechanics where the q's are the generalized coordinates and the Qs are the generalized forces.
[19]Henceforth we will use the notation $\dot{x}$ to mean dx/dt.

So, from the power theorem, we have

$$F^{ext}\dot{x}_3 = \frac{\partial \psi}{\partial x_1^e}\dot{x}_1^e + \frac{\partial \psi}{\partial x_2^e}\dot{x}_2^e + F_1^d \dot{x}_1^p + F_2^d \dot{x}_2^p. \tag{2.44}$$

(6) **Connectivity information is represented as constraint equations:** We clearly have more DOFs than necessary (we have 7 because we have not used any of the information related to the connections between the parts), but that is easily remedied. We simply represent the arrangement of the parts of the system by writing down the relationship between the velocities of all the parts. Specifically, we have the following constraints:

(a) The velocity of joint 1 is equal to the velocity of the dashpot 1: $\dot{x}_1 = \dot{x}_1^p$
(b) The sum of the relative velocities of dashpot 1 and spring 1 must be equal to the velocity of the joint 2: $\dot{x}_1^p + \dot{x}_1^e - \dot{x}_2 = 0$
(c) The spring 2 and dashpot 2 are in parallel so their relative velocities must be the same: $\dot{x}_2^e - \dot{x}_2^p = 0$
(d) The sum of the relative velocities of dashpot 1, spring 1 and spring 2 is equal to the displacement of the joint 3: $\dot{x}_1^e + \dot{x}_1^p + \dot{x}_2^e = \dot{x}_3$

Note that *the No. of DOFs (7) − No. of constraints (4) = No. of joints (3)*. This is an important check to see if we have listed all the constraints.

(7) **Elimination by substitution:** Now, use the equations of the constraints to rewrite (2.44) by eliminating all but three kinematical variables and thus rewrite it in terms of whichever three degrees of freedom that you want. This step allows us to rewrite the power theorem in terms of a "minimal" set of variables (in this case 3). For example, if we use the constraint equations to eliminate all the variables other than $\dot{x}_1, \dot{x}_2$ and $\dot{x}_3$ we will get

$$\dot{x}_1^p = \dot{x}_1, \quad \dot{x}_2^p = \dot{x}_3 - \dot{x}_2, \quad \dot{x}_1^e = \dot{x}_2 - \dot{x}_1, \quad \dot{x}_2^e = \dot{x}_3 - \dot{x}_2. \tag{2.45}$$

Now substituting the above expressions into equation (2.44), and grouping terms, we will get

$$F_{ext}\dot{x}_3 = \sum_i Q_i \dot{x}_i \Rightarrow (F_{ext}-\psi_{,2}-F_2^d)\dot{x}_3+(F_2^d-\psi_{,1}+\psi_{,2})\dot{x}_2+(\psi_{,1}-F_1^d)\dot{x}_1 = 0,$$

where $\psi_{,i} := \partial \psi / \partial x_i^e$.

(8) Once we have the form (2.46), we can read off the equations of motion as the coefficients of the velocities set to zero, i.e. we get

$$\begin{aligned} F_{ext} - \psi_{,2} - F_2^d &= 0 \\ F_2^d - \psi_{,1} + \psi_{,2} &= 0 \\ \psi_{,1} - F_1^d &= 0 \end{aligned} \tag{2.46}$$

(9) Thus, *for spring dashpot systems without inertia, we can find the equations of motion by simply writing the power theorem in terms of a minimal set of variables and then reading off the coefficients.*

We want you to familiarize yourself with this process by doing the following exercise:

⊙ 2.10 (Exercise:). Flow of Mechanical Power

For the spring-mass-damper systems shown in Fig. 2.2(a–c), obtain the equations of motion from the power theorem and the constraints by following the procedure outlined above, assuming that the springs are all linear with the same stiffness k and all dampers being linear with viscous coefficient, c.

Hint: Obtain an expression for the strain energy and external power supply first. Do not substitute for the dashpot forces until after you have obtained the equations of motion. This is vital.

At this stage, dear student, you might be frustrated by the number of different ways that we got to these equations; Even though that we got in a much simpler way by just applying Newton's laws to each joint. We will try to address this issue and answer some of your questions in the following subsections.

2.4.1 *Frequently asked questions and clarifications*

Based on our experience in teaching this material to students at Texas A&M University and IIT Madras, we found it useful to discuss this energy approach as a question and answer session.

Q: What did we gain by finding the equations of motion by using the power theorem?:

For a simple one-dimensional spring dashpot system, we gained NOTHING. It IS simpler to use Newton's laws at the joints. HOWEVER, *what we gained is generalizability. The kinematical variables you introduce and their classification into "joints, springs and dashpots", allow you to specify what elements are present. The constraints you use represent connectivity (or compatibility between the elements). These two ideas can be generalized all the way to continua, and thus we are introducing you to a common framework*. In two or higher dimensions, this approach is much simpler than using Newton's laws.

To elaborate, what was the main advantage of using Newton's laws and free body diagrams? We could see what we were doing and *what you see is what you get*. However, this is very limiting since *what you see is all you've got!*. You can't draw free body diagrams with vector forces for complicated 3-D structures containing hundreds of elements, but you can easily find the power which is a scalar.

We can also consider much more general scenarios that can't even be expressed as simple spring mass systems. For example if we have constraints of the form

$$\sum_{i=1}^{n} A_{\alpha i} \dot{q}_i = 0, (\alpha = 1, \ldots, m) \tag{2.47}$$

We may not even be able to visualize it especially if the matrix $A_{\alpha i}$ changes with time. Nevertheless, we can obtain equations of motion by using the constraint equations to eliminate all but $n - m$ variables. We will later see that even more general constraints (could be nonlinear in $\dot{q}_i$) can be handled effortlessly.

Similarly, the energy function and the external forces can be any function of q_i, not just a sum of terms each depending on only one of the q's. This will come in very handy in more than one dimension where the constraints can be much more difficult. Later, we will also see that it provides a framework for extending "spring-dashpot" mechanics to develop thermomechanical constitutive equations for continua. This is one of the major advantages.

Thus, the power theorem approach is used to see the commonality of approaches between the spring-dashpot systems and the constitutive equations to be dealt with in the later chapters.

Q: Why did we choose x_i's as the coordinates to retain in equation (2.46)? What will happen if I choose some other set of coordinates—say the x^e's or the x^p's?

Excellent question. You can choose any set of three independent variables to retain—but you have to have 3 independent coordinates, not less. You will get different combinations of the equations of motion depending upon which coordinates you choose. The following exercise demonstrates this point:

⊙ **2.11 (Exercise:). Different Variables imply different forms for the equations of motion**
Using $\dot{x}_1^p, \dot{x}_2^p$ and $\dot{x}_1^e$ as your variables, show that the power theorem (2.44) reduces to

$$(F - \psi_{,1})\dot{x}_1^p + (F - \psi_{,2} - F_2^d)\dot{x}_2^p + (F - F_1^d)\dot{x}_1^p = 0. \tag{2.48}$$

What are the equations of motion now. Prove that they are indeed correct by directly applying Newton's Laws.
Can you use x_1^p, x_2^e and x_3^p as the independent variables. If not, why not?

There is actually some benefit in choosing your independent variables wisely. The reason is that we eventually want differential equations of the type $\dot{x} = f(x,t)$. It is easier to obtain this if you chose the dashpot variables x_i^p and whatever additional variables necessary to obtain the equations. The reason is evident from the exercise above. You can see that the equations of motion are

$$\psi_{,1} = F(t),\ F_1^d = F(t),\ F_2^d = F - \psi_{,2}. \tag{2.49}$$

Notice that in all the above equations, the right-hand side is only a function of the state variables and/or time (since ψ is a function of the state variables alone) whereas the left-hand side is a function of the velocities. All velocity variables show up on the left-hand side. For example, for the viscous dashpots, if we had $\dot{x}_1^p = f(F_1^d)$, and $\dot{x}_2^p = f(F_2^d)$, then the governing differential equations are

$$\dot{x}_1^p = f(F_1^d) = \frac{1}{C_{01}}F_1^d = F/C_{01}, \quad \dot{x}_2^p = f(F_2^d) = \frac{1}{C_{23}}F_2^d = (F - \psi_{,2})/C_{23} \quad (2.50)$$

So a general guiding principle is that

Try to make sure that all the dashpot variables are retained as independent variables and as many of the spring variables as possible are eliminated. This will allow us to set the ODEs in a straightforward way as we shall see in the next section.

The next question is very important

Q: Why did we treat the dashpot forces separately?
The answer lies in the fact that power of the spring forces can be replaced by the rate of increase strain energy function but there is no "strain energy function" for dashpot forces.

Q: Just like the spring forces are obtained from a work potential, can we obtain the dissipative forces from a dissipative potential?
This is a very important question and we will address it in detail here.

The fact that the spring forces are obtained from a work potential is due to the observation that there is no perpetual motion machine of the first kind: we cannot get work out of a spring by simply cycling it back and forth at a constant temperature. There is not such a nice clean statement for dissipative forces.

All that is required of the dissipative forces is that they do not generate mechanical power but only consume it, i.e. if F_i^d are the dissipative forces, then we only need to ensure that dissipation be non-negative, i.e.

$$\zeta := \sum_{i=1}^{n} F_i^d \dot{x}_i^p \geq 0 \quad (2.51)$$

As long as the above criterion is met, we are free to specify the dissipative forces in any manner we like. So really speaking, the second law of thermodynamics gives us considerable leeway in choosing dissipative forces.

However if we actually look at all the dissipative forces that we have encountered, they generally tend to have a fairly special structure.

For the vast majority of dissipative mechanisms that we have encountered (including well-known laws / principles such as "Fourier's law of heat conduction", "Darcy's law" "Fick's law", "power law viscosity" etc.) it is possible, as a working hypothesis, to assume something "like" a dissipative potential. The argument in favor of assuming something about the dissipative potential is not nearly as strong or universally proclaimed as with the work potential for springs[20].

To understand the nature of this dissipation potential, let us first examine the work potential assumption. The statement is as follows "There is a function ψ which is a function of the elastic variables x_i^e such that the spring force is equal to the gradient of ψ, in consequence, we also have $\mathbf{F} \cdot \mathbf{v} = \dot\psi$".

Now let us try something like that for a dashpot. We begin with the dissipative force associated with a dashpot. It is better to look at this in two dimensions rather than in 1-D [21]. Recall that in 2-D, we had, for a simple dashpot

$$\mathbf{F}^d = \phi \mathbf{v}, \phi = \phi(\|\mathbf{v}\|) \Rightarrow \zeta(\mathbf{v}) := \mathbf{F}^d \cdot \mathbf{v} = \phi(\|\mathbf{v}\|) \mathbf{v} \cdot \mathbf{v} = \zeta(\|\mathbf{v}\|) \quad (2.52)$$

If you now differentiate ζ with respect to $\mathbf{v}$, you will see that the *direction of* $\mathbf{F}^d$ *is along the gradient of* ζ *with respect to* $\mathbf{v}$, i.e.

$$\mathbf{F}^d = \lambda \frac{\partial \zeta}{\partial \mathbf{v}}. \quad (2.53)$$

However, the magnitude of $\mathbf{F}^d$ is NOT equal to that of the gradient of ζ. But we know that the magnitude of $\mathbf{F}^d$ should be such that

$$\mathbf{F}^d \cdot \mathbf{v} = \zeta. \quad (2.54)$$

Combining (2.53) and (2.54), we arrive at

$$\mathbf{F}^d = \lambda \frac{\partial \zeta}{\partial \mathbf{v}}, \quad \lambda = \frac{\zeta}{\mathbf{v} \cdot \partial \zeta / \partial \mathbf{v}}. \quad (2.55)$$

You can verify for yourself that this actually satisfies (2.54).

Now what happens for dry friction? We will show that it works for dry friction too: Consider $\zeta = \kappa \|\mathbf{v}\|$ where κ is a constant. Now, applying (2.55) we see that

$$\mathbf{F}^d = \lambda \frac{\partial \zeta}{\partial \mathbf{v}} = \lambda \kappa \frac{\mathbf{v}}{\|\mathbf{v}\|} = \phi \mathbf{v}, \phi := \lambda \kappa. \quad (2.56)$$

Now how about ϕ? we can obtain it by setting

$$\mathbf{F}^d \cdot \mathbf{v} = \zeta \Rightarrow \phi \mathbf{v} \cdot \mathbf{v} = \kappa \|\mathbf{v}\| \Rightarrow \phi = \frac{\kappa}{\|\mathbf{v}\|} \quad (2.57)$$

We thus end up with

$$\mathbf{F}^d = \kappa \frac{\mathbf{v}}{\|\mathbf{v}\|} \quad (2.58)$$

Clearly, this is the 3 dimensional generalization of the equation (2.32) for friction that we obtained for friction in 1-D. $\mathbf{F}^d$ is indeterminate if $\mathbf{v}$ is zero and we can see that $\|\mathbf{F}^d\| = \kappa$ if $\mathbf{v} \neq 0$. The friction condition is also met.

So, our assumption about the dissipative forces is not too bad after all! So based on this rather "weak" justification and in an analogy with the springs, we have the following two assumptions for springs and dashpots respectively:

---

[20]In other words, we are too "chicken" to state the existence of a dissipation potential is a fundamental thermodynamical principle because we are not at all sure. Nevertheless, this assumption works out very well for such a wide range of cases (including for plasticity), that it is well worth investigating as we will see later.

[21]In 1-D every dashpot and every spring has a potential so there is nothing to prove.

- There is a work potential ψ such that the conservative forces are given by

$$F_i^c = \partial \psi / \partial x_i^e \Rightarrow \sum_i F_i^c \dot{x}_i^e = \dot{\psi} \qquad (2.59)$$

- There is a dissipative potential ζ which is always non-negative, and is such that the dissipative forces are given by

$$F_i^d = \lambda \frac{\partial \zeta}{\partial \dot{x}_i^p}, \quad \lambda = \sum_i \frac{\zeta}{\dot{x}_i^p \cdot \partial \zeta / \partial \dot{x}_i^p} \Rightarrow \sum_i F_i^d \dot{x}_i^p = \zeta. \qquad (2.60)$$

Q: Why do we need to assume that ζ is a function of $\dot{q}_i$?
It has been observed that a dissipative element dissipates mechanical work only when it moves, i.e. when q_i *changes* (imagine a dashpot with ends fixed, does it dissipate energy?). So, we must have $\zeta = 0$ if $\dot{q}_i = 0$. Therefore, ζ depends upon $\dot{q}_i$.

Q: In classical mechanics we can obtain the equilibrium configurations using variational principles such as Castigliano's theorems. Can we obtain evolution equations in a similar manner? Generally speaking, whenever we have an assumption that something is related to the gradient of a scalar it is possible to state it as an extremum principle. For example the fact that the spring forces are equal to the gradient of the work potential can be stated in the form of the assumption that "the equilibrium configuration of a system is one that extremizes the strain energy function" (which is the well known Castigliano's first theorem). The main attraction of such variational principles is that they give us a feeling that we get "something" for "nothing". We simply have some scalar function, we minimize it and voila, we have vector equations of equilibrium. We will also succumb to this attraction and show that, it is indeed possible to obtain evolution equations from an extremum principle.

If we allow for the dissipative forces to be obtained from a "dissipation potential", we are now in a position to make a very bold "variational principle" that will allow us to incorporate the power theorem, the constraints as well as the equations of motion into a single overarching statement. We will use notation that are consistent with dynamics notations for generalized coordinates and forces (see [Ziegler (1963); Rajagopal and Srinivasa (2004c)] for detailed analysis in a general thermomechanical case):

> **THE MAXIMUM RATE OF DISSIPATION HYPOTHESIS (MRDH)**
>
> Consider an inertialess system that is subject to known external forces. Let the degrees of freedom of the system be q_i and the external forces be $Q_i^{ext}(t)$ (many of them may be zero) so that the power supply is $\sum_i^n Q_i^{ext} \dot{q}_i$. Let the kinematical variables be subject to m constraints of the form $f_\alpha(q_i, \dot{q}_i, t) = 0$ ($\alpha = 1, \ldots, m$). If the constraints f_α are independent of $\dot{q}^i$, we differentiate them once to get it to the form required for this theorem. Its response is characterized by two functions: the work potential $\psi(q_i)$ and the dissipative potential $\zeta(q_i, \dot{q}_i)$ which are related by the power theorem,
>
> $$\sum_i Q^i(t)\dot{q}_i = \dot{\psi} + \zeta = \frac{\partial \psi}{\partial q_i}\dot{q}_i + \zeta(q_i, \dot{q}_i) \qquad (2.61)$$
>
> *The state of the system evolves in such a way that among all possible values of $\dot{q}_i$ which are consistent with the kinematical constraints and the power theorem, the actual value of $\dot{q}_i$ will maximize the dissipation potential ζ*

Note that the requirement of the second law is that "among all the possible values of $\dot{q}_i$, the allowable values satisfy the constraints and render the rate of dissipation be positive but not necessarily be maximum. Thus, this hypothesis is much stronger.

Q: In my previous experiences, when we have used extremum principles, we have only varied q_i. Why are we varying $\dot{q}_i$? Also is this hypothesis really necessary?

Good question. What is our intent? We are trying to find out HOW the state evolves. To be more precise, let us say that time at t, I am at a state $q_i = q_i^0$. Now I want to find out where I should be at time $t + \triangle t$. In other words, I need to find out what should $\triangle q_i$ be, i.e. how large a step to take and in which direction. What the non-negativity of the rate of dissipation tells me is that "you can take any step in any direction as long as you dissipate work in the process". So, discrete constitutive equation of the form

$$\triangle q_i = f_i(q_j, Q_j^{ext}, \ (j = 1, \ldots, n)) \qquad (2.62)$$

which says that, "if you are at state q_i and the external forces are Q_i^{ext} at this instant, then take a step $\triangle q_i$ as specified.". This will be allowed as long as the mechanical work dissipated in the step is non-negative, i.e. so long as $\sum_i \triangle q_i Q_i \geq 0$.

So, we can say: "if every step leads to negative dissipation, then don't go anywhere". The system is at rest, this is equilibrium.

The maximum dissipation hypothesis makes a statement that is much stronger than this. It says "of all the possible steps take a step in the direction that maximizes the dissipation of mechanical work and which satisfies all the constraints." You could think of it as "a principle of laziness".

This kind of a variational principle is called an "incremental variational principle" and is similar in spirit to the well-known Gauss' principle of least constraint.

Embedded in this assumption is the statement that "if every step from the current state leads to negative dissipation, then don't go anywhere. Since not going anywhere leads to zero dissipation which is the maximum possible (since all other steps lead to negative dissipation)". The maximum rate of dissipation criterion is the continuous limit of this discrete time stepping.

Let us reiterate that it is NOT necessary to assume that the rate of dissipation in maximized. All that is necessary is to choose something of the form (2.62) which will ensure that ζ is non-negative. However, maximizing ζ is sufficient to provide an evolution equation that satisfies the criterion that ζ be non-negative. Furthermore, a large class of material response relations (including all the plasticity constitutive equations that we are going to investigate) actually satisfy the maximum dissipation hypothesis.

Q: Show me with an example how to use this variational hypothesis

Let us start with a simple case and consider the spring dashpot system of Fig. 2.2(a) redrawn here for convenience

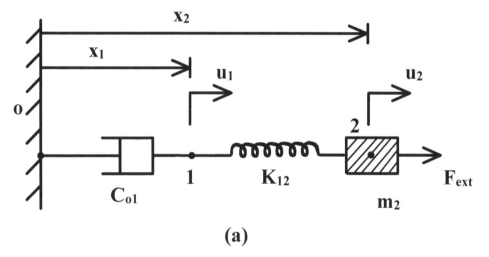

(a)

(1) The degrees of freedom are : $u_1, u_2, x_1^e = (x_2 - x_1), x_1^p = x_1$ reflecting the fact that there are two joints, 1 and 2 with displacements u_1 and u_2, a spring with a displacement, x_1^e and a dashpot with a displacement, x_1^p. These are the state variables and we want to find their evolution.

(2) The energy and dissipation potentials are $\psi(x_1^e)$ and $\zeta(\dot{x}_1^p)$. We don't have to

worry about the precise form of these as long as ζ is non-negative.
(3) The power theorem becomes $F^{ext}\dot{u}_2 = \psi_{,1}\dot{x}_1^e + \zeta(\dot{x}_1^p)$. We can use F instead of F^{ext} for this example since only one F^{ext} exists.
(4) There are two constraints representing the connectivity of the joints: $\dot{u}_1 - \dot{x}_1^p = 0$ and $\dot{u}_2 - \dot{x}_1^p - \dot{x}_1^e = 0$.
(5) Now to find the equations of motion we simply have to maximize ζ over all values of $\dot{u}_1, \dot{u}_2, \dot{x}_1^e, \dot{x}_1^p$ that satisfy the constraints. This is the tricky part. However, have no fear. There is a remarkable procedure that is attributed to the great mechanician and mathematician Lagrange (called the Lagrange multiplier method) that will allow us to do this maximization[22]. It will also allow us to introduce this important method to you.

The Lagrange multiplier method for constrained Maximization:
The trick is simple: if you want to find the extrema of a function $f(\mathbf{x})$ subject to a side condition of the form $g(\mathbf{x}) = 0$, simply form a new function with the original function augmented by $g(\mathbf{x})$ as follows: $h(\mathbf{x}, \lambda) = f(\mathbf{x}) + \lambda g(\mathbf{x})$. Then the extrema of this augmented function (with respect to $\mathbf{x}$ and λ) are

$$\frac{\partial h}{\partial \mathbf{x}} = \frac{\partial f}{\partial \mathbf{x}} - \lambda \frac{\partial g}{\partial \mathbf{x}} = 0 \Rightarrow \frac{\partial f}{\partial \mathbf{x}} = \lambda \frac{\partial g}{\partial \mathbf{x}},$$
$$\frac{\partial h}{\partial \lambda} = g(\mathbf{x}) = 0. \quad (2.63)$$

Thus, the extrema of the augmented function (without any constraints) are the same as that of the original function with constraints built in! The parameter λ is called the Lagrange multiplier. So remember: "augment the function with the constraint equations and just do regular extremization"—nothing could be simpler than this!

(6) In our case, the augmented function is

$$h(\dot{u}_1, \dot{u}_2, \dot{x}_1^e, \dot{x}_1^p, \lambda_1, \lambda_2, \lambda_3) = \zeta(\dot{x}_1^p) + \lambda_1(F\dot{u}_2 - \psi_{,1}\dot{x}_1^e - \zeta(\dot{x}_1^p)) \\ + \lambda_2(\dot{u}_1 - \dot{x}_1^p) + \lambda_3(\dot{u}_2 - \dot{x}_1^p - \dot{x}_1^e). \quad (2.64)$$

Notice that the augmented function is composed of (1) the function to be maximized (which is the dissipation potential), (2) the equations representing the power theorem and (3) the equations representing the kinematical constraints. Observe that we now need three Lagrange multipliers ($\lambda_1, \lambda_2, \lambda_3$) —one for the power theorem and one for each constraint. Now simply maximize this function

---

[22] This is an amazing, elegant, versatile and wonderful method and, as far as we know, this is the only method that actually gives us something for nothing! The trick is so easy but the result is so useful. Just type Lagrange Multiplier on Google and see how many hits you get!

over all values of $\dot{u}_1, \dot{u}_2, \dot{x}_1^e, \dot{x}_1^p, \lambda_1, \lambda_2, \lambda_3$ and get the following set of equations:

$$\frac{\partial h}{\partial \dot{u}_1} = 0 \Rightarrow \lambda_2 = 0$$

$$\frac{\partial h}{\partial \dot{u}_2} = 0 \Rightarrow \lambda_1 F + \lambda_3 = 0$$

$$\frac{\partial h}{\partial \dot{x}_1^e} = 0 \Rightarrow \lambda_1 \psi_{,1} + \lambda_3 = 0$$

$$\frac{\partial h}{\partial \dot{x}_1^p} = 0 \Rightarrow (1 - \lambda_1)\frac{\partial \zeta}{\partial \dot{x}_1^p} - \lambda_2 - \lambda_3 = 0 \quad (2.65)$$

$$\frac{\partial h}{\partial \lambda_1} = 0 \Rightarrow F\dot{u}_2 - \psi_{,1}\dot{x}_1^e - \zeta(\dot{x}_1^p) = 0$$

$$\frac{\partial h}{\partial \lambda_2} = 0 \Rightarrow \dot{u}_1 - \dot{x}_1^p = 0$$

$$\frac{\partial h}{\partial \lambda_3} = 0 \Rightarrow \dot{u}_2 - \dot{x}_1^p - \dot{x}_1^e = 0.$$

Note that this provides 7 equations for the 7 unknowns, but what a mess! This is the price you pay for introducing Lagrange multipliers—you have to solve many more equations[23]!

We can clean this up very nicely if we systematically eliminate the Lagrange multipliers. This will be your next exercise:

⊙ **2.12 (Exercise:). Eliminating the Lagrange multipliers**
Simplify the above set by first dividing each equation by λ_1 and eliminating λ_3/λ_1 from the second, third and fourth equations of (2.65) and show that we get

$$\psi_{,1} = F, \quad \mu\frac{\partial \zeta}{\partial \dot{x}_1^p} = F, \quad \mu = (\lambda_1 - 1)/(\lambda_1). \quad (2.66)$$

Show by directly using Newton's laws that these equations are the equations of motion with the dissipative force given by $F^d = \mu\partial\zeta/\partial\dot{x}_1^p$.
Also show by using the fifth equation that

$$\mu = \frac{\zeta}{(\dot{x}_1^p)(\partial\zeta/\partial\dot{x}_1^p)}. \quad (2.67)$$

In other words, we obtained the equations of motion together with the constitutive equations for the dashpot and for the spring.

---
[23]This is an example of the Free Lunch theorem as in "there is no free lunch":-)))

Q: It appears that we have made our procedure more and more complex rather than simplifying it; do we need to go through this maximization process every time?

Have no fear, we would not go through all of this if there was no payoff: remember this book is meant to be a practical guide! This means that we are going to include techniques only if they have practical utility.

The main point of doing this maximization procedure is to formalize the evolution equation generating procedure. We will not need to do any maximization; once we find the state variables, the energy and dissipation potentials and the constraints, we can *directly write down the equations of motion for a inertialess dissipative system without doing any further calculations!* This is the big payoff.

We don't really need to go through this maximization process every time. The structure of the constraints is so simple that the answer can be written down straightaway, we will state the result below:

EQUATIONS OF MOTION FOR AN INERTIALESS SYSTEM

Consider an inertialess lumped parameter system that is subject to known external forces. Let the degrees of freedom of the system be q_i and the external forces be $Q_i^{ext}(t)$ (many of them may be zero) so that the power supply is $\sum_i^n Q_i^{ext} \dot{q}_i$. Let the kinematical variables be subject to m constraints of the form $f_\alpha(q_i, t, \dot{q}_i) = 0$ ($\alpha = 1, \ldots, m$). Let the work potential be $\psi(q_i)$ and the dissipation potential be $\zeta(q_i, \dot{q}_i)$. Then the equations of motion are given by

$$Q_i^{ext} - \frac{\partial \psi}{\partial q_i} - \mu \frac{\partial \zeta}{\partial \dot{q}_i} - \sum_\alpha \mu_\alpha \frac{\partial f_\alpha}{\partial \dot{q}_i} = 0, \tag{2.68}$$

where μ_α are obtained by satisfying the equations of constraints and

$$\mu = \frac{\zeta - \sum_\alpha \mu_\alpha \sum_i \dot{q}_i \partial f_\alpha / \partial \dot{q}_i}{\sum_i \dot{q}_i \partial \zeta / \partial \dot{q}_i}. \tag{2.69}$$

Further, if f_α are *homogeneous functions of* $\dot{q}_i$, i.e. if they each satisfy

$$f_\alpha(q_i, \lambda \dot{q}_i, t) = \lambda f_\alpha(q_i, \dot{q}_i, t) = 0 \text{ for all } \lambda, \tag{2.70}$$

then μ simplifies to

$$\mu = \frac{\zeta}{\sum_i \dot{q}_i \partial \zeta / \partial \dot{q}_i}. \tag{2.71}$$

This is a major result: Note that the equations of motion are given as a direct formula applicable for ANY inertialess dissipative system!

⊙ 2.13 (Exercise:). Obtaining the general form for the equations of motion

Show that the results (2.68), (2.69) and (2.71) are true.
Hint: Just apply the method of Lagrange multipliers to obtain (2.68). Then, multiply the equations of motion by $\dot{q}_i$ and sum; use the power theorem to simplify and then solve for μ. The last equation is tricky. You need to show that for homogeneous functions, the summation on the numerator of (2.69) is zero. Look up "homogeneous functions" on Wikipedia. You might be especially interested in "Euler's theorem".
Show further that if ζ itself is a homogeneous function, then μ is a constant.

Q: Wait a minute here. You stated this rather general result by just doing one dimensional models. Does it still work for 2 or 3 dimensional models?

True, we have extrapolated from purely one dimensional considerations, in the true spirit of physics[24]. However, we can easily verify it in 2 dimensions as the next example shows:

Consider a spring dashpot system shown in Fig. 2.5. Note that the spring 3 and the external forces are always along the x direction. We will now set up and solve this problem using the results (2.68) and (2.69). For definiteness, let us assume that the springs and dashpots are linear.

(1) As indicated in Fig. 2.5, the kinematical variables are $q_1 = r_1$, $q_2 = r_2$, $q_3 = r_3$, $q_4 = r_4$, $q_5 = \theta_1$, $q_6 = \theta_2$, $q_7 = \theta_3$, $q_8 = \theta_4$.

(2) The work potential is
$$\psi = \frac{1}{2}k_1(r_1 - r_1^0)^2 + \frac{1}{2}k_2(r_2 - r_2^0)^2 + \frac{1}{2}k_3(r_3 - r_3^0)^2. \quad (2.72)$$

(3) The dissipation potential is also assumed to be quadratic since we are going to assume that the dashpot is a linear dashpot.
$$\zeta = c\dot{r}_4^2. \quad (2.73)$$

(4) The external forces are F^{ext} acting at the point 3. The power supplied by it is
$$F^{ext}\frac{d}{dt}(r_1 \sin\theta_1 + r_3) = F^{ext}(\dot{r}_1 \sin\theta_1 + r_1\cos\theta_1\dot{\theta}_1 + \dot{r}_3) \quad (2.74)$$
so that the generalized external forces (which are the coefficients of the $\dot{q}_i$'s)[25] are $Q_1 = F^{ext}\sin\theta_1$, $Q_2 = 0$, $Q_3 = F^{ext}$, $Q_4 = 0$, $Q_5 = F^{ext}r_1\cos\theta_1$ $Q_6 = Q_7 = Q_8 = 0$.

---

[24] where else can you speculate on the motion of planets and stars and galaxies just from observing the falling of an apple?
[25] This is an important point: if you want to find the generalized force, write down the power supplied in terms of the generalized coordinates and find the coefficients of the rate of the generalized coordinates.

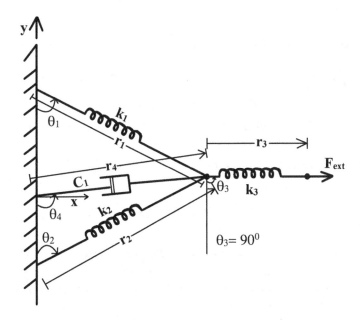

Fig. 2.5 A 2 dimensional spring-dashpot system with two springs and a dashpot on a plane.

(5) The constraints are[26]

$$\begin{aligned}
g_1 &= r_1 \sin\theta_1 - r_2 \sin\theta_2 = 0 \\
&\Rightarrow f_1 = \dot{g}_1 = \sin\theta_1 \dot{r}_1 + r_1 \cos\theta_1 \dot{\theta}_1 - \sin\theta_2 \dot{r}_2 - r_2 \cos\theta_2 \dot{\theta}_2 = 0, \\
g_2 &= r_2 \sin\theta_2 - r_4 \sin\theta_4 = 0 \\
&\Rightarrow f_2 = \dot{g}_2 = \sin\theta_2 \dot{r}_2 + r_2 \cos\theta_2 \dot{\theta}_2 - \sin\theta_4 \dot{r}_4 - r_4 \cos\theta_4 \dot{\theta}_4 = 0, \\
g_3 &= r_1 \cos\theta_1 + r_2 \cos\theta_2 = a_1 \\
&\Rightarrow f_3 = \dot{g}_3 = \cos\theta_1 \dot{r}_1 - r_1 \sin\theta_1 \dot{\theta}_1 + \cos\theta_2 \dot{r}_2 - r_2 \sin\theta_2 \dot{\theta}_2 = 0 \\
g_4 &= r_2 \cos\theta_2 + r_4 \cos\theta_4 = a_2 \\
&\Rightarrow f_4 = \dot{g}_4 = \cos\theta_2 \dot{r}_2 - r_2 \sin\theta_2 \dot{\theta}_2 + \cos\theta_4 \dot{r}_4 - r_4 \sin\theta_4 \dot{\theta}_4 = 0.
\end{aligned} \quad (2.75)$$

Note that *No. of DOFs − No. of constraints = No. of equations of motion*.

Observe also that the constraint functions, f_α, are homogeneous in the generalized velocities. By now you have probably noticed that not all the generalized coordinates have the same dimension. Some of them are lengths and some are angles and we can freely mix them. Now by using (2.68), we can easily write out the equations of equilibrium as follows:

---

[26] We hope you figure out the constraints by yourself: Note that they are statements to the effect that "spring 1 and 2 are connected so the x and y coordinates of the connection must match". We urge you to write down in words what each constraint below means. Also, note that constraints on the geometry are differentiated to get us constraints on the generalized velocities. It is easier to enforce the latter constraints in a computer program, rather than the former.

⊙ **2.14 (Exercise:). Equations of motion**
Apply (2.68) and obtain the following equations (μ_1, μ_2, μ_3 and μ_4 are the Lagrange multipliers for the four constraints, g_i, $i = 1, 2, 3, 4$).

$$0 = F^{ext}\sin\theta_1 - k_1(r_1 - r_1^0) - \mu_1\sin\theta_1 - \mu_3\cos\theta_1,$$
$$0 = -k_2(r_2 - r_2^0) + \mu_1\sin\theta_2 - \mu_2\sin\theta_2 - \mu_3\cos\theta_2 - \mu_4\cos\theta_2,$$
$$0 = F^{ext} - k_3(r_3 - r_3^0),$$
$$0 = -2\mu c\dot{r}_4 + \mu_2\sin\theta_4 + \mu_4\cos\theta_4, \quad (2.76)$$
$$0 = F^{ext}r_1\cos\theta_1 - \mu_1 r_1\cos\theta_1 + \mu_3 r_1\sin\theta_1,$$
$$0 = -\mu_1 r_2\cos\theta_2 + \mu_2 r_2\cos\theta_2 - \mu_3 r_2\sin\theta_2 - \mu_4 r_2\sin\theta_2,$$
$$0 = \mu_2 r_4\cos\theta_4 + \mu_4 r_4\sin\theta_4 = 0.$$

By using Newton's laws at the two joints, show that the equations of motion are indeed correct. Also show that the Lagrangian multiplier μ associated with the satisfaction of the power theorem takes the value $\mu = 1$. You might notice that we have to eliminate a lot of the Lagrange multipliers in order to get the same form as you would obtain from Newton's Laws, but it will convince you that the procedure works.

Note that there are seven equations and they represent a complex set of equations to solve. The reason for our unduly complex form of these equations is that we did not select "good" variables. It would have been much better to select the coordinates of the joint A and r_3 as the coordinates[27]. There would have been no constraints and the process would have been extremely simple:

⊙ **2.15 (Exercise:). New Coordinates**
Re-do the problem using only the x,y coordinates of the joint A and r_3 as the generalized variables: Note that there are only three variables and no kinematical constraints, so this will make the problem considerably simpler.
Hint: $r_1 = \sqrt{x^2 + (y + c_1)^2}$ and similarly for r_2 and r_4. You don't need any of the θ variables.

Congratulations! if you worked out the previous exercise, you have just done a problem with *nonlinear kinematics*. At least you know how to get the equations of motion. The key thing to remember is to choose your coordinates wisely without too many constraints.

---

[27] It is hard to find the perfect set of coordinates for a given problem. Experience and some trial and error is necessary.

PART B
LUMPED PARAMETER MODELS: THERMO-MECHANICAL PROBLEMS

2.5 Thermal considerations: The Helmholtz and Gibbs potentials

Until now, we have ignored the thermal response of the system and focussed solely on the mechanical response of the system. But consideration of the thermal response of the material is vital for many technological purposes. For example, if a bar is subject to a very rapid impact (such as by a projectile or a "blast") it will deform so rapidly that there will be significant increase in temperature. This, in turn will decrease its yield strength so that its collapse will occur earlier. Modeling phenomena such as this and many more requires a good understanding of thermo-mechanics as it is to this aspect that we turn to next.

We will first incorporate thermal effects and then consider some nonlinear springs and dashpots with temperature dependent response, followed by further consideration of frictional dashpots which require a rather careful treatment. The final part of this chapter includes a discussion of an approach which is based on consideration of forces and not displacements.

The first thing to note is that we need to reinterpret the "potential energy" of a spring. To see why we need to reinterpret the notion of "potential energy of a system," recall that for a purely mechanical system that we have been considering so far, *the potential energy changes only if the springs are stretched*. Thus, in this setting, a rigid link has no potential energy at all. On the other hand, if the temperature of the system is changing and then the internal energy of all parts of the system changes. For example we know that a rigid bar has no strain energy but if you heat it, it can "store" energy and you will be able to recover it when you cool it back. However, not all of it is recoverable as work. So we have to generalize the notion of potential energy to include temperature effects and at the same time introduce the notion that only a part of it is recoverable as work.

In order to do this, we first note that we can generalize some well-known results in classical structural mechanics. We know that for a spring, the work done by the spring when it goes from one state to another is equal to the difference in the potential energy between the two states. *The Helmholtz potential (or the isothermal work function)* of a system is a generalization of this idea (see Fig. 2.6):

> ## THE HELMHOLTZ POTENTIAL
>
> If we change the state of any system from s_1 to s_2 keeping the temperature T constant, then the work supplied to all the conservative components (and hence recoverable) is equal to the difference in the Helmholtz potential, ψ, i.e. if we consider a spring connecting joints i and j, and Q_i^s is the generalized force on the spring at node i, then
>
> $$\psi(s_2, T) - \psi(s_1, T) = \int_{s_1}^{s_2} Q_i^s \dot{q}_i \, dt = \int_{s_1}^{s_2} Q_i^s dq_i \qquad (2.77)$$
>
> where q_i is the generalized displacement. Thus
>
> $$Q_i^s = \left.\frac{\partial \psi(q_i, T)}{\partial q_i}\right|_{T=\text{const}}. \qquad (2.78)$$
>
> Thus, *the maximum amount of work recoverable from a system at a given temperature is equal to the change in the Helmholtz potential at that temperature.*

The above equation is *identical* to the familiar Castigliano's first theorem except for the fact that we have now said that the derivative must be calculated keeping temperature fixed. In other words, *the spring forces in a system at a given temperature is equal to the derivative of the Helmholtz potential with respect to the corresponding displacement at that temperature.*

Just as the Helmholtz potential can be viewed as a generalization of the strain energy function of classical mechanics, the Gibbs potential (also called Gibbs free energy) can be viewed as a generalization of the *complementary strain energy function* of classical mechanics.

To understand the notion of the Helmholtz and Gibbs Potentials, consider a spring whose force displacement relationship at a given constant temperature T is shown in Fig. 2.6. Geometrically, the change in the Helmholtz potential, $\Delta\psi$, of the spring is equal to the area ABC under the curve AC. This is generally written as a function of the displacement AB. On the other hand, the change in the Gibbs potential, ΔG, is the negative of the complementary energy, i.e. it is the negative of the area ADC above the curve. Mathematically, we have

$$\Delta\psi = \psi(q, T) - \psi(0, T) = \int_0^q Q(q)\, dq, \qquad (2.79)$$

$$\Delta G = G(Q, T) - G(0, T) = -\int_0^Q q(Q)\, dQ, \qquad (2.80)$$

where Q and q are the generalized force and displacements respectively.

Now, compared to the physical meaning of $\Delta\psi$ as the maximum work recoverable from a spring between two states at the same temperature, the physical meaning of ΔG is more abstract. One way to think of it is by the following thought

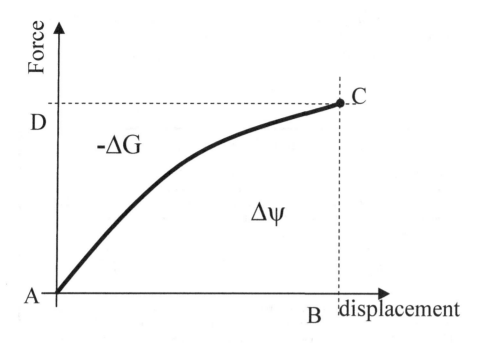

Fig. 2.6 Helmholtz Potential and Gibbs Potential for a force displacement relationship of a spring. Area ABC under the curve AC is the change in Helmholtz Potential and the negative of the area ADC is the change in Gibbs Potential.

experiment: consider a massless linear spring in a vertical plane, if we gradually apply a load to it, increasing its deflection "quasistatically", then we know that the work done by us is $\frac{1}{2}ku^2 = \frac{1}{2}Wu = \frac{1}{2}W^2/k$ where W is the maximum load or weight that is applied to the spring.

On the other hand, consider a process wherein the entire load W is suddenly applied to the spring and we wait until equilibrium is reached. Let us call this process *sudden loading*. The spring oscillates back and forth and eventually settles to the *same equilibrium value as before*. The deflection of the spring is $u = W/k$. The maximum work recoverable from the spring is still $\frac{1}{2}ku^2 = \frac{1}{2}Wu = \frac{1}{2}W^2/k$, however the work done by the weight (since it was applied suddenly, the weight fell a distance u in the presence of gravity), is $Wu = W^2/K$.

Thus in this second example, only half the work done by the weight is recoverable. Where did the remaining energy go? It was dissipated as heat[28].

The Gibbs potential is the *work recoverable from the spring minus the work*

---

[28] Can you visualize the fact that when a sudden load is applied, the weight oscillates back and forth, slowly decaying to equilibrium? In the process, it dissipates $\frac{1}{2}W^2/k$ units of energy.

done by the sudden loading $= \frac{1}{2}ku^2 - Wu = -\frac{1}{2}W^2/K$. This is what we mean by the negative sign of the complementary energy and it is the negative of the energy lost by the sudden load in the process of deforming the spring to its equilibrium configuration.

THE GIBBS POTENTIAL

If we change the state of any system from s_1 (which is at zero force) to s_2 keeping the temperature T constant, and with the same notation as before, the change in the Gibbs potential is defined as the work recoverable from the conservative components minus the work done by the loads on the conservative components as if they were applied as a sudden load, i.e.

$$G(s_2, T) - G(s_1, T) = \sum_i^n \int_{s_1}^{s_2} Q_i^s \dot{q}_i \, dt - Q_i^s(s_2) u_i(s_2) = -\sum_i^n \int_{s_1}^{s_2} q_i \, dQ_i^s \tag{2.81}$$

where q_i and Q_i are the forces and relative displacements associated with the spring connecting nodes i and j of the system. In view of the fact that the Helmholtz potential is the maximum recoverable work, it can be shown that the Gibbs potential at a given temperature is the *minimum work that can be gained by a system in a dead loading process*.

Of course the net work gained in a dead loading process will be negative since we cannot gain positive work by a dead load process, otherwise we would have a perpetual motion machine. The first part of (2.81) is just the work recoverable minus the work of the sudden load, the fact that it is equal to the second comes from integration by parts as you can verify.

Note that, there is a relationship between the Helmholtz and Gibbs Potentials, namely that

$$G = \psi - \sum_{i=1}^n Q_i^s q_i \tag{2.82}$$

$$\Rightarrow G = \psi - \sum_{i=1}^n \frac{\partial \psi}{\partial q_i} q_i; \quad \left(Q_i^s = \left(\frac{\partial \psi}{\partial q_i} \right) \right) \tag{2.83}$$

$$\Rightarrow \psi = G - \sum_{i=1}^n \frac{\partial G}{\partial Q_i^s} Q_i^s; \tag{2.84}$$

The main feature of the Gibbs potential is that it interchanges the roles of the forces and displacements, i.e. rather than obtaining the conservative forces as functions of the displacements, we obtain the displacements as functions of the conservative forces, i.e.

$$q_i = -\frac{\partial G}{\partial Q_i^s}|_{T=\text{const}} \qquad (2.85)$$

The above equation is *Castigliano's second theorem* except for the fact that we have now said that the derivative must be calculated keeping $T fixed$. In other words, *the displacement of a conservative component at a given temperature is equal to the negative of the derivative of the Gibbs potential with respect to the conservative force at that temperature.*

⊙ **2.16 (Exercise:). Helmholtz and Gibbs: the people**
Helmholtz and Gibbs are two of the most important contributors to all aspects of science and philosophy. Write a short essay on the life and contributions of Helmholtz and Gibbs. What struck you as the most interesting aspect of their life? What was Helmholtz'z connection with biology (if any)? What was Gibbs connection with vector notation?

2.5.1 *Extension of the power theorem*

By multiplying (2.78) by v_{ij}, we have the following important result:

POWER THEOREM-THERMOMECHANICAL GENERALIZATION

Consider a system that at a given instant of time is at a particular temperature. The net power supplied to the conservative elements at that instant temperature = rate of change of the Helmholtz potential at that temperature, i.e.

$$P_{spring} = \frac{d\psi}{dt}|_{T=\text{const.}} \qquad (2.86)$$

⊙ **2.17 (Exercise:). Power Theorem: Gibbs potential form**
Show that

$$P_{spring} = \frac{d}{dt}\left(G - \sum_{i=1}^{n} \frac{\partial G}{\partial Q_i^s} Q_i^s\right) \qquad (2.87)$$

Hint: Use (2.84) in (2.86)

The reason that the Helmholtz potential depends upon the temperature is due to the fact that the spring stiffness may depend upon the temperature. Indeed, for almost all metals, the elastic modulus decreases with increasing temperature, whereas for polymers, the modulus can increase with temperature.

> ⊙ 2.18 (Exercise:). Change in strain energy with temperature
>
> *A spring which has a temperature dependent stiffness of the form $k(T) = (1000 - 2T) N/m$ where T is the absolute temperature. What is its strain energy as a function of temperature and displacement?.*
>
> *It is stretched by 0.1 m from its relaxed state at room temperature (300K) and held fixed. How much work is recoverable from it at 300K?*
>
> *Now spring is fixed (after being stretched by 0.1 m) and the temperature is raised to 350K. How much work is recoverable from it now?*
>
> *Discuss what happens to the lost/gained work. At what temperature will the spring become slack. Is this the same as thermal expansion of the spring? If not, why not?*
>
> ANS: $\psi = 1/2(1000 - 2T)u^2$ N-m. *At 300K, the recoverable work is 2N-m. At 350K, the recoverable work is only 1.5N-m. Where did the 1 N-m of work go?*

The key point to notice is that *in a thermomechanical process, the recoverable work and hence the strain energy or the Helmholtz potential is NOT CONSERVED* as you might have discovered from the previous example. If the process were *isothermal* then the Helmholtz potential would be conserved and you would be able to get out whatever you put in.

We also wish to emphasize that *the strain energy is NOT the same as the total energy stored in the material*, since it does not account for heating or cooling of the material and focuses only on the work done. If you recall from the first law of thermodynamics, the change in the total energy stored in a material is the sum of the net work done and the net heat supplied— and we have been counting only the work done.

The Helmholtz potential deals with only the available work at any given temperature of the system. To illustrate this point, consider a rigid bar that is held fixed and simply heated. It is obvious that, as such, the bar is incapable of doing work since it cannot move, but we would agree that its energy increases. This simple example illustrates that the "strain energy" is not the same as total energy in a system.

The total energy including both mechanical and thermal effects but excluding the kinetic energy of the system is called the *internal energy* ∃ of the system[29]. If you consult a classical thermodynamics textbook (e.g. Callen), you will see that

---

[29] we are using the symbol ∃ for the internal energy because it is a backwards E and is also the mathematical symbol for "exists". Since our current vague idea is that all that exists is a form of "energy", we think it is an appropriate symbol for it. You can pronounce it as "E".

the Helmholtz potential is related to the internal energy by[30]

$$\beth = \psi - T\frac{\partial \psi}{\partial T} \tag{2.88}$$

Thus, the internal energy of a system is NOT the Helmholtz potential or "isothermal work function", but can be obtained from it through (2.88).

It can also be shown that (although it is not obvious)

$$\beth = G - \sum_{i=1}^{n}\frac{\partial G}{\partial Q_i^s}Q_i^s - T\frac{\partial G}{\partial T} \tag{2.89}$$

In order to familiarize yourself with the Gibbs potential, we will consider the example of a stiffening spring whose force displacement relationship is $F = ku^3$. Then, we have $u = (F/k)^{1/3}$ and $\psi = \frac{ku^4}{4} = \frac{k}{4}(\frac{F}{k})^{4/3}$. Thus, the Gibbs potential for the spring is

$$G = \frac{k}{4}(\frac{F}{k})^{4/3} - F(\frac{F}{k})^{1/3} = -\frac{3F^{4/3}}{4k^{1/3}} \tag{2.90}$$

We can also obtain the same result by $G = -\int_0^F u(F)dF$. You can verify by yourself whether this gives the same answer.

> ⊙ **2.19 (Exercise:). Simple example for the internal energy of a spring**
> Consider the spring from the previous exercise. Assume that its Helmholtz potential is given by $\psi = \frac{1}{2}(1000 - 2T)u^2$ where u is the displacement. Find an expression for its internal energy. We already saw that the work recoverable from the spring changes when it is heated so that when it gets to 500K, no work is recoverable from it. What happens to the internal energy as a function of temperature? Is the answer surprising? Now can you answer what happened to all the work that was supplied to the spring at room temperature?

We are now ready to extend the power theorem to a general process that includes heating.

---

[30] At this stage we are not ready to plunge into a lot of thermodynamical explanations as to WHY the expression (2.88) is true; If you accept it as the fundamental statement that we need in thermodynamics, then everything else becomes straightforward. We confess that we don't have a simple rationalization for (2.88) as yet.

> ## LAW OF CONSERVATION OF ENERGY
>
> Consider a thermomechanical lumped parameter system with mechanical degrees of freedom q_i and external generalized forces Q_i so that the total mechanical power supplied is $\sum_i Q_i \dot{q}_i$. Assume that the temperature of the system is T and that the Helmholtz potential (or isothermal work function) is $\psi(q_i, T)$. Then the **law of conservation of energy** can be stated as "the rate of increase of internal energy is the sum of the power supplied and the heat supplied", i.e.
>
> $$\dot{\exists}(q_i, T) = \sum_i Q_i \dot{q}_i + H \qquad (2.91)$$
>
> where $\exists(q_i, T) = \psi - T \partial \psi / \partial T$ is the *internal energy of the system* and H is the rate of heating of the system from external sources.

We can consider the following thought process to understand the relationship between the isothermal work function or Helmholtz potential ψ and the internal energy $\exists$. Let us consider a spring such as the one that we considered earlier.

At 300K, we deform it from its unstretched configuration; What happens to the work that you do?
Contrary to what you might think, not all the work that you do is actually "stored" in the spring during the process; only the part $\psi - T \partial \psi / \partial T$ gets stored. The rest of it gets out of the system in the form of heat. You might think "Wait a minute, this is a spring, right? if it loses the work that I do as heat, then how do I get it back?"—therein lies the mystery of thermodynamics!

To understand this paradoxical situation, think of the following analogy: you go to deposit some cash (work) into your account in a new bank called the "Spring Bank Ltd–All your deposits are safe®". The bank credits it to your account. Do you think that the bank actually "stores" your money? It just credits your account (which is the Helmholtz potential) but does not actually store it. It just keeps a certain part of it and "lends" the rest. With an actual bank, the lending is at certain interest and it earns money. The "spring bank" actually lends it out (exchanges heat to the surroundings) to absolutely safe borrowers at zero percent interest.

If you want to withdraw the money "slowly" (i.e. isothermally) the bank can actually call in the "loans" and pays you back and you are none the wiser. You don't care whether the bank has stored it or lent it. On the other hand, if you want to withdraw all your cash at once (adiabatically), i.e. if there is a "run" on the bank, you can get only a part of your money back—the money that is actually stored. This is what happens in real materials.

On the other hand, if you prevent the bank from loaning money, i.e. seal off the external heat supply, then whatever work you do will be "stored" as internal energy. If you control the system so that the temperature is constant, then the work that you do is NOT stored as internal energy but part of it is given away as heat.

Thus, even though a system may undergo an isothermal process, it can still absorb or reject energy to its surroundings in the form of heat. So, in some sense we may say that changes $T\frac{\partial \psi}{\partial T}$ due to deformation at a given temperature T is the energy lost as heat by the system during an isothermal process, i.e. it is the money "lent" out by the bank. Note that since the temperature is constant, this energy lost depends only upon the change in the displacements of the components of the system.

To make this point more precise, if we differentiate the internal energy expression (2.88) with respect to time and simplify the resulting expression, we will get

$$\dot{\beth} = \frac{d}{dt}\left(\psi - T\frac{\partial \psi}{\partial T}\right) = \sum_i \left(\frac{\partial \psi}{\partial q_i}\dot{q}_i - T\frac{\partial^2 \psi}{\partial T \partial q_i}\dot{q}_i - T\frac{\partial^2 \psi}{\partial T^2}\dot{T}\right) \qquad (2.92)$$

Thus, a change in the internal energy of a material is due to three effects. We will try to understand the origins of each of the three terms on the right-hand side.

- The first term, $\sum_i \dot{q}_i \partial \psi / \partial q_i$, is just the power supplied to the springs (since $\partial \psi / \partial q_i$ is the force on the springs). This is the purely "mechanical" term
  The next two terms are the interesting ones:
- The second term, $\sum_i -T\dot{q}_i \partial^2 \psi / \partial T \partial q_i$ is due to the fact that the spring force is dependent on the temperature, and is due to the *mechanical effects of heating* (i.e. the effect of thermal expansion and changes in modulus due to temperature).

  The first two terms are only present if the system is moving (notice that they both have a velocity term, $\dot{q}_i$). It is the loss/gain of mechanical power of the springs due to the change in the temperature.
- The last term, $-T\frac{\partial^2 \psi}{\partial T^2}\dot{T}$, is purely due to the increase in temperature and is present even if no motion occurs and is thus the purely "thermal" term. Not surprisingly, this is the term that is related to the "heat capacity" of the material.

As an example, for the spring introduced in the previous two exercises, if we substitute $\psi = (1000 - 2T)u^2$ into the right-hand side of (2.92) we see that

$$\beth = \psi - T\frac{\partial \psi}{\partial T} = 500u^2 \qquad (2.93)$$

The material is bizarre—its internal energy is independent of the temperature. If you don't stretch the spring but merely heat it, its internal energy will NOT increase! Moreover, we find that the rate of change of the internal energy is given by

$$\dot{\beth} = \frac{d}{dt}(500u^2) = (1000 - 2T)u\dot{u} + 2Tu\dot{u} = 1000u\dot{u} \qquad (2.94)$$

We see that irrespective of what temperature we stretch the spring, the internal energy increases at the same rate! Notice that the power supplied to the spring is only $F_{spr}v = (\partial \psi / \partial u)\dot{u} = 1000u\dot{u} - 2Tu\dot{u}$. Thus the power supplied is *less* than the actual increase in the internal energy of the spring! What a strange material!

The spring "soaks up" the extra $2T u \dot{u}$ units of power from the surroundings or from the dissipative components which allow generation of heat. While the spring is certainly strange, it is not unphysical—this is a very crude approximation for a rubber band or a typical elastomer.

⊙ 2.20 (Exercise:). The rate of change of the internal energy
For future reference, show that we can rewrite the equation (2.92) in the form

$$\dot{\exists} - \sum_i \frac{\partial \psi}{\partial q_i} \dot{q}_i \big|_{T \text{ fixed}} = -T \frac{d}{dt}\left(\frac{\partial \psi}{\partial T}\right) \quad (2.95)$$

Since the left-hand side is the rate of increase of the internal energy minus the power supplied to the springs, we may interpret the right-hand side of the above equation as the increase the internal energy due to heating by internal friction as well as due to external sources.

In order to proceed further, we now define thermal quantities that will be crucial for future developments:

2.5.1.1 *Some crucial results from thermodynamics*

Apart from the spring forces (see equation (2.78)), there are two quantities of interest that can be obtained from the derivative of the Helmholtz potential with respect to the temperature. They are both related to changing temperatures; they are the *entropy S* and *heat capacity at constant displacements*[31], C_v. These two quantities are defined by

$$S = -\frac{\partial \psi}{\partial T}\big|_{q_i}, \quad C_v = -T\frac{\partial^2 \psi}{\partial T^2}\big|_{q_i} \quad (2.96)$$

In terms of the Gibbs potential,

$$S = -\frac{\partial G}{\partial T}\big|_{Q_i}, \quad C_p = -T\frac{\partial^2 G}{\partial T^2}\big|_{Q_i} \quad (2.97)$$

It is not within the scope of this chapter to elaborate on the above results[32]. They are a part of the fundamental relationships in thermodynamics that we shall take for granted and we shall use them. However we cannot pass by without mentioning that the "entropy" of the system is based on the idea that the lumped mass system is composed of a large number of "particles" whose state is given by their individual positions and velocities. Let us assume that we have a lumped parameter system that is made of a 1000 microscopic particles. Then, the actual state of the system is represented by 6000 DOFs (3 components of position and 3 components of velocity for each particle). But let us say we have used only a few "average"

---
[31] This is the generalization of the heat capacity at constant volume.
[32] see e.g., [Callen (1985)] for a very detailed discussion

quantities q_i to represent all the DOFs. Thus there is a tremendous loss of information regarding the state of the system due to the "averaging" process. This loss of information is quantified by the entropy.

The main conceptual idea is the following:

STATISTICAL OR INFORMATION ENTROPY

Consider a system composed of M microscopic particles but being described by only N coordinates q_i, $(i = 1, \ldots, N)$ (M is assumed to be much larger than N), the q_i being "smoothed" or "averaged" positions of the M microscopic particles. Clearly many states (positions and velocities of the microscopic particles) all have the same average values q_i, since the function relating the q_i's to the microscopic positions is clearly not one to one. Assume that the macroscopic kinetic energy of the whole body is zero. Then,
(a) The temperature T is a measure of the total kinetic energy of all the microscopic particles.
(b) The entropy is the negative logarithm of the total number of different possible states (positions and velocities) of the microscopic particles which all have the same "averaged" values q_i, i.e. it measures "redundancy" or "uncertainty" about the actual microscopic state variables corresponding to any given value of the state variables q_i.

The fact that the entropy is the derivative of the Helmholtz potential with respect to the temperature is not at all easy to demonstrate and we won't even attempt to do it[33]. Just accept it as a fundamental property of the Helmholtz potential for now. We will revisit this in a later chapter where we will explore it in (slightly) greater detail.

By using (2.96) in (2.88) we get a familiar equation: $\beth = \psi + TS$. Furthermore, we can write (2.95) in a familiar form (using (2.78))

$$\dot{\beth} - \sum_i Q_i^c \dot{q}_i = T\dot{S} \tag{2.98}$$

where $Q_i^c = \partial \psi / \partial q_i$ are the generalized conservative forces.

If the above equation seems unfamiliar, "multiply by dt" and rewrite it as

$$d\beth - Q_i^c dq_i = TdS \tag{2.99}$$

This is a "TdS" relation, $dU - pdv = TdS$ that you might have seen in a beginning course in thermodynamics with p and v replaced by the generalized coordinates and forces.

---

[33] See for example, the excellent chapters on statistical mechanics in the second half of [Callen (1985)]. This is a very readable account and a great place to start if you want to understand this fascinating subject.

2.5.2 Finding the Helmholtz and Gibbs potentials for a system

Up to now we have considered only very simple systems where the strain energy was quadratic etc. A question arises: is it possible to find the Helmholtz potential for a system if we don't know a-priori what its response is like?

The answer to this question is that in principle, *yes*, provided you don't have any frictional components or their effects are negligible. If we have only viscous dampers, their effects can be reduced to as small a value as you want by changing the state of the system "quasistatically", this has the additional benefit of being able to keep the temperature fixed and eliminating inertial effects. If you are able to do this, then:

(1) Fix a beginning mechanical state (positions of all the parts). We will start from this state for all our mechanical measurements.
(2) At each temperature T find the work done in moving the system quasistatically from the reference state to any given state. We will get a function $f(s, T)$. For example, for the spring problem, the mechanical state is given by u and $f(u, T) = \frac{1}{2}(1000 - 2T)u^2$. This is sufficient for all mechanical calculations. We need to consider the thermal effects next.
(3) Fix the system in its reference mechanical state and find the energy required to heat it from zero Kelvin[34]. Let it be $\bar{h}(T)$.
(4) Then, the Helmholtz potential is

$$\psi = f(q_i, T) - \int_0^T \frac{\bar{h}(T)}{T} dT \qquad (2.100)$$

For the spring, if we assume that $\bar{h}(T) = CT \ln T$ where C is a constant, we will get

$$\psi = \frac{1}{2}(1000 - 20T)u^2 - CT(\ln T - 1) \qquad (2.101)$$

One might wonder why the integration of $\bar{h}(T)/T$. If we do this, we can then show that $\exists = \psi - T\partial\psi/\partial T$ will work also for purely thermal processes. The above procedure typically works only for simple springs, and for a few other "ideal" cases although the final answer is completely general.

Following the same procedure as with the Helmholtz potential, we now describe a thought experiment to find the Gibbs function for a lumped parameter system:

(1) Fix a beginning mechanical state as a state of zero force on all the bars. We will start from this state for all our mechanical measurements.
(2) Now, at each temperature T find the complementary energy in moving the system quasi-statically from the reference force free state to any given state,

---
[34] Before anyone complains that zero Kelvin is not possible, please remember that this is only a thought experiment. If that does not satisfy you, then start from some known temperature and "approach" zero, i.e. get to 0^+.

i.e. we find $-\int_0^F u(F)dF$. We will get a function $g(F,T)$. This is sufficient for all mechanical calculations. We need to consider thermal effects next. This can be done by applying various sudden loads on the body and finding the force to equilibrium deflection relationships as functions of the sudden loading.

(3) Keep the material in a stress free state and find the energy required to heat it from zero Kelvin to any other temperature. Let it be $\bar{h}(T)$. This experiment, in contrast to the one described for the Helmholtz potential, can be done by using a Differential Scanning Calorimeter[35]. Let us call this function $\bar{h}(T)$.

(4) Then, the Gibbs potential is

$$G = g(F,T) - \int_0^T \frac{\bar{h}(T)}{T} dT \qquad (2.102)$$

Q: What if the model that I am investigating has many mechanical degrees of freedom?

You have three options:

(1) The statistical mechanics option: Go to a physics expert and ask them if there is a way to use "statistical mechanics" to get a form for the Helmholtz potential. In statistics, the Helmholtz potential is related to something called the "partition function". Most probably they will tell you that it is too difficult.

(2) Mix and match: use known Helmholtz potentials and mix and match them: for example, if you want a constant heat capacity and a spring-like response then try $\psi = C(T - T\ln T) + f(q_i)$. This will completely separate the thermal and mechanical response, the entropy is purely a function of the temperature while the spring force is a function of the generalized coordinates alone. On the other hand, if you want something which includes thermal expansion, try $\psi = C(T - T\ln T) + f(q_i - \alpha_i T)$. Now you will see that the thermal and mechanical responses are coupled—the entropy will depend upon the mechanical variables q_i also.

(3) The final option is to empirically write down a Helmholtz potential function and try to match parameters[36].

---

[35] well... almost; nobody has done one starting from zero Kelvin yet, but this is a thought experiment.

[36] This is called system identification in the control systems literature and is extremely common there. Generally, in the continuum modeling community, this has not been developed as well as in the controls community and we still favor ad-hoc methods versus using the tools of parameter estimation. We urge researchers in the field to rectify this defect.

⊙ **2.21 (Exercise:). Common Helmholtz potentials**
Consider the following examples of Helmholtz potentials for a spring system, (where u is the displacement associated with the system). Using (2.88) and (2.96) find the expression for the internal energy, the conservative force, the entropy and the heat capacity. For each case, discuss whether the mechanical force is related to the internal energy or the entropy, or both.

(A) $\psi(u,T) = 1/2ku^2 + CT(1 - \ln T)$ (2.103)

(B) $\psi(u,T) = 1/2kTu^4 + CT(1 - \ln T)$ (2.104)

(C) $\psi(u,T) = 1/2k(e^{-\alpha T})u^2 + CT(1 - \ln T)$ (2.105)

(D) $\psi(u,T) = 1/2k(1+T^2)(u - \alpha T) + CT(1 - \ln T)$ (2.106)

Notice that, in the above example, you obtained a constant heat capacity for (A) and (B) but not for (C). Also notice that for the example (B), the internal energy depends ONLY on the temperature. Thus if the spring in (B) is stretched under *isothermal* conditions, its internal energy WILL NOT CHANGE! (How is this possible? Where does the work done on the spring "go" if it is not "inside" the spring?). Moreover, in the example (B), conservative forces are NOT linear in displacements.

⊙ **2.22 (Exercise:). Helmholtz and Gibbs Potentials**
A general nonlinear spring is given by $F = ku^n$. Obtain the mechanical part of the Helmholtz and Gibbs potentials.
Ans: The Helmholtz potential is $ku^{n+1}/(n+1)$, the Gibbs potential is $-n(F/k)^{(n+1/n)}/(n+1)$.

2.5.3 Determination of the equations of motion and the temperature of the system

We finally turn to the determination of the evolution of the temperature of a system. The starting point for this is the law of conservation of energy that was stated earlier but which we are restating since it is so important:

The Law of Conservation of Energy

Consider a lumped mass system whose mechanical degrees of freedom are q_i. Let its Helmholtz potential be $\psi(q_i, T)$ where T is the temperature. Then *the rate of increase in internal energy = mechanical Power supplied + heating through boundaries + heating through other sources*, i.e.

$$\frac{d\exists}{dt} := \frac{d}{dt}\left(\psi - T\frac{\partial \psi}{\partial T}\right) = \sum_i Q_i^{ext} \dot{q}_i + h + R, \qquad (2.107)$$

where, $\exists$, is the internal energy of the system, $\sum_i Q_i^{ext} \dot{q}_i$, is the external power supplied, h is the rate of energy gained by conduction from the surroundings and R, is the rate of heating (or cooling) by other sources (such as radiation, internal electrical heating etc. that are not explicitly accounted for as part of the system).

Now by expanding the left-hand side of (2.107), and grouping terms and combining above result with (2.98) we get the following central result:

THE HEAT EQUATION

$$T\frac{dS}{dt} := -T\frac{d}{dt}\left(\frac{\partial \psi}{\partial T}\right) = \sum_i \left(Q_i - \frac{\partial \psi}{\partial q_i}\right)\dot{q}_i + h + R = \zeta + h + R \qquad (2.108)$$

where

$$\zeta = \sum_i \left(Q_i - \frac{\partial \psi}{\partial q_i}\right)\dot{q}_i = \sum_i Q_i^{ext}\dot{q}_i - \dot{\psi}|_{T=const} \qquad (2.109)$$

is the dissipation potential.

We want you to observe that the only difference between the above equation and (2.61) is that ψ is now a function of T. Thus, (2.109) *is a generalization of the purely mechanical power theorem (2.61) and allows us to extend the maximum rate of dissipation hypothesis to thermomechanical systems also.*

The right-hand side of (2.108) contains terms that all "produce" heat: power supplied to the dissipative components, conduction from the surroundings and radiation. However, the left-hand side, rather than being just the heat capacity times the rate of temperature increase, turns out to be temperature times the rate of entropy change. Again, to make contact with a classical thermodynamics text, we multiply the equation by "dt" and get $dS = dQ/T$, where dQ is the sum total of all the different mechanisms of heating. Thus we can use the following working hypothesis : "heating a body" means increasing its entropy. "cooling a body" means decreasing its entropy. Just like "extension" of a spring means increasing its length and contraction is decreasing its length[37]. To be more precise, we will define "heating" to mean *supplying energy to the body with the intent of increasing its entropy*. On the other hand, "making something hotter" is to increase its temperature. In our view, these are not the same. In this view, supply of mechanical power to the body will cause it to "heat" the body, if there is internal friction. So will the "external heating".

The Second Law of Thermodynamics: Strong Form

You are perhaps aware that the second law of thermodynamics says something to the effect that we can convert work into heat and not vice-versa or perhaps that the entropy always increases. This is not quite correct: what you need to say is that during any process, the change in entropy of the system plus *that of its surroundings* is non-negative. The statement that we want is the following

---

Second Law of Thermodynamics

It is impossible to "cool" a thermomechanical body (i.e. lower its entropy) through work alone. Thus, a body can be "heated" by work alone—in order to cool a body you need some form of heat transfer.

For our case the amount of entropy produced by the mechanical processes is the power dissipated divided by the temperature, and this must be non-negative, i.e.

$$dS_{irr,mech} = \frac{\partial \zeta}{\partial T} \geq 0 \qquad (2.110)$$

Thus, the satisfaction of the second law requires that the power supplied to the dissipative components should always be non-negative.

Thus, external sources of heat can either heat or cool a body. But internal friction can only heat it. Since the dissipation potential ζ is the rate at which the body is heated by work, the second law only demands that only those processes are allowed for which ζ is positive. We now have the complete set up to obtain the thermomechanical equations for a lumped mass system.

---

[37] We want you to think about it before you erupt in howls of indignation saying that heating and cooling have to do with the temperature and not the entropy. Consider the example of a pot of boiling water on a stove. Everyone would agree that we are "heating" a boiling pot, but its temperature does not change- what changes is the entropy!

> **THE MECHANICAL DISSIPATION EQUATION**
>
> The starting point for choosing constitutive equations for thermomechanical springs and dashpots is the mechanical dissipation equation which states that: *At any given instant, the power supplied by the external forces to a system minus the rate of increase in recoverable work keeping the temperature at that instant fixed must be equal to the rate of mechanical dissipation*, i.e. in terms of the Helmholtz potential,
>
> $$\sum_{i=1}^{n} Q_i^{ext} \dot{q}_i = \dot{\psi}\,|_{T=const} + \zeta. \qquad (2.111)$$

The mechanical dissipation equation in terms of the Gibbs Potential, takes a more complex form: We assume that each of the generalized forces in the spring dashpot network is classified as either a conservative generalized force (or spring force) Q^s, a dissipative generalized force Q^d or an external generalized force Q^{ext}. Then the Gibbs potential is a function only of the conservative generalized forces and the temperature and

$$\sum_{i=1}^{n} Q_i^{ext} \dot{q}_i = \sum_{i=1}^{n} Q_i^s \frac{d}{dt}\left(\frac{\partial G}{\partial Q_i^s}\right)\,|_{T=const} + \zeta. \qquad (2.112)$$

It should be obvious to you that the rate of dissipation ζ can be written in terms of the dissipative forces as $\zeta = \sum_{i=1}^{n} Q_i^d \dot{q}_i$.

Now we are ready to use these notions to develop the differential equations for a variety of spring dashpot systems. Before we continue, we make the following important point: A number of systems that we will consider in simple structural applications can be reduced to a single mechanical degree of freedom system. For this case, the way to obtain the equations of motion is very simple and direct.

> **ONE DEGREE OF FREEDOM SYSTEMS**
>
> If the model can be reduced to a one degree of freedom system, then the power theorem either in the form (2.111) or (2.112) (with n=1), together with either a constitutive equation for ζ or a constitutive equation for the dissipative force Q^d delivers the equations of motion. Nothing more is needed!

Thus, for a one DOF system, there is no difficulty. It is only when we have multiple degrees of freedom with constraints and whatnot do real difficulties arise. For these cases, we have the following result,

Thermomechanical Equations for an Inertialess System

Consider an inertialess lumped parameter system that is subject to known external forces. Let the mechanical degrees of freedom of the system be q_i and the external forces by $Q_i^{ext}(t)$ (many of them may be zero) so that the mechanical power supply is $\sum_i^n Q_i^{ext} q_i$. Let the kinematical variables be subject to m constraints of the form $f_\alpha(q_i, T, t)\dot{q}_i = 0$ ($\alpha = 1, \ldots, m$) where T is the temperature. Let the isothermal work potential be $\psi(q_i, T)$ and the dissipation potential be $\zeta(T, q_i, \dot{q}_i)$. Then the internal energy and the entropy are given by

$$S = -\frac{\partial \psi}{\partial T}, \quad \exists = \psi + TS = \psi - T\frac{\partial \psi}{\partial T}. \qquad (2.113)$$

Any evolution equation for the variables $\dot{q}_i$ of the form $\dot{q}_i = f_i(q_j, T, Q_j^{ext}, (j = 1, n))$ is acceptable so long as the right-hand side of (2.109) turns out to be non-negative and all the constraints are met.

If one demands that the rate of dissipation ζ be maximum, then kinetic equations are given by

$$Q_i^{ext} - \frac{\partial \psi}{\partial q_i} - \mu \frac{\partial \zeta}{\partial \dot{q}_i} - \sum_\alpha \mu_\alpha \frac{\partial f_\alpha}{\partial \dot{q}_i} = 0, \qquad (2.114)$$

where μ_α are obtained by satisfying the equations of constraints and the mechanical power dissipation equations

$$Q_i^{ext} \dot{q}_i = \dot{\psi}|_{T=const.} + \zeta \qquad (2.115)$$

and is explicitly given by

$$\mu = \frac{\zeta - \sum_\alpha \mu_\alpha \sum_i \dot{q}_i \partial f_\alpha / \partial \dot{q}_i}{\sum_i \dot{q}_i \partial \zeta / \partial \dot{q}_i}. \qquad (2.116)$$

The equation for the determination of the temperature is

$$T\frac{dS}{dt} = \zeta + h + R, \qquad (2.117)$$

where $h = \bar{h}(T_\infty - T)$ is the heat supply by the surround medium (with $\bar{h}$ being the heat transfer coefficient) and R is the heat supply by other sources such as radiation, electrical heating etc. However, *the heat equation in the form above is not convenient to use, although it is useful for interpretation of the results. It is MUCH more convenient to use the original energy equation*, i.e.

$$\frac{d\exists}{dt} = \frac{\partial \exists}{\partial T}\dot{T} + \sum_i \frac{\partial \exists}{\partial q_i}\dot{q}_i = \sum_i Q_i^{ext}\dot{q}_i + h + R. \qquad (2.118)$$

Q: You have stated this whole thing in many different ways. What is the bottom line? What are the equations to be used for a thermomechanical system?

We have tried to show how to go from the familiar mechanical systems to thermomechanical systems that you might not be that familiar with. Hence the repetition of the equations with slight differences. The bottom line is: *use equations (2.2)–(2.7) in the summary for any thermomechanical lumped mass system. If there is no temperature involved, then ignore (2.7)*.

Q: Do I HAVE to use the maximum dissipation hypothesis? What if I am uncomfortable with it and just want to use some other set of equations?

For simple cases, it is quite easy to simply use Newton's laws at the joints to obtain the equations. For more general models (with weird degrees of freedom and complicated constraints) read the summary at the beginning of the chapter and it will show you that there are at least 3 ways of doing this, only one of which is the MRDH route. Pick any way that you find comfortable.

Q: Show me with an example how to get this set of equations for a specific model

We are now in a position to obtain the governing equations for the thermomechanical mass-spring-damper system that was introduced in Fig. 2.2a and which we have reproduced here for convenience.

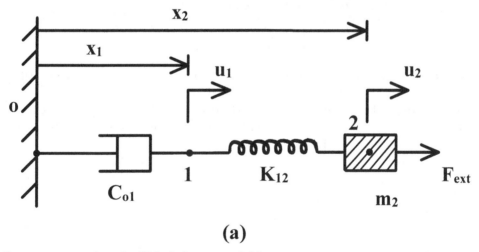

(a)

Let us assume that the Helmholtz potential is

$$\psi = \frac{1}{2}(a - bT)(x_e - cT)^2 + DT(1 - \ln T) \quad J, \tag{2.119}$$

where a is the spring modulus at zero Kelvin, b is the coefficient of the temperature dependence of the modulus, c is the coefficient of thermal expansion, D is a constant related to the heat capacity and x_e is the spring displacement.

This is an innocuous looking form for the Helmholtz potential and you might be tempted to interpret it as "mechanical energy+thermal energy"—but you would be wrong! Remember, the Helmholtz potential is NOT the energy. If the temperature is held fixed, then it does represent the recoverable work. But it is not the energy "stored" in the system. Shortly we will see what is the energy of the system.

The above form for the Helmholtz potential implies that the entropy is

$$S = -\partial \psi/\partial T = D \ln T + \frac{1}{2}(2ac - 3bcT + bx_e)(xe - cT) \; J/K. \tag{2.120}$$

Then the internal energy is

$$\exists = \psi + TS = DT + \frac{1}{2}(x_e - cT)(ax_e + acT - 2bcT^2) \; J. \tag{2.121}$$

Note that the internal energy is not simply the thermal + mechanical energy! Since both the terms involve temperature! Note that it doesn't even have the nice "quadratic form" form that we have become so used to.

The heat capacity for this system is

$$\begin{aligned} C_v = \frac{\partial \exists}{\partial T} &= -T \partial^2 \psi / \partial T^2 \\ &= T \partial S/\partial T = D + cT(-ac + 3bcT - 2bx_e) \; J/K. \end{aligned} \tag{2.122}$$

Again note that the heat capacity has not only the D term but also other temperature and extension dependent terms.

Now let us assume that the rate of dissipation for the system is $\zeta = f(T)(\dot{x} - \dot{x}_e)^2$. Then by using (2.114), the equations of motion become

$$\begin{aligned} F^{ext} = F_{spring} &= (a - bT)(x_e - cT), \\ F_{ext} = F_{dashpot} &= 2\mu \; f(T)(\dot{x} - \dot{x}_e). \end{aligned} \tag{2.123}$$

The value of μ is given by (2.116) and happens to be equal to $1/2$ (as you can verify by yourself). You could have obtained this by directly applying Newton's laws, as long as you realized that the spring force is the derivative of the Helmholtz potential with respect to x_e. Also we avoided needless and silly constraints by simply using $x - x_e$ instead of a new variable.

The heat equation is $TdS/dT = \zeta + Q + r$. As mentioned earlier, it is NOT convenient for MATLAB applications. It is much better to use the energy equation (2.118). This turns out to be

$$\dot{\exists} = C_v \dot{T} + \frac{\partial \exists}{\partial x_e}\dot{x}_e - F^{ext}\dot{x} = H + R \tag{2.124}$$

where $\partial \exists/\partial x_e = ax_e - bcT^2$. Note that it has been written in a special way: the left-hand side has all the rates of the state variables; the right-hand side has the heating terms.

If we assume that there is no radiative heating and we can use "Newton's Law of cooling" then $R = 0$, $H = -h(T - T_\infty)$, where T_∞ is the temperature of the surroundings. The governing equations are then (2.123) and (2.124) with the heat capacity given by (2.122).

We hope you got the idea for how the equations are generated. Essentially, the sequence of operations are :

$$\psi \to S = -\frac{\partial \psi}{\partial T} \to \beth = \psi + TS \to C_v = \frac{\partial \beth}{\partial T} \to \frac{\partial \beth}{\partial q_i}. \quad (2.125)$$

If you have this list then you are all set to solve the equations.

The question is how to solve it: The next section discusses this point.

2.6 Solving the thermo-mechanical evolution equations for an inertialess system

Until now we have blithely obtained equation after equation for lumped mass system with scant regard for their eventual solution. We now have to pay the piper, i.e. we have developed the equations, now we have to solve them. Our solution methodology is simple: Convert everything into a set of first order differential equations of the form $\mathbb{A}(\mathbf{q}, t)\dot{\mathbf{q}} = \mathbf{f}(\mathbf{q}, t)$ and use canned ODE solvers in MATLAB or develop your own techniques[38]. So, as far as our modeling world is concerned *everything is a differential equation*.

We introduce the concept with the spring-dashpot thermomechanical system considered in the last section. To be specific, let us assume that $a = 1000$, $b = 0.1$ $c = 10^{-4}$ and $D = 30$. Recall there are three unknown variables $x(t), x_e(t)$ and $T(t)$ and there are three equations (2.123a), (2.123b) and (2.124).

Note that while (2.123)b and (2.124) are in the form of ordinary differential equations, (2.123)a is clearly not an ODE. It is an algebraic equation. How do we convert it into a differential equation? Simple. We just differentiate it and get

$$\dot{F}^{ext} = (1000 - 0.1T)\dot{x}_e + \left[0.1(x_e - 10^{-4}T) - (0.1 - 10^{-5}T)\right]\dot{T} \quad (2.126)$$

Now we are ready: We can rewrite the original differential equations in matrix form as

$$\begin{pmatrix} 0 & 1000 - 0.1T & A_{13} \\ 0.01T^3 & -0.01T^3 & 0 \\ F^{ext} & -F^{ext} & -30 \end{pmatrix} \begin{pmatrix} \dot{x} \\ \dot{x}_e \\ \dot{T} \end{pmatrix} = \begin{pmatrix} \dot{F}^{ext} \\ F^{ext} \\ h(T - T_\infty) \end{pmatrix} \quad (2.127)$$

where due to lack of space we have introduced the notation $A_{13} = 0.1(x_e - 10^{-4}T) - (0.1 - 10^{-5}T)$.

Once you have it in this form, you are now ready to invoke MATLAB: MATLAB's ode solvers can solve ODEs of the form $M(t, x)\dot{x} = f(t, x)$ just call ODE23 or ODE45 and you are done. Now, what are the initial conditions? You

---

[38] The book titled Numerical recipes in C has excellent chapters on how to solve ODES without a lot of jargon and theorems and proofs. In fact we recommend it as an OUTSTANDING source for learning all about numerical strategies even if you don't write your own code. Think of it as an encyclopedia of numerical techniques.

must specify x, x_e and T at time $t = 0$. Please note that you can't set x_e to be anything you want, it has to satisfy $F^{ext} = (1000 - 0.1T)(x_e - 0.0001T)$ at the initial time. Subsequently the first differential equation will ensure its satisfaction at all other times.

To get the MATLAB code to do this in another way, see authors' websites listed in the preface.

We have now seen that, it is possible to obtain lumped parameter models based on the use of a Helmholtz potential and a kinetic equation to systematically develop models for a variety of inelastic materials. Of course, our models have been zero dimensional but in the subsequent chapters we will develop lumped parameter models for higher dimensional systems. Before leaving this chapter, however, it is important to introduce the other potential that can be used in lieu of the Helmholtz potential. In some cases, (for example, large deformation plasticity such as that found in metal forming), this approach has certain distinct advantages.

2.7 Lumped parameter models: Gibbs potential approach

We will end this chapter by using a Gibbs potential approach to obtain the equations of motion for a one degree of freedom inertialess system. In order to exploit the simplifications that result from using the Gibbs potential we make the following observation: *If any two components are in series, the force flow is the same in both, so we can combine them into one component with both conservative and dissipative elements.* Since two springs in series or two dashpots in series can be trivially reduced to one, the fundamental unit in the Gibbs formulation is a spring and dashpot in series as shown in Fig. 2.7. In this example, note that the same force flows through both the spring and the dashpot. So we introduce only two nodes, one for the ground and the other for the mass and so, we have a one degree of freedom system. In this case, the equations of motion can be obtained directly from the power theorem.

(1) Let us assume that Gibbs potential for the spring dashpot system is

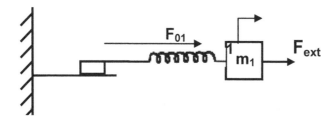

Fig. 2.7 A simple lumped mass system with a dry friction component and a spring in series. Note that there are only two nodes: The ground and the mass since the force is the same for the spring and the dashpot.

$-(F_{01})^2/k(T) + \bar{h}(T)$ (we have assumed that the spring is a linear spring with temperature dependent modulus).

(2) The displacement of the spring is $u_{spr} = F_{01}/k(T)$, and hence the velocity of the dashpot is the difference between the velocity of the combination and that of the spring, i.e.

$$v_{\text{dashpot}} = v_{01} - \frac{d}{dt}(F_{01}/k(T)). \tag{2.128}$$

Now the velocity of the dashpot must be a function of the force F_{01} on the dashpot, so that, for a linear dashpot, we will get a constitutive equation of the from

$$v_{01} - \frac{d}{dt}(F_{01}/k(T)) = (F_{01})/C, \tag{2.129}$$

where C is the damping coefficient. This is all we need from the constitutive equation side.

(3) Equation (2.129) can be augmented by the equation of motion of the mass so that we have

$$m\frac{dv_{01}}{dt} = -F_{10} + F^{ext}, \quad v_{01} - \frac{d}{dt}(F_{01}/k(T)) = (F_{01})/C \tag{2.130}$$

as the two governing equations for the mechanical behavior, together with

$$-T\frac{d}{dt}\frac{\partial G}{\partial T} = T\frac{F_{01}^2}{K(T)^2}\frac{dK}{dT} - T\frac{dh}{dT} = F_0 1(v_{01} - \frac{d}{dt}(F_{01}/k(T))) + h(T - T_\infty) \tag{2.131}$$

as the heat equation. (You can verify this by yourself by using (2.89).)

As you can see, the Gibbs formulation uses only the internal forces and does not involve anything other than the velocities of the masses.

A natural question may arise as to the necessity of going through such an elaborate exercise for a fairly simple problem whose governing equations can be written by inspection. A fair question: The answer lies in the introduction of the methodology and philosophy of the approach. The core steps in the Gibbs potential approach are (1) writing down the governing balance laws (2) introducing the Gibbs potential into the energy equation (3) identifying the elastic parts of the response (4) finding out which variables require kinetic laws and how to state the kinetic laws in such a way as to satisfy the second law of thermodynamics. These steps are the same for all types of models and will be repeated for different types. On a philosophical note, observe that the Gibbs formulation requires only terms (forces and velocities) in the current state and do not need any information about where things are measured from, i.e. past history information. This is a great advantage for "flow type" formulations that are appropriate for finite deformation problems and for metal forming applications.

2.8 Projects and exercises

(1) State, in words, what is meant by "spring stiffness" and "coefficient of thermal expansion". Consider a material whose Gibbs potential is a general quadratic expression of the form $-F^2/k(T) - \alpha(T)F + f(T)$ where $k(T), \alpha(T)$ and $f(T)$ are general functions of temperature. What are the forms for the Helmholtz potential and the internal energy of the body. Are the following statements true? (1) $k(T)$ is the elastic stiffness of the body. (2) $\alpha(T)$ is related to the coefficient of thermal expansion. Explain how you would check the veracity of the above statements.

(2) Show that for a spring whose Gibbs potential is of the form $-\frac{3F^{4/3}}{4k^{1/3}} + f(T)$, the internal energy is of the form

$$\exists = \frac{k}{4}\frac{F^{4/3}}{k} - T\frac{df}{dT} \qquad (2.132)$$

Also find the form of the internal energy for a spring whose Gibbs potential is of the form $-n(F/k)^{(n+1/n)}/(n+1)$.

(3) Show that

$$\frac{d\exists}{dt} = -T\frac{d}{dt}\left(\frac{\partial G}{\partial T}\right) - F^c_{ij}\frac{d}{dt}\left(\frac{\partial G}{\partial F^c_{ij}}\right) \qquad (2.133)$$

Hint: This is a brute force computation, involving use of the chain rule.

(4) Write a MATLAB program to simulate the response of the system for a sinusoidal force, shown in Fig. 2.2a, assuming that the damper is a dry friction damper. Assume suitable values for the constants (for example, if you choose k to be numerically equal to the modulus of a metal and α to be equal to the yield strength, and m to be the density of the material, then the displacement will be the strain and you will get an idea as to how the material deforms). Does the answer depend upon the frequency of oscillation?

(5) **Simplified model for a frictional system**
Consider the spring mass damper system in Fig. 2.2a. Assume that the dashpot is a nonlinear dashpot whose kinetic equation is given by $F_{10} = C_{10}v_{10}|v_{10}|^{1/n}, n = 20$. Write a MATLAB Program to simulate the response of the system subject to a sinusoidal force. How does the response change with frequency. Show that over a certain range of frequencies, the response mimics that of a frictionally damped system. (Assume suitable values for k, m and c.) What are the units of c.

(6) **Other models**
Repeat Exercise 2 for the other models in Fig. 2.2.

(7) **Massless response**
In each of the situations, we have always assumed that the mass of the system is non zero. Considerable simplification can be realized for quasistatic processes by neglecting the mass. Derive the simplified evolution equations for the systems in Fig. 2.2 after setting their masses to zero.

Chapter 3

Inelastic Response of Truss Elements

Learning Objectives

By learning this material, you should be able to

(1) find the forces in a simple, statically determinate truss

(2) derive the balance laws for mass, momentum and energy for a bar in a truss

(3) use a spring dashpot LPM to develop the constitutive equations for a bar under tension or compression

(4) formulate constitutive equations for the hardening functions

(5) obtain expressions for the rate of total and plastic strains from the knowledge of the force on the bar, the Gibbs potential and the hardening laws, i.e. for force control

(6) obtain expressions for the rate of total and plastic strains from the knowledge of the strain in the bar, the Gibbs potential and the hardening laws, i.e. for displacement control

(7) use these to find out how much plastic strain is "accumulated" per cycle. Use the Coffin-Mason rule to find out fatigue life for low cycle fatigue.

> **CHAPTER SYNOPSIS**
>
> In carrying out analysis with trusses and cables which are statically determinate, the procedure is to find the axial forces in the truss member, find the critical element(s), and then use a lumped parameter model to analyze the inelastic response of the truss. If you are interested in low cycle fatigue, then you will need a hardening model for the cyclic response as well as a fatigue failure criterion such as the Coffin Manson rule. In this chapter you will see how these ideas can be implemented.
>
> If the truss is statically indeterminate, the analysis is much more involved. One way to do this is to replace ALL the bars in a truss with spring-dashpot equivalents and then carry out a LPM analysis as in equation (2.5) on p.63.

Chapter roadmap

A convenient way to learn the material in this chapter is to combine it with the engineering applications in Chapter 5. Our experience has been that many students have forgotten truss analysis and so the first exercise in Chapter 5 introduces a real world example utilizing a statically indeterminate Pratt Truss and carries out an elastic analysis to ensure safety against yield, and static failure without consideration of post yield behavior. This can then be followed by section 3.2 which introduces a very simple statically determinate truss with a requirement for low cycle fatigue. Section 3.3 introduces students to continuum mechanics in one dimension, introducing ideas of stretch and strain measures and balance laws and their reduction to local form —all in one dimension. It could be used as a basis for introducing students to continuum mechanics if necessary. On the other hand, if the aim is to strictly deal with structural mechanics, you can go directly to section 3.2.3.1, where a Gibbs potential based approach is used to introduce the spring-dashpot constitutive relations. This is followed by discussion of hardening behavior and the governing differential equations are obtained. Next we introduce the Bauschinger effect and introduce the notion of ratchetting and shakedown as well as a model that incorporates these effects. A sample cyclic loading problem is then solved using a spring dashpot model with hardening and Bauschinger effects and the number of cycles to failure is calculated using the Coffin Manson rule with parameters selected to fit Aluminum.

Instructors, if they choose, can then use the SMA bracing problem in Chapter 5 as the other end of the spectrum where dynamic loading is also included. We kept these two examples in a separate chapter since we did not want them to be necessarily combined. To find more examples, see authors' websites listed in the preface.

3.1 An example problem

You have been asked to estimate the life of an aluminum structure under a sinusoidally varying load. Perhaps, the structure is a mounting bracket for a motor in a non-critical application and you have been asked to find out if any possible imbalance is dangerous. You have simplified and modeled the structure as a truss shown in Fig. 3.1, where the length of the bar AB is $1m$. The load on the structure is a cyclic load, whose frequency is low compared to the natural frequency of the truss, so that you can neglect the inertia of the truss in your calculations.

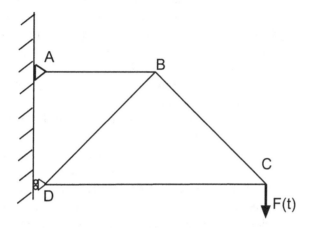

Fig. 3.1 A structure modeled as a truss ABCD. The joints have been approximated as pin joints. The bar CD is twice the length of the bar AB and the distance AD=AB. The bar is subjected to a vertical force F(t) which is approximated as a sinusoidally varying force.

Given the slow cyclic loads on the truss, your task is to find out answers to the questions such as "how many cycles will the truss last? Which bar is likely to fail? Are temperature changes due to internal mechanical heating important? If not, when will they become important?", etc.

There are two fundamentally different types of measurements that you can make to determine the loading on the truss. You can either measure or calculate the response of the truss, or you can measure the displacement of the joints of the truss. The former is a problem with known loads while the latter is a problem with known displacements.

Since the truss is a statically determinate structure, the forces in the bars are easy to find from elementary considerations, so that load control problems are easy. Thus, given the load $F(t)$, by the method of sections, you find that bar AB is the one subjected to the maximum force and that the force on the bar AB is $2F(t)$. So, you decide to concentrate your attention on bar AB as the one that is likely to decide the fate of the structure and its lifespan (with an assumption that all the bars have the same cross-sectional area and made of the same material).

Next, you need to decide on some criterion for failure. For this you should be aware of what "failure" means in this particular context. In general, failure means loss of functionality of the part. In the case of the bracket truss in question, this could mean: (i) the bracket breaks as soon as the load is applied—a case referred to as static strength failure. This is somewhat simpler to guard against since as long as the stress is below the ultimate tensile strength of the material, it will not break. (ii) The bracket may deform so much that the motor cannot be operated: this could be due to inadequate stiffness (imagine a bracket made out of rubber bands) or because the material yields and undergoes permanent deformation. The second situation will also imply failure even though the bracket is not broken. This is more complicated and requires deflection analysis. Even here, if you ensure that the stress is below the yield stress of the material, then you can perform an elastic analysis of the structure and determine if the stiffness is sufficient. If not, the structure has to be braced properly to increase stiffness.

A third and more complicated kind of failure and the one to guard against, is due to gradual micro-yielding that occurs due to repeated loading. This is called the *fatigue* or the *cyclic loading failure* and macroscopically, will have the appearance of a sudden catastrophic strength failure, even if the stresses are nowhere near the ultimate tensile strength of the material (see [Hertzberg (1976)] Chapter 12 for a detailed metallurgical description of this phenomenon together with some photographs). Here, the failure originates at minute imperfections on the surface of the bar and accumulates microscopically, cycle by cycle even though macroscopically, the material seems to be elastic. Problems such as these are a classic 0 times ∞ type problem—meaning that *per cycle* the damage to the structure is negligible and practically invisible, but over thousands or millions of cycles, the damage can accumulate, leading to a fatigue crack. Generally, such problems are, at present, being tackled primarily by means of statistical methods that are beyond the scope of this book. However, a special case is that of low-cycle fatigue, i.e. damage to the structure accumulates and the structure fails in a few hundred to a few thousand cycles. Here, there is measurable permanent deformation but not so much that the structure is non-functional. We can then use plasticity models to estimate the amplitude and extent of plastic deformation. Then the number of cycles that the structure can tolerate can be determined empirically with the central assumption that *the damage is due to the "accumulation" of plastic strain*. A commonly used rule for estimating life under these circumstances is the Coffin-Manson Law.

In summary, in order to simulate the low cycle fatigue response of a truss or cable system whose members are subjected to axial loads alone, you will need to

(1) Create a LPM for the truss member in question.
(2) Derive the evolution equations for the plastic deformation of the truss member.
(3) Find the relevant parameters from available resources.
(4) Solve it for the cyclic loading in question and simulate the response.
(5) Decide upon a failure criterion and then use the solution together with the

failure criteria to make decisions on when and how the truss is likely to fail and,

(6) Determine what (if any) remedial measures are necessary.

The first two steps represent *the mathematical modeling* steps. Step 3 is called the *model validation* step. Step 4 is the *simulation* step and Steps 5 and 6 represent the *engineering decision making* steps.

Problems like these, involving cables and trusses are very common and quite easy to set up and solve. Many practical cases with dissipative wires (such as superelastic SMA wires) are used as damping tendons in structures. Given the importance of this task and the fact that procedure for modeling these one-dimensional problems are illustrative of the general procedure, we will take this opportunity to introduce you to continuum mechanics in one dimension. For this case, the problem is simple enough even to solve by a finite deformation theory and we will do so.

3.2 Inelastic bars under axial loading

Having discussed these issues, we will idealize the bars of a truss as a long narrow bar of Length L which is capable of sustaining only axial loads that may be either tensile or compressive. Transverse loads, moments and torques are completely ignored. Such a bar is called a one dimensional continuum.

3.2.1 *Task 1: Make a preliminary list of the variables of interest*

It is clear that the input variables (or variables whose values we know) for the statically determinate truss shown in Fig. 3.1 are the force at the joint C and the ambient temperature. The output variables are the amount of accumulated "damage" as a function of location, the displacements and the temperature along the bar. In order to set up the governing equations, it may become necessary to define additional variables along the length of the bar (such as the total and inelastic strains), although we are not directly interested in these variables for the problem at hand.

3.2.2 *Task 2: Make simplifying assumptions about the response*

Simplification of the kinematical and thermal parameters

Let us begin by considering a truss member or cable shown in Fig. 3.1. There are several ways to model such a body. It is clear that for our purpose, since the cross-sectional dimensions of the body are very small compared to the length, we will assume that all properties and variables are uniform across the cross-section and vary only along the length of the truss member.

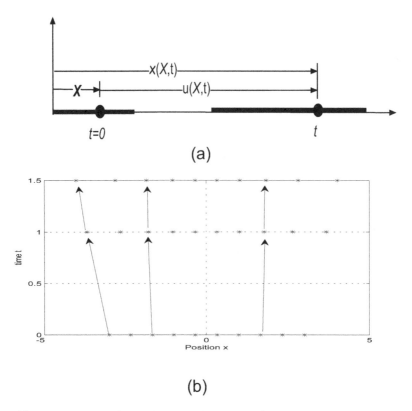

Fig. 3.2 (a) Motion of a bar (a one dimensional continuum) along the x-axis. The location of a point on the bar at time $t = 0$ and at a different time is shown. The displacement depends upon the position and time as shown. (b) The location of a set of initially equispaced particles are shown as functions of time. The displacement of a particle is $u = t^2 X^3/40$. The arrows represent the way in which the particles move. Note that the spacing between particles change as time goes on.

Simplification of the properties:

We will assume that none of the material properties are dependent on the temperature. This may not be a good assumption for cyclic loading since variation of temperature due to internal mechanical heating may get to be large enough to "weaken" the material. After we solve the problem, we can come back and verify whether the assumption is reasonable.

Simplification of the thermal coupling

To begin with, we will ignore temperature variations and see if we can get away with this simplifying assumption. With wires, cables and truss-members, this would usually be a good assumption given the large surface to volume ratio so that any temperature rise is quickly conducted away[1]. However, if the cyclic loading were sufficiently rapid, then internal mechanical heating could become an issue. Thus, the thermal and mechanical properties are uncoupled and we can solve them separately. We will have to investigate this very carefully a-posteriori and make suitable modifications.

3.2.3 Task 3: Modeling of the response of the truss

A "one-dimensional continuum" is a set of particles arranged along a segment of the real number line such that there is a one-to-one correspondence between the numbers in the segment and the particles. Roughly, there are an uncountably infinite number of particles with no "gap" between them. The main difference between continuum mechanics and particle mechanics is that in continuum mechanics, the particles are "smeared out" and we only apply all the laws of physics to *chunks* of the body in question—never to the individual particles. We then use the "magic of calculus," i.e. limits to obtain rules that are valid at every location in the body. Please remember: We first consider a finite chunk, apply physical principles and then take limits; It is very hard to directly justify the final equations that we get in terms of physics.

Thus, the development of the model for this truss is made up of 3 steps: (1) a description of the kinematics of the particles (the kinematical laws) (2) The balance laws in their local form, in other words, laws of motion of "infinitesimal chunks" of the material (the kinetic laws) (3) Choosing a LPM for the "infinitesimal chunks" (the constitutive laws). We will follow this same principle for beams as well as for three-dimensional continua, except that it gets increasingly more complicated as we go towards full 3D.

kinematics of a one-dimensional continuum

To begin the modeling, we need to label the particles of the bar (see Fig. 3.2). Our most convenient choice is to choose the location X of the particles at time $t = 0$ as the labels. At a subsequent time, due to the forces on the bar, the bar will stretch and the locations will change. Let us imagine the particles to be standing in a queue. If different particles move differently, then the separation between the particles in the queue will change. The displacements of the particles[2] are given by $u = u(X, t)$ so that the new locations x are given by $x = X + u(X, t)$. Thus, $x = x(X, t)$ means that "x is the location at time t of the particle that used to be at X at time $t = 0$".

---

[1] This is an assumption that has to be carefully verified and in fact appears to be a poor assumption for many shape memory alloys and polymers since their properties are very sensitive to temperature. For these cases, one MUST solve for the temperature profile also.

[2] Capital X stands for the initial location and small x stands for the current location.

⊙ **3.1 (Exercise:). Verbal descriptions**
Give a verbal description of $u = u(X, t)$

The stretch and deformation gradient:
Consider two particles A and B (see Fig. 3.3) that have their initial locations X_1 and X_2 along the bar. Due to the forces on the bar, they will be displaced and their new locations are x_1 and x_2 at time t respectively. The ratio of the final separation to the initial separation between them, i.e. $(x_2 - x_1)/(X_2 - X_1)$ is called the *average stretch ratio*.

If we use a mathematical limiting process, we can then replace the differences with infinitesimals and we will get the *local stretch ratio* or simply "stretch" as

$$\lambda := \lim_{\Delta X \to 0} \left. \frac{\Delta x}{\Delta X} \right|_{t=const} := \frac{\partial x}{\partial X}. \tag{3.1}$$

In other words, the stretch λ is the *slope of the $x - X$ curve at a given time t*.

The linear or engineering strain ε, is defined as $\varepsilon = \lambda - 1$ while the so-called *true or logarithmic strain* is defined as $e = ln(\lambda)$. It can be shown that for λ between 0.89 and 1.12, i.e. for engineering strains between 12% and -11%, the error, or the difference between the values of the engineering strain, ε, and the true strain e is less than 6%.

The variation of position with time is the velocity, which for the bar, can be a function of both position and time. The acceleration is then the rate of change of velocity, i.e.

$$v(X, t) = \frac{\partial x}{\partial t} \Big|_{X \text{ fixed}}, \quad a(X, t) = \frac{\partial v}{\partial t} \Big|_{X \text{ fixed}} \tag{3.2}$$

This is all the kinematics we need for our purpose.

In order to clarify these points, consider a motion of a bar which is defined by $x = t^2 X^3/40$. If we consider equispaced particles at time $t = 0$, their subsequent locations at time $t = 1$ and $t = 1.5$ are shown in Fig. 3.2b. Note that the particles are moving further apart; a question might arise in your head that "if the particles move farther apart, won't gaps open up?". The answer is one of the mysteries of "continua" — no matter how much you "stretch" the number line you will never get gaps[3].

Note also that the particles that are farther from the origin are moving faster. This becomes evident when we calculate the stretch, the velocity and acceleration as

$$\lambda = \frac{\partial x}{\partial X} = \frac{3}{40} t^2 X^2, \quad v = \frac{\partial x}{\partial t} = \frac{1}{20} t X^3, \quad a = \frac{\partial^2 x}{\partial t^2} = \frac{1}{20} X^3. \tag{3.3}$$

---
[3] If you want to learn more, take a math class on real analysis as we did and you too can give clever answers like us! :-)))

Note that the acceleration of every particle (i.e. for every fixed X) is constant but different particles have different values.

> ⊙ **3.2 (Exercise:). Extension, velocity and acceleration**
> *Given that the displacement of the particles in a rod is $x(X,t) = 4(1 - \cos(t))\sin(3X)$, find the engineering strain, the logarithmic strain, the velocity and the acceleration as functions of X and t. Consider two points on the bar, one at $X_1 = 1$ and the other at $X_2 = 2$. Describe their motion relative to each other in words. Plot their relative positions (i.e. x_1-x_2) as functions of time and see if it agrees with your description.*
> *Hint: Use (3.1) and (3.2).*

Balance Laws

Every body that obeys classical physics is assumed to satisfy four fundamental laws, namely the conservation of mass, momentum, angular momentum and energy. For the one-dimensional continuum in question, since the bar is not rotating and since we already used global angular momentum balance (when we invoked the method of sections) we do not have to worry about angular momentum balance for each chunk of the bar; so that leaves us with the task of ensuring that the remaining three laws are satisfied for every "part" or "section" or "chunk" of the bar. This sounds like a daunting task since it is possible to imagine cutting up the bar into small slices and demanding that each slice needs to be individually examined. But have no fear—calculus to the rescue. To help you develop a mental picture of the process, imagine that Δx means "slice of x" and dx means "sliver of x".

We will illustrate the mathematical representation of these laws in full detail in this one-dimensional case. Since the process is similar in the higher dimensions, in the subsequent sections, we will not derive these laws but simply state them instead, referring the reader to suitable texts for the derivations expecting that the reader will be able to follow these derivations based on the discussions here.

Mass Conservation

So, we consider the same two particles A and B (see Fig. 3.3a) and examine the "chunk" or "slice" of material between A and B. Let the linear density (mass per unit length) be equal to ϱ_0 at time $t = 0$. What will be its linear density at time t? If the bar gets longer so that A and B separate, you might expect the linear density (mass per unit length) to decrease. How did you arrive at this conclusion?

You have automatically invoked the law of conservation of mass, i.e. "the amount of material between A and B is the same no matter how much we stretch it out". This is true only if the body is a closed system (i.e. it does not take in or give out mass). Here, we are considering just a simple metallic bar so we can safely say that the mass of the material between A and B is the same even when the bar is

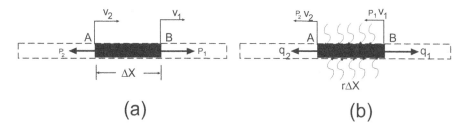

Fig. 3.3 (a) Free body diagram of a small element of the beam of length $\triangle X$, with the axial force shown at either end. The velocities of the two ends A and B are indicated. (b) The same element with the flow of mechanical power PV and heat flux q into the system. The direction of the arrows indicate the direction of energy flow. Note that there is lateral heating due to the possible temperature difference with the surroundings.

deformed, since the mass is density times the distance between A and B, the mass per unit length after deformation is given by

$$\varrho \approx \frac{\text{mass}}{\text{current length}} = \frac{\varrho_0(X_2 - X_1)}{(x_2 - x_1)} \qquad (3.4)$$

as A and B get close together by taking the limit so that the "slice" of material becomes a "sliver" and by using the definition for the stretch, we will get

$$\varrho(X,t) = \varrho_0 \frac{\partial X}{\partial x} = \frac{\varrho_0}{\lambda} \qquad (3.5)$$

Thus, if we know the stretch λ and the initial density ϱ_0 we can always find the linear density at any other time.

Momentum Conservation

Next, we consider the free body diagram (FBD) of the chunk AB (see Fig. 3.3a). The axial forces on the ends of the chunk AB are P_1 and P_2, both being drawn in tension. The assumption of a truss type system precludes any other forces on the bar. By applying Newton's laws to the chunk and noting that the mass of the chunk is $\varrho_0(X_2 - X_1)$, we get

$$(+\rightarrow)ma = (\varrho_0(X_2 - X_1)) \frac{\partial v}{\partial t} = P_2 - P_1 \qquad (3.6)$$

Dividing throughout by $\triangle X := (X_2 - X_1)$ and taking limits as $X_2 \to X_1$, we get

$$\varrho_0 \frac{\partial v}{\partial t} = \frac{\partial P}{\partial X} \qquad (3.7)$$

This is the differential equation that represents momentum conservation. Of course, if there is no acceleration then the equation becomes trivial.

Energy Conservation

We repeat the procedure for energy conservation. We assume that the internal energy (or adiabatic potential energy) per unit mass is $\exists$ so that the total energy of the chunk AB (potential + kinetic) is $\varrho_0(X_2 - X_1)(\exists + v^2/2)$.

The law of conservation of energy states that the rate of change of this energy must equal the rate of working by the forces applied at the ends of the bar, the rate of heating or cooling by convection, r per unit length of the lateral surfaces and the rate of heating from conduction along the bar. Thus, by looking at Fig. 3.3b, we get

$$\frac{d}{dt}\left[\varrho_0(X_2 - X_1)(\exists + v^2/2)\right] = (P_1 v_1 - P_2 v_2) + r(X_2 - X_1) - (q_2 - q_1) \quad (3.8)$$

As usual, upon dividing by $(X_2 - X_1)$ and noting that the right-hand side can be written as $((Pv)_2 - (Pv)_1)/(X_2 - X_1)$ and $(q_2 - q_1)/(X_2 - X_1)$, and taking limits we will get

$$\frac{d}{dt}\left[\varrho_0(\exists + v^2/2)\right] = r + \frac{\partial(Pv)}{\partial X} - \frac{\partial q}{\partial X}. \quad (3.9)$$

The above equation can be simplified by expanding the left-hand side as well as the second term of the right-hand side (and using the fact that ϱ_0 is a constant) to get

$$\varrho_0 \frac{d\exists}{dt} + \varrho_0 v \frac{dv}{dt} = r + v\frac{\partial P}{\partial X} + P\frac{\partial v}{\partial X} - \frac{\partial q}{\partial X}. \quad (3.10)$$

Now using the balance of linear momentum (3.5), we get

$$\varrho_0 \frac{d\exists}{dt} = r + P\frac{\partial v}{\partial X} - \frac{\partial q}{\partial X} \quad (3.11)$$

Finally, we use the fact that *the gradient of the velocity is equal to the rate of change of the stretch*, i.e.

$$\frac{\partial v}{\partial X} = \frac{\partial^2 x}{\partial X \partial t} = \frac{\partial}{\partial t}(\partial x/\partial X) = \frac{d\lambda}{dt} = \frac{d\epsilon}{dt}. \quad (3.12)$$

In other words, *the derivative of the velocity with respect to position X is equal to the time derivative of the engineering strain*. Using this, we finally get the so-called *energy equation*

$$\varrho_0 \frac{d\exists}{dt} = r - \frac{\partial q}{\partial X} + P\frac{d\epsilon}{dt}. \quad (3.13)$$

In words, the above equation states that:
the rate of change of the internal energy equals the rate of external heating due to convection (the r term) and conduction (the $\partial q/\partial X$ term) plus the axial force times the strain rate.

The first two terms are obvious. The additional term $P\dot\epsilon$ may not be and so we will explain it further. The key point is that if you apply forces on a *rigid* body, the work done will become the kinetic energy of the rigid body. On the other hand not all the work done on a *deformable or stretchable* body raises its kinetic energy. Only a part of it does. The remaining part goes to raise its internal potential energy. It is this part which is called the *internal working* or *work of deformation* and is equal to $P\dot{\epsilon}$.

The energy equation (3.8) is the first among the balance laws to connect the external stimulus to the system (heating and working on the system) with the internal changes in the system, i.e. the change in the internal energy of the system.

Next, we define the axial stress in the bar to be $\sigma = P/A_0$, where A_0 is the reference cross-sectional area of the bar. This is the engineering stress[4].

Simplification for the Truss in Fig. 3.1: For the current problem, we have already found that the forces at points A and B are equal to each other and that their value is $2F(t)$. Then, by neglecting the inertial terms on the left-hand side of (3.7), we obtain

$$\frac{\partial P}{\partial X} = 0, P(0,t) = P(L,t) = 2F(t) \Rightarrow P(X,t) = 2F(t). \quad (3.14)$$

Thus, not unexpectedly, the equation of equilibrium suggests that the axial force is constant throughout the length of the bar. If the area of the bar is constant, then this implies that the axial stress throughout the bar is also constant, i.e. $\sigma(t) = P(t)/A_0 = 2F(t)/A_0$. Thus, if we know the force on the bar as a function of time, then we can find the stress in the bar. Such a problem is called a *stress controlled problem*.

This result is somewhat unusual in the sense that the stress in the bar was determined without regard to the response of the bar[5]. This allows us to distinguish consideration of the constitutive laws from consideration of the momentum conservation laws and considerably simplifies the solution. In most problems, the equations resulting from momentum conservation cannot be solved without constitutive assumptions, leading to coupled equations. We will see these kinds of problems in later sections.

3.2.3.1 *The equivalent LPM and the Gibbs potential*

Now that we have the form of the energy equation, we are in a position to develop constitutive laws for the truss. The idea is as follows: we break up the bar into "slivers" of length dx, and assume that each sliver is like a LPM. In view of our

---

[4] This is also referred to as the *First Piola Kirchhoff* stress. Perhaps, a more descriptive name would be "referential force intensity". But, in the interests of consistency with the rest of the world, we will call it the *engineering stress*.

[5] Actually, this is not strictly true, since we assumed that the bars of the truss are nearly rigid so that angles and lengths are nearly the same as the original lengths. To see this point, imagine that all the bars are made of rubber bands; would you still expect the bars to be of equal length when a force is applied?

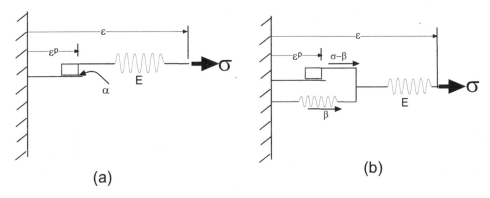

Fig. 3.4 A simple lumped mass model of an elastoplastic bar. (a) A simple spring and frictional dashpot in series. The maximum force that can be sustained by the dashpot is κ. (b) A model that is capable of exhibiting the "Bauschinger effect", through the use of an additional spring as shown. In case (b), the force in the second spring is α and the force in the frictional dashpot is thus reduced to $\sigma - \alpha$. The total and plastic strains are also shown.

assumption that all "slivers" are identical in their properties and responses and the result (3.14) that the axial forces on each sliver is the same, it should be clear to you that *each sliver is identical to any other sliver and it is enough to consider a single "typical" sliver and figure out its response.* The key to figuring out an equivalent LPM is the energy conservation equation (3.13). If you compare this with (2.91), we see that the term $P d\epsilon/dt$ in (3.13) is the equivalent of $\sum_i Q_i \dot{q}_i$ in (2.91). Based on this, we will identify ϵ as the generalized displacement q and $\sigma = P/A_0$ as the generalized force, **Q**.[6]

We will now model the individual slivers of the bar as a spring-frictional dashpot systems as shown in Fig. 3.4(a). The spring represents the elasticity of the bar and the dashpot in this case is a frictional dashpot to represent the permanent deformation of the bar. Note that the force on the LPM is equivalent to the stress in the bar and that the displacement of the right end of the dashpot is equivalent to the engineering strain ε in the bar[7].

Later, we will also consider the alternative and more realistic (but more complicated) model in Fig. 3.4(b) and we will see how the answers differ.

We will model this as a single degree of freedom system and will use a Gibbs potential approach to show you the utility of the Gibbs potential. Note that the generalized force on the spring, Q^s, is the stress σ. So our Gibbs potential is given by $G(\sigma, T)$. Recalling from Chapter 2 that the mechanical part of the Gibbs potential

---

[6] We can use P itself as the generalized force but then the results will depend upon the area.

[7] Although for the current problem, the mechanical model is shown with the displacements being additive, we will not necessarily make such an assumption. This illustrates one of the dangers of a purely mechanistic view of such models. They are meant to serve only as aids to reasoning and are not meant to be taken literally. In a general nonlinear context, we can only say that the total stretch of the element is some (possibly nonlinear) function of the stretches of the individual elements. This is also a constitutive assumption. An additive function is just one of the possibilities albeit a very useful one.

is the negative of the isothermal complementary energy function, for our case, we assume that the isothermal complementary energy per unit length of the bar is quadratic (since we assume that the spring is linear) in the applied force, so that

$$G = -\frac{\sigma^2}{2E} + CT(1 - \ln T). \tag{3.15}$$

The first term is the familiar complementary energy (with a negative sign) where E is the Young's modulus. The second term is related to the heat capacity of the slivers of the bar. Note that we have assumed Young's modulus to be a constant. For example, for an aluminum bar with a density of $2.7 gm/cc$, modulus of $70 GPa$, and a heat capacity of about 900 J/kg/K, the form for the Gibbs potential per mm^3 of the rod is given by,

$$G = \left(-\frac{\sigma^2}{140000} + 2.43 x 10^{-3} T(1 - \ln T)\right) \ J/mm; \quad \sigma \ \text{in MPa} \tag{3.16}$$

Further, we know that the internal energy *per unit length* of the bar is given by (see (2.89))

$$\varrho_0 U = A_0 (G - T\frac{\partial G}{\partial T} - \sigma \frac{\partial G}{\partial \sigma}), \tag{3.17}$$

where A_0 is the cross sectional area of the bar.

⊙ **3.3 (Exercise:). Heat equation in 1 dimension**
Substitute (3.17) into (3.13) and show that we get the following heat equation

$$-A_0 T \frac{d}{dt}\left(\frac{\partial G}{\partial T}\right) = r - \frac{\partial q}{\partial X} + A_0 \sigma \frac{d}{dt}\left(\varepsilon + \frac{\partial G}{\partial \sigma}\right). \tag{3.18}$$

The second term on the right of the above equation (3.18) is *the mechanical power dissipated (or lost irrecoverably) by the rod due to internal viscous and frictional damping processes*, so that our internal dissipation equation together with the second law of thermodynamics becomes

$$\sigma \frac{d}{dt}\left(\varepsilon + \frac{\partial G}{\partial \sigma}\right) = \xi \geq 0 \tag{3.19}$$

The above equation is the cornerstone of constitutive modeling, being the starting point for subsequent analysis.

Power dissipated by the dissipative components and identification of inelastic parts of the strain rate: Now that we have a form (3.15) for the Gibbs potential, we substitute it into (3.19), and find that the mechanical power dissipated by the frictional component of the model can be given explicitly in terms of the stresses and strains as

$$\sigma(\dot{\varepsilon} - \frac{\dot{\sigma}}{E}) = \xi \geq 0 \tag{3.20}$$

> ⊙ **3.4 (Exercise:). Power dissipated by internal friction**
> *Verify the above result* directly by drawing the FBD for the frictional element in Fig. 3.4a and using the fact that the displacements of the two components are additive.
> *Hint:* The force through the dashpot is σ and its displacement is $\varepsilon - \sigma/E$.

For simplicity of representation and in view of the mechanical analogy, we define the *inelastic strain rate* as

$$\dot{\varepsilon}^p = \dot{\varepsilon} - \frac{\dot{\sigma}}{E}. \tag{3.21}$$

Note that this is the velocity of the dashpot.

Loading criteria and the origin of permanent deformation: By substituting (3.21) into (3.20), we see that, from the second law of thermodynamics

$$\sigma \dot{\varepsilon}^p \geq 0, \tag{3.22}$$

i.e. the second law of thermodynamics implies that $\dot{\varepsilon}^p$ must have the same sign as σ, thus, if σ is positive, then only, increase in ε^p is allowed whereas if σ is negative only decrease in ε^p is allowed. This "one way" or "ratchetting" feature of elastoplastic materials is critical to their response. Thus, if we cycle the material only with positive forces, (i.e. only in tension) there will be an accumulation of tensile plastic strain or permanent strain, leading to eventual failure of the material.

To complete the process, we need to supply a constitutive law for the dashpot that relates the inelastic strain rate (the generalized velocity of the dashpot) to the stress (the generalized force on the dashpot) and it is to this task that we turn next:

Formulation of the Inelastic Laws: This is perhaps the most complicated part of the analysis which needs careful study. Since we want to model rate independent elastoplastic behavior, we are going to use a friction model. Thus, we assume that the frictional component will hold until the magnitude of the stress σ on it reaches a critical value κ and will slide subsequently (see equation (2.34) on p.42). For our case, we take the "yield strength" of the 6061 Aluminum alloy to be $150 MPa$ (N/mm^2). This will allow us to immediately make the following prediction:

> ### ONSET OF YIELD
>
> Based on our truss calculations, we can conclude that if the stress on bar AB exceeds 150 MPa, then it will begin to undergo permanent shape change. If we assume that the cross sectional area of the bar AB is 156.16 mm^2, then, since the axial force is 2F, we can conclude that if F reaches $150 \times 156.16/2 = 11.7 kN$, the truss will begin to yield.

Beyond yielding, we have the following constitutive laws

$$\sigma \dot{\varepsilon}^p \geq 0; \quad \dot{\varepsilon}^p = 0 \Rightarrow \sigma^2 - \kappa^2 <= 0, \dot{\varepsilon}^p \neq 0 \Rightarrow \sigma^2 - \kappa^2 = 0; \quad (3.23)$$

This is not a simple constitutive law and is presented in an implicit form. In words, the equations mean the following:

(1) Plastic deformation is dissipative.
(2) If there is no plastic deformation, then the load cannot exceed κ.
(3) During plastic deformation, the load has to be equal to κ.

In view of the above conditions, the central task becomes the following:
Given the load $\sigma(t)$, find $\dot{\varepsilon}^p$ that satisfies the criteria (3.23).

We will now show that, if κ is constant, the model predicts that the truss will collapse as soon as the stress reaches κ (i.e. there is no unique solution for $\dot{\varepsilon}^p$).

To see this, let us assume that κ is indeed constant. Then,

(1) If $\sigma^2 < \kappa^2$ then only $\dot{\varepsilon}^p = 0$ is possible and we get an elastic response.
(2) It is impossible for σ^2 to be greater than κ^2.
(3) If $\sigma^2 = \kappa^2$ then as long as it remains at that value *any value of $\dot{\varepsilon}^p$ which has the same sign as σ is allowed, since all the criteria are met.* To see this, we consider the case when $\dot{\varepsilon}^p \neq 0$, then we have $\sigma \dot{\varepsilon}^p > 0$ which means that σ and $\dot{\varepsilon}^p$ must have the same sign. Further, we note that since as long as $\dot{\varepsilon}^p \neq 0$, we have $\sigma^2 - \kappa^2 = 0 \Rightarrow \sigma \dot{\sigma} = 0 \Rightarrow \dot{\sigma} = 0$. Thus, as long as $\dot{\varepsilon}^p$ does not vanish, σ is constant. Thus, we note that all the conditions in (3.23) can be met by any value of $\dot{\varepsilon}^p$ which is of the same sign as σ as long as $\|\sigma\| = \kappa$.

To understand this unexpected result, note that the LPM is a spring+friction model. Notice from the spring-dashpot model that if the force on the dashpot equals κ, the dashpot moves freely and the force cannot be increased further, whereas if the force on the dashpot is below κ the dashpot does not move at all! This represents *ideal plasticity*. We will return to this model again when we discuss prescribed displacements rather than prescribed forces on the two ends A and B. We will then find that the model is perfectly well behaved.

Thus, according to this model, the maximum load supportable by the truss is $11.7kN$. Below $11.7kN$ there is no "damage" ($\dot{\varepsilon}^p = 0$) to the bar whereas as soon as the force reaches $11.7 kN$, the bar's life is over (since $\dot{\varepsilon}^p$ is effectively infinite). In spite of this, useful conclusions can be drawn. Maximum load calculations of this kind fall under the category of "the theory of the plastic hinge" and are a very useful tool for estimating static strength of structures and their subsequent failure modes. Hence, while this model is suitable for finding the maximum load that the truss is capable of withstanding, it is unsuitable for simulating any kind of life prediction calculation.

There is life after yielding: In reality, for a material like aluminum, the material has a nonlinear stress-strain curve as shown in Fig. 3.5.

In order to simulate this kind of hardening behavior, i.e. increasing frictional resistance, *we need to have a frictional dashpot with variable resistance.* Such models are called *hardening models* and the prescription for the variation of the frictional resistance is called a *hardening law*. We are, thus, led naturally to the next step in our modeling.

Work hardening and its origin: Plastically deforming a metal at room temperature is referred to as plastic working or cold working (if the room temperature is less than half the melting point). During this process, a large number of dislocations (a kind of defect in the crystal structure) are formed which make it increasingly difficult to subsequently deform the material, i.e. which increase the resistance to further plastic deformation. Generally speaking, this is a very complex process involving the interaction of a very large number of dislocations. However, for the purposes of a macroscopic model, it is usual to assume that *the resistance to further deformation is a function of the total amount of "damage" done to the material.* Thus, we need a parameter to represent "how much damage has been done to the material". For this purpose, a parameter called the "equivalent plastic strain" is introduced. It is defined as

$$s = \int_0^t \|\dot{\varepsilon}^p\| \, dt \quad \Rightarrow \quad \dot{s} = \|\dot{\varepsilon}^p\|. \tag{3.24}$$

This parameter is a measure of how much "sliding" has occurred. Notice that this parameter can only increase with time and will never decrease. So, even if the bar is alternately compressed and extended, the value of s will go on accumulating. We could consider this as an internal "time" or "damage" parameter. We now stipulate that the resistance of the material is a monotonically increasing function of the "equivalent plastic strain", i.e. stipulate that

$$\dot{\kappa} = \Theta(s,\kappa)\dot{s} = \Theta(s,\kappa)\|\dot{\varepsilon}^p\|, \Theta(s,\kappa) \geq 0. \tag{3.25}$$

The function $\Theta(s,\kappa)$ is referred to as the *hardening modulus or hardening rate*. In principle, the differential equation (3.25) can be integrated to give $\kappa = \kappa(s)$ and in many cases, this is how it is presented. However, the form (3.25) is convenient for numerical implementation and is much more versatile. Some commonly used

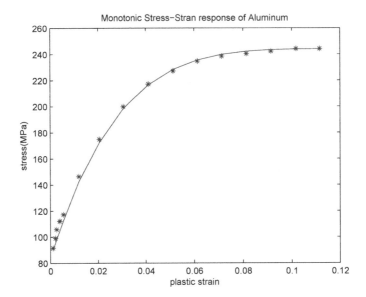

Fig. 3.5 Experimental stress-strain curve of an aluminum rod. The plastic strain versus the stress beyond yield has been plotted. The hardening law given by $\dot{\kappa} = \Theta(s,\kappa)\dot{s} = 100(244-\kappa)^{0.8}\dot{s}$, is an excellent fit to the data as is evident.

hardening functions, $\Theta(s,\kappa)$, are:

$$\text{Linear hardening:} \quad \Theta = C_1 \Rightarrow \kappa = C_1 s \tag{3.26}$$

$$\text{Swift Law:} \quad \Theta = mC_2(s_0+s)^{m-1} \Rightarrow \kappa = C_0 + C_2(s_0+s)^m \tag{3.27}$$

$$\text{Voce Law:} \quad \Theta = -C_4 C_5 e^{-C_5 s} \Rightarrow \kappa = C_3 + C_4 e^{-C_5 s}. \tag{3.28}$$

In the above equation, the hardening parameter, κ is also integrated and shown as a function of s. While laws such as the above are commonly used, with the advent of modern software packages, we have found it easier to work directly with the hardening modulus $\Theta(s,\kappa)$. For many metals such as aluminum and copper, where a sharp "yield point" is not apparent, a function of the form

$$\Theta(s,\kappa) = a_0(a_1 - \kappa)^{a_2} \tag{3.29}$$

seems to fit available experimental data very well. For example, for a sample of aluminum for which experimental monotonic stress-strain curve was available, the equation $\dot{\kappa} = \Theta(s,\kappa)\dot{s} = 100(244-\kappa)^{0.8}\dot{s}$, along with $\kappa(s=0) = 95 MPa$ seems to fit the data extremely well up to a strain of about 0.1 (10%) as seen in Fig. 3.5.

Note that the form (3.29) does not involve s at all. There are very good physical reasons for assuming forms for Θ that do not involve s and so we will try to follow this as far as possible.

⊙ 3.5 (Exercise:). Power law hardening

Solve the differential equation (3.25) with Θ given by (3.29) in closed form (by using separation of variables) and obtain an expression for $\kappa(s)$. Compare the result with Swift hardening law. What does the constant a_1 represent? How about a_0?

Ans: The form is similar to the Swift law, although not the same. a_0 is related to the initial hardening rate and a_1 is the maximum resistance to permanent deformation.

Evolution equations for variable resistance (or hardening models):
Now that we have the form (3.25) with a suitably chosen law for $\Theta(s, \kappa)$, we are now ready to find the value of $\dot{\varepsilon}^p$. We will find that, unlike the case when $\kappa = const.$ we will be able to solve for $\dot{\varepsilon}^p$ for a given $\sigma(t)$.

Thus, we have to solve the following problem:
Given $\sigma(t)$, find $\dot{\varepsilon}^p$ such that the following conditions are met:

$$\sigma \dot{\varepsilon}^p \geq 0, \quad \dot{\varepsilon}^p = 0 \Rightarrow \sigma^2 - \kappa^2 \leq 0, \quad \dot{\varepsilon}^p \neq 0 \Rightarrow \sigma^2 = \kappa^2 \tag{3.30}$$

where the evolution equation for κ is given by (3.25). This is not a simple equation to solve. As we go further along this book, we will encounter this equation many times and we will invent increasingly general ways to solve it.

Here, we find $\dot{\varepsilon}^p$ in the following way: Notice that the first of (3.23) implies that $\dot{\varepsilon}^p$ must have the same sign as σ. Furthermore, since $\|\sigma\| = \kappa > 0$ for plastic flow, the sign of $\dot{\varepsilon}^p$ is the same as the sign of σ which is equal to $\|\sigma\|/\sigma = \kappa/\sigma$. So, we arrive at

$$\dot{\varepsilon}^p \neq 0 \Rightarrow \dot{\varepsilon}^p = \frac{\kappa}{\sigma} \|\dot{\varepsilon}^p\| = \frac{\kappa}{\sigma} \dot{s} \tag{3.31}$$

where $\dot{s}$ is given by (3.24).

⊙ 3.6 (Exercise:). Calculation of the plastic strain rate

By using $(3.23)_3$ and differentiating $\sigma^2 - \kappa^2$ with respect to time and using (3.25) and after some simplification, obtain the following expression for $\dot{\varepsilon}^p$:

$$\dot{\varepsilon}^p \neq 0 \Rightarrow \dot{\varepsilon}^p = \frac{\dot{\sigma}}{\Theta(s, \kappa)}. \tag{3.32}$$

Hint: $\sigma \dot{\sigma} = \kappa \dot{\kappa} = \kappa \Theta \dot{s} \Rightarrow \dot{s} = \sigma \dot{\sigma}/\kappa \Theta$.

Now the above solution is only possible if it also meets the second law requirement that $\sigma \dot{\varepsilon}^p > 0$. Since $\Theta(s, \kappa) > 0$, we then find that $\dot{\varepsilon}^p > 0$, only if $\sigma \dot{\sigma} > 0$. This is called a *loading criterion*.

Combining all of the above considerations, together with (3.23), we arrive at the following kinetic law for $\dot{\varepsilon}^p$

$$\dot{\varepsilon}^p = \begin{cases} \frac{\dot{\sigma}}{\Theta(s,\kappa)}, & \text{if } \sigma^2 = \kappa^2, \sigma \dot{\sigma} > 0; \\ 0, & \text{otherwise.} \end{cases} \tag{3.33}$$

In words, the above equation says that *the rate of plastic strain is the stress rate divided by the hardening modulus if the stress rate and stress have the same sign. Otherwise, it is zero and the response is elastic.* Here, at last, is the sought after kinetic law.

STRAIN HARDENING BAR UNDER PRESCRIBED STRESS RATE

We summarize the complete procedure for finding the equations for a bar in a statically determinate truss subject to a load $F(t)$.

(1) Find the forces on all the bars and find the bar with the largest force. For the truss in Fig. 3.1, this is the bar AB with a force of $2F(t)$ on it.

(2) Divide by the cross sectional area of the bar to obtain the stress $\sigma(t)$ on the bar; For the bar AB, the cross sectional area is A_0, so that the stress is $\sigma(t) = F(t)/A_0$. Since we need the stress rate, it is actually much simple to specify $\dot{F}(t)$ and pretend that the differential equation for calculating the stress is

$$\dot{\sigma} = \dot{F}/A_0 = a(t) \tag{3.34}$$

where $a(t)$ can be prescribed to be a known function of t and reflects the applied load.

(3) Look up the modulus of elasticity E and choose a suitable hardening law $\Theta(s, \kappa)$, for the material in question. This is a hard problem and requires quite a bit of research and curve fitting. For the case of Aluminum at room temperature, we have found that $E = 70GPa = 70000MPa$ and $\Theta(s, \kappa) = 100(244 - \kappa)^{0.8}$ fits the data very well.

(4) Assume initial conditions for ε, ε^p, s and κ; We have $\varepsilon^p(t=0) = 0$, $s = 0$, $\kappa(t=0) = \kappa_0 = 95MPa$.

(5) Solve the following set of differential equations for the strain, plastic strain, the equivalent plastic strain and the resistance κ :

$$\text{Stress}: \dot{\sigma} = a(t) \tag{3.35}$$

$$\text{Plastic flow equations (from (3.33))}: \dot{\varepsilon}^p = \begin{cases} \frac{\dot{\sigma}}{\Theta(s,\kappa)}, & \text{if } \sigma^2 = \kappa^2, \sigma\dot{\sigma} > 0; \\ 0, & \text{otherwise}. \end{cases} \tag{3.36}$$

$$\text{Equivalent plastic strain (from (3.24))}: \dot{s} = \|\dot{\varepsilon}^p\| \tag{3.37}$$

$$\text{Hardening law (from (3.25))}: \dot{\kappa} = \Theta(s,\kappa)\dot{s} \tag{3.38}$$

$$\text{Total strain rate (from (3.21))}: \dot{\varepsilon} = \dot{\varepsilon}^p + \frac{\dot{\sigma}}{E} \tag{3.39}$$

Why are we spending so much time in setting up a set of five differential equations? Because, my dear readers, ordinary differential equations are our "friends" not our "enemies"—especially with the advent of MATLAB. You can solve this set of ODEs in a jiffy on MATLAB and gain valuable insight! If you are familiar with MATLAB, then try it out for yourself before looking up our solution. If not, please put this book down and spend some time learning to solve ODEs on MATLAB. There are excellent primers on the subject and we promise that if you are familiar with programming, you can learn this in a day [8].

⊙ **3.7 (Exercise:). MATLAB program to solve the evolution equations**
Write a MATLAB program to solve the above set of equations (use ODE23, it is a low accuracy solver that is quite forgiving of discontinuities etc. and sufficient for our purpose. ODE45 may sometimes give trouble because of the discontinuities in $\dot{\varepsilon}^p$).

We will assume that our input is

$$\dot{\sigma} = \begin{cases} 0.05 \ MPa/s, & \text{if } t < 500s; \\ 17\cos(0.1(t-500)), & \text{if } t \geq 500s. \end{cases} \quad (3.40)$$

The applied stress is shown as a function of time in Fig. 3.6.

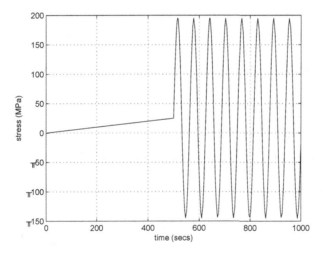

Fig. 3.6 The applied cyclic loading on the truss. Initially the load is increased linearly and then cycled back and forth (simulating a piece of vibrating machinery).

For a complete computer program for the simulation of this, see authors' websites listed in the preface.

---
[8]here are some of the popular ones: [Pratap (2005); Gilat (2007); Chapra (2006)].

The resulting stress-strain response and the strain and plastic arc length with time are shown in Fig. 3.7:

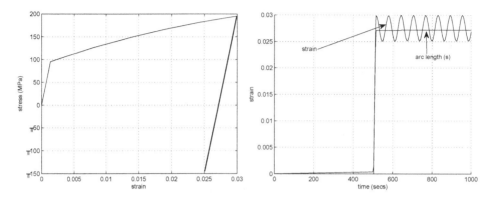

Fig. 3.7 The simulated response of the Al rod subjected to the cyclic load of Fig. 3.6. Note that the rod quickly stabilizes to a new elastic response with a much higher yield value. Moreover, there is no accumulation of "damage", i.e. the value of s does not increase!

The response shows that the rod simply becomes "harder"—there is an initial permanent change of shape but then it subsequently responds elastically. We see that this model again predicts that there is no accumulating damage in this case!

3.3 Prescribed displacements

Compared to stress cycling, loading with strain control is much more "damaging" to the material.

Compared to the case of prescribed loading, which is a relatively straightforward problem for a statically determinate truss since the load on the bars can be obtained independently of the constitutive equations for the bars, it is not simple to find the displacements of the bars given the displacement at some of the joints. We shall see that such problems will need to be solved when dealing with statically indeterminate structures and we will take it up as a case study in the next section. Here, we content ourselves with the following problem: Rather than knowing the stress σ on the bar, if we know the strain $\varepsilon(t)$, how do we find the plastic strain rate?

Notice that (3.33) is still valid. However, we don't know $\dot{\sigma}$. We can use (3.21) to solve for $\dot{\sigma}$ and get the following equation

$$\dot{\varepsilon}^p = \frac{\dot{\sigma}}{\Theta(s,\kappa)} = \frac{E(\dot{\varepsilon} - \dot{\varepsilon}^p)}{\Theta(s,\kappa)}. \tag{3.41}$$

The above equation can be solved for $\dot{\varepsilon}^p$ to give

$$\dot{\varepsilon}^p = \frac{E\dot{\varepsilon}}{\Theta(s,\kappa) + E}. \tag{3.42}$$

Note that, (3.42) is solvable as long as $\Theta(s, \kappa) > -E$. This is in contrast to the case of applied loads where (3.33) was solvable only if $\Theta(s, \kappa)$ was positive. Thus the prescribed displacement loading will allow us to solve for cases where the material may *soften*, thus considerably enhancing the scope of applications. Furthermore, as long as the denominator of (3.42) is positive, we see that $\dot{\varepsilon}^p$ and $\dot{\varepsilon}$ have the same sign. Thus, the condition that $\sigma \dot{\varepsilon}^p > 0$ now reduces to the condition that $\sigma \dot{\varepsilon} > 0$.

BAR UNDER PRESCRIBED STRAIN RATE

We now summarize the procedure for finding the equations for a bar with known strain rate $\dot{\varepsilon}(t)$:

(1) Look up the modulus of elasticity E and choose a suitable hardening law $\Theta(s, \kappa)$, for the material in question. Unlike the case of prescribed load, we are able to deal with softening materials ($\Theta(s, \kappa) < 0$) as well as perfectly plastic materials ($\Theta(s, \kappa) = 0$). The only restriction being that $\Theta(s, \kappa) > -E$. In other words, as long as the softening is greater than the negative of the modulus, we can find the plastic strain.

(2) Assume initial conditions for ε, ε^p, s and κ; We have $\varepsilon^p(t=0) = 0$, $s = 0$, $\kappa(t=0) = \kappa_0$.

(3) Solve the following set of differential equations for the strain, plastic strain, the equivalent plastic strain and the resistance κ:

$$\text{Plasticity equations (from (3.33))}: \dot{\varepsilon}^p = \begin{cases} \frac{E\dot{\varepsilon}}{E+\Theta(s,\kappa)}, & \text{if } \sigma^2 = \kappa^2, \sigma\dot{\varepsilon} > 0; \\ 0, & \text{otherwise,} \end{cases} \quad (3.43)$$

$$\text{Equivalent plastic strain (from (3.24))}: \dot{s} = \|\dot{\varepsilon}^p\|, \quad (3.44)$$

$$\text{Hardening law (from (3.25))}: \dot{\kappa} = \Theta(s, \kappa)\dot{s}, \quad (3.45)$$

$$\text{Stress rate (from (3.21))}: \dot{\sigma} = E(\dot{\varepsilon} - \dot{\varepsilon}^p). \quad (3.46)$$

⊙ **3.8 (Exercise:). MATLAB program for bar under prescribed strain rates**

Modify the given program to find the response of a bar under a prescribed cyclic strain of the form

$$\dot{\varepsilon}(t) = 0.001 \sin(0.1t) \quad (3.47)$$

Does the response of the bar become elastic? Does the damage accumulate now?

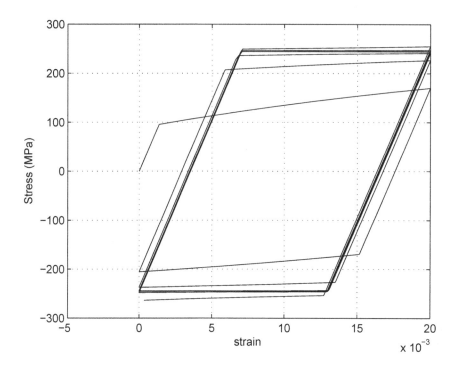

Fig. 3.8 Response of a hardening bar under cyclic straining. Note the hysteresis loop.

Ans: The response function for cyclic loading prescribed above is shown in Fig. 3.8.

3.4 The Bauschinger effect

The previous section assumed that the behavior of a material under monotonic loading is similar to that under cyclic loading. Cyclic loading poses certain demanding challenges on models because of the fact that very small microscopic changes in the internal structure accumulates over a number of cycles and causes eventual cracking. It is a technologically important issue. However, most current methods use a fairly conservative definition of failure and have models based on the notion of a "fatigue limit" and ways of counting how many cycles have occurred. These models have limited predictive capability and are based on the notion of infinite life, i.e. the predictable phenomenon is that if the conditions for the applicability of the model are satisfied (stresses much smaller than the 0.2% offset yield strength and periodic loading) then one can design the structural member such that there is no likelihood of failure.

On the other hand, materials subjected to large loads (compared to the yield

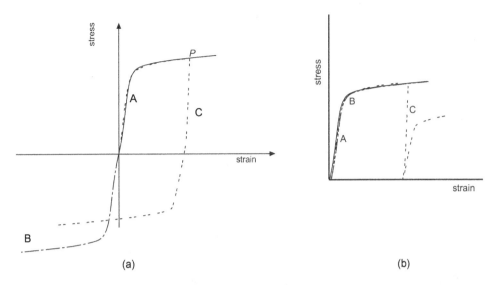

Fig. 3.9 History dependence of elastoplastic metals. (a) The response of a metal to a simple tension test (curve A), a simple compression test (curve B) and a test where the material is extended to point P, then compressed and then extended again. (b)To quantify the history dependence, we plot the absolute value of the stress against the absolute value of the logarithmic strain . Note that the pure tension and compression curves A and B practically coincide, but the curve C is actually below curves A and B. In other words, the metal "remembers" that it was cycled through compression.

strength) and somewhat more random loads such as that experienced during earthquakes etc. cannot be modeled this way. A simple illustration of this phenomenon of "path dependence" is illustrated by the following thought experiments (see [Bate and Wilson (1986)]). Consider three identical samples (A, B, C) of a metal. Sample A is subject to pure monotonic tension, sample B to pure monotonic compression and sample C is subject to an initial monotonic tension followed by compression. The three stress strain curves are shown in Fig. 3.9. If we now plot the magnitude of the stress versus the magnitude of the logarithmic strain, we see that the curves for A and B now coincide. But the curve for C is different from the original curve . We notice that the magnitude of the yield stress in compression is "decreased" by the prior tensile load. This asymmetry between tension and compression, which is path dependent, is called the *Bauschinger effect* ([Bauschinger (1881)]).

Observations have shown that stress strain curve is extremely complex for a general loading but certain general features can be observed. Typically, materials that are subject to cyclic loading, exhibit the following behavior:

(1) **Ratchetting:**

If the sample is subject to cyclic loading with controlled stress, (Fig. 3.10a), there may be a progressive increase in the strain from one cycle to the next continuing until the material actually fails. This phenomenon is called *ratchetting*.

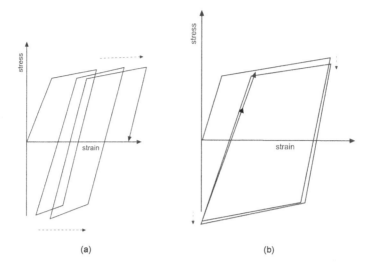

Fig. 3.10 Cyclic loading of a typical metal showing (a) The phenomenon of *ratchetting* where the maximum strain continues to increase with each cycle and (b) *Mean stress relaxation* where the maximum stress decreases from cycle to cycle.

(2) **Mean stress relaxation:**
   If the sample is subject to cyclic strain between two values of the strain, the maximum stress slowly reduces. This is called *mean stress relaxation* (Fig. 3.10b).

(3) **Shakedown:**
   In some materials, when subject to continuous cyclic loading, the stress-strain graph may show an asymptotic limit (somewhat like a "limit cycle" in dynamics). Such a limiting behavior is called *shakedown*. This is a desirable property since further failure will not occur. If the limiting "cycle" is a line and not a loop, then it is called *elastic-shakedown* (see Fig. 3.11b). If the limiting cycle is a closed hysteretic cycle, this is called *plastic shakedown* and the cyclic response is hysteretic (see Fig. 3.11d).

Most of the behavior of materials undergoing cyclic loading are at least approximately describable in terms of the Bauschinger effect. The basic idea behind this phenomenon is explained by means of examples given in a forthcoming section.

3.4.1 *Modeling the Bauschinger effect*

In order to arrive at a suitable, physically realistic simulation of the Bauschinger effect we can develop the following simple lumped model consisting of a spring in parallel with a frictional dashpot, the two being in series with another spring (see Fig. 3.4b) ([Prager (1975)]).

⊙ **3.9 (Exercise:). Evolution laws for the Bauschinger effect**
Consider the model shown in Fig. 3.4b. Assume that the spring constant E is the elastic modulus and K, the Bauschinger effect. Using the concepts introduced in Chapter 2, show that the mechanical part of the Gibbs potential for the material is:

$$G = -\sigma^2/2E - \alpha^2/2K \quad (3.48)$$

where α is the stress in the Bauschinger spring and is referred to as the back stress. Show also from the spring dashpot model in Fig. 3.4b, that

$$\dot{\varepsilon}^p = \dot{\varepsilon} - \dot{\sigma}/E = \dot{\alpha}/K, \quad (3.49)$$

where ε and ε^p represent the displacements of the two springs.

Using the above, we can show that the equation for the mechanical dissipation becomes

$$(\sigma - \alpha)\dot{\varepsilon}^p = \xi \geq 0. \quad (3.50)$$

Thus, with the use of a new spring in the model, the effective force on the dashpot is decreased from σ to $\sigma - \alpha$. The equations (3.49) and (3.50), are the replacements for (3.33) and (3.20) respectively. For this case, the response of the dashpot is given by

$$\dot{\varepsilon}^p = 0 \Rightarrow (\sigma - \alpha)^2 - \kappa^2 < 0, \dot{\varepsilon}^p \neq 0 \Rightarrow (\sigma - \alpha)^2 - \kappa^2 = 0 \quad (3.51)$$

⊙ **3.10 (Exercise:). Equation for plastic strain rate**
By using (3.49)–(3.51) and following the procedure that lead up to (3.36), show that the equation for the plastic strain rate is obtained as

$$\dot{\varepsilon}^p = \begin{cases} \left(\frac{\dot{\sigma}}{K+\Theta(s,\kappa)}\right), & \text{if } (\sigma-\alpha)^2 = \kappa^2, (\sigma-\alpha)\dot{\sigma} > 0; \\ 0, & \text{otherwise}, \end{cases} \quad (3.52)$$

where, as usual, the hardening function is given by $\dot{\kappa} = \Theta(s,\kappa)\dot{s}$.
Hint: First note that the second law (3.50) requires that $sign(\dot{\varepsilon}^p) = sign(\sigma - \alpha)$. Next differentiate $(\sigma - \alpha)^2 = \kappa^2$, and show using the condition that $\overline{(\sigma - \alpha)} = \Theta(s)\dot{\varepsilon}^p$. Finally, solve for $\dot{\varepsilon}^p$ using the second of (3.49). Make sure you derive the condition $(\sigma - \alpha)\dot{\sigma} > 0$ too.

The strain control version of the Bauschinger model mentioned above can be obtained by substituting $\dot{\sigma} = E(\dot{\varepsilon} - \dot{\varepsilon}^p)$ into (3.52) and resolving for $\dot{\varepsilon}^p$ to obtain,

$$\dot{\varepsilon}^p = \begin{cases} \frac{E\dot{\varepsilon}}{(E+K+\Theta(s,\kappa))}, & \text{if } (\sigma-\alpha)^2 = \kappa^2, (\sigma-\alpha)\dot{\varepsilon} > 0; \\ 0, & \text{otherwise}. \end{cases} \quad (3.53)$$

3.5 An example cyclic loading problem

To illustrate the use of the equations thus far derived for the Bauschinger effect, consider an Aluminum rod ($E = 70GPa$, $\sigma_y \approx 100MPa$).[9] Now that we have included the Bauschinger effect, the hardening law required to match the experimental data for aluminum will be different. Let us assume that the hardening law is given by $\Theta(s,\kappa) = 95(175 - \kappa)$ (in MPa), and further that the stiffness of the "Bauschinger spring" is $K = 1000MPa = 1GPa$. Note that, for aluminum, the elastic modulus is about $70GPa$ so that the "Bauschinger spring" has 1/70th the value of the elastic modulus. If you want to fit more realistic data, then you will need data regarding cyclic loading of the material.

3.5.1 *Stress cycling*

Let us assume that the static stress on the bar is $25MPa$ and that due to the dynamic load, the stress varies in a sinusoidal manner from about $190MPa$ to about $-140MPa$. Of course, this is a huge variation, and the bar will fail very quickly. But we want to illustrate the response of the material in a clear way without having to compute thousands of cycles. We thus assume that the stress rate is given by

$$\dot{\sigma} = \begin{cases} 0.05, & t < 500s; \\ 17\cos(0.1(t-500)), & t > 500s. \end{cases} \quad (3.54)$$

By integrating the above equation, we obtain the stress versus time graph shown in Fig. 3.11a. We can write a computer program to solve for the relevant parameters, by solving the differential equations given by (3.52), together with the elastic response equation for the strain $\dot{\varepsilon} = (\dot{\sigma}/E + \dot{\varepsilon}^p)$, the equivalent plastic strain equation $\dot{s} = \|\dot{\varepsilon}^p\|$ and the hardening law $\dot{\kappa} = \Theta(s,\kappa)\dot{s}$. (To get the MATLAB code for this, see authors' websites in the preface.)

When we do this, we can plot the stress versus strain (see Fig. 3.11b). We immediately note that after the first few cycles, the hysteresis loop associated with the plastic deformation decreases rapidly, vanishing eventually! This is because the material hardens rapidly so that resistance to plastic flow increases and, after a few cycles, there will be no subsequent plastic flow. If we plot the "accumulated plastic strain" or equivalent plastic strain (see Fig. 3.12a), we clearly see that after the first few cycles, there is no more plastic strain. Thus, we may be assured that for this load, if the material does not fail in the first few cycles, there is no worry of immediate failure over the next several thousand cycles and low cycle fatigue is not a concern. Of course, one has to calculate the life of the bracket for high cycle fatigue, but this is a different problem that does not involve plasticity calculations

---

[9]The yield strength given here is assumed to be based on a departure from linearity measurement. This is more suitable for modeling cyclic loading, since it will enable us to capture small changes in plastic strain, rather than the 0.2% offset (which is closer to $210MPa$ which is more useful for static loading). The ultimate tensile strength is about $250MPa$.

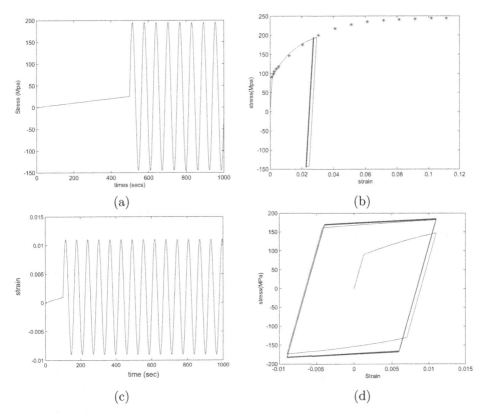

Fig. 3.11 Cyclic response of an aluminum bar, (a), (b) represent response to stress cycling while (c), (d) are the strain cycling response. Note that in (b), experimental monotonic loading response data is also shown for comparison. Note also that the response for strain cycling is significantly different from that of stress cycling. The Bauschinger effect is more important for stress cycling than for strain cycling.

but the use of empirical correlations. So we will not address them here, referring the reader to one of several excellent books on fatigue life estimation (see [Suresh (1998); Gijsbertus (2006)] for example). Since the strains are of the order of 3%, a deflection analysis is probably necessary to verify that the bracket is not deformed too much to affect the function.

3.5.2 *Strain cycling*

The situation is far more serious when the strain in the bar is the control parameter and it cycles. Let us consider this problem next. Again, for the aluminum bar with the properties listed above, we assume that the strain in the bar cycles sinusoidally

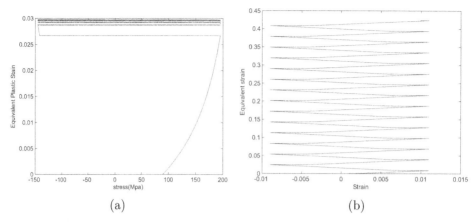

Fig. 3.12 Plot of the accumulated plastic strain versus applied stress or strain, showing a gradual build up of equivalent strain. (a) Stress cycling: Note that initially, the plastic strain accumulates quickly, but subsequently, there is no more accumulation and the response becomes elastic. (b) Strain cycling: Note the continual accumulation of strain cycle by cycle. There is no saturation as in the case of stress cycling.

between 1.2% and −0.8%, i.e. the strain rate is given by

$$\dot{\epsilon} = \begin{cases} 0.00001, & t < 100s; \\ 0.001\cos(0.1(t-100)), & t \geq 100s. \end{cases} \quad (3.55)$$

By integrating the above equation, we obtain Fig. 3.11c showing the strain versus time. Notice that the strain levels are quite small but the strain amplitude is quite large compared to the stress cycling.

In order to estimate the life of the part, we need to solve (3.53), together with the elastic response equation $\dot{\sigma} = E(\dot{\varepsilon} - \dot{\varepsilon}^p)$, the equivalent plastic strain equation $\dot{s} = \|\dot{\varepsilon}^p\|$ and the hardening law $\dot{\kappa} = \Theta(s,\kappa)\dot{s}$. This is just a set of ordinary differential equations that can be programmed using ODE23 in MATLAB and solved for any strain loading (to get MATLAB code for this, see authors' websites listed in the preface). When this is programmed and solved, we see from Fig. 3.11d that now, while the hysteresis loop stabilizes *it does not vanish*, i.e. we continually accumulate plastic strain cycle after cycle. This becomes evident in Fig. 3.12(b) where the "accumulated plastic strain" is plotted as a function of the strain. We see that, after an initial transient period, the "damage per cycle" (taken in equivalent plastic strain units) stabilizes to around 0.0268/cycle.

We can use this for life estimation as follows: If we have an idea of how much damage the material can tolerate (say a damage that is equivalent to an accumulated strain of 60%) then the life is given by $.6/\triangle s = N$ cycles. We cannot predict how much "damage", in equivalent strain units, a material can tolerate—this has to be found by experimentation. The most commonly used rule for low cycle fatigue is the Coffin-Manson rule and is given by the following equation:

$$N = \left\{\frac{\Delta s}{e_f}\right\}^{\frac{1}{c}} \qquad (3.56)$$

where N is the number of cycles to failure, Δs is the "equivalent strain amplitude" which is half the accumulated plastic strain per cycle, e_f and c are empirical constants. e_f is called the "fatigue ductility coefficient" and is the plastic strain required to cause failure in a single cycle. c is the "fatigue ductility exponent". For aluminum, the values quoted are $e_f = 0.105$ and $c = -0.495$.

Using these numbers, we can estimate the cycles to failure as about 15 cycles. This is very low cycle fatigue indeed. You might think, that is too few cycles. True. We have chosen very high loads to demonstrate the fact that there will be a stable hysteresis loop. With more realistic numbers we can get a much clearer idea. However, the calculation is sensitive to the exact values of the yield strength and hardening functions—things that become very difficult to measure if the strains are very small and also due to the presence of local stress-raisers (flaws, defects, scratches). It is for this reason that calculations such as these are restricted to the low cycle fatigue regime where the influence of these errors are much smaller.

3.6 General loading conditions

Until now, we have considered only two types of loads: known stresses $\sigma = \sigma(t)$ or known strains $\varepsilon = \varepsilon(t)$. We appear to have two different strategies for dealing with each of these cases: for known stresses we use the equations (3.36) whereas, for known strains, we use (3.46), with their extensions (3.52) and (3.53) as appropriate. A question arises as to whether there exists a unified way to treat all of these conditions? The answer is yes. We begin by assuming that neither the stress, nor strain in the bar are prescribed, but only a combination of them in the form

$$a(t)\dot{\sigma} + b(t)\dot{\varepsilon} = f(t) \qquad (3.57)$$

⊙ **3.11 (Exercise:). Special loading cases**
What should $a(t)$ and $b(t)$ be for the cases of (a) stress control and (b) strain control? Can you describe the condition when $a(t) = K$ and $b(t) = -1$. What kind of loading is it? Assume that a spring of constant K is connected to the end of the bar and a known force is applied to the other end of the spring. What is the prescribed loading condition now?
Ans: $a(t) = 1, b(t) = 0$ for stress control. When $a(t) = K$ and $b(t) = -1$ then a loading is through a spring whose other end has a known displacement.

We have to solve for $\sigma(t)$ and $\epsilon(t)$ given the following
(i) general loading in the form of (3.57),
(ii) the elastic constitutive equation in the form $\sigma = E(\varepsilon - \varepsilon^p)$,

(iii) kinetic equations for plastic flow together with hardening and Bauschinger effect, whose form is

$$\dot{\varepsilon}^p = \begin{cases} \frac{E\dot{\varepsilon}}{(E+K+\Theta(s,\kappa))}, & \text{if } (\sigma-\alpha)^2 = \kappa^2,\ (\sigma-\alpha)\dot{\varepsilon} > 0; \\ 0, & \text{otherwise.} \end{cases} \quad (3.58)$$

where

$$\dot{s} = \left\|\dot{\varepsilon}^p\right\|,\ \dot{\kappa} = \Theta(s,\kappa)\dot{s},\ \dot{\alpha} = K\dot{\varepsilon}^p. \quad (3.59)$$

Note that we have rewritten the kinetic equation for plastic flow pretending that we have strain control. We will see that this is a critical idea. The problem with solving this set of equations is the conditional nature of the plastic flow equations above. We need to find a way to figure out which of the conditions are valid. *$\dot{\varepsilon}^p = 0$ as long as we have not satisfied the yield criteria or if we have satisfied it but are stepping back into the elastic regime.* To find out whether or not $\dot{\varepsilon}^p = 0$, we will take the "dumbest possible guess" and just pretend that it is indeed zero and see if we can work the problem. To implement this strategy, we first differentiate the elasticity law and get

$$\dot{\sigma} = E(\dot{\varepsilon} - \dot{\varepsilon}^p) \quad (3.60)$$

Note that *if $\dot{\varepsilon}^p = 0$ then we can solve (3.60) and (3.57) very easily* and get

$$(a(t)E + b(t))\dot{\varepsilon} = f(t) \Rightarrow \dot{\varepsilon} = \frac{f(t)}{b(t) + Ea(t)} \quad (3.61)$$

and we are home free! Let us be greedy and *always hope that $\dot{\varepsilon}^p = 0$* and calculate $\dot{\varepsilon}$ by using (3.61). Now, how do we know if we are correct? If you look at the kinetic equation (3.58), then it is clear that our guess is correct either if $(\sigma - \alpha)^2 < \kappa^2$, i.e. the material has not yet yielded[10], or $(\sigma - \alpha)\dot{\varepsilon} = (\sigma - \alpha)f(t)/(b(t) + Ea(t)) < 0$ i.e. we will look back into the elastic regime. Thus, the strategy is based on the following observation.

> A foolproof way to figure out if $\dot{\varepsilon}^p$ is nonzero or not, first check if we are on the yield surface, i.e. find $(\sigma - \alpha)^2 - \kappa^2$. If it is less than zero, then $\dot{\varepsilon}^p = 0$. If it is equal to zero, assume that $\dot{\varepsilon}^p = 0$, find the strain rate $\dot{\varepsilon} = f(t)/(b(t) + Ea(t))$ and see if for this strain rate, we step out of the yield surface, i.e. if $(\sigma - \alpha)\dot{\varepsilon} = (\sigma - \alpha)f(t)/(b(t) + Ea(t)) > 0$! If we do, then we have to go back and find the strain rate by assuming $\dot{\varepsilon}^p \neq 0$.

Thus, we have the following strategy:

General strategy to solve for the response of an elastoplastic material under general loading conditions:

In order to solve for the elastoplastic response subject to general loading conditions (3.57), we can set it up as a set of ordinary differential equations of the form

---

[10] In which case it is not a guess at all, if the material is not at yield, it would be foolish to guess that $\dot{\varepsilon}^p$ is not equal to zero.

$M(y,t)\dot{y} = f(y,t)$ where $y = \{\sigma, \varepsilon, \varepsilon^p, s, \kappa\}$. The form has been deliberately chosen to resemble the standard form that is solvable in MATLAB.

We need to set up the matrices $M(y,t)$ and the vector $\mathbf{f}(y,t)$. The matrix equation is of the form

$$M(y,t)\dot{y} = \begin{pmatrix} a(t) & b(t) & 0 & 0 & 0 \\ 1 & -E & E & 0 & 0 \\ 0 & M_{32} & 1 & 0 & 0 \\ 0 & 0 & 0 & 1 & 0 \\ 0 & 0 & 0 & -\Theta(s,\kappa) & 1 \end{pmatrix} \begin{pmatrix} \dot{\sigma} \\ \dot{\varepsilon} \\ \dot{\varepsilon}^p \\ \dot{s} \\ \dot{\kappa} \end{pmatrix} = \begin{pmatrix} f(t) \\ 0 \\ 0 \\ h_4(t) \\ 0 \end{pmatrix} \quad (3.62)$$

The equations are written as follows: the first row represents the external loading conditions. The second row are the elasticity relations in the rate form, i.e. (3.60), the third equation is the critical one. We need to discuss it in detail. We note that the equation is of the form $M_{32}\dot{\varepsilon} + \dot{\varepsilon}^p = 0$. Note that, if we choose $M_{32} = 0$ we get $\dot{\varepsilon}^p = 0$. In view of (3.58) we set

$$M_{32} = \begin{cases} -\frac{E}{(E+K+\Theta(s,\kappa))}, & \text{if } (\sigma - K\varepsilon^p)^2 = \kappa^2 \text{ and } (\sigma - K\varepsilon^p)\frac{f(t)}{a(t)E+b(t)} > 0; \\ 0, & \text{otherwise.} \end{cases}$$
(3.63)

Note that the right-hand side of (3.63) depends only on known quantities at any time step. How did this happen? The key is that, as we discussed earlier, the condition $(\sigma - \alpha)\dot{\varepsilon} > 0$ has been replaced by $(\sigma - \alpha)(f(t)/(a(t)E + b(t))) > 0$.

⊙ **3.12 (Exercise:). Explicit solution for the first three equations**
Show that the first three equations in (3.62) can be solved for $\dot{\sigma}, \dot{\varepsilon}$ and $\dot{\varepsilon}^p$ to give

$$\dot{\sigma} = \frac{E(1 - M_{32})f(t)}{b(t) + a(t)E(1 - M_{32})}.$$

$$\dot{\varepsilon} = \frac{f(t)}{b(t) + a(t)E(1 - M_{32})}. \quad (3.64)$$

$$\dot{\varepsilon}^p = \frac{M_{32}f(t)}{b(t) + a(t)E(1 - M_{32})}.$$

Hint: Write out the three equations for the three unknowns; notice that it is just a simple set of linear equations that can be solved by hand.

Now that we have the three equations, it is easy to see that since $\dot{s} = \|\dot{\varepsilon}^p\|$, and upon using $(3.64)_3$ we obtain,

$$h_4(t) = \|\dot{\varepsilon}^p\| = \left\| \frac{M_{32}f(t)}{b(t) + a(t)E(1 - M_{32})} \right\| \quad (3.65)$$

We can, in principle use the ODE Solver ODE23, to solve the set of equations

(3.62) with M_{32} given by (3.63) and $h_4(t)$ given by (3.65).

⊙ **3.13 (Exercise:). Functions to be specified for solution**
What input functions and constants should be specified for the set of ODEs to be solvable?
Hint: We need 3 functions of time representing the loading, one function of the state variables representing the hardening function and two other constants, apart from five initial conditions for the ODE.

While it is possible to solve it with ODE23, if you try it out, you may find that the final graph is not very smooth (this depends upon the version of MATLAB that you are using). The solution may have many fine "bumps". The reason for this is due to just one term: M_{32}. If you observe equation (3.63) closely, you will see that it has a conditional statement, it is *not a continuous function of time*. This causes a standard general purpose ODE solver to have problems. We will develop special purpose numerical solution techniques for these kinds of conditional differential equations in a subsequent chapter. For now, we will live with this problem.

The general loading condition that we have introduced here is a prototype for all our future solutions to boundary value problems. A heuristic argument for this is provided here. Consider a "real life" situation like a plate with a hole in it. We know that the force on a part of the boundary and the rest of the boundary be fixed, say. Now we pick some generic material point inside the plate: we don't know either the stress or the strain at this location. So how do we find out whether it is going to yield? We will see later that what we know is some linear combination of stresses and strains of ALL the points (through the equilibrium and compatibility equations). From this, by following a procedure that is similar to the one outlined above, we can find the response of the material.

3.7 Summary

(1) The modeling of elastoplastic properties of a statically determinate truss system is composed of two steps:

   (a) finding the forces in the truss (using method of joints, for example) and,
   (b) solving for the strains and the plastic strains using an elastoplastic model.

(2) The elastoplastic model is based on a spring-frictional dashpot LPM, and, for us to solve it under prescribed force conditions, requires that the resistance of the friction element to increase with deformation (strain hardening).

(3) In order to model the strain hardening phenomena, we introduce the notion of "accumulated" or equivalent plastic strain defined by $\dot{s} = \|\dot{\varepsilon}^p\|$. We then assume that the resistance to deformation κ evolves as $\dot{\kappa} = \Theta(s,\kappa)\dot{s}$, the form for $\Theta(s,\kappa)$ is called a hardening law.

(4) Once the hardening law is known, the process of solving for the strain and plastic strain is given by (3.36).

(5) One can derive a similar expression for strain control (see equations (3.46)).
(6) Unlike the stress control case, the strain control case can be solved even for negative values of $\Theta(s, \kappa)$. This feature is critical for modeling high strength alloys like maraging steel.
(7) In most cases, for cyclic loading, the yield stress in the reverse direction is lower than that in the forward direction. This is called the "Bauschinger Effect".
(8) A working hypothesis for this effect is that there are internal stresses that are built in during the forward motion that are released during the reverse motion thus reducing the stress needed for the backward motion.
(9) The Bauschinger effect can be modeled (to a first approximation) by adding a spring in parallel to the frictional element. The resulting equations to be solved are listed in (3.52) for stress control and (3.53) for strain control.
(10) The graphs that result for different sinusoidal loads or displacements show characteristic features such as cyclic stabilization and ratchetting.
(11) For strain control loading, one can use this along with the Coffin-Manson Rule for life prediction.
(12) For a general loading, we can write it as a set of ODEs, after some manipulation. These ODEs can, in principle be solved by a package such as ODE45.
(13) The key point about any general loading is that **we find out if further plastic straining occurs by assuming that $\dot{\varepsilon}^p = 0$ and examining if we step OUT of the yield surface. If we do, then we have to go back and assume that further plastic straining does take place.**

Life estimation under cyclic loading of frames is a very important topic and there is a vast literature on the subject. The articles by [Mughrabi (1977); Mughrabi and Christ (1997)] surveys the experimental evidence for a number of cyclic loading experiments on various steel alloys. Many kinematic hardening models other than the rather simple ones used here can be adopted; For further reading on this matter, see e.g., [Drucker and Tachau (1945); Hardesty et al. (1946); Chaboche (1989b, 1991); Mughrabi (1993); de Andres et al. (1999); Bache et al. (2001); Ding et al. (2002); Voyiadjis and Sivakumar (1991, 1994)] which represent a small sample of the various approaches to modeling this phenomenon spanning 50 years. It is clear that this phenomenon of fatigue and failure prediction will keep many engineers and researchers busy for many years to come[11].

---

[11] Offering some "job security" in trying economic times:-))

Chapter 4

Elastoplastic Beams: An Introduction to a Boundary Value Problem

Learning Objectives

By learning this material, you should be able to do the following:

(1) Write down the equilibrium equations for a Bernoulli-Euler beam.

(2) Write down the assumptions regarding the deformation of Bernoulli-Euler beam and use them to obtain the form of the displacement and the strain.

(3) Derive the equations connecting the total strain and the plastic strain for statically determinate beams under thermal loads.

(4) Implement the iterative solution process (in a spreadsheet application) to find the stress distribution.

(5) Implement the solution scheme for other statically determinate beams.

(6) Derive the governing differential equations for a statically indeterminate beam.

(7) Describe the iterative solution method that was employed for statically indeterminate beams.

> **CHAPTER SYNOPSIS**
>
> This chapter introduces the first spatially varying deformation involving inelasticity and illustrates the interplay between the equations of equilibrium and the constitutive equations. We also show how thermal expansion effects can be dealt with. One of the key aspects of beam theory is that, apart from the practical significance, it illustrates a wide range of procedures. For example, the fact that the terms that show up in the governing equations of equilibrium are obtained by integration is connected with the "weak formulation" of the governing equations—an approach used in finite element methods. Moreover, in polycrystalline plasticity and other areas of micromechanics, it is common to develop constitutive equations at a micro level and then average the quantities to obtain the macroscopic behavior. In the beam problem, the macroscopic variables are the forces, moments and the centerline deflections while the microscopic quantities are the displacements and stresses across the cross-section.

Chapter roadmap
The approach to beams in this chapter follows that for trusses very closely. As before, we begin the chapter by introducing a simplified engineering task and a description of the importance of simplifying the geometry and the applied loads in order to convert an engineering task into a mathematically precise albeit much simplified problem. We then introduce the steps in the analysis process beginning with making a list of variables and the simplifying assumptions. The modeling process begins with the kinematics of the beam followed by balance laws and then the constitutive laws. Here, unlike the truss problem, we go through the balance laws very quickly but the constitutive equations are more involved since they involve integrating the kinematical variables across the thickness of the beam. For the case of elastoplastic beams, this cannot be done exactly and a numerical scheme is required. We then follow it up with the setting up and solving a statically determinate problem and then, a statically indeterminate problem. If you are learning this material by yourself, then you should implement a MATLAB program for the statically indeterminate problem following the algorithm given in the book. If you are an instructor, you can help the students set this program up. The statically indeterminate beam has all the elements of a fully 3-D inelasticity problem and is well worth doing as the final project in a structural inelasticity class. The approach to beams used here is based on a chapter on thermal loads in beams in [Mendelson (1983)], an excellent book on clever numerical techniques [1] that has several excellent practical solutions (especially for torsion and plane strain problems). We have revisited these problems with modern tools such as MATLAB. We urge the reader to consult that book to see how to non-dimensionalize elastoplasticity problems and to set up numerical schemes.

---

[1] written in the days before the dominance (tyranny? ;-)) of canned finite element programs

4.1 An example task

Assume that you are a design engineer for a company that is involved in the design and deployment of a frame[2] that is subject to thermal as well as mechanical loads. Your client would like to increase the load on the beam without redesigning it and is requesting your recommendation as to whether this is feasible and safe.

There are other similar situations where you may need to find out how much load a structure (let us say a building or a machine frame) will withstand and if it fails, which of the parts of the structure will begin to collapse and how. Perhaps the structure is subject to mechanical and thermal loads (heating) at the same time. Such situations arise in "fail safe" designs where you might need not only to assess whether it will fail but also how it fails. In many cases this is a dynamic event (imagine the frame of a car collapsing) and a careful dynamic analysis will be needed, but for our present case, we shall imagine slow failure due to a combination of mechanical and thermal effects to illustrate the procedure.

Your specific task in this case is to model the beam subject to a 4-point bending load as shown in Fig. 4.1 and to find out the maximum load carrying capacity of this beam. You also need to find out the region where the plastic strain occurs and how it spreads as the bending moment increases. Based on the solution, you need to make recommendations to your client as to the condition of the beam, the safety factor and what (if any) changes need to be made.

4.2 Modeling of a thermoelastoplastic beam

Thus, the tasks that you have to do can be listed as follows:

(1) develop a suitable model
(2) find the material parameters
(3) convert the task into a mathematically tractable problem with physically realistic properties
(4) solve the resulting equations
(5) extract the relevant variables and *make a recommendation/decision based on your numerical results.*

As an engineer, a major part of your task will involve the formulation of the problem by making the appropriate simplifying assumptions and for making design or retrofitting recommendations based on the calculations and your experience.

In order to do this, you have to make a number of decisions and simplifications regarding (1) the geometry (2) the external loading and displacement boundary conditions and (3) a reasonable constitutive model to use. The complexity of the resulting boundary value problem (BVP) and the ease of obtaining a solution de-

---

[2]What is the difference between a truss and a frame? A frame is made up of bars that can sustain bending and is connected in such a way that moments can be transmitted through the bars.

pends upon the choices made here. Part of the choice is guided by the data and time available and the criticality of the part in question. But first, it is wise to come up with a simple model to get an estimate on what is likely to happen to the structure. This involves making a number of simplifying assumptions that will ease the solution process. It will also help decide if a further analysis is necessary or warranted. For example, if this is a critical part that is crucial to the structure in question, it might be worth your while to develop a very detailed model and to spend some resources in obtaining a detailed constitutive equation.

Let us discuss the various steps involved in the formulation of the problem:

Idealization of the geometry: This is perhaps one of the most productive areas for simplification since, as you will see later, it will reduce the problem to the solution of a set of ordinary differential equations (which can be accomplished with readily available software) rather than partial differential equations (which will usually require significant additional effort on your part). We have a wide range of geometrical models to choose from (1) for structural entities where one of the dimensions is much larger than the other, we can either choose truss models (that allow only axial load as seen in the last chapter) or beam models (that allow bending loads also), (2) If two dimensions are much larger than the third, then it is possible to choose a membrane, a plate, or a shell model (including curved membrane-shells etc.) and (3) only if the geometry is complex or if you are really interested in the

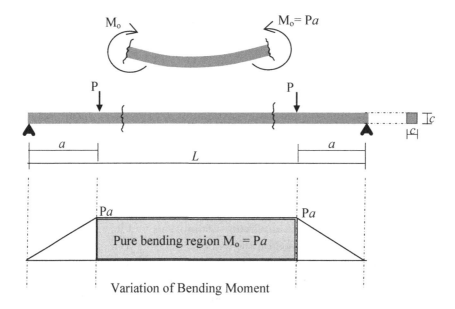

Fig. 4.1 A beam with a set of concentrated loads of same magnitude acting two points as shown. Notice that the region between the two loads is a pure bending moment region as shown in bending moment diagram.

phenomena that occur at very small length scales will you need a full three dimensional analysis. Even in such cases, you could simplify the geometry by considering certain dimensions to be infinite (layers, half spaces etc.) and/or removing features that are not essential. This process is called defeaturing and it must be done in a judicious manner.

The governing equations for such bodies have been very well developed for elastic structures and have been widely used (see [Zenkour (1999)] for a survey of beam models). However, geometrical simplifications for elastoplastic bodies are challenging due to the need to track the spread of the plastic strain across the thickness.

Boundary and external loading conditions: In many cases, geometry alone is not the deciding factor. If one wants to study the effects of thermal loading on the body, then it may be simple enough to assume that the body is a rigid conductor (for the purposes of calculating the temperature field) and since the resulting equation is a Laplace or Poisson type equation for the steady state temperature distribution, it may be relatively easy to find the temperature field (especially since these equations have been well studied and the techniques to solve them are well developed). Major simplifications of the geometry are not necessary.

Let us say that based on our experience and the geometry of the member, we have decided to model the body as a **beam** for the purposes of mechanical loading. We next need to decide what kind of loading there is on the beam. Here, one needs to exercise some judgement since, in many cases it is not completely clear as to how to decide whether some joint is a "pin type assembly" or a "built in or welded assembly". Typically, we would then decide to make the simplest choice, i.e. choose the joint to be one that will give us a statically determinate structure unless it is very clear that it is not a suitable choice.

Constitutive Model: We choose the simplest possible constitutive equation that will capture the phenomena in question. For slow monotonic loading, it might suffice to consider perfect plasticity or a simple hardening model and to ignore temperature effects on the yield strength. For rapid cyclic loading, a much more detailed model is usually necessary since even minor differences between the model and the actual response might accumulate over many cycles so as to ruin your predictions.

Finally, temperature effects on the strain (through thermal expansion) is usually significant even when the temperature differences may have negligible effect on the yield strength etc. In most cases, (unless rapid and large deformations are involved) the effect of plastic dissipation on the temperature may be safely ignored for a quasistatic loading condition, especially in a preliminary calculation. It is important to check whether these assumptions are indeed valid after the calculation has been done. This seemingly obvious point can sometimes be lost in the focus on obtaining solutions.

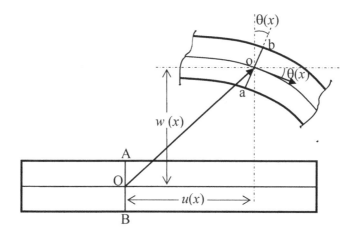

Fig. 4.2 An exaggerated view of the deflection of a beam showing the deformation of a typical point on the centerline of the beam and the deformation of all the points on the cross-section in such a way that (1) the cross-section is undistorted and (2) the cross-section remains normal to the centerline. Thus the line AOB becomes the line aob making an angle $\theta = -w(x),_x$ with the centerline.

4.3 The elastoplastic beam

Having discussed these issues, we now assume that the idealization step is complete and the engineering task has been reduced to the following problem:

We have simplified the geometry to that of a thin beam of length L, and square cross section of side $2c$ with $L >> c$, as shown in Fig. 4.1.

4.3.1 Task 1: Make a preliminary list of variables of interest

This is the first task to be carried out. This will enable us to see what parameters need to be found out in this problem. The input or control variables are
(1) the force or moment applied to the bounding surfaces of the beam
(2) the temperature of the surroundings, and,
(3) the fixity conditions and the temperature at the ends.

At any location along the beam we need to find the following set of parameters (output variables)
(1) the bending moment and hence the stresses,
(2) the deflection of the beam,
(3) the extent of plastic deformation,
(4) the stress distribution, and,
(5) the temperature distribution.

In order to obtain these quantities, it may become necessary to introduce additional variables. We will revisit this when we discuss specific models.

4.3.2 Task 2: Make simplifying assumptions on the response

Let us describe some simplifying assumptions first;

Simplification of the temperature field

We will assume that the lateral sides are exposed and that the temperature varies only along the length and not across the thickness of the beam so that we could say that we will only be interested in axial heat conduction.

Simplification of deformation field

We will assume that the beam undergoes small deformations and that they satisfy the Bernoulli-Euler beam assumptions, namely that
(1) plane sections remain plane and normal to the neutral axis which is the line connecting the centroids of the cross-sectional plane;
(2) the planes themselves suffer no distortion so that only axial distortions are present[3].

Considerations of inertial effects

We shall assume that we are not interested in dynamic problems (such as impact) and so we can consider the process to be "quasistatic" and therefore, ignore inertial and rate effects.

Temperature dependence of properties

We will also neglect temperature effects on the material properties especially, moduli (this would depend upon whether it is a metal or polymer and how close it is to a transition point) but we want to include the coefficient of thermal expansion, especially, for beams that are confined at either end.

Effect of internal heating

We assume small quasistatic deformations and that, for the purpose of the thermal problem we will ignore (1) the heating due to plastic dissipation and (2) the deformation of the body. If we do not do this, we will have a "coupled problem", i.e. temperature fields and deformation fields will need to be solved simultaneously.

Items (4) and (5) above are key to "decoupling" the mechanical and thermal effects. This, together with the assumption of small deformations allows us to treat the heat conduction as an independent problem. Assumptions like the above have to be made for every problem that we deal with. By making them explicit, we can then test them when the solutions are obtained to verify their range of applicability

---

[3][Zenkour (1999)] discusses a number of other possible assumptions for different beams in the context of elasticity.

4.3.3 Task 3: Modeling the response of the beam

We now arrive at the modeling stage. Here for each kind of geometry we need to develop suitable balance laws for mass, linear momentum, angular momentum and energy. These are the "hard constraints" imposed by classical physics and restricts our choice of what we can do. The balance laws alone are insufficient to find all the parameters of interest in the material. We need additional equations (reflecting the behavior of the specific material at hand) to "close" the set of mathematical equations. These equations are the constitutive equations.

Kinematics of the beam

Based on the Euler-Bernoulli assumptions listed earlier, and with reference to Fig. 4.2, we assume that the displacement of any particle at a location x along the centerline of the bar is $\mathbf{u}_0(x) = u(x)\mathbf{i}_1 + w(x)\mathbf{i}_2$ as marked in Fig. 4.2. Notice that unlike the case of the truss, there are two components to the displacements. Now what about particles that do NOT lie on the centerline? Consider any line such as AB in Fig. 4.2 that is originally straight and parallel to the y-axis. We will assume that after the beam bends, the line becomes ab which is still straight, but not parallel to the y-axis, instead it is assumed to be *normal to the distorted centerline*. The angle which it makes with the y-axis is θ.

DISPLACEMENTS AND STRAINS IN A SIMPLE BEAM

Now based on these geometrical considerations, and by elementary calculations which, *for small deformations*, implies that $\tan(\theta)$ is the slope of the centerline, we can show that any particle whose position in the undistorted configuration (x, y) is given by

$$\mathbf{u}(x,y) = (u(x) + y\theta(x))\mathbf{i}_1 + w(x)\mathbf{i}_2, \quad \theta = -w(x)_{,x}. \tag{4.1}$$

A routine calculation of the strain $\varepsilon_{xx} = u_{x,x}$ then gives

$$e := \varepsilon_{xx} = u_{,x} + \theta_{,x}\, y := e_0(x) + \varrho(x)y \tag{4.2}$$

where e represents the xx component of the strain ε, $e_0(x)$ represents the axial strain of the centerline and $\varrho(x) = -w(x)_{,x}$ is the curvature of the centerline.

The above assumption indicates that the axial strain varies linearly with the cross-sectional height. All other components of the strain are neglected in this approximation.

It might seem somewhat strange at first sight to introduce kinematic assumptions about the beam as part of the model. The reason for doing so is that assumptions such as equation (4.1) may be viewed as constraints on the deformations possible in the body and may be considered as generalizations of constraints as incompressibility or inextensibility that represent material behavior. For example, if we were modeling a sandwich beam with a soft core, it would be unreasonable to persist in using equation (4.1) since there would be substantial shear deformations in the core.

The balance laws

As with the case of the truss, we take a "chunk" or "slice" of the beam and draw its free body diagram (FBD) as shown[4] in Fig. 4.3. With reference to Fig. 4.3, M, is the internal moment about the z-axis, P, is the internal axial force, V, the internal shear force, $n(x)$ the transverse external force applied, $t(x)$, the transverse shear force applied to the beam, h, the heat flux and, r, the rate of heating or cooling from the lateral sides.

From the FBD, we conclude the following:

(1) <u>Mass balance:</u>
Since we are assuming a "closed system", i.e. closed to the mass flow and since the deformations are assumed to be small, we shall apply all the laws to the undeformed state since the mass balance simply means that the density is independent of time. We shall assume that density is the same throughout the body.

(2) <u>Linear and angular momentum balance:</u>
Ignoring inertia, and referring to Fig. 4.3b, these equations imply that the sum of the forces and the sum of the moments around any point P must be zero, i.e.

$$+\rightarrow : \sum \mathbf{F}_x = 0 \Rightarrow (P + dP) - P + t(x)dx = 0 \Rightarrow \frac{dP}{dx} + t(x) = 0, \quad (4.3)$$

$$+\uparrow : \sum \mathbf{F}_y = 0 \Rightarrow (V + dV) - V + n(x)dx = 0 \Rightarrow \frac{dV}{dx} + n(x) = 0, \quad (4.4)$$

$$+\curvearrowright_{left} : \sum M = 0 \Rightarrow (M + dM) - M + Vdx = 0 \Rightarrow \frac{dM}{dx} + V(x) = 0. \quad (4.5)$$

(3) <u>Energy balance:</u> Assuming steady state and referring to Fig. 4.3(b), we note that the net heat outflow must be equal to the rate of heat generated in the body or that lost from the lateral surfaces. This means that

$$(h + dh) - h + rdx = 0 \Rightarrow \frac{dh}{dx} + r = 0 \quad (4.6)$$

---

[4] We apologize for the rather crowded figure because we wanted to show both the mechanical and thermal stimuli to the beam in the same figure. All the wavy lines represent thermal stimuli.

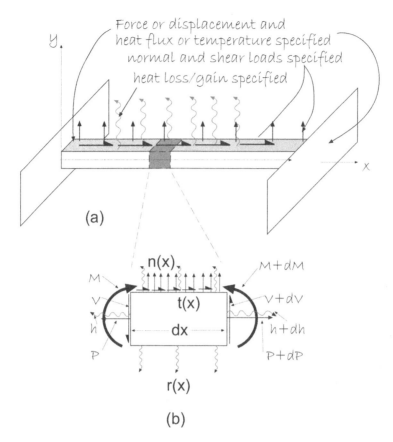

Fig. 4.3 (a) A perspective view of an elastoplastic beam, connected to walls (i.e. the rest of the structure). (b) A free body diagram (FBD) of a small section of the beam showing all the internal and external forces with energy flow showing all the various heat transfer processes.

In obtaining the above equation, note that we have ignored the stress power or internal mechanical working, which is quite appropriate for steady state problems. Moreover, the energy equation becomes drastically simplified since all time derivative terms including that of the internal energy vanish.

(4) <u>Boundary conditions</u>: We assume that at the boundaries, the beam is subjected to a pure moment M_0, that the axial and shear forces at the ends vanish and that the hot end is at a temperature T_h, i.e. $T|_{x=0} = T_h$ and the cold end the temperature is T_c, i.e. $T|_{x=L} = T_c$. We also assume that the lateral surfaces are traction free and the ambient temperature surrounding the beam is T_0. In a more general situation, we might have the requirement that the ends are built in or some other such boundary condition. We will actually deal with these cases later.

Notice that in contrast to our elaborate treatment of truss balance laws, we are being very brief with beams. This is by design since we want you to fill in the details by yourself. Remember: "you can't build muscles by watching an exercise video."

Constitutive equations

Form of the Complementary energy function (or Gibbs Potential)

Now, as we did in the chapter on one-dimensional inelasticity, we assume that the *Gibbs potential* has up to quadratic terms in the stress in the beam, i.e. we assume that

$$G(\sigma, T) = -\sigma^2/E - \alpha(T - T_0)\sigma + f(T) \tag{4.7}$$

where σ is the axial stress, E is Young's modulus (assumed to be constant), α is the coefficient of linear thermal expansion, T is the temperature of the particle and T_0 is the temperature of the surroundings and $f(T)$, the thermal free energy.

A couple of points to note: in the truss chapter, the stress in the beam was assumed to be constant and hence the axial force was simply $P = \sigma A_0$ where A_0 was the cross-sectional area of the bar. Not so here. When we have bending, we are forced to assume that *the stress varies from point to point in the beam; it is a function of both x and y*. Please resist the temptation to say that "Of course, I know it! $\sigma = My/I$!". Please note that for elasto-plastic beams, the stress is NOT a linear function of y and is NOT given by My/I! The formula is true only for elastic beams under small deformation.

⊙ **4.1 (Exercise:). Relationship between the axial stress, the axial force and the bending moment**
Show that

$$P(x) = 2c \int_{-c}^{c} \sigma(x,y)\, dy, \quad M(x) = -2c \int_{-c}^{c} y\sigma(x.y)\, dy. \tag{4.8}$$

The negative sign for the moment is due to the fact that we have assumed that the moment will make the beam curve UP. What happens if σ is an even function of y?

Moving along and following the same pattern as for trusses, we get the equation for the mechanical dissipation as (*See Chapter 3 for details*)

$$\sigma(\dot{e} - \dot{\sigma}/E - \alpha\dot{T}) = \xi. \tag{4.9}$$

If we now define the terms in the bracket on the left-hand side of (4.9) as $\dot{e}_p$ (this is nothing more than a definition at this point. Its physical significance will become evident shortly), then we have

$$\dot{e}_p := \dot{e} - \dot{\sigma}/E - \alpha\dot{T} \tag{4.10}$$

which can be immediately integrated to reveal that the total axial strain is the sum of the elastic strain, the plastic strain and the thermal strain, i.e.

$$e = e_e + e_T + e_p = \frac{\sigma}{E} + \alpha(T - T_0) + e_p \tag{4.11}$$

Then e_p is the *plastic strain or residual strain* in the x direction, i.e. the strain remaining in the material when the stress is zero and the temperature is the same as the ambient.

Note that, in this thermodynamic approach, the decomposition of the strain etc. follows directly from the assumption of the Gibbs Potential. Thus, the first term on the right-hand side of (4.11) is a statement of Hooke's law, which in this case is derived from the Gibbs potential. Moreover, the second law of thermodynamics will be satisfied if we choose the kinetic equation for $\dot{e}_p$ such that

$$\sigma \dot{e}_p \geq 0 \tag{4.12}$$

It is easy to see that if we choose $\dot{e}_p$ to be any function of the form $\mu\sigma$ where μ is non-negative, then the above requirement is automatically met. This is precisely what the one dimensional friction element does (see the chapter on one dimensional inelasticity).

It is easy to solve for σ and rewrite (4.11) as

$$\sigma = E(e_e) = E(e - e_T - e_p) = E(e - \alpha(T - T_a) - e_p). \tag{4.13}$$

For convenience, we shall define the *mechanical strain* as $e_m = e - e_T = e_e + e_p$ and we shall see that changes in the mechanical strain are what are really important for the problem.

We can use these results to connect the stresses, axial forces and moments to the deflection of the centerline.

⊙ **4.2 (Exercise:). Relationship between axial forces, moments, and the axial deflections**
Substitute (4.13) and (4.2) into (4.8), and show that

$$\sigma = E(u_{,x} + \theta_{,x} y) - E(e_p + \alpha(T - T_a))$$

$$P = 4c^2 E u_{,x} - 2cE \int_c^c e_p\, dy - 2cE\alpha \int_c^c (T - T_a)\, dy = P_e - P_p - P_T$$

$$M = EI\theta_{,x} - 2cE \int_c^c y e_p\, dy - 2cE\alpha \int_c^c (T - T_a)y\, dy = M_e - M_p - M_T, \tag{4.14}$$

where $I = 4c^4/3$ is the area moment of inertia and $\theta_{,x} = w_{,xx}$ is the curvature. Hint: Functions that are odd in y integrate out to zero.

The above equations show that the axial force and bending moment can be considered as the sum of the usual elastic formula for the axial force and bending moment together with corrections for plasticity and thermal expansion.

Note that while the y dependence of the total strain is known (from equation (4.2)), we do not yet know the y dependence of e_T (since we do not know the temperature field) or e_p (since we haven't yet solved the evolution equations) and we need equations to determine them. A key point to observe here is that while the total strain is linear in the y coordinate (as seen from (4.2)), the same need not be true for the thermal or plastic strains. The point is that *for a given distribution of thermal and plastic strains, the stress distribution will adjust itself in such a way that the total strain varies linearly across the cross section.* Since the temperature field can be determined (in view of the assumptions) a-priori, there is no leeway in the thermal strains. So, the trick is to find a distribution of elastic and plastic strains in such a way that the equations of equilibrium, elastic constitutive equations and the flow rule are satisfied.

Equation for the plastic strain rate and its integration: The final step in the material modeling process is the prescription of the constitutive law for $\dot{e}_p$ in such a way that the second law is met and which reflects the response of the beam. In the current context we will assume that the material model is akin to that of a spring and friction element model. By now, you should be familiar with the form for the frictional response, so we will simply state it as follows

$$\dot{e}_p = \phi\sigma;\ \phi > 0,\ f(\sigma) = \sigma^2 - \kappa^2 \leq 0, \phi\, f = 0. \tag{4.15}$$

⊙ **4.3 (Exercise:). Reduction of the Kuhn Tucker condition and the loading criteria**
Assume that κ is constant, and that the temperature is not a function of time and show that the flow rule (4.15) reduces to

$$\dot{\varepsilon}^p = \dot{e}^p_{xx} = \begin{cases} \dot{e} & \text{if } |\sigma| = \kappa \text{ and } \sigma\dot{e} > 0, \\ 0 & \text{otherwise} \end{cases} \tag{4.16}$$

Hint: Differentiate the yield function and then use the constitutive equation (4.13) to solve for ϕ. What happened to the e_T term? What would happen if the temperature distribution depends upon time?
Ans: Since the temperature is steady, the thermal strains do not change with time so that changes in e are the same as changes in e_m, the mechanical strain.

For the case of *monotonic loading*, i.e. where at each point (x, y) in the beam the magnitude of the strain e keeps on increasing, we can solve explicitly for the plastic strain by integrating (4.16) and getting $e_p = e + const$. The constant of integration needs careful treatment however.

⊙ **4.4 (Exercise:). Initial conditions for the calculation of the plastic strain**
Since the equation (4.16) is an ODE, what are the initial conditions for it? Hint: Use the elastic response and the yield condition to find the condition for first yield.
<u>Ans</u>: $e_p = 0$ when $E(e - e_T) \leq \pm\kappa$. Why are there two possibilities?

It is possible to integrate the expression for the plastic strain (assuming monotonic loading) and get two solutions:

$$\text{monotonic extension: } e - e_T > \frac{\kappa}{E}, \dot{e} > 0 \Rightarrow \varepsilon^p = e - e_T - \frac{\kappa}{E}$$

$$\text{monotonic compression: } e - e_T < -\frac{\kappa}{E}, \dot{e} - E_T < 0 \Rightarrow \varepsilon^p = e - e_T + \frac{\kappa}{E}. \quad (4.17)$$

Thus, for the monotonic loading case considered here, we were able to obtain an explicit expression for the plastic strain in terms of the total strain.

The equations (4.11), (4.13) and (4.17) are the main equations that define the mechanical response of the beam under monotonic loading.

The thermal response:

Up to now, we have said pretty little about the thermal response of the material other than to say that *for steady state, the temperature field is fixed*. The question is, how do we find the temperature in the beam?

Fortunately, compared to the constitutive equations for the mechanical response, the equations for the thermal response are relatively simple. We need constitutive equations for the heat flux h and the transverse heat loss r. For the first, we assume Fourier's Law which says that the heat flux is proportional to the negative of the temperature gradient, so that

$$h = -kA\frac{dT}{dx} \quad (4.18)$$

where k is the thermal conductivity per unit area (with units $W/m^2/K$) and A is the cross-sectional area.

For the transverse heat loss, we shall assume "Newton's Law of Cooling" according to which, the rate of cooling is proportional to the difference in temperature between the beam and the ambient atmosphere, i.e.

$$r = \hat{h}P_l(T - T_0) \quad (4.19)$$

where $\hat{h}$ is the heat transfer coefficient per unit lateral area and P_l is the perimeter of the cross-section. The value of $\hat{h}$ depends not only upon the material of the beam but also upon ambient conditions (such as the rate of air flow etc.)[5].

---

[5] If you remember your heat transfer classes you might recall some empirical formulas for $\hat{h}$ involving various powers of the Prandtl, Nusselt and other numbers.

Thus, we now list the governing differential equations and boundary conditions for this model:

<u>*Balance Laws*</u>:

$$\frac{dP}{dx} + \bar{t} = 0; \quad \frac{dV}{dx} + \bar{q} = 0; \quad \frac{dM}{dx} + V = 0; \quad -\frac{dh}{dx} + r = 0$$

$$P(x) = 2c \int_{-c}^{c} \sigma \, dy$$

$$M(x) = -2c \int_{-c}^{c} \sigma y \, dy$$

<u>*Constitutive equations*</u>:

$$e = u_{,x} + y\theta_{,x}$$
$$\sigma = E(e - e_p - \alpha T)$$
$$\dot{\varepsilon}^p = \dot{\varepsilon}^p{}_{xx} = \begin{cases} \dot{e} & \text{if } |\sigma| = \kappa_0 \text{ and } \sigma\dot{e} > 0, \\ 0 & \text{otherwise} \end{cases}$$
$$r = \hat{h} P_l (T - T_0), h = -kA\, dT/dX$$

We do not include the boundary conditions here, since we can then consider different problems with different boundary conditions in the same way.

4.3.4 Task 4: Formulate the solution strategy

4.3.4.1 Statically Determinate Problems

The governing equations given above are a set of rather complicated integro-differential equations and are generally not solvable by analytical means. However, with the advent of software tools like MATLAB, they can be easily discretized and numerically solved. There are two types of discretization that are necessary: one is the discretization along the length of the bar needed for the numerical solutions of the equilibrium equations; the other is the discretization through the thickness to carry out the numerical integration through the thickness. In general, for statically indeterminate problems these are coupled and the solution is quite involved. However, for a statically determinate problem, the problem becomes much easier since the equilibrium equations can be solved immediately (without the constitutive equations) and then the constitutive equations can be solved separately.

We will demonstrate the core solution strategy by means of the problem that was described in the beginning of this chapter, namely a beam under a simple bending moment M_0 at either end. This is a trivial statically determinate problem whose solution is given by $M(x) = M_0$, $V = 0$ and $P = 0$ irrespective of the temperature field, the specimen geometry or material (assuming the body undergoes small deformations).

Now, in order to find the constitutive response of the material, i.e. the deflected shape of the beam, the curvature etc. we need to find some parameters:

As a first example, consider a steel beam of length $L = 1m$, and with a **square cross section** of side $0.10m$, i.e. $\bar{h} = c = 0.05m$ subject to pure bending moments at both ends and that both ends are at a temperature of $T = 323K$ whereas the ambient temperature is $303K$. We need to find out the load carrying capacity (i.e. the maximum bending moment possible in the beam before "failure" occurs), the way in which the plastic zone, (i.e. the region with nonzero plastic strain occurs) and the deflection of the beam.

The properties of steel are as follows: Young's Modulus: 210GPa, Yield Strength: 170 MPa, Tensile Strength:300 MPa, Conductivity: 50W/m/K, coefficient of thermal expansion: 13×10^{-6}/K. A rough estimate of the heat transfer coefficient is $30W/m^2/K$.

Notice that steel under these conditions is a strain hardening material. But we would like to make a conservative estimate of the load carrying capacity in a fairly simple manner. For this reason, we will make a drastic simplification and use an ideal plasticity model, so that we shall assume that $\kappa = 210 MPa$.

Before we do the calculations, it is always a good idea to non-dimensionalize all the quantities so that it will be easy to compare different values; So we let

$$s = \frac{\sigma}{\kappa}; \quad H = \frac{E}{\kappa} = 0.001; \quad t = T/T_0. \quad (4.20)$$

4.3.5 Problem 1: No axial confinement

Our first problem assumes that at $x = 0$ and $x = L$ there is no axial load P. Also, no tangential tractions $\bar{t}$ are prescribed and there is no normal load n. Under these conditions, the differential equations of equilibrium can be solved trivially to give $P = V = 0$, $M = M_0$, where M_0 is the applied bending moment.

We are now ready to carry out the calculations:

Finding the Temperature distribution:

⊙ **4.5 (Exercise:). Solving the heat equation**
Show that the temperature distribution in the bar is given by

$$T(x) - T_a = 20 \frac{\cosh \sqrt{hP_l/(2kA)}x}{\cosh \sqrt{hP_l/(2kA)}L} \qquad (4.21)$$

Hint: Substitute (4.18) and (4.19) into (4.6), solve the resulting second order linear ODE in x together with the boundary conditions.

Finding the displacement field as a function of the plastic strain field:

⊙ **4.6 (Exercise:). Axial stretch and curvature**
Use (4.2) (4.13), (4.8) and (4.21) and show that, if $P = 0$ then

$$e_0(x) := u_{,x} = \frac{1}{2c} \int_{-c}^{c} e_p \, dy + \alpha(T(x) - T_0) \qquad (4.22)$$

$$\varrho(x) := \theta_{,x} = \frac{M_0}{EI} + \frac{3}{2c^3} \int_{-c}^{c} y e_p \, dy \qquad (4.23)$$

where I is the cross sectional moment of the area, e_0 is the axial strain and ϱ is the curvature.

The above equations clearly show that once the plastic strain is known, the total strain can be found by first finding $e_0(x)$ and $\varrho(x)$ and then using (4.2). Note that *if there is no variation of temperature across the thickness of the beam, the curvature is independent of the temperature.*

If we define the *mechanical strain* $e_m = e - \alpha(T - T_0)$ it should be clear that $\sigma = E(e_m - e_p)$ and σ is also independent of x. Moreover, from (4.22) and (4.23). Thus, our main task becomes one of finding e_p.

Iterative process for finding the plastic strain field

We now gradually increase M_0 and find the value of M_0 at which yield first begins.

⊙ **4.7 (Exercise:). Initiation of yield**
Show that the value of M at which yielding begins is given by

$$M_y = \kappa I/c \qquad (4.24)$$

and that the yielding begins along the entire upper and lower face of the beam, the top in compression and the bottom in tension. What is the magnitude of the curvature $\varrho = \theta_{,x}$ at the beginning of yield?

The above condition means that yielding is going to occur along the entire length of the beam so that as the moment increases, the plastic strain also increases along the entire length of the beam. In other words, *for this problem, the plastic strain will only be a function of y and will be independent of x, and hence by (4.22) and (4.23) so will the mechanical strain!*

This further means that we don't have to integrate the plastic strain for each location along the bar, but we can simply do so at one cross-section along the length. All other locations will be the same. With this in mind we proceed with the next step.

Let us increase M_0 in steps of δM so that $M_0^i = M_y + i\delta M$ with δM be equal to $0.1 M_y$.

FINDING PLASTIC STRAIN DISTRIBUTION: An Iterative Procedure

1. Choose n locations $y_j, j = 1, \ldots, n$ along the normal to the axis of the beam. We will use the notation $[y]$ for these n values of y. Let $e_p^i([y])$ represent the vector plastic strains of length n along these n locations corresponding to the ith increment of loading. Since the plastic strains are zero until yielding occurs, we shall begin with $M_0 = M_y$

initialization:

2. Set $i = 1$, $e_p^0([y]) = 0$, $M^0 = M_y$

elastic update:

3. Set $e_p([y]) = e_p^{i-1}([y])$, $M^i = M^{i-1} + \delta M_y$
4. Set $e_{m0} = \frac{1}{2c} \int_{-c}^{c} e_p \, dy$, where the integration is carried out numerically.
5. Set $\varrho = \frac{M^i}{EI} + \frac{3}{2c^3} \int_{-c}^{c} y e_p \, dy$ again using numerical integration.
6. Find $e_m([y]) = e_{m0} + \varrho [y]$

plastic correction:

7. Find $e_{pnew}([y])$ from (4.17) with e_m from item 5.
8. Convergence check: find $E = \|e_p([y]) - e_{pnew}([y])\|$ and set $e_p = e_{pnew}$
9. If $E \geq Tol$, go to 3.
10. If $E < Tol$, set (i=i+1) and go to 2.

For the current case, this process was implemented in MATLAB (see authors' websites listed in the preface) and the result obtained is shown as Moment-Curvature graph in Fig. 4.4. The plastic strain distribution is shown in Fig. 4.5. The graph shows the features such an initial yield moment, M_{init} followed by a hardening type of response of the beam. A point to note is that it is not possible to go on increasing

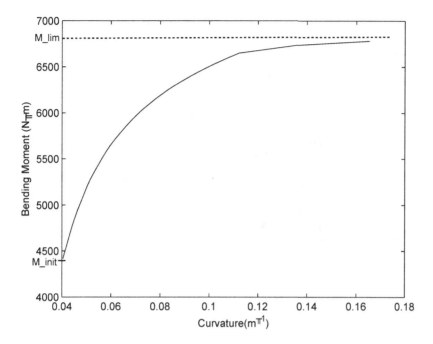

Fig. 4.4 Graph of the Moment versus curvature obtained for a material that satisfies ideal plastic behavior. Note that the moment curvature relationship exhibits "hardening", i.e. increase in the moment beyond yielding. It, however, saturates upon substantial amount of bending of the beam.

the bending moment since it saturates at a certain value M_{lim}. This is the value at which large scale plastic flow begins and the material behaves like a "plastic hinge". The value at which this occurs is around $6500 N-m$ for this model.

This is the key point for the decision making that was introduced at the beginning of the chapter: The material will not fail (i.e. go fully plastic) as long as the bending moment is less than $6500 N-m$ and the factor of safety is $M_0/6500$ where M_0 is the applied moment. This is a conservative estimate in many ways: First, we have constrained the deformation by demanding that the through-thickness strain be linear. Generally speaking, such constraints on the system will result in an overestimate of the load for a given deformation. Second, we have not used the ultimate tensile strength of the material so that the actual failure load may be somewhat higher. If further loading occurs, there might be localization and sudden collapse.

Note also that the thermal strain had no effect on the response whatsoever. This is because, the bar was free to expand and the thermal strains only affect the displacement and not the mechanical strains (and hence the axial force and bending moments). As we shall see presently, as soon as the beam is prevented from expanding, the thermal strains will begin to play a role.

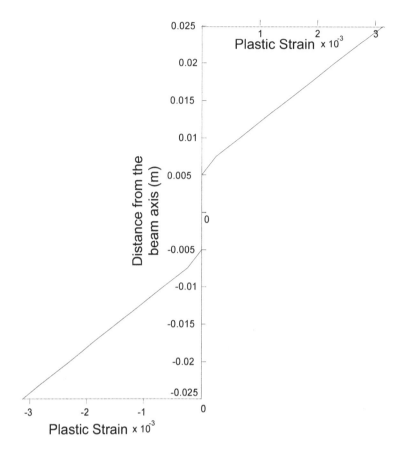

Fig. 4.5 Graph of the plastic strain distribution along the cross-section of the beam. Note the central core where plastic strain has not taken place.

4.3.6 Problem 2: Beam with axial confinement

We now consider the case when the axial displacements are confined. Again the process is the same, except that now, the axial force is unknown.

⊙ **4.8 (Exercise:). Axial Force and Bending Moment**
Show that the axial force is constant (but unknown) throughout the beam and that the bending moment is constant at M_0. How do we find the value of the axial force?

> **PLASTIC STRAIN DISTRIBUTION: AXIAL CONFINEMENT**
>
> 1. Choose n locations $y_i, i = 1, \ldots, n$ perpendicular to the axis of the beam and m locations, $x_j, j = 1, \ldots, m$ along the axes of the beam. This will form a grid of points on the beam. We will use the notation $[x, y]$ for the $n \times m$ points whose coordinates are formed by this grid (can be done with MATLAB).
> 2. Set $i = 1$, $e_p^0([x, y]) = 0, M^0 = M_y$
> 3. Set $e_p([x, y]) = e_p^{i-1}([x, y])$, $M^i = M^{i-1} + \delta M_y$
> 4. Set $\bar{e}_0[x] = \frac{1}{2c} \int_{-c}^{c} e_p[x, y]\, dy + \alpha(T[x] - T_a)$, where the integration is carried out numerically and $T - Ta$ is given by (4.21).
> 5. By using the fact that the axial displacement $u(L/2) = 0$, we can show that $P = -A * (\int_{L/2}^{L/2} \bar{e}_0 dx)$. (This integration can also be done numerically.)
> 6. Find the mechanical strain $e_{m0} = P/A + \frac{1}{2c}\int_{-c}^{c} e_p\, dy$; (Why?)
> 7. Set $\varrho[x, y] = \frac{M^i}{EI} + \frac{3}{2c^3}\int_{-c}^{c} y e_p\, dy$ again using numerical integration.
> 8. Find $e_m([x, y]) = e_{m0}[x, y] + \varrho[x, y]$.
> 9. Find $e_{pnew}([x, y])$ from (4.17) with e_m from item 6.
> 10. Convergence check: find $E = \|e_p([x, y]) - e_{pnew}([x, y])\|$; set $e_p = e_{pnew}$
> 11. If $E \geq Tol$ go to 4.
> 12. If $E < Tol$, set (i=i+1) and go to 3.

Compared with the previous version, this is more complicated since now everything varies with x as well as y and so this problem is an order of magnitude more computer intensive than the previous one. But it is not much harder. It just requires more numerical integration!

4.3.7 *Statically indeterminate problems / general beam problems*

The previous cases all dealt with the solution of statically determinate beam problems. The purpose was to illustrate the fact that it is possible to solve beam equations with relative simplicity for many practical applications by separating the equations of equilibrium and the constitutive equations. This will not always be possible, however, when we come across statically indeterminate beams.

So the general statement of the conditions for a thermo-elasto plastic beam problem to be in equilibrium under monotonic loading conditions is as follows:

We are given a beam that is assumed to satisfy the Bernoulli-Euler hypothesis, i.e. one whose displacement, strain field, plastic strain field and temperature fields are given by

$$\begin{aligned}
\mathbf{u}(x,y) &= (u(x) - yw(x),_x)\mathbf{i}_1 + w(x)\mathbf{i}_2, \\
e(x,y) &= u,_x - w,_{xx} y \\
\varepsilon^p &= \varepsilon^p(x,y) \\
T &= T(x,y)
\end{aligned} \quad (4.25)$$

where, $T(x,y)$ is known a priori.

We need to find the three scalar functions $\mathbb{U} = \{u(x), w(x), \varepsilon^p(x,y)\}$. These scalar functions satisfy

(1) The equations of equilibrium, i.e. (4.3), (4.4) and (4.5) with known lateral forces $t(x)$ and $n(x)$.

(2) Each end of the beam (say the end at $x = x_0$) can satisfy one of a variety of different boundary conditions such as

$$\begin{aligned}
\text{Simply Supported: } & u(x_0) = 0, \; w(x_0) = 0, \; M(x_0) = 0. \\
\text{Built in Support: } & u(x_0) = 0, \; w(x_0) = 0, \; w,_x(x_0) = 0. \\
\text{Free end: } & P(x_0) = 0, \; V(x_0) = 0, \; M(x_0) = 0. \\
\text{Forced end: } & P(x_0) = P_0, \; V(x_0) = V_0, \; M(x_0) = M_0. \\
\text{Spring supported: } & P(x_0) = 0, \; V(x,0) = ku(x_0), \; M(x_0) = 0.
\end{aligned} \quad (4.26)$$

(3) The constitutive equations, given by

$$\begin{aligned}
\text{Axial Force: } P &= AE\left(u(x),_x - \frac{1}{2c}\int_{-c}^{c}\{\varepsilon^p(x,y) + \alpha T(x,y)\}\,dy\right) \\
&= P_e + P_p + P_T \\
\text{Bending Moment: } M &= -EI\left(w(x),_{xx} - \frac{1}{c^2}\int_{-c}^{c}\{\varepsilon^p(x,y) + \alpha T(x,y)\}y\,dy\right) \\
&= M_e + M_p + M_T.
\end{aligned}$$
(4.27)

Note that the axial force and the bending moment are each given as the sum of three parts, one from the displacement gradient (purely elastic), one from the plastic strain (which we will call P_p and M_p) and one from the thermal strain (which we will call P_T and M_T). Note that the thermal strains do not vary with time for steady state temperature distributions and is known a-priori.

The equations for plastic flow are given by

$$\dot{e}_p = \begin{cases} \dot{e}, & \text{if } \sigma^2 - \kappa^2 = 0 \text{ and } \sigma\dot{e} > 0; \\ 0, & \text{otherwise.} \end{cases} \quad (4.28)$$

The above problem can be split into two problems as shown below:

(1) The displacement problem:

Let us "pretend" that we have guessed the value of $e_p(x, y)$ and since we know the temperature distribution, we know $e_T(x, y)$. We can solve for $u(x)$ and $w(x)$ from the equilibrium equations (4.3)–(4.5), the elastic constitutive equations (4.27) and the boundary conditions (4.26).

⊙ **4.9 (Exercise:). The Displacement equations**
Substitute (4.27) into (4.3)–(4.5) and eliminate the shear force to get

$$AE\frac{d^2u}{dx^2} = -t(x) + \frac{dP_p}{dx} + \frac{dP_T}{dx} := a(x)$$
$$EI\frac{d^4w}{dx^4} = \frac{dn}{dx} + \frac{d^2M_p}{dx^2} + \frac{d^2M_T}{dx^2} := b(x) \quad (4.29)$$

These equations, together with the boundary conditions (4.26) can be solved simply by integrating twice (to get $u(x)$) or four times (to get $w(x)$). If we know the value of e_p only at the $n \times m$ grid points, we have to carry out a bunch of numerical quadrature procedures to do this.

(2) The plastic flow problem:

Now that we found $u(x)$ and $w(x)$, we can find the strain from (4.25) and the plastic strain using the evolution equation for the plastic strain equation (4.28). Notice that this is a nonlinear ODE for ε^p and is actually a transcendental equation for hardening materials discussed earlier in Chapter 3. Thus, only numerical solutions are usually possible so that *the value of the plastic strain can be calculated only at certain discrete points in the beam and not everywhere*, i.e. we can only find $e_p(x_i, y_i), (i = 1, \ldots, N)$.

Solution procedure

The displacement and plastic flow split explained above suggests the following iterative scheme: (1) given the temperature $T(x, y)$, the boundary forces, $\bar{t}(x)$ and the transverse tractions, $\bar{q}(x)$:
STEP 1: Choose N points $(x_i, y_i), (i = 1, \ldots, N)$ distributed throughout the beam (typically as a grid) where the plastic strain will be calculated.
STEP 2: Guess $e_p(x_i, y_i)$. If the load is applied in a sequence of steps, then use the value of e_p from the previous load step as the beginning guess. At the start of the whole procedure assume zero plastic strains.
STEP 3: Find $M_P + M_T$ and $P_P + P_T$ at each point x_j along the axis of the beam

by numerically integrating the second and third terms in equation (4.27).

STEP 4: Using the values of STEP 3, the tractions $\bar{t}(x), \bar{n}(x)$, and appropriate boundary conditions, use (4.29) to find $u(x)$ and $w(x)$. This involves a simple numerical integration of these two equations with respect to x. If the boundary conditions do not involve $w(x)$, it is enough to find $w(x),_x$.

STEP 5: Find the mechanical strain $e_m(x_i, y_i) = e(x_i, y_i) - e_T(x_i, y_i)$ using the temperature distribution and the values of $u(x)$ and $w(x)$.

STEP 6: Using these values of strain, calculate a new value of plastic strain $e_p(x_i, y_i)$.

STEP 7: Check for convergence and if not converged, go to STEP 3.

The solution scheme shown here is extremely simple to implement and is quite robust, converging even for softening materials (see Fig. 4.6). But alas, it is not a very practical scheme due to its poor rates of convergence. However, if you have a reasonably small problem and unlimited computer time, this is a useful starting point for practical schemes.

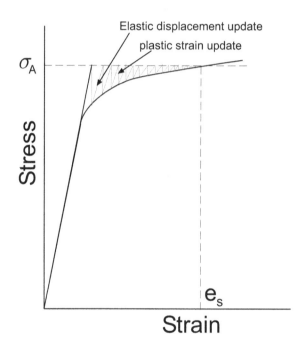

Fig. 4.6 This figure schematically demonstrates the iterative solution of a general elastoplastic problem using the elastic and plastic split procedure. The external load applied is σ_A and the value of the strain corresponding to this load is e_s. The slant lines represent the use of the elastic solution to find the displacements, while the vertical lines represent the use of the flow rule to find the plastic strain for monotonic loading.

4.4 Summary

(1) Classical beam models can be used successfully for many practical structural mechanics problems
(2) The Bernoulli-Euler beams are particularly simple since in case of statically determinate beams, the calculations drastically simplify and one can decouple the equilibrium equations from the constitutive equations
(3) The governing equations are obtained by considering the force and moment equilibrium for a small chunk of the beam
(4) The thermal problem can be solved by assuming that the beam is a rigid conductor.
(5) The constitutive equations proceed from the assumption that total strain is the sum of the elastic plastic and thermal strains and these adjust themselves in such a way that the axial strain $e = e_{xx}$ is linear in y
(6) For monotonic loading the flow rule and the Kuhn Tucker conditions can be explicitly integrated to give an equation for the determination of the plastic strain
(7) The general procedure to solve such a problem is outlined in the last section and involves an iterative scheme wherein we solve for the displacements for given plastic and thermal strains and then use the strains to obtain a new value for the plastic strain
(8) The solution for the plastic strain usually involves the numerical solution of a transcendental equation for monotonic loading and hence, even for the simplest loading, one has to do a numerical solution of the whole problem

For further reading on the subject of elastoplastic beams, there are a number of classic works on limit loads of frames etc. (see e.g., [Drucker and Prager (1952)], [Onat and Prager (1953)] - for arches, [Onat and Shield (1953)]- for combined bending and torsion, [Prager and Rozvany (1975)]-for location of supports, [Prager (1976)]- for design of trusses and [Winter et al. (1948)]–for a discussion of limit design on beams). Elastoplastic beams have a lot in common with sandwich and other "layered beam" structures since in both cases, we are interested in the evolution of properties through the thickness (see e.g., [Manoach and Karagiozova (1993); Seyman and Esendemir (2002)]). See also [Lu and Sayir (1982)] for a singular perturbation analysis of initial plastic zones in beams.

4.5 Projects

(1) Find the plastic strain field for (i) a cantilever beam and (ii) a simply supported beam with uniform distributed load of the same dimensions and as the beam considered here. This is a statically determinate problem but the moment varies with location. So you have to integrate across the thickness at several locations in order to plot the variation of plastic strain.

(2) Write a numerical scheme for solving for the plastic strain as a function of the total strain and temperature for monotonic loading.
(3) Do exercise 1 for cyclic loading. You need to go back to the Kuhn Tucker conditions to do this. This is a challenging task.
(4) Write a MATLAB code to solve a general elasto plastic beam problem. The inputs would be (1) geometry and elastic constants (2) the coefficient of volume expansion (3) function file names indicating temperature distributions, the hardening laws (4) loading file names with lateral tractions as functions of x and time so that time stepping can be done.

Chapter 5

Simple Problems

Learning Objectives

After going through this chapter, you should be able to do the following:

(1) Simplify an application problem into an idealized problem without much information loss with regard to the engineering design decision to be made.

(2) Identify the appropriate characteristics to be modeled to answer the relevant design questions that arise in the application.

(3) Apply the principles learnt in the earlier chapters to come up with a lumped parameter model for the material.

(4) Simplify the structural model for analysis.

(5) Identify the critical loading conditions that need to be analyzed.

(6) Perform analysis of the application by writing a simple MATLAB code, and,

(7) Extract the important numbers from the analysis results necessary for making the design decision and conclusions.

> **CHAPTER SYNOPSIS**
>
> In this chapter, we select some simple real life applications involving inelasticity and illustrate how one goes about modeling, analyzing and solving the problem from the point of view of the design question asked.
>
> The first application shows how one can introduce apparent hardening in the load-deflection behavior of a truss consisting of members assumed to behave elastic-perfectly plastically. The hardening is introduced indirectly by introducing an indeterminacy in the truss structure useful for additional factor of safety against collapse of the truss system.
>
> The second application is related to damping. It is illustrated how a modern material such as shape memory alloy can be modeled and analyzed using the material learnt so far in this book. It also illustrates how one would go about analyzing the dynamic behavior of a braced frame for seismic excitations.

Chapter roadmap

The two problems that are solved in this chapter are simple engineering applications that connect to the principles learn in the previous chapters, especially, chapters on lumped parameter models and one-dimensional plasticity.

We graduate from a simple statically determinate truss that was discussed in the one-dimensional inelasticity chapter, to a statically indeterminate Pratt Truss to show how we carry out analysis and answer the question of how safe the truss is against yield. We will see that complex post yield behavior (which is very important here) arises because of the static indeterminacy in the truss. This example is introduced in Section 5.1.

The second example of an SMA braced frame introduced in Section 5.2 uses concepts discussed in Chapters 1 and 2 first to establish a lumped parameter model useful for analysis. The exercise also shows how one can be innovative in choosing an appropriate LPM that depicts the idealized behavior of the SMA material. Then an elaborate discussion is provided for how one goes about the analysis of dynamic loading conditions.

In this chapter, we have limited ourselves to providing two examples. More examples will be added as and when they are done in the authors' websites listed in the preface.

5.1 Case Study I: Rehabilitation of a crane girder

A crane girder is designed, fabricated and erected, that carries a teeming crane of span 15 meters. Sometime after the crane girder was erected, there is a requirement for the crane wheel loads to be now revised upwards considerably due to customer needs. It was decided to use the crane girder as the top chord of a Pratt-like truss with compression verticals and tension diagonals. For safety, a bottom chord has been introduced that introduces an indeterminacy in the truss. The web members and bottom chord are made of W sections in order to bring in truss action on the plane of the truss (see Fig. 5.1). The out-of-plane bending that may arise due to the eccentricity in the wheel loads with respect to the plane of the crane girder may be neglected. The problem at hand is to find out the factor of safety against the following two factors
1. The initial yield of the truss and
2. Ultimate collapse of the truss

In addition, it should be made sure that the truss does not undergo more than a permissible deflection of 60mm in order to avoid difficulties related to crane operation under excessive deflections.

The truss structure fabricated is shown in Fig. 5.1. All connections are made by bolting to gusset plates that are strong and stiff enough to provide near rigidity and do not fail even under the collapse load conditions. The crane's wheels transfer load directly on the top girder. There are five wheels each carrying roughly 400kN and spaced equally as shown in the figure. An impact factor of 1.25 should be taken into account for the quasi-static analysis as the magnification factor for dynamic conditions. For all practical purposes, for the truss analysis, the five wheel loads can be combined into a single load $P = 2500 kN$ acting at the joint above the verticals BE and CF. The steel is of grade with a minimum yield strength of 300MPa.

Elastic Analysis

Considering the loads acting at B and C to be the critical condition, and denoting the loads to be P, the reactions at the two supports can be obtained as P each at A and D. The skeletal diagram of the truss with these forces marked is shown in Fig. 5.2(a). The forces in the members are positive if they are tensile. Each of the member forces will be represented by the two joints they connect (for example, member AB connects joints A and B).

As can be seen, the truss is symmetric about the center. This establishes that

$$AB = CD; \quad BF = CE; \quad BE = CF; \quad and \quad AE = DF. \tag{5.1}$$

Establishing equilibrium at joints A, B and at E, we obtain,

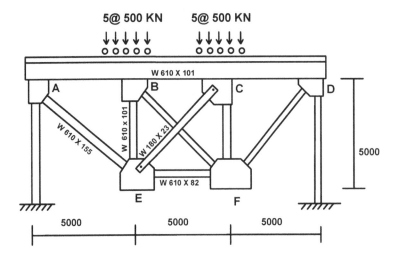

Fig. 5.1 Refurbished Pratt Truss with different W-section members. Crane loading is also shown.

$$AE = \sqrt{2}P,$$
$$AB = -P,$$
$$\sqrt{2}BC + BF = -\sqrt{2}P, \quad (5.2)$$
$$\sqrt{2}BE + BF = \sqrt{2}P, \quad \text{and}$$
$$\sqrt{2}EF + BF = \sqrt{2}P.$$

By considering symmetry, it is enough that we focus on members AB, AE, BC, BE, BF and EF (a total of 6 member force unknowns). The truss has one degree of indeterminacy and therefore, the above 5 equations obtained from equilibrium are inadequate to solve for the member forces. The additional equation needs to be obtained from a kinematic compatibility condition.

Consider the support point A to be the reference with respect to which we can consider horizontal and vertical movements of the joints B, C, E and F. The displacements of the joints are indicated in Fig. 5.2(b). In the following, we analyze the compatibility of the member elongations to obtain the additional kinematic compatibility equation necessary to solve for the member forces.

By symmetry, we have,

$$u_E = -u_F \quad \text{and}$$
$$v_E = v_F. \quad (5.3)$$

Also,

$$u_A = v_A = 0 \quad (5.4)$$

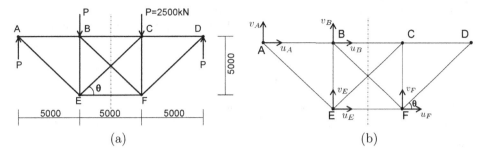

Fig. 5.2 (a) Skeletal diagram of the Pratt Truss with member forces and external forces marked. (b) Joint displacements of the truss along the horizontal and vertical directions.

The bar elongations can now be found out to be, in conjunction with the above conditions,

$$\delta_{AB} = u_B - u_A = u_B,$$
$$\delta_{BE} = v_B - v_E,$$
$$\delta_{BF} = u_F\sin\theta - v_F\cos\theta - (u_B\sin\theta - v_B\cos\theta)$$
$$= \frac{1}{\sqrt{2}}\{-u_E - v_E - u_B + v_B\} \text{ and,}$$
$$\delta_{EF} = u_F - u_E = -2u_E.$$

Eliminating the joint displacements from the above equations, we obtain,

$$\sqrt{2}\delta_{BF} = \delta_{EF}/2 - \delta_{AB} + \delta_{BE}, \qquad (5.5)$$

In order to use the above compatibility equation to find the member forces, we need to first find the stiffness of the individual members. Using the stiffness values of the members, it is possible, in the elastic analysis, to convert the above equation (5.5) into an equation relating the member forces.

Member stiffnesses:

Members AB, BC, and CD are made of steel sections W610×101 with a minimum yield strength of 300 MPa with a cross-sectional area of 12900mm^2 while the diagonal bracing members BF and CE are made of steel sections W310×21 with a cross-sectional area of 9480mm^2. The end diagonal bracing member AE is made of a heavy steel section W610×82 with a cross-sectional area of 19800mm^2. The bottom chord member EF is made of a section W610×82 with a cross-sectional area of 10500mm^2.

The elastic axial stiffness of the members can be computed as,

$$k_{AB} = \frac{EA_{AB}}{L}; \; k_{EF} = \frac{EA_{EF}}{L}; \; k_{BE} = \frac{EA_{BE}}{L}; \; k_{BF} = \frac{EA_{BF}}{(\sqrt{2}L)}; \qquad (5.6)$$

where E is the elastic modulus of steel (taken as 200GPa) and L, the single span length of 5 meters.

Given the forces in the members and their respective stiffness, we can now substitute the deflections in equation (5.5) with the respective force by stiffness expressions. Eliminating E/L that appears in common in the equation, we obtain,

$$2\frac{BF}{A_{BF}} = \frac{1}{2}\frac{EF}{A_{EF}} - \frac{AB}{A_{AB}} + \frac{BE}{A_{BE}} \quad (5.7)$$

⊙ 5.1 (Exercise:). Elastic behavior : member forces
Show that solving for the forces assuming that all the members behave elastically, we obtain,

$$BF = 142kN; \ EF = 2400kN; \ BC = BE = -2601kN; \ AE = 3536kN, \quad (5.8)$$

and the corresponding stresses in the members are,

$$\sigma_{BF} = 53MPa; \quad \sigma_{EF} = 228.5MPa;$$
$$\sigma_{BC} = \sigma_{BE} = -201.6MPa; \quad and, \quad \sigma_{AE} = 178.5MPa.$$

Show that the highest stress is in member EF, and, is within the yield limit.

⊙ 5.2 (Exercise:). Initial yield
In order to find the factor of safety against first yield, the load on the truss is to be increased. Show that the load to first yield, $P_Y = 3283kN$ and hence, the factor of safety against the first yield is 1.31. At this load, the member EF reaches the yield limit.

There is some more reserve capacity in the truss even if we assume that the member EF ceases to take any additional load (perfectly plastic behavior). Let's find the additional factor of safety available before the eventual collapse of the truss structure.

⊙ 5.3 (Exercise:). Indeterminate to determinate truss
Show that the truss now becomes a determinate truss with the bottom chord member EF resisting a constant member load of $300MPa$.
Also show that the factor of safety against the ultimate collapse is 1.39, a shade higher than the factor of safety against the first yield and the ultimate load is $P_{ult} = 3472kN$. The member forces at this state are given by:

$$BF = 455kN; \quad BC = BE = -3794kN; \quad EF = 3150kN;$$
$$AB = -3472kN; \quad and \quad AE = 4910kN.$$

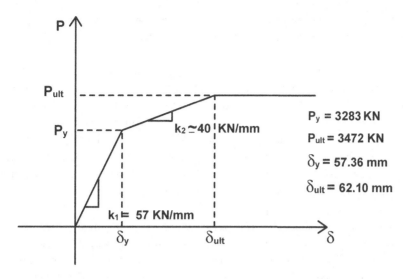

Fig. 5.3 Load-deflection plot of the truss structure. Note that initial yielding takes place at P_y and the truss becomes a mechanism when the crane load reaches P_{ult}.

Since this is an application where it is important to avoid excessive deflections, if the permissible deflection of the rail is 60mm, say, we need to find out the factor of safety against deflection.

The deflection analysis can be carried out in two stages, the first stage which is elastic is a straightforward analysis.

⊙ **5.4 (Exercise:). Check for deflection**
Show that the deflection at yield is 57.36mm. Further analysis can be carried out assuming no additional resistance offered by the bottom chord member for deflection beyond yield. This can be found to be an additional deflection of 4.737mm. The load-deflection plot is shown in Fig. 5.3.

The factor of safety against excessive deflection of 60mm can be found to be 1.35 and the load to this excessive deflection limit is $P = 3388 kN$.

It is possible to extend this analysis to hardening behavior also.

5.2 Case Study II: Passive Damping of a frame structure using super-elastic shape memory alloys (SMA) bracings

Seismic loads are the most critical loads in framed structures. Several structures in the high intensity seismic zones need upgrading for withstanding seismic loads.

There are several possibilities of mitigating the effect of earthquake loads. In passive damping, there are two modes that are usually adopted. One is the seismic isolation and the other is the energy absorption. While the first mode provides a means of shunting the seismic mechanical work through the isolation systems, the second one provides damping. Damping converts the mechanical energy transferred to the structure to heat by means of dissipation and to the other forms of energy. Many times a hybrid means is adopted.

In general, it has been shown that the passive energy dissipation and re-centering devices are effective for the prevention of inter-storey drifts in framed buildings during the seismic excitation. Permanent deformations are noticed to occur requiring replacement of the device after a seismic event. In such cases, SMA braces offer possibilities of limiting the inter storey drifts with reasonable energy dissipation capabilities.

Shape memory alloys in their superelastic state provide good dissipative capabilities comparable to metals undergoing plastic deformation with large hysteretic loops but at the same time behave like pseudoelastic materials in terms of return to the original configuration recovering the large deformations upon release of loading. Therefore, there are many benefits to using SMAs in their superelastic state:

1. Energy absorption with large hysteresis,
2. Near zero residual deformations upon unloading. This is often called the *recentering* property in relation to structural damping,
3. Good low and high cycle fatigue properties,
4. Corrosion resistance, and
5. Ductility due to the formation of stress plateaus that help limit the force transfer through the SMA member.

Attempts have been made in the past, to use SMA-based passive dampers to absorb energy for mitigation of damage that structures undergo, during strong motions such as earthquakes [Dolce and Cardone (2001); Tamai and Kitagawa (2002); Auricchio and Desroches (2006)]. Currently, efforts are on to make these dampers much more practical by a combination of an appropriate choice of SMA materials in order to maximize re-centering and energy dissipation actions of SMA wires [Baratta and Corbi (2002); Desroches and Smith (2004); Dolce and Cardone (2001); Saadat et al. (2002)]. Pre-tensioned SMA wires [Desroches and Smith (2004); Dolce and Cardone (2001)] or SMA wires arranged in multiple loops around the cylindrical posts are some forms in which SMA elements have been incorporated in structures. Their high initial stiffness helps absorb service loads or light earthquakes. Seismic SMA based devices also stiffen in case of unforeseen strong earthquakes (when they

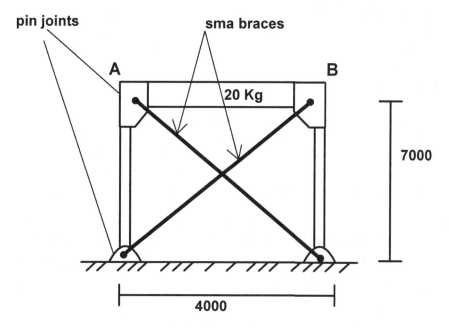

Fig. 5.4 A simple SMA braced single storey frame. For illustration purposes, all the constraints at the joints A and B are released. The mass of the slab is taken to 20kg.

reach pure martensite compositions) and hence assure good displacement control. The stiffness of the SMA wires varies with the displacement of the structure due to a continual change in the fractional populations of martensite and austenite which are two different states that the material can exist in. Thus, the chances of the structure resonating with the earth quake loads are minimized. The ambient air temperature and proper convective environment to take away heat produced during damping action is critical for proper functioning of the SMA and therefore, should be considered while designing the braces. It is reported in the literature [Bartera and Giacchetti (2004); Martinelli et al. (1996)], that SMA braces can damp up to 3 times better than normal steel bracings.

As an illustration to establish the seismic vibration control using SMAs in a building under earthquake loadings, a simple SMA braced single storey frame is analyzed in this study. The frame has a single bay and the cross SMA braces are installed parallel to the direction of motion (see Fig. 5.4 for details of the frame dimensions etc.).

For all practical purposes, it may be assumed that the mass is concentrated on the floors of a framed building structure and therefore, in this example, we will assume that the mass is concentrated on the floor slab and beam supported by the columns of the frame. In addition, in our analysis, we will assume that the floor slab is relatively rigid compared to the columns and that all stabilizing force / damping force is offered by the SMA braces. Stabilized behavior of SMAs is assumed to reduce the behavior sensitivity to repeated loading conditions and rate effects. Strands of 10 SMA wires of 1mm diameter are used for SMA braces. The material used is a 50.4% Ni-Ti composition in its fully annealed condition and heat treated in its stranded configuration.

The response of the bracing material under uniaxial tension is shown in Fig. 5.5 from which the properties can be identified.

The performance measures related to the damping and residual drift associated with the floor have to be assessed. In typical engineering applications related to seismic excitations, equivalent viscous damping is used to measure the energy dissipation provided by a system. The equivalent viscous damping refers to the energy dissipated per cycle divided by the product of 4π and the strain energy for a complete cycle.

The behavior of SMA material is to be tested under cyclic loading conditions, particularly with respect to loading rates caused by a seismic excitation. As a first step to understanding, a loading-unloading cyclic test under tension was conducted on the SMA wire that constitutes a single wire of the strands of the bracings. The results of the test are shown in Fig. 5.5. These results are idealized and used for extracting the essential mechanical properties such as forward and reverse transformation stresses, σ_{Ms} and σ_{As} respectively, austenite and martensite elastic moduli, E_A and E_M, respectively, and the total transformation strain ε_T. Phase transformation occurs from the austenite state to the martensite state at the forward transformation stress and the other way back to the austenite state at the reverse transformation. Notice that the slope of the initial linear response of the austenite state is at least about 3 times that of the slope of the linear response after the complete transformation has occurred. In the intermediate stage of transformation, rule of mixtures can be used to obtain the effective elastic modulus.

An idealized behavior is used in the analysis (see Fig. 5.5. The dotted lines represent the idealized response). As can be seen later in this chapter, this simplifies the analysis greatly.

5.2.1 *Modeling the material response*

From the material response, idealized by the dotted line, the following are observed.
- The material involves a phase transformation going from austenite to martensite upon stressing and reverse while unloading. Once the transformation initiates at the onset of transformation stress, supply of more energy goes toward the latent energy of transformation leading to further transformation. The process is assumed to be isothermal. Any change in temperature will lead to changes in both the stress for onset of transformation and the latent energy of transformation.
- There is a large hysteresis seen in the experimental response. This is also reflected in the idealization (see Fig. 5.7).
- There is a small amount of kinematic hardening (almost linear) taking place

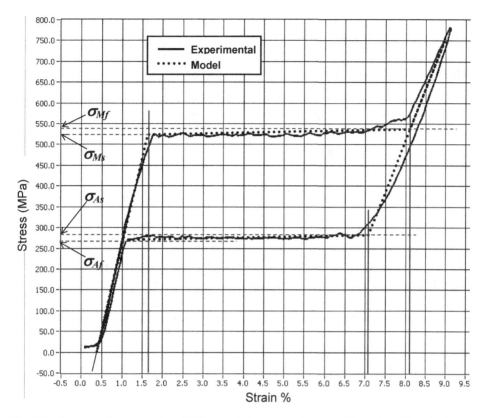

Fig. 5.5 Stress-strain graph of an SMA material tested under uniaxial tension. The idealization is shown as a dotted line. The appropriate values measured are also shown.

during phase transformation.
- Upon loading from zero stress state, the material behaves like a linear elastic material till the stress reaches a particular value, σ_{Ms}. Upon further stressing, phase transformation occurs from austenite to martensite till it completely transforms into martensite at a stress of σ_{Mf}. Upon full transformation to martensite, the material again behaves more or less like a linear elastic material.
- Upon unloading, the reverse transformation initiates at a much lower stress, σ_{As}. Further reverse transformation occurs while unloading further till a stress level of σ_{Af} is reached when the reverse transformation is complete and the material is fully austenite again. On further unloading, it behaves like a linear elastic material.
- The slope of the initial part (elastic modulus of austenite, E_A) is different from the slope of fully martensite curve (elastic modulus of martensite E_M). E_A is much larger than E_M.

There are a number of models of SMA structural elements under different loading conditions (see e.g., [Lim and McDowell (2002)] for cyclic loading). For our illustrative purposes, we can identify the model that could represent the behavior to be consisting of the following elements:

(1) A spring representing the elastic modulus of a combination of austenite and martensite. Simple rule of mixtures can be used to define the elastic modulus at the intermediate stage of transformation. Using the volume fraction of the martensite, ξ, the spring modulus, E can be taken to be

$$E = \xi E_M + (1 - \xi) E_A, \tag{5.9}$$

where, E_A represents the elastic modulus of the pure austenite phase while, E_M, the elastic modulus of the pure martensite phase.

(2) A friction element representing the phase transformation with a threshold effective stress of σ_y. The sliding is equivalent analogously to the transformation strain, ε_T. When in fully austenite state, the transformation strain, $\varepsilon_T = 0$, while at fully martensite state, the "sliding" stops with a maximum transformation strain of $\varepsilon_T = \varepsilon_T^{max}$. Thus, the transformation strain in between can be expressed by the volume fraction as,

$$\varepsilon_T = \xi \varepsilon_T^{max}. \tag{5.10}$$

(3) A spring element attached in parallel to the friction element, representing the Bauschinger hardening with a modulus, h, related to the kinematic hardening taking place during the phase transformation.

In the above, it should be noted that the stress that acts on the assembly of the above elements is "shifted" (see Fig. 5.6), i.e. there is an effective stress $\bar{\sigma}$ which is given in terms of the stress acting, σ and the stress due to latent heat effects, σ_A by,

$$\bar{\sigma} = \sigma - \sigma_A; \quad \bar{\varepsilon} = \varepsilon \tag{5.11}$$

Thus, the stress in the spring E can be calculated as,

$$\bar{\sigma} = E(\varepsilon - \varepsilon_T) \tag{5.12}$$

See Fig. 5.6 for a representative illustration of the assembly. This assembly also provides for the necessary hysteresis exhibited by the material.

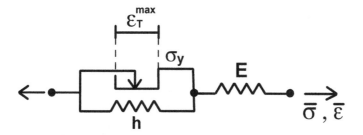

Fig. 5.6 A mechanistic understanding of the material using the assembly of simple response elements such as friction block and springs.

⊙ 5.5 (Exercise:). Evolution equations for the SMA material

Show that the response of the material model represented by the assembly shown in Fig. 5.6 simulates the idealized response (dotted line in the experimental response graph in Fig. 5.5, also in Fig. 5.7), i.e. show that the evolution equations for the variables, the transformation strain, ε_T, that occurs due to the sliding of the friction element and the backstress, β, that appears as stress in the Bauschinger element h for the material can be written as,

$$\dot{\varepsilon}_T = \mu \bar{E} \dot{\varepsilon} \text{ and,}$$
$$\dot{\beta} = h \dot{\varepsilon}_T, \tag{5.13}$$

where $\bar{E}$ is given by

$$\bar{E} = E/\left[E + (E_A - E_M)(\varepsilon - \varepsilon_T)/\varepsilon_T^{max} + h\right], \tag{5.14}$$

and μ is a switch parameter that defines sliding / no sliding conditions. $\mu = 0$ indicates no sliding of the frictional dashpot and $\mu = 1$ indicates sliding that occurs when the following sliding conditions (or transformation conditions) are satisfied:

$$\|\bar{\sigma} - \beta\| = \sigma_y; \quad (\bar{\sigma} - \beta)\dot{\varepsilon} > 0; \quad 0 \le \xi \le 1 \tag{5.15}$$

Note that there are three conditions, the 'yield' condition, the loading condition and the volume fraction condition.

⊙ 5.6 (Exercise:). Material Constants

Show that the values of the various constants can be evaluated to be (see Fig. 5.7):

$$E_A = 41.6 GPa; \quad E_M = 24.1 GPa;$$
$$\sigma_y = 130 MPa \quad \sigma_A = 390 MPa;$$
$$\epsilon_T^{max} = 0.055; \quad h = 364 MPa$$

Also show that the values of the stresses, σ_{Ms}, σ_{Mf}, σ_{As}, and σ_{Af} shown in Fig. 5.7 can be obtained using the above constants.

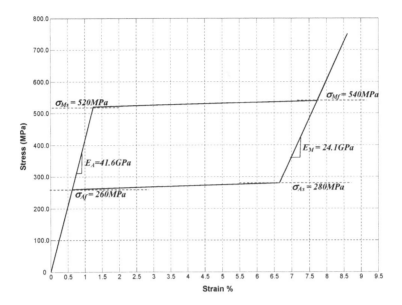

Fig. 5.7 Stress-strain response of the model in Fig. 5.6 for a loading-unloading cycle.

5.2.2 Modeling the SMA braced frame

With the model for SMA material available, we are now ready to model and analyze the SMA braced frame for its response under different dynamic loading conditions.

Before analyzing, we will introduce some assumptions and simplifications to help obtain an idealized model of the SMA braced frame that is simple and readily usable for the dynamic analysis:

- Since the SMA braces are made of strands of SMA wires, they resist only tension, and therefore, assuming that the initial condition is a natural state of the braces, at any given time, only one of the SMA braces will be active in taking the load rendering the frame a determinate structure.
- Columns and beam are assumed to be rigid. In addition, columns are assumed to be of negligible mass compared to the beam. The mass of the system is assumed to be concentrated in the beam and is constant with time.
- Stabilized behavior of the SMA material is assumed, so that the brace has less behavioral sensitivity to repeated loading conditions and rate effects.

The details of the idealized frame taken up for the analysis are shown in Fig. 5.4 and Fig. 5.8. The SMA braces are made up of 10 NiTi SMA wires of 1 mm diameter that take only tensile loads. The material model developed in the previous section is used to model the stress-strain behavior of the SMA material that the brace is made of.

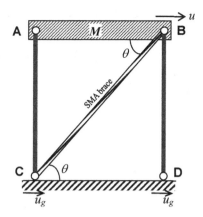

Fig. 5.8 Idealized frame. The idealization is done for the case when SMA brace AC is active. The horizontal movement of the beam is the only degree of freedom. The ground motion u_g is also shown. Mass is assumed to be concentrated in the rigid beam BC.

5.2.2.1 SMA brace system

Another simplification has to do with the SMA braces that connect the bases of the frame to the edges of the beam. We are interested in the horizontal displacements of the beam and the damping effectiveness of the braces with respect to these displacements. Given that the mass of the system is assumed to be concentrated at the center of the beam and that the SMA braces take only tensile axial loads, we seek to find out if the SMA brace can be idealized into a simple spring-frictional dashpot element system with spring stiffness, K with a total elongation of δ.

⊙ **5.7 (Exercise:). Equivalent spring-frictional dashpot assembly for the SMA brace system**
Assuming that the response of the SMA brace is homogeneous all along its length, show that it can be modeled using a similar spring-friction dashpot assembly as the material model shown in Fig. 5.6 simply by replacing the spring constants by stiffness constants, the stress related variables by member force related variables and the strain quantities by axial elongation, δ related quantities. The spring-dashpot model for the SMA brace is shown in Fig. 5.9.
Thus, the various constants for the SMA brace system can be obtained as,

$$K = E(A_b/L_b) = \xi K_M + (1-\xi)K_A; \quad \xi = \delta_T/\delta_T^{max}.$$
$$K_A = E_A(A_b/L_b) = 40.53 kN/m; \quad K_M = E_M(A_b/L_b) = 23.48 kN/m;$$
$$k = \sigma_y A_b = 1020.5N; \quad A = \sigma_A A_b = 3061.5N;$$
$$\delta_T^{max} = \epsilon_T^{max} L_b = 0.443m; \quad \bar{h} = h A_b = 2857.4N.$$

A_b *and* L_b *are the total cross sectional area and the length of the braces. The effective elastic modulus E is given by equation (5.9).*

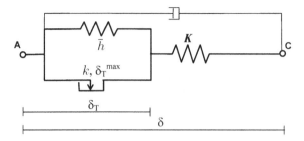

Fig. 5.9 Mechanical analogy in terms of spring-dashpot assembly for the SMA brace for dynamic analysis. This model is based on the SMA material model chosen. Note that a small dashpot has been added in addition to introduce internal damping effects.

5.2.2.2 Reduction of the frame into a single DOF system

Since only one SMA brace is active at any given time, we can idealize the frame structure as a pinned system of rigid bodies and the SMA brace system assembly as shown in Fig. 5.8. The columns, AC and BD and the beam AB are assumed to be rigid. The mass, M, is assumed to be concentrated at the beam, AB. The SMA brace, BC, is the only deforming active member when the top of the frame moves in the horizontal direction. When $u - u_g$ is positive, the SMA brace BC is active while the SMA brace AD, (not shown in the figure) is active when $u - u_g$ is negative. The ground to which the hinges A and B are attached, undergoes seismic horizontal displacements, u_g, while the beam AB undergoes a horizontal displacement, u (the small secondary vertical displacements that the beam may undergo due to the rotation of the columns are neglected in this analysis).

⊙ **5.8 (Exercise:). Governing equation of equilibrium**
Show that the equilibrium in the horizontal direction at B leads to:

$$-F_{spr}\cos\theta = M\ddot{u}, \qquad (5.16)$$

where F_{spr} is the member force in the SMA brace.
Show also that the elongation in the SMA brace, δ, can be determined geometrically for small deformations as (see Fig. 5.8), $\delta = (u - u_g)\cos\theta$.
Given that the SMA brace undergoes transformation elongation of δ_T, the force in the SMA brace can be determined as $F_{spr} = K(\delta - \delta_T)$ and thus, the governing equation of equilibrium can be written as,

$$\ddot{u} = -K/M\left((u - u_g)\cos\theta - \delta_T\right)\cos\theta. \qquad (5.17)$$

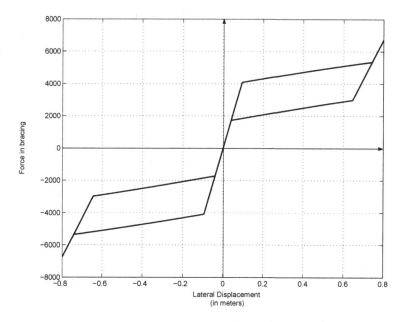

Fig. 5.10 Load-deflection response of the braced frame - SMA bracing. Note that the response shows hysteresis in both the directions of the loading.

The above equation can be simplified and written as,

$$\ddot{u} + \left(K/M\cos^2\theta\right)u = \left(K/M\cos^2\theta\right)u_g - (K/M\cos\theta)\,\delta_T, \quad \text{or}$$
$$\ddot{u} + \omega_n^2 u = F(t)$$

where $F(t) = \omega_n^2(u_g(t) + \delta_T/\cos\theta)$, and, $\omega_n = \sqrt{K/M}\cos\theta$.

The quasi-static response of the frame subject to a displacement u can be obtained simply by finding the resistance offered by the SMA brace to the horizontal beam displacement, u. The resistance-displacement curve of the frame obtained is similar to the stress-strain curve of the SMA material making up the brace. This can be seen in Fig. 5.10. As expected, the resistance-displacement curve shows a perfect symmetry in response to the cyclic horizontal displacements, u.

5.2.2.3 Response of a non-SMA brace

Before finding the response of the above discussed SMA brace frame, we will try and understand the characteristics of a non-SMA brace (consisting of just a spring) that has similar elastic properties as the SMA brace. The SMA brace switches from a spring stiffness of K_A to a lower stiffness of K_M in the process of transformation from austenite to martensite state.

One of the important phenomena that needs to be taken care of during a seismic excitation is the resonance that could take place which may cause excessive deformations. The typical seismic excitations consist of higher power density frequencies in the range of 1 to 2 Hz. In general, while designing, one should make sure that the natural frequency of the structure does not fall near this range of seismic excitation, especially for structures to be built in earthquake prone zones.

The equations of motion for a simple undamped spring mass system can be written as,

$$\ddot{u} + \omega_n^2 u = F(t). \tag{5.18}$$

The natural frequencies of the structure considered here for austenite and martensite state are found to be $\omega_n^A = \sqrt{(K_A/M)}\cos\theta \approx 3.5 Hz$ and $\omega_n^M = \sqrt{(K_M/M)}\cos\theta \approx 2.7 Hz$. This implies that the resonant frequency of 1-2Hz is almost avoided.

Let us consider a typical excitation to be of a simple sinusoidal form, say, $F(t) = F_0 \sin\omega t$. The solution for the system response to this excitation can be obtained as an addition of the complementary solution, u_c, and the particular integral, u_p, i.e.

$$u(t) = u_c + u_p = C\cos(\omega_n t - \phi) + \frac{F_0/m}{\omega_n^2 - \omega^2}\sin\omega t \tag{5.19}$$

where ϕ is the initial phase angle.

As you can see, the solution is an addition of two sinusoidal functions. Though it sounds simple, just the addition of two sinusoidal functions $\sin\omega_n t + a\sin\omega t$ may look far from simple. Figure 5.11a shows the addition of two sine functions with the frequencies close to each other. In this case, the graph display the classic "beats" phenomenon.. Addition of two sine functions, one frequency that is an order of magnitude higher compared to the other shows disturbance like patterns at the peaks as shown in Fig. 5.11b.

Consider an asymptotically decaying excitation of $F(t) = F_0 e^{-0.5t}\sin t$. Notice that the excitation becomes almost zero in 15 seconds (see Fig. 5.12a). However, for such an excitation, the system response does not die down even after the excitation has stopped (see Fig 5.12b). However, in a real life situation we don't see a spring-mass system oscillating forever. Thus, in the real life situation, *there is no such thing as pure spring system when it comes to dynamic response* since every system modeled has some amount of internal damping eventually bringing the system down to a static condition upon removal of an excitation. In this sense, in order to simulate the real response, *every spring should accompany a damper*. With a weak damper introduced in parallel to the spring, the system response for the decaying excitation is shown in Fig 5.12b. One can see that the complementary part of the solution dies off for a weak damper as one would expect from a real life dynamic situation.

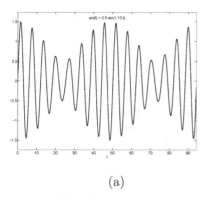
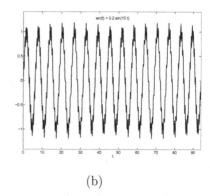

(a) (b)

Fig. 5.11 Addition of two simple sine functions. Note that we obtain far from simple looking functions! (a) shows beats like formation upon addition of sine functions with frequencies close to each other, while (b) shows disturbance like patterns at the peaks due to the addition of a high frequency sine function.

There is no pure spring

EVERY SPRING SHOULD BE ACCOMPANIED BY A WEAK DAMPER TO SIMULATE THE INTERNAL LOSSES IN THE DYNAMICS OF A REAL LIFE SYSTEM

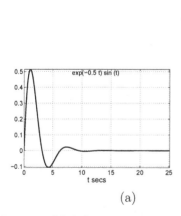
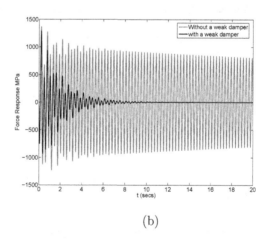

(a) (b)

Fig. 5.12 (a) A decaying excitation. Notice that the excitation goes almost to zero in 15 seconds or so. (b) The response of a spring with and without a weak damper to the decaying excitation is shown. Note that the spring without a damper continues to oscillate indefinitely.

Therefore, in this analysis of frame with an SMA brace system, a weak damper is introduced in parallel to represent the internal losses (in the structure including joints) that knocks off the complementary part of the response after an initial transience. This is very important to obtain a realistic dynamic response of the system.

5.2.3 Dynamic analysis of the SMA braced system

We are now ready to perform the dynamic analysis of the frame with the SMA brace as a nonlinear spring mass system (Fig. 5.8). A weak damper is introduced, as discussed in the previous section, apart from the SMA brace system model described in the aforementioned discussions.

We will now focus on writing the evolution equations and equations of state of the entire frame + SMA brace system. For simulation purposes, it is useful to list all the state variables and their related evolution with time. This helps in setting up the appropriate integration that can be carried out using the ODE solver ODE23 available in MATLAB package.

⊙ **5.9 (Exercise:). State equations for the SMA braced frame**
The state variables thus chosen are: u, v, δ_T, β and F_{spr}. The last three variables represent the state of SMA brace. Note that these variables are not an independent set. δ_T represents the internal variable of the SMA brace system related to the permanent displacement occurring due to the transformation.

Show that the following set of equations define the state of the frame for a given displacement rate, v: $v = \dot{u}$.

$$\dot{v} = -(K/M)\cos^2\theta\left((u - u_g) - \delta_T/\cos\theta\right),$$
$$\dot{u} = v,$$
$$F_{spr} = K\cos\theta(u - u_g - \delta_T/\cos\theta), \tag{5.20}$$
$$\dot{\delta}_T = \mu\bar{K}\dot{\delta} = \mu\bar{K}\cos\theta(v - \dot{u}_g) \quad \text{with} \quad \bar{K} = \bar{E}A_b/L_b \quad \text{and,}$$
$$\dot{\bar{\beta}} = \bar{h}\dot{\delta}_T.$$

u_g *is the ground motion as shown in Fig. 5.8. $\mu = 0$ for no sliding condition of the frictional dashpot in the SMA brace system and $\mu = 1$, if sliding conditions are met. $\bar{\beta}$ is the force in the spring $\bar{h}$ as shown in Fig. 5.9. δ_T is the transformation extension in the SMA brace as shown in Fig. 5.9.*

Notice that the SMA system gives rise to a complex nonlinear system of equations. The above set of equations can now be solved[1] given the ground motion u_g

---
[1] thanks to MATLAB:-))

to obtain the response of the SMA braced frame. In the following, simulations are performed for some cases of ground motion typical of the seismic excitations. For a sample MATLAB code, see authors' websites listed in the preface.

5.2.4 Simulation studies

The damping characteristics of the frame system are observed by simulating the response corresponding to different types of excitations. The results are also compared with the response of a non-SMA braced frame to highlight the amount of damping obtained for an SMA braced frame. Three loading conditions are considered: sinusoidally varying ground motion, depleting sinusoidal ground motion and a small duration sinusoidally varying ground motion. The results of the above modeling are presented in the following:

5.2.4.1 Sinusoidally varying ground motion

The frame behavior is first analyzed for the sinusoidal ground motion, u_g, given by:

$$u_g = u_{g0}\sin(\omega t). \tag{5.21}$$

The amplitude u_{g0} is taken to be 0.25m in order to invoke realistic damping. The frequency $(\omega/2\pi)$ was taken to be 2.1Hz. This frequency is well below the resonant frequency of the frame with SMA brace in its austenite state. The response of the non-SMA brace system is calculated for reference.

The results are presented as variation of displacements of the top joint relative to the ground with respect to time in Fig. 5.13. The force versus displacement response of the two braces - the SMA brace and the non-SMA brace are shown in Figs. 5.14 and 5.15. Hysteretic loops in the response of SMA brace dampens the vibration to a steady state. Though the displacements are almost the same for both the simple spring and the SMA braced response as shown in Fig. 5.13, the force induced in the frame due to the ground motion can be seen to be comparatively reduced for the SMA braced frame as shown in Fig. 5.16.

5.2.4.2 Decaying sinusoidal ground motion

During earthquakes, it is reasonable to assume that an oscillating decaying ground motion occurs that affects the structure over and beyond the duration of the earthquake. The frame is thus analyzed for a decaying sinusoidal ground motion of:

$$u_g = u_{g0}e^{-0.5t}\sin(\omega t), \tag{5.22}$$

where the amplitude and time period are taken to be the same as before. As in the response to sinusoidal excitation, it can be observed from Fig. 5.17 that the SMA

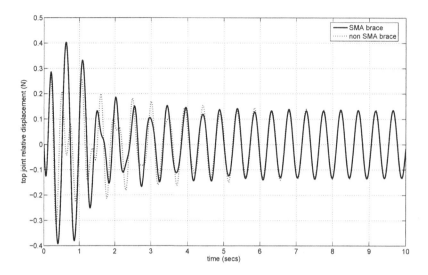

Fig. 5.13 Time variation of displacement of the top of the frame.

cuts out the higher forces that may get transferred to the frame. However, initial displacements are higher for SMA braced frame compared to the non-SMA braced frame. The non-SMA braced frame oscillates longer with a higher amplitude than the SMA braced frame (see Fig. 5.18). The peak forces in the SMA braced frame are cut off and the appropriate energy is absorbed by the SMA braces. It can be observed from the time-force diagram in Fig. 5.17 that internal damping in the structure will be enough to damp out the low level oscillations eventually (beyond 10 seconds).

5.2.4.3 Short duration ground motion

In general, strong ground motions occur over a fairly short period of time just like an impulse load. It is important to analyze the damping provided by the SMA braces for such a short duration motion. The short duration displacements at the ground, u_g as a function of time can be represented by the following expressions:

$$u_g = \begin{pmatrix} u_{g0} e^{-0.5t} \sin(\omega t) & \text{for } t \leq 1s \\ 0 & \text{for } t > 1s \end{pmatrix} \quad (5.23)$$

Here a single half sine wave is applied over the one second period of impulse load. Therefore, $\omega = 0.25$ is taken. The response of the structure to such a load is shown in Fig. 5.19. The amplitude of the force decreases with each cycle and eventually dies down to zero. For a non-SMA bracing, the amplitude of the force

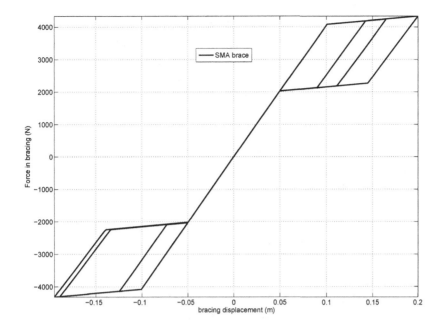

Fig. 5.14 Force-displacement response of the brace under a sinusoidally varying ground motion - SMA bracing.

transferred to the frame can be seen to be higher since there is no damping in the ordinary bracing.

> ⊙ **5.10 (Exercise:). Equivalent damping ratio**
> *Show that the equivalent damping ratio for SMA braces can be found to be around 6% provided the displacements involve full activation of the superelastic response of the SMA braces.*
> *(Damping ratio is defined as the ratio of the mechanical energy dissipated or absorbed to mechanical energy supplied per cycle.)*

Summarizing and concluding from the analysis results, we find that while SMA braces do help in damping out motions with advantages of frequency independent damping and isolation of the frame from transfer of higher loads, there is a downside in terms of "softening" of the SMA brace (when the load reaches the "plateau region" of the stress strain curve", leading to higher initial displacements for short duration excitations. The re-centering ability (zero residual strain for a complete

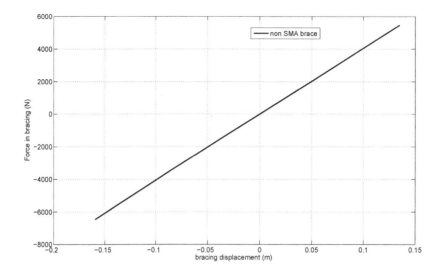

Fig. 5.15 Force-displacement response of the frame under sinusoidally varying ground motion - ordinary bracing. Note that the response is linear and as can be seen the bracing transfers more load to the structure in comparison to the SMA braced frame.

cycle of loading-unloading) is definitely attractive. However, this property can only be used beyond the initial transformation force or stress of the SMA bracing. One way to address this issue is to introduce an initial prestressing in the SMA braces to realize activation of the transformation at much lower loads. Even in such cases, with large displacements generally not permitted in realistic frames, the amount of damping that can be realized is quite small, so the use of SMA braces may not be very good for such applications.

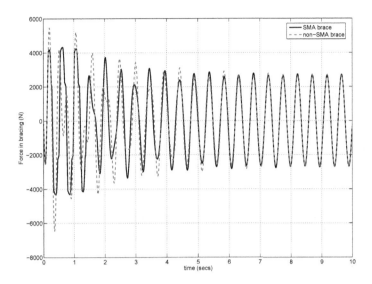

Fig. 5.16   Time variation of force transferred to the frame from the bracings.

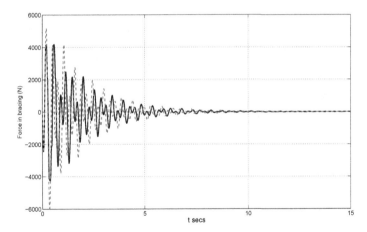

Fig. 5.17   Time variation of force transferred to the frame from the bracings for a ground motion that reduces asymptotically with time.

176 *Inelasticity of Materials: An engineering approach and a practical guide*

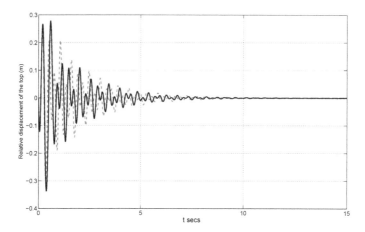

Fig. 5.18 Time-variation of the displacements at the top of the frame due to a ground motion that reduces asymptotically with time.

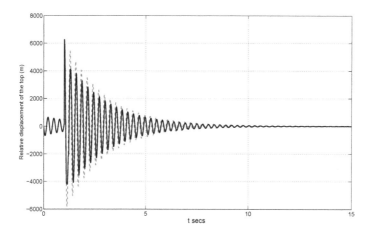

Fig. 5.19 Time variation of force transferred to the frame from the bracings for a ground motion that reduces with asymptotically with time.

PART 2
INELASTICITY OF CONTINUA
Small Deformations

Chapter 6

Introduction to Small Deformation Plasticity

Learning Objectives

By learning this material, you should be able to do the following:

(1) Write down the set of constitutive equations that correspond to J2 plasticity with and without hardening.

(2) Describe the underlying assumptions that are made in arriving at this model.

(3) List a few applications where such a model might be applicable.

(4) Find the form of the hardening function from uniaxial tension experiments.

(5) Write down the loading criteria for materials with the von Mises yield condition ("J2 yield condition").

(6) Calculate the tangent modulus for this model.

(7) Obtain a criterion to decide whether or not the material is currently hardening, softening or perfectly plastic.

(8) Write a MATLAB program to simulate the response of a material subjected to a homogeneous loading condition.

CHAPTER SYNOPSIS

This is the first chapter of part II of the book dealing with 3-D continua undergoing small deformations. The central result is that the applied loading can be written in matrix form as $\mathbb{A}\dot{\varepsilon} + \mathbb{B}\dot{\sigma} = \mathbf{f}(t)$, while the evolution laws can be written as $\dot{\sigma} = \mathcal{K}\dot{\varepsilon}$ where $\mathcal{K}$ is called the tangent modulus. It is then possible to use MATLAB to solve for the response of the material subjected to any homogeneous deformation history. The main result is that the tangent modulus is of the form

$$\mathcal{K} = \mathbb{C}\left(\mathbb{I} - \frac{4\mu\boldsymbol{\tau}\otimes\boldsymbol{\tau}}{(4\mu+\Theta)\kappa^2}\right), \tag{6.1}$$

where $\mathbb{C}$ is the isotropic elastic modulus, $\mathbb{I}$ is the six by six identity tensor, $\boldsymbol{\tau}$ is the deviatoric stress, μ is the shear modulus, κ is the threshold stress and Θ is the hardening function.

Prerequisites

In order to learn the material in this chapter, you need to be familiar with (a) strain, velocity gradients and some elements of small deformation kinematics of continua, (b) stress and equations of motion, (c) indicial notation, and, (d) eigenvalues, eigenvectors and the Π plane.

Chapter roadmap

The chapter begins with the generalization of a frictional dashpot to three-dimensional continuum mechanics which leads to the notion of a "rigid plastic material". This model is then extended by adding some elastic response leading to a three-dimensional continuum mechanics equivalent of a spring and dashpot in series. The constitutive laws for this model are then carefully examined and an explicit expression is obtained for the Kuhn-Tucker or consistency parameter ϕ in terms of the so-called *strain space loading function* $\hat{g}$. This is then followed by obtaining an expression for the tangent modulus $\mathcal{K}$. The ways in which combinations of stresses and strains can be specified for homogeneous deformation are then introduced and an iterative scheme for determining whether or not plastic flow occurs.

The resulting set of differential equations is solved for combination of tension and torsion of a thin walled tube. Instructors can use this problem to introduce the students to the response of elastic-perfectly plastic materials to homogeneous deformations.

This is then followed by a discussion of hardening and the tangent modulus is re-derived in this case. The chapter ends with the simulation of the combined tension and torsion of thin walled tubes.

6.1 An example task

You are working for an engineering consulting company (or the design and analysis support group) and a client wants to know what is the range of load bearing capacity or maybe the low cycle fatigue life for a particular part. The part is made up of aluminum, which is very ductile under operating conditions.

You might want to do a full scale FEM analysis of this part if there is time and need. But before you embark on such an analysis, you might want to get a rough idea of the applied loads. Generally, it is important to have some kind of a range of loads as a way to check if the answers are right. Also, it is important to have a qualitative idea as to what is likely to happen as a way to compare your solution with your intuition. Often, there is a tendency to just "throw it at the computer" and believe whatever it says as the last word. Remember computers do what you *tell* them to do and not what you *want* them to do–garbage in ⇔ garbage out!

Once you have found an approximate solution (using a simple LPM from Chapter 2), you need to decide whether you want to do a full scale FEM analysis. If you choose to do so, then the question arises as to what kind of material property information you will need. Clearly, at a minimum you will need the elastic moduli, and the yield strength. What would you use? The lower yield strength? The ultimate tensile strength (UTS)? Some average? This depends upon what you mean by failure of the part. For example, the easiest thing to do is to say that the beginning of yield is failure; in this case only an elastic analysis is needed and this is relatively straightforward to do.

But this is a very conservative estimate and may lead to unacceptably bulky or expensive solutions. You may want to use the UTS as your criterion, but this may correspond to rather large strains and excessive deformations and in many cases getting too close to the actual failure may lead to an unpredictable behavior. So, one possibility is to assume a certain permanent strain criterion at which point the structural member is so distorted that (1) the part would be considered as failed functionally or, (2) the part has distorted so much that your analysis based on small strains becomes useless anyway. So, a strategy might be to assume some small plastic strain as your acceptable limit of failure, find the corresponding yield stress and then use that as the failure criterion. A typical value and the one that is widely used is the "proof stress". This is the stress needed to cause 0.2% permanent strain. This is schematically illustrated in Fig. 6.1:

You should always perform a post calculation check to see if the solution is acceptable, meets the criteria set and meets your intuition. This is why you need approximate solutions and some idea as to what to expect BEFORE you start the rigorous calculations. That way, if the answer does not agree with your expectation, you now have the chance to learn something. Remember, *Experience is what you*

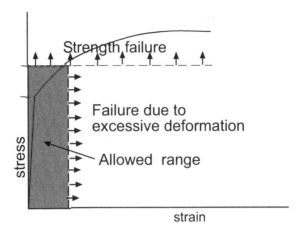

Fig. 6.1 A schematic stress strain response indicating allowable operating range and the types of failure indicated.

get when you don't get what you want.

So, let us assume that for aluminum part in question, you have no access to a sample of the material. What do you do? Let us say that you choose to look at the "Atlas of Stress Strain Diagrams" [International (2002)], find the aluminum that has properties that are nearest to your material and use that. For example, for a particular aluminum alloy, you get a stress-strain diagram as in Fig. 6.8 and you hope to extract information from it. But, this graph is entirely one-dimensional. How can you build a three-dimensional model with only this much experimental data? This chapter is focused on developing a very simple inelasticity model that you could use. We will address the following questions:

(1) What is the simplest three-dimensional model that can use the information provided by a tension test?
(2) What can we use to develop some intuition as to how materials behave after beginning to yield?
(3) How do we find bounds for the allowable loads?

6.2 The J2 rigid-plasticity model

As with the case of the truss and the beam, there are three major parts to the modeling: (1) kinematics, (2) the balance laws and (3) the constitutive model.

6.2.1 *3-D kinematics*

The aluminum part is now considered to be a 3-D continuum, i.e. it occupies a three-dimensional region of space with every particle uniquely associated with a corresponding location in space with a smooth boundary etc.—you get the picture! The displacement of the particle and its velocity are given by

$$\mathbf{u} = \mathbf{u}(\mathbf{x}, t), \mathbf{v} = \mathbf{v}(\mathbf{x}, t) \qquad (6.2)$$

where $\mathbf{x}$ is the location of the particle at time t. As you might know already, the linearized strain and the rate of deformation tensor are defined by

$$\boldsymbol{\varepsilon} = \frac{1}{2}\left(\nabla \mathbf{u} + (\nabla \mathbf{u})^t\right), \mathbf{D} = \frac{1}{2}\left(\nabla \mathbf{v} + (\nabla \mathbf{v})^t\right). \qquad (6.3)$$

This is all the kinematics we are going to do. You might notice that we are getting increasingly brief with regard to these steps since we hope that you are embarking on this part of the book after having taken a course in elasticity. If you haven't, this will be hard for you and you ought to go back and do elasticity first.

6.2.2   3-D equations of motion

If we consider a 3-D "chunk" of metal and consider the equations of motion, we will get the following equations:

$$\text{div}\boldsymbol{\sigma} = \varrho\dot{\mathbf{v}}, \quad \boldsymbol{\sigma} = \boldsymbol{\sigma}^t. \qquad (6.4)$$

Again we expect you to know what div means and what the components of $\boldsymbol{\sigma}$ mean.

6.2.3   Constitutive laws: The J2 rigid plasticity equations

The formal study of permanent deformations of solids due to metal working operations began with a paper by Henri Edouard Tresca [Tresca (1864)] a French mechanical engineer and number 3 on the list of 72 people who were said to have played a crucial role in the building of the Eiffel tower.

Hitherto, the work of blacksmiths was mostly by experience and was shrouded in secrecy since they were the makers of swords and axes and hence formed the "defense industry" of that time. They were the ones who knew how to work the high carbon steel called Wootz steel [Verhoeven (1987)][1] by carefully working the grains to make the legendary ultra sharp and ultra tough Damascus swords of repute [Verhoeven et al. (1998)]. These swords had a characteristic wavy pattern called "damask" and recent thought is that they contained carbon nanotubes and fine wires of cementite (see [Reibold et al. (2006)])[2]. Wootz steel aroused great curiosity in the western world and with the British occupation in the 17th century, were subject to serious scientific investigation by the royal society ([Pearson (1795)].

---

[1] A special steel which has abundant ultra hard metallic carbides and was invented in India around 300BC; the technique to make this steel was widely used in India and Sri Lanka (see [Juleff (1996)]). It is speculated that the word Wootz is a corruption of the word *ukku* in Tamil and Kannada and other south Indian languages, perhaps meaning steel.

[2] Ha! we have connected up this work with the current buzzword "nanotechnology". Goes to show that there is nothing "really" new under the sun:-)))

Getting back to more mundane matters, the first model for the response of a ductile metal that we are going to introduce is the 3-D continuum analogue of a frictional dashpot. This is the simplest plasticity model that can explain some features of the response of metals. The original "J2 plasticity model" was apparently developed simultaneously by St.Venant, Levy and von Mises based on an analogy to fluid flow, with some modifications to account for the yielding behavior.

A simple three-dimensional dashpot:
In order to understand the development of inelastic models, it is the simplest to begin with a three-dimensional viscous dashpot. Recall that for a one-dimensional viscous dashpot, the constitutive law is of the form $v = \eta F$ where v is the velocity (i.e. the rate of increase of the end-to-end separation), F is the force and η is the reciprocal of the viscosity, C. We now generalize these two ideas: (1) the velocity is proportional to the force and, (2) it is in the same direction. Our first generalization is to a three-dimensional dashpot where we will demand that the velocity vector be proportional to the force vector, i.e.

$$\mathbf{v} = \eta \mathbf{F}. \tag{6.5}$$

A generalization of the linear dashpot leads to the celebrated "Navier-Stokes fluid" or "linear viscous fluid". In modeling, modifications are necessary for two distinct properties of these fluids:

(1) These fluids are incompressible, and
(2) they "flow" under shear, i.e. undergo large deformations when subject to any nonzero shear stress.

In order to generalize the velocity and the force to a 3-D continuum, we substitute the *rate of deformation*, $\mathbf{D}$ for the velocity (the rate of deformation is the symmetric part of the velocity gradient) and the *deviatoric stress*, $\boldsymbol{\tau}$ for the force. Thus, we generalize the equation (6.5) to:

$$\mathbf{D} = \eta \boldsymbol{\tau},$$
$$\text{where } \mathbf{D}_{ij} = \frac{1}{2}(\mathbf{v}_{i,j} + \mathbf{v}_{j,i}) = \eta \boldsymbol{\tau}_{ij}, \tag{6.6}$$

$1/\eta$ is the viscosity and $\boldsymbol{\tau}$ is the deviatoric part of the stress. Notice that the first of the two properties is automatically satisfied by the above equation since $\boldsymbol{\tau}$ is deviatoric (or, in other words, the trace of $\mathbf{D}$ will come out to be zero). Also, note the second law condition that $\boldsymbol{\tau} \cdot \mathbf{D} \geq 0$ is satisfied as long as we choose $\eta > 0$.

Thus, *the Navier-Stokes fluid is the continuum generalization of a linear viscous dashpot*. Now, what about a frictional dashpot? For the one-dimensional frictional dashpot, we had the condition that $v = 0$ as long as $\|F\| < k$, *i.e. no sliding unless the force is sufficient to overcome friction* and then $v \neq 0 \Rightarrow \|F\| = k$, i.e. when it is sliding, the force has to be equal to the maximum frictional force. A naive three-dimensional extension of this would be $\|\mathbf{v}\| = 0$ if $\|\mathbf{F}\| < k$ and $\|\mathbf{v}\| \neq 0 \Rightarrow \|\mathbf{F}\| = k$. But this is NOT enough. The key point is that just like for a viscous dashpot, we need to specify not only the magnitude of the velocity, but also the direction. As you might expect, the simplest choice would be that the direction of the sliding be the same as that of the force, i.e.

$\mathbf{v} = \phi\mathbf{F}$, the body will slide in the direction of the force,

$\|\mathbf{F}\| < k \Rightarrow \phi = 0$, no sliding when force is less than critical value, (6.7)

$\phi \neq 0 \Rightarrow \|\mathbf{F}\| = k$, when sliding, force is maximum possible.

Note that in the above equation ϕ plays the role of the reciprocal of the viscosity, η, in (6.6). However, it is NOT a constant. Nevertheless, the criterion that $\mathbf{F} \cdot \mathbf{v} \geq 0$ will be satisfied so long as $\phi \geq 0$. In view of this, the above set of equations (6.7) can be written in a much more convenient, compact and generalizable form called the *Kuhn-Tucker form*

$$\mathbf{v} = \phi\mathbf{F},\ \phi \geq 0,\ \|\mathbf{F}\| - k \leq 0,\ \phi(\|\mathbf{F}\| - k) = 0. \quad (6.8)$$

⊙ **6.1 (Exercise:). Kuhn-Tucker Form**
Show that the form (6.8) is equivalent to the conditions (6.7). Show also that, $\phi \geq 0$ is needed to ensure that $\mathbf{F} \cdot \mathbf{v} \geq 0$, i.e. it is a statement of the second law of thermodynamics for this system. It is equivalent to demanding that the viscosity be positive. Hint: Use the fact that multiplication is equivalent to the logical operation "or" i.e. a=0 or b=0 is the same as a*b=0.

Now, to deal with a plastic solid, St. Venant and von Mises [Mises (1913)] modified the above structure by assuming that
- The solid is rigid if the yield stress is not reached.

- Once yielding begins, the solid is incompressible and the rate of deformation is proportional to the deviatoric stress (similar to equation (6.7)). They, thus, assumed that

$$\mathbf{D} = \phi\boldsymbol{\tau},\ \phi \geq 0, \quad (6.9)$$

where ϕ which has the units of inverse viscosity, i.e. $1/Poise$ is not a constant. (This theory will be henceforth referred to in this book as the *viscous fluid theory*.)

- Although they originally used the condition that the plastic solid "flows" whenever the maximum shear stress reaches a critical value (called the Tresca criterion [Tresca (1864)]), von Mises [Mises (1913)] suggested that it was simpler to use the notion that the second invariant of the deviatoric stress (i.e.

$J2 = \boldsymbol{\tau}\cdot\boldsymbol{\tau} = \tau_{ij}\tau_{ij}$), which is a measure of the "magnitude of the shear stress", reaches a critical value (a suggestion that was apparently also in the work of Maxwell). This is popularly called the *von Mises criterion*. We will call this criterion as the *J2 criterion* in this book[3]). Thus we will assume that

$$\mathbf{D} = \phi\boldsymbol{\tau}$$
$$\phi = 0 \Rightarrow \boldsymbol{\tau}\cdot\boldsymbol{\tau} = \tau_{ij}\tau_{ij} < \kappa^2, \text{ i.e. rigid behavior,} \qquad (6.10)$$
$$\phi > 0 \Rightarrow \tau_{ij}\tau_{ij} = \kappa^2.$$

YIELD FUNCTION AND SURFACE

The function $f(\boldsymbol{\sigma}) = J2 = \boldsymbol{\tau}\cdot\boldsymbol{\tau} = \tau_{ij}\tau_{ij}$ is called the *yield function* and it demarcates the range of stresses beyond which plastic flow occurs. The boundary where plastic flow begins is given by $f(\boldsymbol{\sigma}) = \kappa^2$. This equation can be considered to represent a smooth closed surface in stress space (much as the function $x^2 + y^2 + z^2 = r^2$ represents the surface of a sphere of radius r) which is called the *Yield surface*. See Fig. 6.2. In Chapter 8, we will generalize this notion further.

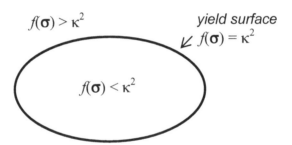

Fig. 6.2 A schematic representation of how the equation $\boldsymbol{\tau}\cdot\boldsymbol{\tau} = \tau_{ij}\tau_{ij} = \kappa^2$ represents the yield surface which delineates the stresses at which plastic flow begins.

The value of ϕ is not prescribed separately but must be deduced from the above conditions. This is the trickiest part of the constitutive equations and makes plasticity somewhat more challenging to understand than most other constitutive theories for dissipative processes.

From a materials modeling point of view, the constitutive relations (6.10) have a number of features that are extremely useful:

(1) It requires only one material constant–the yield stress κ;
(2) It lends itself to be used as an idealization that allows us to develop intuition about plasticity;

---

[3]The reason for "$J2$", is that $\boldsymbol{\tau}\cdot\boldsymbol{\tau}$ is the 2nd invariant of the deviatoric stress and is referred to as $J2$ in the linear algebra literature.

(3) it allows for many exact solutions especially in plane strain.
(4) Note that a constitutive equation for ϕ is NOT necessary. This puzzling aspect will be explored shortly.

Once the velocity field is given, the equations (6.10) are sufficient to determine the stress field in those regions of the body **where the velocity gradient is not zero**. This may not be entirely obvious at first sight. However, this will become evident when we try to calculate ϕ. To do this, we first take the inner product of **D** with itself and then use (6.10a) and get

$$\mathbf{D}\cdot\mathbf{D} = \mathbf{D}_{ij}\mathbf{D}_{ij} = \phi^2 \tau_{ij}\tau_{ij}. \tag{6.11}$$

Now if **D** is not zero, we know that $\tau_{ij}\tau_{ij} = \kappa^2$ and so we get the following important result for a rigid plastic material:

$$\phi = \frac{\sqrt{\mathbf{D}_{ij}\mathbf{D}_{ij}}}{\kappa} = \frac{\|\mathbf{D}\|}{\kappa}, \tag{6.12}$$

i.e. the "inverse viscosity" ϕ is proportional to the magnitude of **D**. Thus, as **D** goes to zero, the viscosity tends to infinity and the material behaves like a rigid body. This is in contrast to most viscous fluids where the viscosity tends to a finite value as **D** tends to zero. Furthermore, the linear dependence of ϕ on the magnitude of **D** also implies that it does not explicitly depend upon how fast the material deforms, i.e. the material response is *rate independent*. No other form for viscosity has this feature. To see this, substitute (6.12) into (6.9) and you will get

$$\tau_{ij} = \frac{\kappa}{\|\mathbf{D}\|}\mathbf{D}_{ij}, \tag{6.13}$$

if we now increase the velocity by a certain constant factor $\bar{c}$ so that $\mathbf{v}' = \bar{c}\mathbf{v}$ then $\mathbf{D}' = \bar{c}\mathbf{D}$, but *the stress remains the same since the constant factor $\bar{c}$ cancels out in (6.13)*.

Thus, if you compare (6.13) with the Newtonian viscous fluid model (6.6), we see that the "viscosity" of the plastic solid is not a constant but roughly "inversely proportional" to the rate of deformation. Now we can see that as the rate of deformation goes to zero, the viscosity goes to infinity and the material becomes "rigid". Note, however, that *the stress does not go to infinity when **D** goes to zero, rather it becomes indeterminate*. This is one of the principal difficulties with the rigid plasticity model and reduces its usefulness for structural applications where plastic flow is generally localized (i.e. the bulk of the material is not yielding).

6.2.4 *Standard form for rigid plastic (Kuhn-Tucker form)*

We will now rewrite equations (6.9) and (6.10), in a slightly different form which will be our *standard form for rigid plastic materials*, which we will follow in all our subsequent developments.

> **J2-RIGID PLASTICITY MODEL**
>
> (1) **Constitutive or state variables:** τ_{ij} & D_{ij}.
> (2) **Yield function:** $f(\sigma) = \tau \cdot \tau - \kappa^2 \leq 0$.
> (3) **Elastic Response:** The body is rigid if $f(\sigma) < 0$.
> (4) **Flow Rule or kinetics:** $\mathbf{D} = \phi \tau$.
> (5) **Kuhn-Tucker Conditions:** $\phi \geq 0, f \leq 0, \phi f = 0$.
> (6) **Value of ϕ:** $\phi > 0 \Rightarrow f = 0$ and $\phi = \|\mathbf{D}\|/\kappa$.

6.2.5 Example problems

We can now use the constitutive equation (6.13) to calculate the stress once the velocity field is known. Two examples are given below.

Example 1:
Consider the velocity field corresponding to simple shearing, i.e.

$$\mathbf{v} = ky\mathbf{i} = \begin{Bmatrix} ky \\ 0 \\ 0 \end{Bmatrix} \tag{6.14}$$

where k is a constant representing the rate of shear (see Fig. 6.3).

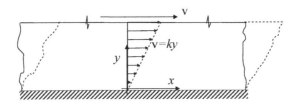

Fig. 6.3 Simple shearing with a velocity field $\mathbf{v} = ky\mathbf{i}$. k is the shearing rate.

We can calculate the rate of deformation tensor $\mathbf{D}$ as

$$\mathbf{D} = \frac{1}{2}(\mathbf{v}_{i,j} + \mathbf{v}_{j,i}) = \begin{pmatrix} 0 & k/2 & 0 \\ k/2 & 0 & 0 \\ 0 & 0 & 0 \end{pmatrix}. \tag{6.15}$$

Now a routine calculation using (6.13) shows us that the deviatoric stress is given by

$$\tau_{ij} = \frac{\kappa}{\|\mathbf{D}\|} D_{ij} = \kappa/\sqrt{2} \begin{pmatrix} 0 & 1 & 0 \\ 1 & 0 & 0 \\ 0 & 0 & 0 \end{pmatrix}. \tag{6.16}$$

Thus, the shear stress is a constant of magnitude $\tau = \kappa/\sqrt{2} = \tau_y$, irrespective of the shear rate. It is easy to see that the constant shear stress satisfies the balance of linear momentum when inertia is ignored.

Example 2:
Consider a flow of the form
$$\mathbf{v} = kx\,\mathbf{i} - \frac{1}{2}ky\,\mathbf{j} - \frac{1}{2}kz\,\mathbf{k}. \tag{6.17}$$
This represents a uniaxial volume preserving stretching - extension combined with a lateral shrinkage so that the volume is preserved. For this deformation we can calculate the rate of deformation tensor $\mathbf{D}$ as (see Fig. 6.4)
$$\mathbf{D} = \frac{k}{2}\begin{pmatrix} 2 & 0 & 0 \\ 0 & -1 & 0 \\ 0 & 0 & -1 \end{pmatrix} \tag{6.18}$$

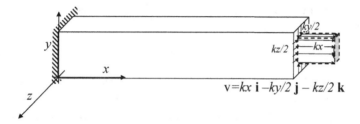

Fig. 6.4 Uniaxial incompressible stretching - extension in x leads to volume preserving contraction in y and z directions during the deformation.

and the deviatoric stress, $\boldsymbol{\tau}$, is given by
$$\boldsymbol{\tau} = \frac{\kappa}{\sqrt{6}}\begin{pmatrix} 2 & 0 & 0 \\ 0 & -1 & 0 \\ 0 & 0 & -1 \end{pmatrix}. \tag{6.19}$$

> ⊙ **6.2 (Exercise:). Rigid plasticity example**
> *Find the stress field for a flow field of the form $\mathbf{v} = ay^2\mathbf{i}$. Does the stress satisfy the equations of motion (you can neglect inertia if you like because we will mostly be interested in very slow flow)?*
> *Hint: Remember that only the deviatoric stresses are specified from the constitutive equations. Ignore inertia and see if you can solve for the pressure by using the equations of equilibrium.*

This model is very useful for studying situations where there is a widespread plastic flow (such as for example, a soft material like lead or solder or when a material is extruded or deformed by large amounts).

A number of exact solutions have been developed based on this model, especially for plane strain. In all these cases, the solution is obtained by a judicious choice of flow field combined with rigid body motion in such a way that the equations of equilibrium are satisfied. Prominent among them is the *slip line field theory*. In this theory, the system is reduced to a set of first order hyperbolic equations that are solved by using the method of characteristics. We will not get into that approach here. Many excellent books are available on this subject. The interested reader may refer to [Chakrabarthy (2000)] for a comprehensive modern account of the subject.

The model given in equation (6.13) is extremely simple and easy to understand and so modifications of these have found wide application in the metal forming industry. On the other hand, the model is not useful for structural problems because, when **D** goes to zero, the stress cannot be calculated. This is a major drawback of this model. One modification that is done routinely in the metal forming industry is to remove the indeterminacy of the stress when the rate of deformation is small by replacing the "rigid" region with the response of a very viscous fluid, i.e. convert it into a non-Newtonian fluid mechanics problem with a non-constant viscosity, and write it in the form

$$\tau_{ij} = \sqrt{\kappa} f(\|\mathbf{D}\|) \mathbf{D}_{ij}, \tag{6.20}$$

where, for example, $f(x) = x^\alpha$. As alpha approaches -1, the response approaches that of a rigid plastic model. There are some issues with the predictions of such models however. The primary issue being that we have replaced a solid with a fluid. Thus, we don't recover elastic behavior at all when the stresses are small, rather the material slowly "creeps" for any nonzero stress. Nevertheless such models have been implemented in many FEM programs for metal forming problems.

6.3 The J2 elasto-plasticity model

The key ingredient that we would like for an elastoplastic model that would be suitable for structural problems is to ensure that *when there is no yielding, we should get an elastic response*. This model was the first truly elastoplastic model that was developed. It was developed for plane strain by Prandtl [Prandtl (1924)] and later by Reuss [Reuss (1939)] and is usually referred to as the *Prandtl-Reuss theory*. We will use the descriptive name *small strain J2 elastoplasticity* for this model.

One could view this model as a way to fix the difficulties associated with the J2-rigid plastic model, especially with regard to the problems of non-uniqueness and the inability to find the stresses in the rigid regions. Small strain J2-elastoplasticity addresses the problems associated with the rigid plastic model in the following way:

(1) the deformations considered are restricted to **small deformations** only. This restriction is quite suitable for problems in structural mechanics.

(2) **Constitutive Variables**: Borrowing ideas from the 1-D response (i.e. the spring-frictional dashpot model), it is assumed that the total strain tensor ε is the sum of an "elastic strain tensor" ε_e and "plastic strain tensor" ε_p, i.e.

$$\varepsilon = \varepsilon_e + \varepsilon_p. \tag{6.21}$$

Thus, a sample list of constitutive variables are: the total strain ε, the plastic strain ε_p and the stress σ. In other words, this is the 3-D continuum analogue of a spring and dashpot in series.

(3) **Plastic Incompressibility**: Borrowing ideas from fluid mechanics, we assume that $\varepsilon_{ii}^p = 0$, i.e. the permanent deformation leaves the volume of the body unchanged. This is an empirical observation which has been justified by experiments for a class of materials [Bridgman (1964)]. But there are a number of material for which this is not true. Furthermore, this does NOT mean that the material is incompressible; only that there is no permanent volume change.

(4) **Elastic Range**: We assume that for given values of the plastic variables (in this case the plastic strain), the stress must always be such that the constraint $f(\sigma) \leq 0$ is satisfied. For the constraint function, called the *yield function*, we use the same function as in the J2 rigid plastic model, i.e. the constraint is given by

$$f(\sigma) = \frac{\|\tau\|}{\kappa} - 1 = \frac{\sqrt{\tau_{ij}\tau_{ij}}}{\kappa} - 1 \leq 0. \tag{6.22}$$

Please note that the constraint above can be written in many different ways, all of which are equivalent:

$$f(\sigma) = \|\tau\| - \kappa \leq 0, \tag{6.23}$$

or,

$$f(\sigma) = \tau \cdot \tau - \kappa^2 = \tau_{ij}\tau_{ij} - \kappa^2 \leq 0. \tag{6.24}$$

(5) **Elastic Response**: Assume that the stress is given as an isotropic linear function of the elastic strain, i.e. $\sigma = \mathbb{C}(\varepsilon - \varepsilon_p) = \mathbb{C}\varepsilon_e$ where $\mathbb{C}$ is the standard isotropic stiffness matrix, i.e.

$$\tau_{ij} = 2\mu(\gamma - \varepsilon_p)_{ij} \quad \text{and} \quad p = K\varepsilon_{ii}. \tag{6.25}$$

where $\gamma = \varepsilon - (1/3)\varepsilon_{ii}\mathbf{I}$ is the deviatoric strain, $p = (1/3)\sigma_{ii}$ is the mean normal stress, μ, the shear modulus and K, the bulk modulus. Note that the above equation (6.25) is just a restatement of the usual isotropic elastic response function, i.e. it is a restatement of

$$\sigma_{ij} = \lambda \delta_{ij}\varepsilon_{kk} + 2\mu\varepsilon_{ij}, \quad \lambda = \frac{E\nu}{(1+\nu)(1-2\nu)}, \quad \mu = \frac{E}{2(1+\nu)}, \tag{6.26}$$

where E and ν are the elastic modulus and the Poisson's ratio respectively. These are the quantities that are usually listed. We can write stress-strain relation in terms of E and ν as,

$$\sigma_{ij} = \frac{E}{1+\nu}\left(\frac{\nu}{1-2\nu}\delta_{ij}\varepsilon_{kk}^e + \varepsilon_{ij}^e\right)$$

$$\Rightarrow \mathbb{C}_{ijkl} = \frac{E}{1+\nu}\left(\frac{\nu}{1-2\nu}\delta_{ij}\delta_{kl} + \delta_{ik}\delta_{jl}\right), \tag{6.27}$$

where $\varepsilon_e = \varepsilon - \varepsilon_p$.

(6) **Flow Rule**: We again borrow the result from the theory used in equation (6.10) and assume that

$$\dot{\boldsymbol{\varepsilon}}_p = \phi\boldsymbol{\tau}, \quad \phi \geq 0, \quad f \leq 0, \quad \phi f = 0, \tag{6.28}$$

where f is the yield function given by (6.24). Note that the flow rule is IDENTICAL to that of the J2 rigid plasticity model (6.10) except that we use, $\dot{\boldsymbol{\varepsilon}}_p$, rather than $\mathbf{D}$.

There are a number of equivalent alternative forms for this flow rule. The form listed above is the one that is directly related to the viscous fluid theory. It is convenient for manipulation but not very convenient for physical or geometrical interpretation of ϕ.

A form that is convenient (and widely used) is

$$\dot{\boldsymbol{\varepsilon}}_p = \phi \frac{\boldsymbol{\tau}}{\|\boldsymbol{\tau}\|} = \phi \frac{\boldsymbol{\tau}}{\kappa}, \tag{6.29}$$

where the yield condition has been used to replace $\|\boldsymbol{\tau}\|$ with κ. This form can also be interpreted as related to the gradient of the yield function in the sense that

$$\dot{\boldsymbol{\varepsilon}}_p = \phi \frac{\partial f}{\partial \boldsymbol{\tau}}, \tag{6.30}$$

where f is given by (6.22). Thus, this form provides a geometrical interpretation of $\dot{\boldsymbol{\varepsilon}}_p$ since the gradient of f with respect to $\boldsymbol{\tau}$ is the normal to the yield surface $f(\boldsymbol{\tau}) = 0$. For this reason (6.30) is usually called *the normality rule*. More on this will be discussed in Chapter 9.

In this form, if we take the magnitude of the tensors on both sides, i.e. if we calculate $\|\dot{\boldsymbol{\varepsilon}}_p\| = \sqrt{\dot{\varepsilon}^p_{ij}\dot{\varepsilon}^p_{ij}}$, we find that $\phi = \|\dot{\boldsymbol{\varepsilon}}_p\|$. In other words, ϕ is the same as the norm of the plastic strain rate[4] and is usually called *the equivalent plastic strain rate*.

Yet another alternative (and equivalent form) is

$$\dot{\boldsymbol{\varepsilon}}_p = \bar{\phi}\frac{\boldsymbol{\tau}}{\kappa\|\boldsymbol{\tau}\|} = \bar{\phi}\frac{\boldsymbol{\tau}}{\kappa^2}, \tag{6.31}$$

where the yield condition has been used to replace $\|\boldsymbol{\tau}\|$ with κ. In this case, by taking the dot product $\boldsymbol{\tau} \cdot \dot{\boldsymbol{\varepsilon}}_p = \tau_{ij}\dot{\varepsilon}^p_{ij}$ and using the yield function we get $\bar{\phi} = \tau_{ij}\dot{\varepsilon}^p_{ij}$, i.e. in this version, $\bar{\phi}$ is the rate of *plastic work*, and you could argue that it should be non-negative since we expect that whatever internal work is done due to plastic deformation should be dissipated and the second law would suggest that it should be non-negative. In all of these cases the *Kuhn-Tucker conditions* are the same.

---

[4] Actually we get $\phi = \pm \|\dot{\boldsymbol{\varepsilon}}_p\|$ and we use the fact that $\phi \geq 0$ to conclude that $\phi = \|\dot{\boldsymbol{\varepsilon}}_p\|$.

(7) **Kuhn-Tucker conditions**: As usual, we assume that the value of ϕ or $\bar{\phi}$ is given by the *Kuhn-Tucker conditions*, i.e.

$$\phi \geq 0, \quad f(\boldsymbol{\sigma}, \kappa) \leq 0, \quad \phi f = 0. \tag{6.32}$$

In all the forthcoming discussions, we will use the J2 plasticity model given by the following equations unless otherwise specified:

J2-ELASTO PLASTICITY MODEL

(1) **State variables**: ε, ε_p, and $\boldsymbol{\sigma}$.
(2) **Yield function**: $f(\boldsymbol{\sigma}) = \boldsymbol{\tau} \cdot \boldsymbol{\tau} - \kappa^2 \leq 0$.
(3) **Elastic Response**: $\boldsymbol{\sigma} = \mathbb{C}(\varepsilon - \varepsilon_p) \Rightarrow \boldsymbol{\tau} = 2\mu(\boldsymbol{\gamma} - \varepsilon_p)$, $p = 3Ke$.
(4) **Flow Rule or kinetics**: $\dot{\varepsilon}_p = \phi \boldsymbol{\tau}$.
(5) **Kuhn-Tucker Conditions**: $\phi \geq 0, f \leq 0, \phi f = 0$.

$$\tag{6.33}$$

The model described above is a minimalist model in that it requires only three constants E, ν and κ. The first two are just the elastic constants and the third is related to the yield strength of the material. These are typically listed in any reasonable book on materials science and is also widely available on the web. However, ***κ is NOT the same as the yield strength, σ_y of the material.***

To understand this, let us revert to the one-dimensional case discussed in Chapter 3 on trusses, in the light of the 3-D constitutive equations we have developed. Thus, if we subject a bar to axial forces along the $\mathbf{i}_i$ direction, we have, $\boldsymbol{\sigma} = \sigma \mathbf{i}_i \otimes \mathbf{i}_i$, representing uniaxial tension along the x_1 direction. Let the yield stress be $\sigma_y = 100 MPa$ say. How is this number related to the value of κ in equation (6.22)? To see this, we must evaluate $\boldsymbol{\tau} \cdot \boldsymbol{\tau}$ for the case when $\boldsymbol{\sigma} = \sigma_y \mathbf{i}_i \otimes \mathbf{i}_i$, i.e. when the material is just at yield.

⊙ **6.3 (Exercise:). Relationship between κ and σ_y**
Show that $\sigma_y = \sqrt{3/2}\kappa$.
Hint: First show that $\boldsymbol{\tau} = (1/3)\sigma(2\mathbf{i}_i \otimes \mathbf{i}_i - \mathbf{i}_j \otimes \mathbf{i}_j - \mathbf{i}_k \otimes \mathbf{i}_k)$.

Thus if the yield stress is reported as 100 MPa, then the value of κ to be used will be $\sqrt{2/3} \times 100 = 81.64$ MPa.

Typical values for some commonly occurring materials are listed here.

Material	E (GPa)	ν	σ_y (MPa)	κ (MPa)
Steel (MS)	210	0.29	300	245
Aluminum	70	0.33	100	82
Copper	110	0.343	33.3	27
Red brass	115	0.307	270	220

Of the numbers listed above, **the elastic constants are very "robust"**, i.e. they don't change drastically by "processing" the material and changing its microstructure (mainly metals). On the other hand, the *yield strength, i.e. the value of the stress at which permanent deformation begins, is very sensitive to the exact microstructure*. So, in using the values for κ, it is essential to know what kind of microstructure is present in the sample. The values reported above for the yield stress are usually called the "0.2% yield strength" or simply "proof strength". It is the stress level at which a permanent strain of about 0.2% is seen. This is quite suitable for most structural analysis purposes. We will use these numbers for the subsequent work.

6.3.1 The value of ϕ and the "tangent modulus"

We are now in a position to find the value of the consistency parameter ϕ and obtain the evolution laws.

Let us assume that the part that you need to analyze is a thin walled steel tube (ratio of inner to outer diameter greater than 0.8) that is subject to various forces (tension, torsion etc.). You do not have time to test it carefully and you have decided to use the J2 plasticity model given by equations in (6.33). Let us further assume that the loading on the tube corresponds to a homogeneous deformation and we are interested in finding the stresses and the accumulated damage in the material.

VECTOR NOTATION FOR SECOND ORDER TENSORS

From now on, we will use the notation $[\cdot]$ to represent column vectors and $(\cdot)$ to represent matrices.

For convenience of solving, we write any symmetric second order tensor **s** with components s_{ij} in vector form $[\mathbf{s}] = [s_{11}, s_{22}, s_{33}, \sqrt{2}s_{12}, \sqrt{2}s_{13}, \sqrt{2}s_{23}]^t$, the $\sqrt{2}$ terms are used to ensure that $[\mathbf{s}]^t[\mathbf{s}] = s_{ij}s_{ij}$.

Thus, we will state our task as follows:

From the constitutive relations (6.33) obtain a relationship between $[\dot{\sigma}]$ and $[\dot{\varepsilon}]$ of the form

$$[\dot{\sigma}] = (\mathcal{K})[\dot{\varepsilon}], \dot{\sigma}_{ij} = \mathcal{K}_{ijkl}\dot{\varepsilon}_{kl}, \qquad (6.34)$$

where $(\mathcal{K})$ is called the *tangent stiffness matrix* and is the generalization of the notion of the tangent to a line on a graph. We expect the tangent to depend upon the state of the material in some complicated manner.

Once we have an explicit expression for $\mathcal{K}$, then we can get the relationship between the unknown stresses and strains with respect to the loading (as a combination of specified stresses and strains, in this case).

Therefore, we need to decide what kind of loading is indicated and how to represent it mathematically. For example, if we want to simulate a simple tension response, we assume that all the stress components other than σ_{11} are zero and that ε_{11} is known. One can see that the loading is given in terms of a combination of specified stresses and strains.

REPRESENTATION OF HOMOGENEOUS DEFORMATIONS

The key point that we are going to make is that *for homogeneous deformations, every possible combination of stress and strain that can be specified can be represented by the following equation*

$$(\mathbb{A})[\dot{\boldsymbol{\sigma}}] + (\mathbb{B})[\dot{\boldsymbol{\varepsilon}}] = [\mathbf{f}(t)], \tag{6.35}$$

where $(\mathbb{A}(t))$ is a six by six matrix representing the combination of components of stress that are known while $(\mathbb{B}(t))$ is a six by six matrix representing the combination of the components of strains that are known, $[\mathbf{f}(t)]$ is a six-term column vector representing the way in which these known components change with time. In all, there are six independent equations for the twelve unknown components of stress and strain. The matrices $(\mathbb{A})$ and $(\mathbb{B})$ allow us to represent any combination of stress and strain that is prescribed.

A few examples are given below:

1. **SIMPLE TENSION**, strain ε_{11} known as a function of time:
For simple tension with lateral sides free of stress, we have

$$\varepsilon_{11} = f(t),\ \sigma_{ij} = 0 \ (i, j \neq 1). \tag{6.36}$$

In our notation, this is represented as

$$\begin{pmatrix} 0&0&0&0&0&0 \\ 0&1&0&0&0&0 \\ 0&0&1&0&0&0 \\ 0&0&0&1&0&0 \\ 0&0&0&0&1&0 \\ 0&0&0&0&0&1 \end{pmatrix} \begin{bmatrix} \dot{\sigma}_{11} \\ \dot{\sigma}_{22} \\ \dot{\sigma}_{33} \\ \sqrt{2}\dot{\sigma}_{12} \\ \sqrt{2}\dot{\sigma}_{13} \\ \sqrt{2}\dot{\sigma}_{23} \end{bmatrix} + \begin{pmatrix} 1&0&0&0&0&0 \\ 0&0&0&0&0&0 \\ 0&0&0&0&0&0 \\ 0&0&0&0&0&0 \\ 0&0&0&0&0&0 \\ 0&0&0&0&0&0 \end{pmatrix} \begin{bmatrix} \dot{\varepsilon}_{11} \\ \dot{\varepsilon}_{22} \\ \dot{\varepsilon}_{33} \\ \sqrt{2}\dot{\varepsilon}_{12} \\ \sqrt{2}\dot{\varepsilon}_{13} \\ \sqrt{2}\dot{\varepsilon}_{23} \end{bmatrix} = \begin{bmatrix} f_1(t) \\ 0 \\ 0 \\ 0 \\ 0 \\ 0 \end{bmatrix}. \tag{6.37}$$

From the above set of equations, you should be able to see that all components of stress other than the first are set to zero (provided that the initial value is chosen to be zero also). The first component of the strain rate is prescribed by $f_1(t)$. This corresponds to simple tension under controlled strain.

2. COMBINED STRETCHING AND SHEARING: Strain control

$$\begin{pmatrix} 0&0&0&0&0&0 \\ 0&1&0&0&0&0 \\ 0&0&1&0&0&0 \\ 0&0&0&0&0&0 \\ 0&0&0&0&1&0 \\ 0&0&0&0&0&1 \end{pmatrix} \begin{bmatrix} \dot{\sigma}_{11} \\ \dot{\sigma}_{22} \\ \dot{\sigma}_{33} \\ \sqrt{2}\dot{\sigma}_{12} \\ \sqrt{2}\dot{\sigma}_{13} \\ \sqrt{2}\dot{\sigma}_{23} \end{bmatrix} + \begin{pmatrix} 1&0&0&0&0&0 \\ 0&0&0&0&0&0 \\ 0&0&0&0&0&0 \\ 0&0&0&1&0&0 \\ 0&0&0&0&0&0 \\ 0&0&0&0&0&0 \end{pmatrix} \begin{bmatrix} \dot{\varepsilon}_{11} \\ \dot{\varepsilon}_{22} \\ \dot{\varepsilon}_{33} \\ \sqrt{2}\dot{\varepsilon}_{12} \\ \sqrt{2}\dot{\varepsilon}_{13} \\ \sqrt{2}\dot{\varepsilon}_{23} \end{bmatrix} = \begin{bmatrix} f_1(t) \\ 0 \\ 0 \\ f_4(t) \\ 0 \\ 0 \end{bmatrix}. \quad (6.38)$$

⊙ **6.4 (Exercise:).**
Which of the components of stress have been set to zero in the above set of equations?

Although this notation appears somewhat cumbersome, it is an extremely useful form that can be implemented easily in MATLAB. There are six equations for the determination of the 12 unknowns (6 components of the stress and 6 components of the strain). If we are able to find six other equations that relate $[\dot{\sigma}]$ to $[\dot{\varepsilon}]$, then will have enough equations for the unknowns. This, however, turns out to be a somewhat challenging task since we have to extract this information from the constitutive relations given in (6.33).

If the tangent stiffness matrix $(\mathcal{K})$ is an explicit expression as in equation (6.34), then we can combine it with (6.35) and get

$$[\dot{\varepsilon}] = ((\mathbb{A})(\mathcal{K}) + (\mathbb{B}))^{-1}[f(t)]. \quad (6.39)$$

Unfortunately, the terms in the matrix $(\mathcal{K})$ depends upon whether or not plastic flow takes place and hence obtaining a form for it is somewhat complicated. The main source of the nonlinear response of the material is in the Kuhn-Tucker parameter ϕ. So the steps involve first finding an explicit representation for ϕ and then obtaining the tangent modulus.

The procedure for obtaining this is common to all plasticity type models and so it is well worth the effort to write it out in a greater detail for this simplest case. For this reason, we have listed it as a step by step exercise in the following.

Finding ϕ from equation (6.33):

⊙ **6.5 (Exercise:). Step 1: Rate form for elastic law**
Differentiate the elastic response relation (6.33a), use (6.33b) and obtain

$$[\dot{\sigma}] = (\mathbb{C})[\dot{\varepsilon} - \phi\tau]. \qquad (6.40)$$

⊙ **6.6 (Exercise:). Step 2: Rate form for the yield function**
Assume $\phi > 0$ so that $f = 0$, differentiate the yield function with respect to time, (6.33a) and obtain

$$\frac{\partial f}{\partial \sigma} \bullet \dot{\sigma} = 0 \Rightarrow 4\mu\tau\bullet(\dot{\gamma} - \phi\tau) = 4\mu\tau\bullet\dot{\gamma} - 4\mu\phi\kappa^2 = 0. \qquad (6.41)$$

Hint: Instead of using the rate form (6.40) it is easier to use $\tau = 2\mu(\gamma - \varepsilon_p)$. The term $\tau\bullet\dot{\gamma}$ plays a crucial role to the whole development. We will represent it by the symbol $\hat{g}$ and we will refer to it as the *strain space loading function*[5].

⊙ **6.7 (Exercise:). Importance of $\hat{g}$**
1. Show that for the yield function given by (6.22), Step 2 can be written as

$$\frac{df}{dt} = \left.\frac{df}{dt}\right|_{\phi=0} + \frac{\partial f}{\partial \varepsilon_p}\bullet\dot{\varepsilon}_p. \qquad (6.42)$$

2. Show further that

$$\left.\frac{df}{dt}\right|_{\phi=0} = \hat{g}. \qquad (6.43)$$

i.e. $\hat{g}$ is the rate of change of the yield function assuming no plastic flow, i.e. pure elasticity.
3. Show that if $f = 0$ and $\hat{g} > 0$ then the Kuhn-Tucker conditions will be violated, unless plastic flow takes place. (Hint: Show that if plastic flow does not take place then we will be at a location where $f > 0$ which is a violation of the Kuhn-Tucker conditions.)

---

[5]The reason for the name "strain space loading function" for $\tau\bullet\dot{\gamma}$ is due to the fact that, in the 1970s, Naghdi and coworkers (in the interests of disclosure, Naghdi was the thesis advisors of one of the coauthors) discovered that the most conceptually appealing way to obtain elasto-plastic constitutive equations was by writing all the equations in terms of strains (see [Naghdi and Trapp (1975c)]). We have not followed that approach here since we want to motivate the forms for the constitutive equations from that of viscous fluids. The approach presented here is a variant of that proposed by Naghdi and coworkers, but considerably less general, and, we hope, is more intuitive. We will revisit this issue later when we introduce finite plasticity; there strain space formulations can be particularly useful. But the connection to that formulation is in the exercises.

⊙ **6.8 (Exercise:). Step 3: Solve (6.41) to find ϕ**
Show from (6.41) that

$$\phi = \frac{\boldsymbol{\tau}\bullet\dot{\boldsymbol{\varepsilon}}}{\kappa^2} = h\hat{g}, \ h = \frac{1}{4\mu\kappa^2}, \ \text{where} \ \hat{g} = 4\mu\left(\boldsymbol{\tau}\bullet\dot{\boldsymbol{\varepsilon}}\right), \tag{6.44}$$

the solution being valid as long as $\phi > 0$, i.e. $\hat{g} > 0$.

You will see this three-step process repeated many times in the subsequent sections and chapters so that you are able to carry out the steps without difficulty.

The final step in the process is to substitute the value of ϕ into (6.40) to obtain

$$\dot{\boldsymbol{\sigma}} = \mathbb{C}(\dot{\boldsymbol{\varepsilon}} - h\boldsymbol{\tau}\hat{g}) = \mathbb{C}(\dot{\boldsymbol{\varepsilon}} - \frac{1}{\kappa^2}\boldsymbol{\tau}(\boldsymbol{\tau}\bullet\dot{\boldsymbol{\gamma}})). \tag{6.45}$$

Note that this form is *almost* the final form since the right-hand side involves $\dot{\boldsymbol{\varepsilon}}$. We can now rewrite it in the required form using a simple vector identity namely, $(\mathbf{a}(\mathbf{b}\bullet\mathbf{c})) = (\mathbf{a}\mathbf{b}^t)[\mathbf{c}] = (\mathbf{a}\otimes\mathbf{b})\mathbf{c}$. Using this, we finally obtain

$$\dot{\boldsymbol{\sigma}} = \mathcal{K}\dot{\boldsymbol{\varepsilon}} = \mathbb{C}(\mathbb{I} - \frac{1}{\kappa^2}\boldsymbol{\tau}\otimes\boldsymbol{\tau})\dot{\boldsymbol{\varepsilon}}, \tag{6.46}$$

where $\mathbb{I}$ is the fourth order symmetric identity tensor. The above equation is so important that we will write it out in the matrix form as,

$$\begin{bmatrix} \dot{\sigma}_{11} \\ \dot{\sigma}_{22} \\ \dot{\sigma}_{33} \\ \sqrt{2}\dot{\sigma}_{12} \\ \sqrt{2}\dot{\sigma}_{13} \\ \sqrt{2}\dot{\sigma}_{23} \end{bmatrix} = \begin{pmatrix} \lambda+2\mu & \lambda & \lambda & 0 & 0 & 0 \\ \lambda & \lambda+2\mu & \lambda & 0 & 0 & 0 \\ \lambda & \lambda & \lambda+2\mu & 0 & 0 & 0 \\ 0 & 0 & 0 & 2\mu & 0 & 0 \\ 0 & 0 & 0 & 0 & 2\mu & 0 \\ 0 & 0 & 0 & 0 & 0 & 2\mu \end{pmatrix} \\ \left\{ \begin{pmatrix} 1 & 0 & 0 & 0 & 0 & 0 \\ 0 & 1 & 0 & 0 & 0 & 0 \\ 0 & 0 & 1 & 0 & 0 & 0 \\ 0 & 0 & 0 & 1 & 0 & 0 \\ 0 & 0 & 0 & 0 & 1 & 0 \\ 0 & 0 & 0 & 0 & 0 & 1 \end{pmatrix} - \frac{1}{\kappa^2} \begin{bmatrix} \tau_{11} \\ \tau_{22} \\ \tau_{33} \\ \sqrt{2}\tau_{12} \\ \sqrt{2}\tau_{23} \\ \sqrt{2}\tau_{13} \end{bmatrix} \begin{bmatrix} \tau_{11} \\ \tau_{22} \\ \tau_{33} \\ \sqrt{2}\tau_{12} \\ \sqrt{2}\tau_{23} \\ \sqrt{2}\tau_{13} \end{bmatrix}^t \right\} \begin{bmatrix} \dot{\varepsilon}_{11} \\ \dot{\varepsilon}_{22} \\ \dot{\varepsilon}_{33} \\ \sqrt{2}\dot{\varepsilon}_{12} \\ \sqrt{2}\dot{\varepsilon}_{13} \\ \sqrt{2}\dot{\varepsilon}_{23} \end{bmatrix} \tag{6.47}$$

The above set of equations represents the six additional equations to augment the six equations in (6.35). However, the story is NOT complete yet. The six equations (6.47) were derived under the *assumption* that $\phi > 0$. How do we know whether this is true or not? A partial answer is provided by equation (6.44). Clearly, ϕ has the same sign as $\hat{g}$. So if we know the sign of $\hat{g}$ then we can decide whether to use (6.47) or not. You might heave a sigh of relief and say "Ok then, just find $\boldsymbol{\tau}\bullet\dot{\boldsymbol{\gamma}}$

and check the sign of ϕ''. Not so fast! We DON'T know $\dot{\gamma}$. All we know is some combination of $\dot{\sigma}$ and $\dot{\varepsilon}$ as given by (6.35)!

The way out of our difficulty is based on an iterative scheme:

FINDING $\mathcal{K}$

Given the loading as described by (6.35) and given σ, ε and t, we find $\dot{\sigma}$ and $\dot{\varepsilon}$ by the following steps:

Step 1: determine whether you are on the yield surface or not by calculating $f(\sigma) = \tau \bullet \tau - \kappa^2$.

Step 2: See if you can get by guessing an elastic response. Thus assume $\phi = 0$. Then $\mathcal{K} = \mathbb{C}$, i.e. the material has an elastic response. Hence

$$[\dot{\varepsilon}] = ((\mathbb{A})(\mathbb{C}) + (\mathbb{B}))^{-1}[f(t)], \quad \dot{\sigma} = (\mathbb{C})[\dot{\varepsilon}]. \tag{6.48}$$

Step 3: See if the guess is right. If $f \not< 0$ and $\hat{g} = 2\mu\tau\bullet\dot{\varepsilon} > 0$, then our guess is wrong. Plastic flow occurs. So, we recalculate the "correct" value of $\dot{\varepsilon}$ through

$$[\dot{\varepsilon}] = ((\mathbb{A})(\mathcal{K}) + (\mathbb{B}))^{-1}[f(t)] \quad \dot{\sigma} = (\mathcal{K})[\dot{\varepsilon}], \tag{6.49}$$

where, now $\mathcal{K}$ is given by (6.47).

Thus, at each time step you have to make two calculations. This process can be easily implemented in MATLAB for homogeneous deformations. However, it should be used with caution since the rate equations are discontinuous, so many ODE solvers have trouble with them. However, ODE23 has been found to be suitable (if not exactly fast) for solution. This cannot be used in general solutions for boundary value problems since it is too slow and inefficient. However, for class room purposes and for the understanding of the method of solution, this is very simple.

Given all these conditions, we can write the value of ϕ in a compact form as

$$\phi = h\langle f \rangle^0 \langle \hat{g} \rangle \tag{6.50}$$

where we have used the Macauley Bracket notation for which

$$\langle x \rangle^n = \begin{cases} 0, & \text{if } x < 0; \\ x^n, & \text{if } x \geq 0, \end{cases} \tag{6.51}$$

so that using this notation, we can write a very compact, explicit form for $\dot{\sigma}$, i.e.

$$\dot{\sigma} = \mathbb{C}\left(\dot{\varepsilon} - h\langle f \rangle^0 \langle \hat{g} \rangle \tau\right), \tag{6.52}$$

where $h = 1/\kappa^2$ and $\hat{g} = \tau\bullet\dot{\gamma}$. This form can be implemented in MATLAB in a relatively simple manner. Thus, in order to use an ODE solver such as ODE23 to simulate the response, we need to implement the following process.

6.3.2 Non-dimensionalization

It is generally beneficial to properly non-dimensionalize all the variables so that we can get improved accuracy in computer simulations and at the same time reduce the number of parameters that are needed for comparison. The standard non-dimensionalization that is suitable for our purpose is to divide all the stresses by the yield stress, σ_y and to divide all the strains by the ratio of the yield stress to the Young's modulus of the material. If we do that, the governing equations will depend only upon *one parameter*: the Poisson's ratio. In other words, *in the current model, the response of any two materials that have the same Poisson's ratio will be the same, apart from a scaling factor.* The governing equations then become

---

NON-DIMENSIONALIZED J2 PLASTICITY MODEL

$$\hat{\boldsymbol{\tau}} = \frac{1}{1+\nu}(\hat{\boldsymbol{\gamma}} - \hat{\boldsymbol{\varepsilon}}_p), \quad p = \frac{e}{1-2\nu}, \quad \text{tr}(\hat{\boldsymbol{\varepsilon}}_p) = 0.$$

$$\dot{\hat{\boldsymbol{\varepsilon}}}_p = \hat{\phi}\hat{\boldsymbol{\tau}}, \qquad (6.53)$$

$$\phi \geq 0, \hat{f} = \hat{\boldsymbol{\tau}} \bullet \hat{\boldsymbol{\tau}} - 2/3 \leq 0, \hat{\phi}\hat{f} = 0.$$

Now consider a thin walled tube subject to uniaxial extension, to uniaxial strain cycling and to tension-torsion cyclic loading are shown in Fig. 6.5. The response of the elastic, perfectly plastic material to such loading is shown in Fig. 6.6 and Fig. 6.7. The applied loads for the tension-torsion cyclic loading are given by equation (6.38) on p.196 with

$$f_1(t) = 4\pi \cos(2\pi t), \; f_2(t) = -2\sqrt{2}\pi \cos(2\pi t) \qquad (6.54)$$

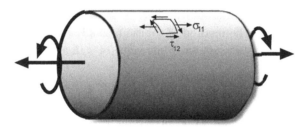

Fig. 6.5 State of stress in a thin walled tube subject to simultaneous tension and torsion. This is a classic experiment since, by varying the amplitude and phase difference between the tension and torsion, one can create complex two-dimensional loading paths. In general, due to the need for stability, the tube may be internally pressurized (this condition is not shown here).

These graphs are generated using a MATLAB code. The code has been written for both non-hardening and hardening cases. For this case which is non-hardening,

a is set to zero (for the MATLAB codes, see authors' websites listed in the preface).

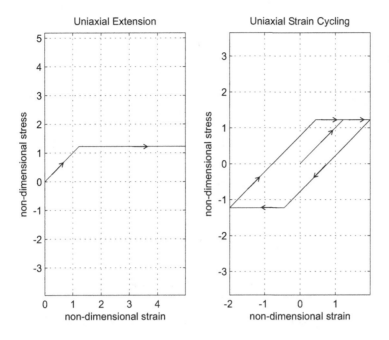

Fig. 6.6 Non-dimensional uniaxial extension and uniaxial cyclic loading response obtained by solving (6.37) together with (6.52).

A key feature of the response that is simulated above is that it is "memoryless", i.e. the previous cyclic loading history of the material does not affect its subsequent response. Thus, the model, while simple to implement (requiring only 3 constants) and extremely useful for structural problems is rather of limited use for studying response under cyclic loading. In the next section, we will consider some improved models for cyclic loading. However, this improved modeling comes at a great expense–a proliferation in the number of parameters.

6.4 Hardening and the plastic arc length

6.4.1 *Definition of hardening, softening and perfectly plastic behavior*

We have up to now formulated the structure of the elastoplastic model for a material based on just three constants. As is demonstrated in Fig. 6.6, this model under uniaxial tension gives a response in which the stress remains constant after the initial elastic increase, i.e. the material does not exhibit "hardening". While this is

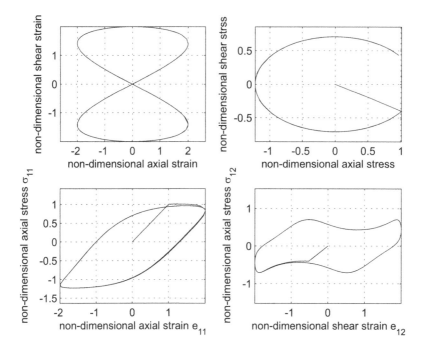

Fig. 6.7 Non-dimensional tension-torsion response obtained by solving (6.38) together with (6.52)

a very crude approximation, nevertheless this has been a very fruitful approximation since many exact solutions have been obtained for this theory (see e.g., [Prager and G (1951)]) for such classical problems as pure bending of a beam, torsion, internal pressurization, bending of wedges, and other special solutions[6].

In casual parlance, "hardening" means that the stress increases with strain even beyond yielding while perfectly plastic means that the stress does not change with strain. In other words, hardening means that the ratio between the stress rate and strain rate must be positive.

How does one generalize this to 3-D? In order to generalize the notion of one-dimensional stress-strain response, we recall that the tangent modulus formulation says that the stress rate is proportional to the strain rate, i.e. in matrix form, $[\dot{\sigma}] = (\mathcal{K})[\dot{\varepsilon}]$ where, in the presence of plastic flow $(\mathcal{K})$ is given by (6.47). Thus when the material undergoes perfectly plastic behavior, we expect that for non zero $[\dot{\varepsilon}]$ we should expect a zero stress rate. This can happen only if the determinant of $\mathcal{K}$ is zero. Otherwise the equation will be invertible. More generally, we can say that *the material is hardening if $\mathcal{K}$ is positive definite, i.e. all its eigenvalues are positive.*

---

[6]A curious feature of the solutions is that they depend actually on incompressibility of the material and not so much on small deformations. This suggests that perhaps we might be able to obtain a number of exact solutions to *finite deformation* problems assuming full incompressibility.

⊙ **6.9 (Exercise:).**
Show that for $\mathcal{K}$ given by equation (6.46) on p.198, the material is not hardening, i.e. at least one of the eigenvalues of $\mathcal{K}$ is non-positive. Specifically, show that $\mathcal{K}$ has one eigenvalue that is zero.
Hint: The eigenvectors of $\mathcal{K}$ are $\boldsymbol{\tau}$ and any tensor that is perpendicular to $\boldsymbol{\tau}$.

Definition of Hardening[7]:
We now define a 3-D version of hardening as follows: We define a quantity Φ by

$$\Phi = \frac{\det(\mathcal{K})}{\det(\mathbb{C})}. \tag{6.55}$$

It is possible to show that if Φ is positive then $\mathcal{K}$ is positive definite. Hence, we stipulate that the elastoplastic material that we have considered here *hardens, softens or is a perfectly plastic at the current state if Φ is greater than, less than or equal to zero*. Thus, since the $\det(\mathcal{K})$ is zero for the model that we have considered so far, $\Phi = 0$. Therefore, it is a non-hardening model or is *perfectly plastic* model. We divide by the determinant of $(\mathbb{C})$ to remove the influence of the elastic constants and note that Φ is non-dimensional.

Since most materials show hardening behavior in a one-dimensional response[8], we still have the task of finding the full model from the one-dimensional response. For this we must model or simulate hardening. The simplest way to think about this is to imagine that hardening is an increase in the "friction force" with deformation. To understand this, let us revert to the one-dimensional case discussed in Chapter 3 on trusses, in light of the 3-D constitutive equations we have developed. Thus we have $\boldsymbol{\sigma} = \sigma \mathbf{i}_i \otimes \mathbf{i}_i$, representing uniaxial tension along the x_1 direction. Let the yield stress be σ_y. Since $\kappa = \sqrt{2/3}\sigma_y$, the increase in value of σ_y during plastic deformation can be reflected in the increase in the value of κ. Recall that in chapter 3, we had introduced the notion of the accumulated plastic strain s as a measure of how much microstructural change had occurred to the material during plastic flow and we stipulated that $\dot{\sigma}_y = f(s, \sigma_y)\dot{s}$ as the "hardening rule". We need an extension of this to the current problem, i.e. we will look for a three-dimensional equivalent for s which will reduce to the correct form in one dimension so that we can match the experimental data. How do we do this?

As the body is deformed, the plastic strain traces out a path in a six-dimensional strain space, (imagine a slug leaving a trail of slime as it crawls along). We will

---
[7] see [Casey and Naghdi (1981, 1983)] for a more in depth discussion of this measure.
[8] For a long time, this was the only type of material response thought to be reasonable until plasticity theory began to be applied to soil mechanics. Typically, rocks and soil (and certain very high strength steels like maraging steel) soften with increasing load and their simulation requires special care (see [Pietruszczak and Mroz (1981, 1983)]).

use the total arc length along this path as the measure of the accumulated plastic strain, i.e.

$$s = \int_0^t \|\dot{\varepsilon}_p\| \, dt \Rightarrow \dot{s} = \|\dot{\varepsilon}_p\|, \tag{6.56}$$

and we will propose that

$$\dot{\kappa} = \Theta(s, \kappa)\dot{s} \tag{6.57}$$

as the generalization of the 1-D hardening law.

How can we find the function $\Theta(s, \kappa)$? Generally, other than for some trivial cases, many forms for $\Theta(s, \kappa)$ can be made to agree with the experimental data. We already saw that $\kappa = \sqrt{2/3}\sigma_y$ for the uniaxial tension. What should s be? Well remember that ε_p is traceless (sum of diagonals of the matrix is zero) so that it will have the form

$$\varepsilon_p = \frac{1}{2}(2\epsilon^p_{11}\mathbf{i}_i \otimes \mathbf{i}_i - \epsilon^p_{11}\mathbf{i}_j \otimes \mathbf{i}_j - \epsilon^p_{11}\mathbf{i}_k \otimes \mathbf{i}_k) \tag{6.58}$$

for uniaxial tension. It is easy to see that in this case $\|\varepsilon_p\| = s = \sqrt{3/2}\,\|\epsilon^p_{11}\|$. Thus we can convert a stress strain diagram in 1-D into a κ-s diagram as follows:

> Given a list of experimental yield stress versus plastic strain values for monotonic yielding, simply plot $\sqrt{2/3}\sigma_y$ versus $\sqrt{3/2}\epsilon^p_{11}$. This is exactly the κ versus s function.

In other words, the x-axis is extended by a factor of $\sqrt{3/2}$ while the y-axis is contracted by the same amount. Thus $\sigma_y \times \epsilon^p_{11} = \kappa \times s$. By using this graph, we can find some suitable forms for $\Theta(s, \kappa)$.

In the materials science literature it is common to plot the *slope* of the stress-strain curve (usually referred to as Θ) against the stress beyond yield. At first sight, this might appear rather unorthodox to a mechanician. However, we must keep in mind two major facts that are taken for granted in the materials science literature. One is that when they say stress, they invariably mean yield stress under monotonic loading conditions, and since they are plotting these graphs for large strains for metals, there is hardly any difference between the total strain and the plastic strain (since the elastic strain is about 0.2% compared to a total strain of 10 or even 100%). Thus, these graphs are very useful as a means of determining the change in the yield strength with plastic arc length.

If we assume that

$$\dot{\kappa} = a(1 - \kappa/b)^c \dot{s}, \tag{6.59}$$

the value b represents the point where a yield stress plateau is reached (i.e. where $\dot{\kappa} = 0$) while a is the slope when $\kappa/b = 0$ and c is a hardening index. For a particular aluminum alloy with a proof stress of about 100MPa and an ultimate stress of about 244 MPa, the values $a = 100, b = 244$ and $c = 0.8$ provide an excellent fit to the

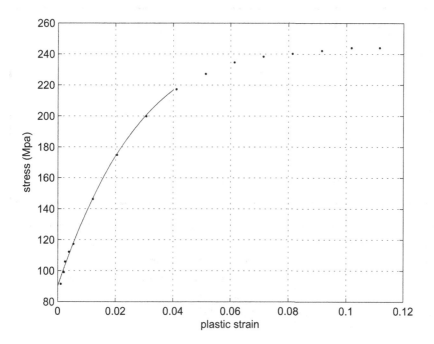

Fig. 6.8 Experimental post yield stress-strain curve of a sample of aluminum with the curve modeled by (6.59) with suitable values for the parameters a, b and c.

reported data as seen in Fig. 6.8.

If we non-dimensionalize κ with its initial value (i.e. $\sqrt{2/3}$ times the proof stress=81.64 MPa) and we non-dimensionalize the plastic arc length with the plastic strain for a yield stress of 100MPa, i.e. 0.2%, then the non-dimensional parameters are $a = 0.1625$, $b = 2.44$ and $c = 0.8$. This non-dimensionalization of the strains by 0.2% and the stress by the corresponding yield stress is very convenient for small strain yielding (i.e. localized yielding) since we are then comparing the actual strains and stresses to the yield stresses and strains. This provides a natural way to compare results for different materials.

In summary, for an aluminum part that you are trying to analyze, you need five constants: The Young's modulus : E=70GPa, the Poisson's ratio $\nu = 0.33$, the initial value of κ (initial condition for equation (6.59)) which is obtained as $\sqrt{2/3}\sigma_y = 81.64$ MPa, and the non-dimensionalized constants a, b, c which we have taken to be ($a = 0.1625$, $b = 2.44$ and $c = 0.8$) [9]. With these values, the governing constitutive equations are

---

[9] If you don't have data other than the proof stress, then assume that a, b and c are zero.

> **HARDENING J2 ELASTO-PLASTICITY MODEL**
>
> Elastic response in rate form: $\dot{\boldsymbol{\sigma}} = \mathbb{C}(\dot{\boldsymbol{\varepsilon}} - \dot{\boldsymbol{\varepsilon}}_p)$.
>
> Isotropic material: $\dot{\boldsymbol{\tau}} = 2\mu(\dot{\boldsymbol{\gamma}} - \dot{\boldsymbol{\gamma}}_p), \dot{p} = 3K(\dot{e})$.
>
> Plastic response: $\dot{\boldsymbol{\varepsilon}}_p = \dot{\boldsymbol{\gamma}}_p = \phi\boldsymbol{\tau}, \dot{\varepsilon}^p_{kk} = 0$.
>
> Yield function: $f(\boldsymbol{\sigma}, \kappa) = \boldsymbol{\tau}\cdot\boldsymbol{\tau} - \kappa^2 \Rightarrow \|\boldsymbol{\tau}\| = \kappa$. (6.60)
>
> Loading conditions: $\phi \geq 0, f \leq 0, \phi f = 0$.
>
> Accumulated plastic strain: $\dot{s} = \|\dot{\boldsymbol{\varepsilon}}_p\| = \phi\kappa$.
>
> Hardening function: $\dot{\kappa} = \Theta\dot{s}, \Theta = a(1 - \kappa/b)^c$.

We have deliberately written it in the form of rate equations since it will be convenient to implement in an ODE solver. Furthermore, comparisons with other yield functions and nonlinear problems will become easier. In order to complete the formulation and try a few simulations to understand the material response, we need to prescribe the loading conditions, i.e. the combination of stress and strain components that are known. We are now ready to simulate any homogeneous deformation.

6.5 Finding the response of the material

In this chapter, we will focus only on homogeneous deformations, relegating boundary value problems to the next chapter. Thus, we assume that the part in question is undergoing a homogeneous deformation (say, it is a thin walled tube if the ratio of the inner to the outer diameter is greater than 0.8), which is under simultaneous tension and torsion (see Fig. 6.5). The prescribed loading is given by a differential equation (in $[\boldsymbol{\sigma}]$ and $[\boldsymbol{\varepsilon}]$) of the form (6.35). We need to find the relationship between $\dot{\boldsymbol{\sigma}}$ and $\dot{\boldsymbol{\varepsilon}}$ that is implied by (6.60). The procedure is identical to the previous case and involves first finding ϕ and then finding the tangent modulus.

⊙ **6.10 (Exercise:). Finding ϕ for the hardening material**
Follow the procedure introduced earlier for finding ϕ (i.e. assume that $\phi > 0$, differentiate the yield function and substitute the rate form of the elastic response) and show that we get

$$\phi = h\hat{g}, \quad \hat{g} = \left.\frac{df}{dt}\right|_{\phi=0} = 4\mu\boldsymbol{\tau}\cdot\dot{\boldsymbol{\varepsilon}}, \quad h = \frac{1}{(\kappa^2)(4\mu + 2\Theta)}, \quad (6.61)$$

where Θ is the hardening function. Note that we recover the original value if $\Theta = 0$. (Hint: When you differentiate the yield function and substitute the elastic response, you should get $4\mu\boldsymbol{\tau}\cdot\dot{\boldsymbol{\varepsilon}} + 2(\boldsymbol{\tau}\cdot\dot{\boldsymbol{\varepsilon}}_p - \kappa\dot{\kappa}) = 0$. The first term is $\hat{g}$. Now substitute for $\dot{\boldsymbol{\varepsilon}}_p$ and $\dot{\kappa}$ and get the result.)

6.5.1 Finding the tangent modulus

Assume that you are loading the material in some fashion. Until you hit the boundary of the elastic domain, no problems occur and you assume that $\dot{\varepsilon}_p = 0$. If you hit the boundary you have to decide whether plastic flow will occur. Recall from Chapter 3, and from the discussion in the previous section that the basic idea for figuring out whether plastic flow occurs or not is that *you assume that no plastic flow occurs and see if you are still inside or on the yield surface. If you are, then continue to the next time step. If not, then backup and recalculate, now assuming that plastic flow does occur.* We have repeated this many times at the risk of overstating the point since this is the most delicate and important element of plasticity modeling.

Thus, the method works as follows: For each time step

(1) First assume that ϕ in (6.60) is zero, (i.e. no plastic flow). Then from (6.60), we compute

$$\dot{\boldsymbol{\sigma}} = \mathbb{C}(\dot{\boldsymbol{\varepsilon}}) \Rightarrow \mathcal{K} = \mathbb{C}, \tag{6.62}$$

and hence from (6.35) and (6.34) we obtain

$$\dot{\boldsymbol{\varepsilon}} = \{(\mathbb{A})(\mathbb{C}) + (\mathbb{B})\}^{-1}[\mathbf{f}]. \tag{6.63}$$

(2) Next we check if, by this assumption, we have violated any of the Kuhn Tucker conditions, i.e. we check two things, *is $f < 0$ or is $f = 0$ and $\dot{f}|_{\phi=0} < 0$?* if either of them is true then we continue to the next time step. Thus, we check

$$\boldsymbol{\tau} \bullet \boldsymbol{\tau} - \kappa^2 < 0? \tag{6.64}$$

If true, then go to the next time step. If not, we check if

$$\hat{g} = \boldsymbol{\tau} \bullet \dot{\boldsymbol{\tau}}|_{\phi=0} = 4\mu(\boldsymbol{\tau} \bullet \dot{\boldsymbol{\varepsilon}}) < 0? \tag{6.65}$$

If this is true, then we go to the next step[10].

(3) If neither (6.64) nor (6.65) are true, then we conclude that $\phi > 0$ and loading takes place. Now we use the conditions (6.60d) and require that $f = 0$ during the time step, i.e. $\dot{f} = 0$. we then have

$$\begin{aligned}\dot{\boldsymbol{\sigma}} &= \mathbb{C}(\dot{\boldsymbol{\varepsilon}} - \dot{\boldsymbol{\varepsilon}}_p) = \mathbb{C}(\dot{\boldsymbol{\varepsilon}} - \phi\boldsymbol{\tau}) \Rightarrow \dot{\boldsymbol{\tau}} = 2\mu(\dot{\boldsymbol{\gamma}} - \phi\boldsymbol{\tau}),\\ \dot{f} &= \boldsymbol{\tau}\dot{\boldsymbol{\tau}} - \kappa\dot{\kappa} = 0,\end{aligned} \tag{6.66}$$

where $\dot{\kappa}$ is given by (6.60f) and ϕ is given by (6.61) p.206.

---

[10] How did we get the expression listed above? Note that for the material in question, $\boldsymbol{\tau} = 2\mu(\boldsymbol{\varepsilon} - \boldsymbol{\varepsilon}_p)$. Differentiate it, keeping $\boldsymbol{\varepsilon}_p$ fixed.

(4) We now have the general result for the tangent modulus when there is plastic flow

$$\dot{\boldsymbol{\sigma}} = \mathbb{C}\left\{\dot{\boldsymbol{\varepsilon}} - \frac{2\mu\boldsymbol{\tau}\bullet\dot{\boldsymbol{\varepsilon}}}{2\mu\kappa^2 + a\kappa^2\left(1 - \kappa/b\right)^c}\boldsymbol{\tau}\right\}$$
$$= \mathbb{C}\left\{\mathbb{I} - \frac{2\mu\boldsymbol{\tau}\otimes\boldsymbol{\tau}}{2\mu\kappa^2 + a\kappa^2\left(1 - \kappa/b\right)^c}\right\}\dot{\boldsymbol{\varepsilon}}. \qquad (6.67)$$

Here $\mathbb{I}$ stands for the fourth order symmetric identity tensor.

We can see that even the simple elasto-plastic model with hardening requires a fair bit of calculation. The key result is (6.67). Computationally, it is generally not very suitable to solve (6.67) since we anyway need a number of intermediate parameters such as ϕ. It is much simpler to keep ϕ as a variable and solve for it at each time step. For completeness, we will list the set of differential equations that we need to solve:

$$\begin{aligned}
\text{Strain rate: } & [\dot{\boldsymbol{\varepsilon}}] = \{(\mathbb{A})(\mathcal{K}) + (\mathbb{B})\}^{-1}[\mathbf{f}(t)], \\
\text{Stress rate: } & [\dot{\boldsymbol{\sigma}}] - (\mathcal{K})[\dot{\boldsymbol{\varepsilon}}] = 0, \\
\text{Plastic arc length: } & \dot{s} - \phi\kappa = 0, \\
\text{Hardening law: } & \dot{\kappa} - (as - b\kappa)^c \dot{s} = 0,
\end{aligned} \qquad (6.68)$$

where

$$\phi = \begin{cases} 0, & \text{if } f < 0 \text{ or } f = 0 \text{ \&} \\ & [\boldsymbol{\tau}]^T\left\{(\mathbb{A})(\mathbb{C}) + (\mathbb{B})\right\}^{-1}[\mathbf{f}] < 0; \\ \left(\frac{2\mu\boldsymbol{\tau}\bullet\dot{\boldsymbol{\varepsilon}}}{2\mu\kappa^2 + (as - b\kappa)^c}\right) & \text{otherwise.} \end{cases} \qquad (6.69)$$

In the above equations $(\mathcal{K}) = (\mathbb{C})$ if $\phi = 0$, and, $(\mathcal{K})$ is given by (6.67) if $\phi > 0$.

These form a system of 14 equations for 14 unknowns ($\boldsymbol{\sigma}, \boldsymbol{\varepsilon}, s, \kappa$). Notice that $\boldsymbol{\varepsilon}_p$ is NOT one of the variables that we solve for, since it can be completely eliminated from the equations. MATLAB can easily handle this using ODE23 (including the fact that ϕ is given by a discontinuous function). A MATLAB code can be written for this 14 equations ODE set is solved for (for a sample code, see authors' websites in the preface). Figures 6.9 and 6.10 show the material response for a uniaxial extension and a cyclic uniaxial loading for a material with hardening. The second solution is for a cyclic tension torsion with the tension and torsion out of phase[11]. In both cases, note the fact that the material has "memory", the previous cycles do influence the response of the material. This is a consequence of the hardening introduced in the previous section. Compare the results with [Naghdi and Nikkel (1984, 1986); Khan and Parikh (1986); Khan et al. (2007); Khan and Cheng (1996)].

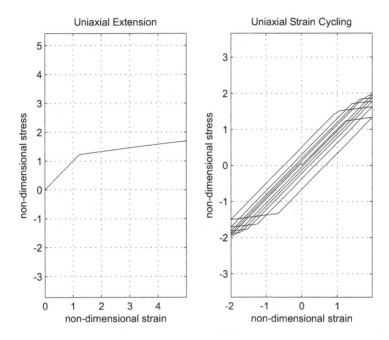

Fig. 6.9 A thin walled tube of aluminum subject to monotonic and cyclic tension. Note the fact that the material hardens and that the cyclic response shows the "memory", i.e. the influence of the previous cycles. Compare with Fig. 6.6 to see the differences.

Before we end this chapter, we note that we have only considered homogeneous deformations. In the next chapter, we will introduce boundary value problems where we have to solve the equations of equilibrium together with the plasticity laws.

6.6 Summary

(1) The response of a plastic material was originally obtained as a generalization of that of a viscous fluid by modifying the inverse of the viscosity (see equations (6.10)).
(2) This model is very simple, requires only one parameter (the yield strength) and allows for the calculation of the stress at any point where the symmetric part of the velocity gradient is not zero (see equations (6.13)).
(3) The viscosity in this model is inversely proportional to the magnitude of the deformation rate, $\mathbf{D}$, and hence goes to infinity as $\mathbf{D}$ goes to zero. Thus, the stress response is rate independent. We refer to this model as *J2 rigid plasticity*.

---

[11] Out of phase loading implies that when two load parameters are varied, one goes to a maximum when the other is a minimum.

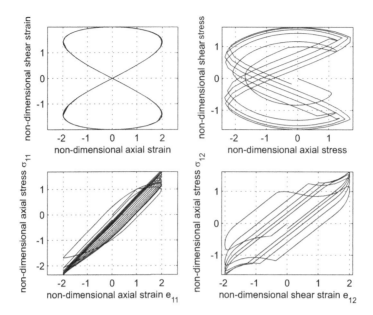

Fig. 6.10 A thin walled tube subject to a combination of tension and torsion. Notice the fact that the axial stress versus strain shows a gradually decreasing hysteresis. Also observe the rather complicated path in the axial stress versus shear stress graph. As the cycling processes, the response becomes increasingly more "elastic", i.e. the hysteresis decreases due to hardening and the stress path begins to resemble the strain path.

(4) Small strain *J2 elasto-plasticity* is a modification of the above structure where the total strain is written as a sum of the elastic and plastic strains, with the elastic strain given by Hooke's law (6.27) while the plastic strain obeys the J2 rigid plasticity laws (6.28) introduced earlier.

(5) It is possible to find an expression for the rate of stress in terms of the rate of strain in such a model. The matrix that connects these two quantities is called the *material tangent stiffness matrix*. It is, however, NOT a function of the current state of the material (i.e. of the current values of $\boldsymbol{\sigma}$ and $\boldsymbol{\varepsilon}$ but depends upon whether plastic flow occurs or not). When the plastic flow does not occur, it is equal to the elastic modulus matrix. When the plastic flow occurs, its explicit form is given by equation (6.46).

(6) Whether or not plastic flow occurs is determined by the value of two scalars: f and $\hat{g}$. Plastic flow occurs only if you are on the yield surface and stepping out, i.e. if $f = 0$ and $\hat{g} = df/dt|_{\phi=0} > 0$.

(7) For a given stress or strain control, these quantities can be written and a simple MATLAB program using ODE23 (the MATLAB ODE solvers) can be written to find the response of these materials.

(8) Hardening or softening is decided based on the eigenvalues of the tangent matrix. If they are all positive then the material is currently hardening. For the

model considered here, there is only one eigenvalue that may change sign. So the nature of the response can be determined by the value of the determinant of the tangent matrix.

(9) It is possible to model some aspects of hardening by using the notion of a plastic arc length (see equation (6.56)) and then write an equation for the rate of change of the yield strength (see equation (6.59)).

(10) The tangent matrix is now of a slightly different form and involves the function Θ (see equation (6.60)).

6.7 Projects

(1) **Isotropic hardening laws**
Find the book "Atlas of stress-strain curves" [International (2002)] and pick three monotonic stress-strain curves. Find a suitable hardening law of the form $\dot{\kappa} = \Theta(s, \kappa)\dot{s}$ that will provide a fit for the data. (Try a law of the form $\Theta(s, \kappa) = a(1 - \kappa/b)^c$.)

(2) **Simulation of complex homogeneous motions**
Write your own MATLAB code to solve the differential equations (6.68) and (6.69) for different kinds of loading such as tension-torsion and internal pressurization. (See [Caulk and Naghdi (1978); Naghdi and Nikkel (1984, 1986)].)

6.8 Homework

(1) **Strain space yield conditions:** Show that for an elastic plastic material, the yield function $f(\tau) = \tau \cdot \tau - \kappa^2$ can be written as $\hat{g}(\varepsilon, \varepsilon_p) = 4\mu^2(\gamma - \varepsilon_p) \cdot (\gamma - \varepsilon_p) - \kappa^2$. This is called a *strain space yield function*. While this is not generally used directly for small deformations, it is very useful for inelasticity of polymers where the strains are large and many microstructural features are naturally related to the strain of the material.

(2) **Emergence of $\hat{g}$:** Show by direct calculation, that $\dot{\hat{g}} = \frac{\partial g}{\partial \varepsilon} \cdot \dot{\varepsilon}$. Thus even when you use yield criteria in terms of stresses, deciding whether plastic flow occurs is based on strains.

(3) **Hardening and softening:** For the tangent modulus defined by (6.67), show that Φ defined by (6.55) becomes $\Phi = \Theta/(2\mu + \Theta)$ Thus the sign of Φ depends upon Θ which is related to the uniaxial hardening response.

Chapter 7

The Boundary Value Problem for J2 Elastoplasticity

Learning Objectives

After studying this chapter, you should be able to

(1) derive the governing equations for a J2 elastoplastic body and simplify them for the case of plane stress and plane strain conditions;

(2) derive the compatibility conditions for the plane stress and plane strain cases and reduce the problem to the one involving the Airy stress function and the biharmonic operator;

(3) solve for the plastic strain for a given total strain for the 3-d and the plane strain case;

(4) obtain the yield surface for plane stress and describe, using geometrical arguments as to why the algorithm that works for 3-D and plane strain would not work for the plane stress case.

CHAPTER SYNOPSIS

In this chapter, we extend the considerations of Chapter 6 to include boundary value problems for small deformation. The resulting equations are like that of elasticity with a body force (which depends upon the gradients of plastic strain). When this is specialized to plane strain or plane stress, and the Airy stress function (A) is introduced, we end up with an inhomogeneous biharmonic equation which is of the form

$$\nabla^4 A = g(x,y), \quad g(x,y) = E(\epsilon^p_{xy,xy} - \epsilon^p_{xx,yy} - \epsilon^p_{yy,xx}) \qquad (7.1)$$

This can be solved by an iterative scheme that cycles between solving an elasticity problem and updating the plastic strains. A finite difference scheme is introduced and the problem of a plate with a square hole is solved.

Chapter roadmap

This is one of the chapters that is mostly elasticity. Our experience has been that students who have taken an elasticity class can completely learn the material in this chapter by themselves. This chapter can be considered as a supplementary material for a class project involving the elastoplastic solution of a plate with a hole using a finite difference scheme.

The chapter begins with the derivation of the equations of equilibrium for an isotropic elastoplastic body. We then specialize it to plane stress and plane strain and show that while the equations of equilibrium are similar, the yield function takes slightly different forms depending upon whether it is plane stress or plane strain. We then introduce the Airy stress function and obtain the governing equations for a plane stress and plane strain.

Since most students are unaware of the boundary conditions that are needed for the Airy function formulation, we introduce these next [1]. We finally end the chapter with a brief description of the finite difference method for the Airy function drawing heavily upon the material in [Timoshenko and Goodier (1970)]. We use this to solve the problem of a plate with a square hole.

---

[1] we follow closely [Sadd (2005)], Chapters 7 and 8, and [Timoshenko and Goodier (1970)].

7.1 The governing equations

We are now in a position to consider the general boundary value problem for a three-dimensional elastoplastic body obeying **J2 Plasticity with isotropic strain hardening**. Unlike the case of the beam problem, we will not *derive* the governing equations but simply state them, leaving the derivation as exercises that can be used as a means to recollect developments in elasticity. It turns out that a surprisingly large number of results that were developed in linear elasticity carry over without modifications to the case of elastoplastic materials. We shall see that the governing equations appear identical to that of an elastic body with a body force (which is a function of the plastic strain). However, analytical solutions are not possible unless we make some simplifying assumptions (such as incompressibility). On the other hand, we can carry out some numerical solutions which will allow us to explore the model in great detail.

7.1.1 3-D case

We begin with a quick summary of the 3-D governing equations in local form. Consider a body $\mathbb{B}$ with boundary $\partial \mathbb{B}$. On one part $\partial \mathbb{B}_u$ of the boundary, displacements are specified while on the remaining part $\partial \mathbb{B}_f$, tractions are specified.

The displacement field in the body is given by $\mathbf{u}(\mathbf{x}) = u_i(x,y,z)\mathbf{i}_i$, and the stresses are given by $\boldsymbol{\sigma} = \sigma_{ij}\mathbf{i}_i \otimes \mathbf{i}_j$.[2]

> ⊙ **7.1 (Exercise:). Local forms of the equations of equilibrium**
> *Draw a small volume element and with the classical arguments found in any elasticity book, show that the local forms of the balance of linear and angular momentum (if we ignore inertial and gravitational effects) are given by*
> $$\text{div } \boldsymbol{\sigma} = 0, \quad \boldsymbol{\sigma} = \boldsymbol{\sigma}^T, \quad \Rightarrow \quad \sigma_{ij,j} = 0, \quad \sigma_{ij} = \sigma_{ji}. \qquad (7.2)$$

Recall from Chapter 6 on J2 plasticity that for a small strain elastoplastic body
1. the total strain is assumed to be the sum the elastic and the plastic strains;
2. the trace of the plastic strain is zero; and,
3. the stress is related to the elastic strain by means of Hooke's law, i.e. that

$$\begin{aligned}
\varepsilon_{ij} &= \frac{1}{2}(u_{i,j} + u_{j,i}) = \varepsilon^e_{ij} + \varepsilon^p_{ij} \; , \\
\varepsilon^p_{ii} &= 0 \; , \\
\sigma_{ij} &= \lambda \delta_{ij}\varepsilon_{kk} + 2\mu\varepsilon_{ij} - 2\mu\varepsilon^p_{ij} \; ,
\end{aligned} \qquad (7.3)$$

---

[2] Segel's book [Segel (2007)] is an excellent and inexpensive resource for a great and short introduction to index notation, and elasticity.

where $\mu = E/2(1+\nu)$ and $\lambda = \mu[2\nu/(1-2\nu)]$ are the Lamé constants. The equations can be easily inverted to write the strain in terms of the stress as

$$\varepsilon_{ij} = \frac{\sigma_{ij}}{2\mu} - \frac{\lambda \sigma_{kk} \delta_{ij}}{2\mu(3\lambda + 2\mu)} + \varepsilon_{ij}^p = \frac{1+\nu}{E}\sigma_{ij} - \frac{\nu}{E}\sigma_{kk}\delta_{ij} + \varepsilon_{ij}^p. \qquad (7.4)$$

⊙ **7.2 (Exercise:). Displacement equations of equilibrium**
Substitute (7.3) into (7.2) and show that the equations of equilibrium can be written as

$$\mu u_{i,jj} + (\lambda + \mu)(u_{j,ij}) = 2\mu \varepsilon_{ij,j}^p \;, \qquad (7.5)$$

which, in direct notation reduces to

$$\mu \nabla^2 \mathbf{u} + (\lambda + \mu)\nabla(\nabla \cdot \mathbf{u}) = 2\mu \nabla \cdot \mathbf{e}^p \;. \qquad (7.6)$$

If we compare this with the Navier's equations of equilibrium for an isotropic elastic body, we see that the elastoplastic boundary value problem is one which looks like an elasticity problem with a nonzero body force. Of course, the problem is that the plastic strain depends upon the stresses and strains through constitutive laws involving the yield stress and hardening parameter so that it cannot be directly solved. Nevertheless, the following iterative scheme can be used to solve an elastoplastic problem:

ITERATIVE SOLUTION OF ELASTOPLASTICITY PROBLEM

Assume that the external boundary conditions are applied in increments $i = 1, 2, 3, 4, \ldots$ beginning from zero. Assume that at the zeroth increment, the plastic strain field is zero. Now for i=1,2,3,4

(1) Assume that the plastic strain field is set as the value at the end of the previous increment
(2) Calculate $2\mu \nabla \cdot \varepsilon_p$.
(3) Apply the ith increment of the boundary conditions and solve (7.6) as you would for any elasticity problem. (This can be done by any numerical scheme that you like.)
(4) Calculate the new strain field $\varepsilon(\mathbf{x}) = 1/2(\nabla \mathbf{u} + \nabla \mathbf{u}^t)$.
(5) Now use the evolution laws of J2 plasticity equation (6.60) on p.206 to update the plastic strains. (This step requires the MATLAB programs of Chapter 6 or other techniques)
(6) If the change in the plastic strain is negligible, $i \leftarrow i+1$ and go to STEP 1.
(7) ELSE, update the plastic strain field and Go to STEP 2. This method of solution allows us to use any elasticity solver to solve small deformation plasticity problems.

The scheme outlined involves two major procedures: (1) solving elasticity problems and, (2) updating plastic strains. If you have ways of doing these two things, then the scheme can be used. However, it is not very efficient but is simple to implement and is quite robust. But trying this on a 3-D problem is usually not advised! Contrary to what you might think, three-dimensional inelasticity problems are still not simple to solve using canned packages. It is well worth your effort to simplify (defeature) the problem so that it can be considered either as plane stress or plane strain. Then you can actually do a lot more. In preparation for this, we will discuss compatibility equations next.

7.1.2 Compatibility equations

A key point to note is that the strain compatibility conditions are conditions that are purely kinematic in nature and are hence independent of the specific constitutive model. Thus, the six strain compatibility conditions are exactly the same as that for elasticity, namely,

$$\varepsilon_{ij,kk} + \varepsilon_{kk,ij} - \varepsilon_{ik,jk} - \varepsilon_{jk,ik} = 0. \tag{7.7}$$

⊙ **7.3 (Exercise:). Compatibility conditions in terms of the stress**
Substitute (7.4) into (7.7) and using (7.2) show that the compatibility conditions in terms of the stress reduces to

$$(1+\nu)\sigma_{ij,kk} + \sigma_{kk,ij} = E\left(\varepsilon^p_{ik,jk} + \varepsilon^p_{jk,ik} - \varepsilon^p_{ij,kk} + \nu\varepsilon^p_{mk,mk}\delta_{ij}\right) \tag{7.8}$$

Comment: This is not easy to do and is a test of your prowess in using the equations of equilibrium and index notation. You can think about it as an exercise in getting familiar with index notation.

For reference in developing solution schemes we summarize here the flow rule, the hardening rule and the Kuhn Tucker conditions for the three-dimensional case as follows:

> **SUMMARY OF FLOW RULE WITH VOCE HARDENING**
>
> $$\text{Yield Condition}: \quad f(\boldsymbol{\tau},\kappa) = \frac{\sqrt{\boldsymbol{\tau}\bullet\boldsymbol{\tau}}}{\kappa} - 1 = 0$$
>
> $$\text{Flow Rule}: \quad \dot{\boldsymbol{\varepsilon}}_p = \bar{\phi}\frac{\boldsymbol{\tau}}{\kappa^2}$$
>
> $$\text{Plastic arc Length}: \quad \dot{s} = \sqrt{3/2}\,\|\dot{\boldsymbol{\varepsilon}}_p\| = \sqrt{3/2}\,\bar{\phi}/\kappa$$
>
> $$\text{Hardening Rule (for Voce hardening)}: \quad \kappa = \sqrt{\frac{2}{3}}(\sigma_{y0} + as + b(1 - e^{-s/s0})).$$
>
> $$\text{Kuhn Tucker Conditions}: \quad \bar{\phi} \geq 0,\ f \leq 0,\ \bar{\phi}f = 0,$$
>
> $$\text{Consistency Condition}: \quad \bar{\phi} > 0 \Rightarrow \dot{f} = 0.$$
>
> (7.9)

7.2 Plane problems

Compared to the solution of three-dimensional problems, the solution to two-dimensional ones are much easier and are also applicable to a wide range of practical problems. This is so because, as was done in elasticity, we can use the Airy stress function to reduce the problem to finding a single scalar function. Thus, the above set of equations can be specialized for the case of plane stress and plane strain and this is the task that we turn to next.

In this regard, the equations and solution procedure for plane strain turn out to be different from that of plane stress due to the fact that the stress component σ_{33} plays an important role in determining the plastic response. The equations for plane stress turns out to have other challenges that must be dealt with separately.

7.2.1 *Plane strain*

We consider a prismatic body of length $2h$. The cross section will be referred to as $\mathbb{S}$ and the lateral surface is referred to as $\partial\mathbb{B}_l$. We attach a coordinate system to the centroid of the body with the z axis along the axis of the body and, for definiteness, the x and y along the principal directions of the cross sectional area moment of inertia[3] as shown in Fig. 7.1.

We assume that

1. The ends $z = \pm h$ are confined between rigid frictionless planes so that $\mathbf{u}_3(x,y,\pm h) = 0$ and $\sigma_{13}(x,y,\pm h) = \sigma_{23}(x,y,\pm h) = 0$.

2. The tractions specified along the lateral boundary $\partial\mathbb{B}_l$ are such that the components of the traction in the z-direction is always zero.

---

[3] Recall that this was the ideal choice of coordinate system for solving torsion and flexure problems in elasticity.

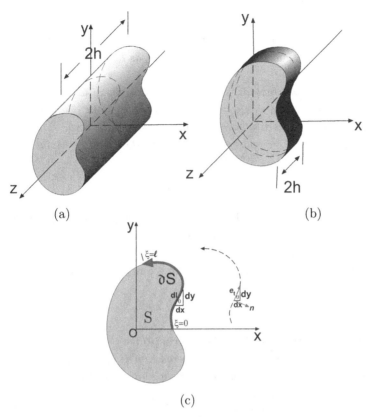

Fig. 7.1 (a) a two-dimensional prismatic body $\mathbb{B}$ whose depth in the z-direction is substantially greater than its cross sectional dimensions and which will be approximated by plane strain, (b) a two-dimensional slab-like body $\mathbb{B}$ whose depth in the z-direction is substantially smaller than its cross sectional dimensions and which will be approximated by plane stress, (c) the cross section marking the tangent, normal vectors as well as showing the relationship with the arc length along the edge. Note that $\sin\theta = dy/dl$, $\cos\theta = dx/dl$ and that the tangent and normal vectors are given by $\mathbf{e}_t = \cos\theta \mathbf{i}_1 + \sin\theta \mathbf{i}_2$ and $\mathbf{n} = -\sin\theta \mathbf{i}_1 + \cos\theta \mathbf{i}_2$.

7.2.1.1 Reduced constitutive equations:

Under these conditions, we can assume that the displacements are given by $\mathbf{u}(x,y,z) = u_1(x,y)\mathbf{i}_1 + u_2(x,y)\mathbf{i}_2$ which we write collectively as $\mathbf{u} = u_\alpha \mathbf{e}_\alpha$ where the Greek index α takes the values $1, 2$. In fact, for future reference, we shall now stipulate that all Greek indices take the values $1, 2$ and all Latin indices take the values $1, 2, 3$. This way we can specialize all the equations for the two-dimensional case without having to constantly remind ourselves as to how far to sum the indices. With the above assumptions, the only nonzero strain components are the planar normal strains $\varepsilon_{11} := u_{1,1}$, $\varepsilon_{22} := u_{2,2}$ and the shear strain $\varepsilon_{12} := \frac{1}{2}(u_{1,2} + u_{2,1})$.

⊙ **7.4 (Exercise:). Nonzero components of the plastic strain**
Assume that $\varepsilon_p = 0$ at time $t = 0$, and show that, as long as $\varepsilon_{13} = \varepsilon_{23} = 0$ so are $\varepsilon^p{}_{13}$ and $\varepsilon^p{}_{23}$.
Hint: Show that if at any instant, $\sigma_{13} = \sigma_{23}$ is zero, then so is $\dot\varepsilon^p_{13}$, and $\dot\varepsilon^p_{23}$. Show also further that if $\dot\varepsilon_{13} = \dot\varepsilon_{23} = 0$ then $\dot\sigma_{13} = \dot\sigma_{23} = 0$. Thus argue that once these quantities are zero at any instant, they remain zero. The tangent modulus $\mathcal{K}$ that you found in equation (6.67) on p.208 would be useful for your purposes.

Thus, for plane problems, the only nonzero plastic strains are $\varepsilon^p{}_{\alpha\beta}$ as well as $\varepsilon^p{}_{33}$. However, in view of the plastic incompressibility condition, we don't have to compute $\varepsilon^p{}_{33}$ separately, but simply use $\varepsilon^p{}_{33} = -(\varepsilon^p{}_{\alpha\alpha})$.

Thus the constitutive equations for the in-plane components of the stress reduce to

$$\sigma_{\alpha\beta} = \lambda\delta_{\alpha\beta}\varepsilon_{\gamma\gamma} + 2\mu\varepsilon_{\alpha\beta} - 2\mu\varepsilon^p{}_{\alpha\beta}. \tag{7.10}$$

⊙ **7.5 (Exercise:). Plane strain elastic response**
Use (7.3) and the plastic incompressibility condition (i.e. that $\varepsilon^p_{kk} = 0$ to find σ_{33} in terms of $\varepsilon_{\alpha\beta}$ and $\varepsilon^p{}_{\alpha\beta}$ and hence find the counterpart of (7.4) for plane strain, i.e. write $\varepsilon_{\alpha\beta}$ in terms of $\sigma_{\alpha\beta}$ and $\varepsilon^p{}_{\alpha\beta}$).
Hint: Rather than use the index notation version of (7.4), write out the terms and solve for σ_{33} in terms of $\sigma_{\alpha\beta}$ and $\varepsilon^p{}_{\alpha\beta}$. You should get

$$\sigma_{33} = \nu(\sigma_{\gamma\gamma} + E(\varepsilon^p{}_{\gamma\gamma})). \tag{7.11}$$

Ans: You should have obtained
$$\varepsilon_{\alpha\beta} = \frac{1+\nu}{E}\sigma_{\alpha\beta} + \varepsilon^p{}_{\alpha\beta} - \nu\left(\frac{1+\nu}{E}\sigma_{\gamma\gamma} + \varepsilon^p{}_{\gamma\gamma}\right)\delta_{\alpha\beta} \tag{7.12}$$

We now turn to the flow rule for the plastic strain rate and note that this involves the *deviatoric stress*. Since the deviatoric stress is obtained by subtracting the mean normal stress, leading to terms that have *normal stress differences of the form* $\sigma_{11} - \sigma_{22}$ etc. This means that the stress component σ_{33} *plays a major role in determining the rate of plastic strain*.

⊙ **7.6 (Exercise:). Role of σ_{33} in the flow rule**
Write out explicitly all the components of plastic strain rate in the flow rule (7.9) in terms of the corresponding components of the stress. Which of the components of $\dot\varepsilon_p$ depend explicitly on σ_{33}?

Balance Laws:

Note that the constitutive equation (7.10) looks remarkably like the constitutive equation (7.3) for the full three dimensional theory. This will also turn out to be true for the displacement equations of equilibrium:

> ⊙ **7.7 (Exercise:). The displacement equations of equilibrium**
> *By using the constitutive equations (7.10) show that the displacement equations of equilibrium reduce to*
> $$\mu u_{\alpha,\beta\beta} + (\lambda + \mu) u_{\beta,\alpha\beta} = 2\mu \varepsilon^P_{\alpha\beta,\beta}. \qquad (7.13)$$
> *Compare with the displacement equations of equilibrium for the full 3-dimensional problem.*
> *Remark: The similarity with the full equations is one of the reasons that many computer codes are tried out in plane strain.*

Thus for plane strain, we solve the above equations (7.13) together with the plastic flow equations (7.9). Note that at every stage we need to actually find σ_{33} and use it in the flow rule.

The situation with plane stress case is quite different as we shall see now.

7.2.2 Plane stress

We again consider a prismatic body of length $2h$ with the cross section will be referred to as $\mathbb{S}$ and the lateral surface is referred to as $\partial \mathbb{B}_l$. We attach a coordinate system to the centroid of the body with the z axis along the axis of the body and, for definiteness, the x and y along the principal directions of the cross sectional area moment of inertia as shown in Fig. 7.1. In this case however, we assume that h *is much smaller than the typical dimensions of the cross section*. We now consider the following boundary conditions

We assume that
1. The ends $z = \pm h$ free of traction so that $\sigma_{13}(x, y, \pm h) = \sigma_{23}(x, y, \pm h) = \sigma_{33}(x, y, \pm h) = 0$
2. The tractions are specified along the lateral boundary $\partial \mathbb{B}_l$ and are such that the components of the traction in the z-direction is always zero.

Compared to the plane strain case, the situation is somewhat less clear here. Since the faces are traction free and the body is not very thick in the z-direction, it would be reasonable to assume that $\sigma_{13}(x, y, z) = \sigma_{23}(x, y, z) = \sigma_{33}(x, y, z) = 0$ and that the other stress components are only functions of x and y. This turns out to violate the stress compatibility conditions even for elasticity. However, for practical purposes, if the thickness is not too great, one can show that the error is not too great with this approximation (see [Timoshenko and Goodier (1970)] for a detailed explanation for the case of elasticity).

One can instead approach it like a "plate" approximation and simply calculate the value of the stresses, displacements etc. averaged across the thickness. In either case, there is a certain approximation associated with this.

Under these conditions, the equilibrium equation in the z- direction is automatically satisfied and we need to satisfy only two equations of equilibrium, i.e. the equations of equilibrium reduce to

$$\sigma_{\alpha\beta,\beta} = 0. \tag{7.14}$$

Given that the only nonzero stresses are $\sigma_{\alpha\beta}$, the *in-plane strain-stress relations* (7.4) reduce to

$$\varepsilon_{\alpha\beta} = \frac{1+\nu}{E}\sigma_{\alpha\beta} - \frac{\nu}{E}\sigma_{\gamma\gamma}\delta_{\alpha\beta} + \varepsilon^p{}_{\alpha\beta}. \tag{7.15}$$

⊙ **7.8 (Exercise:). The Stress-Strain relations**
By inverting the above expression (i.e. by solving for $\sigma_{\alpha\beta}$ in terms of $(\varepsilon_{\alpha\beta} - \varepsilon^p_{\alpha\beta})$), show that the stress strain relations can be written in the form

$$\sigma_{\alpha\beta} = \lambda^*\delta_{\alpha\beta}(\varepsilon_{\gamma\gamma} - \varepsilon^p_{\gamma\gamma}) + 2\mu(\varepsilon_{\alpha\beta} - \varepsilon^p_{\alpha\beta}), \tag{7.16}$$

where $\lambda^* = \frac{2\lambda\mu}{\lambda+2\mu}$
Remark: Note that this form is quite different from that of plane strain where $\varepsilon^p{}_{\gamma\gamma}$ did not appear in the stress-strain equations

Thus the equations of equilibrium are met by the planar stress components which in turn can be written in terms of the in-plane displacements alone. This does not mean that all the other strains and plastic strains are zero. Specifically, the fact that $\sigma_{33} = 0$ implies that u_3 is nonzero. This gives rise to all kinds of problems.

⊙ **7.9 (Exercise:). Out of plane strain displacement relations**
Show that the three strain displacement relations involving the z direction cannot be met in this approximation.
Remark: See the extensive discussion in [Timoshenko and Goodier (1970)] in this regard.

Note that the constitutive equation (7.16) looks remarkably like the constitutive equation (7.3) for the full three-dimensional theory. In fact, so do the displacement equations of equilibrium:

> ⊙ **7.10 (Exercise:). The displacement equations of equilibrium**
> By using the constitutive equations (7.16) show that the displacement equations of equilibrium reduce to
> $$\mu u_{\alpha,\beta\beta} + (\lambda^* + \mu)u_{\beta,\alpha\beta} = \lambda^* \varepsilon^p{}_{\gamma\gamma,\alpha} + 2\mu\varepsilon^p{}_{\alpha\beta,\beta}. \qquad (7.17)$$

Systematic reduction of the flow rule for plane stress

We first need to write the yield condition in terms of the in-plane components of the stress. This is a relatively easy task. We begin by noting that the components of the stress for plane stress forms a 2×2 matrix. So that the state of stress, strain and plastic strain can be conveniently written in the matrix form

$$[\boldsymbol{\sigma}]' = [\sigma_{11}, \sigma_{22}, \sqrt{2}\sigma_{12}]^T,$$
$$[\boldsymbol{\varepsilon}]' = [\varepsilon_{11}, \varepsilon_{22}, \sqrt{2}\varepsilon_{12}]^T, \qquad (7.18)$$
$$[\boldsymbol{\varepsilon}_p]' = [\varepsilon^p_{11}, \varepsilon^p_{22}, \sqrt{2}\varepsilon^p_{12}]^T,$$

where we have used the notation $(\cdot)'$ to indicate a tensor whose components along the z direction have been suppressed. It is important to note that for the stress the suppressed terms are zero but for the total and plastic strains, the suppressed terms are the normal strains in the z direction and *are not zero but can be calculated in terms of the components listed in (7.18)*.

> ⊙ **7.11 (Exercise:). The plane stress elastic modulus**
> Show either from (7.15) or from (7.16) that
> $$[\boldsymbol{\sigma}]' = \begin{bmatrix} \sigma_{11} \\ \sigma_{22} \\ \sqrt{2}\sigma_{12} \end{bmatrix} = \frac{E}{1-\nu^2} \begin{pmatrix} 1 & \nu & 0 \\ \nu & 1 & 0 \\ 0 & 0 & (1-\nu) \end{pmatrix} \begin{bmatrix} (\varepsilon - \varepsilon^p)_{11} \\ (\varepsilon - \varepsilon^p)_{22} \\ \sqrt{2}(\varepsilon - \varepsilon^p)_{12} \end{bmatrix} \qquad (7.19)$$
> $$= (\mathcal{C}_s)[\boldsymbol{\varepsilon} - \boldsymbol{\varepsilon}_p],$$
> where $\mathcal{C}_s$ is the plane stress elastic stiffness matrix.
> What is the corresponding form for plane strain?

In the subsequent developments we will use the matrix notation introduced earlier for all our calculations since this is the form that is most useful and concise.

7.12 (Exercise:). The deviatoric stress

Show that, for plane stress, the deviatoric stress is of the form

$$(\tau) = \frac{1}{3}\begin{pmatrix} 2\sigma_{11} - \sigma_{22} & 3\sigma_{12} & 0 \\ 3\sigma_{12} & 2\sigma_{22} - \sigma_{11} & 0 \\ 0 & 0 & -\sigma_{11} - \sigma_{22} \end{pmatrix} \qquad (7.20)$$

What is the vector notation for $[\tau]'$?

Remark: Note that, even for plane stress, the deviatoric part has a 33 component. Thus, the flow rule will imply that the plastic strain has a 33 component. However, for the plane strain and plane stress cases, this component can be eliminated using the plastic incompressibility condition. This is what simplifies the situation.

7.13 (Exercise:). The 2-D stress deviator
Show that

$$\tau \bullet \tau = \sigma \bullet \tau. \qquad (7.21)$$

Using this, show that for plane stress, the von-Mises yield condition can be written using matrix notation as

$$f(\sigma) = \frac{\sqrt{[\sigma]'^T[\tau]'}}{\kappa} - 1 = \frac{\sqrt{[\sigma]'^T(\mathcal{Y}_p)[\sigma]'}}{\kappa} - 1, \qquad (7.22)$$

where

$$(\mathcal{Y}_p) = \frac{1}{3}\begin{pmatrix} 2 & -1 & 0 \\ -1 & 2 & 0 \\ 0 & 0 & 3 \end{pmatrix} \qquad (7.23)$$

is the *plane stress projection operator*.

Remark: The main point of the above result is that, even though the stress deviator is a 3×3 matrix, the yield function for plane stress can be written in terms of only the planar components. The price you pay is that the yield condition for plane stress is not a circle (as it is for plane strain) anymore but is an ellipse.

When we now consider the plasticity equations, we need to find the plastic arc length parameter (to account for hardening). For the case of plane stress, this is not simply the magnitude of the column vector $[\dot{\varepsilon}_p]'$ since the 33 component has been ignored. However, the plastic incompressibility condition implies that $\dot{\varepsilon}^p_{33} = -(\dot{\varepsilon}^p_{11} + \dot{\varepsilon}^p_{22})$ so that we can find the plastic arc length in terms of the planar components of ε_p.

⊙ **7.14 (Exercise:). The plasticity equations in plane stress**
Show that for plane stress, the flow rule, the arc length parameter and the Kuhn Tucker conditions are given by

$$[\dot{\varepsilon}_p]' = \frac{\bar{\phi}(\mathcal{Y}_p)[\sigma]'}{\kappa^2}$$

$$\dot{s} = \sqrt{2(\dot{\varepsilon}^{p\,2}_{11} + \dot{\varepsilon}^{p\,2}_{22} + \dot{\varepsilon}^{p\,2}_{12} + \dot{\varepsilon}^p_{11}\dot{\varepsilon}^p_{22})} = \sqrt{[\dot{\varepsilon}_p]'^T(\mathcal{Y}_p)^{-1}[\dot{\varepsilon}_p]'} = \frac{\bar{\phi}}{\kappa}, \quad (7.24)$$

$$\bar{\phi} \geq 0, f \leq 0, \bar{\phi}f = 0.$$

Thus, for the plane stress case, we need to solve (7.17) together with (7.24)

⊙ **7.15 (Exercise:). The consistency parameter $\bar{\phi}$**
Differentiate the yield function $(7.24)_1$, use (7.19) and following the usual procedure show that the consistency parameter is given by

$$\bar{\phi} = \frac{\kappa^2[N]^T(\mathcal{C}_s)[\dot{\varepsilon}]'}{\kappa^2[N]^T(\mathcal{C}_s)[N] + d\kappa/ds} \quad (7.25)$$

where

$$[N] = \frac{\partial f}{\partial[\sigma]'} = \frac{1}{\kappa^2}(\mathcal{Y}_p)[\sigma]' \quad (7.26)$$

Remark: This form for $\bar{\phi}$ is much more general than just the case of plane stress and we will see this repeatedly for other yield surfaces later.

This completes the general set of constitutive equations for plane problems. The set of equations that needs to be solved for plane stress are
1. the equations of equilibrium (7.14),
2. the stress strain relations (7.19) and,
3. the plastic flow equations (7.25).

If one were interested in using a displacement based formulation, the equilibrium equations can be replaced by the equivalent set (7.17).

However, for many cases it is quite advantageous to eliminate the displacements from consideration and concentrate only on the stresses and strains and it is to this aspect that we turn to next.

7.3 The stress function and the equations of compatibility

For many two-dimensional problems (plane stress, plane strain, axisymmetric and torsion problems) it is in many cases simpler to use a "stress function" formulation rather than a displacement formulation. Thus, for the plane stress or plane strain problem it is common to use the Airy stress function. Before we go into this formulation for plasticity, let us examine some of the reasons for the use of such stress functions and the reason why some of these seem to have fallen out of favor.

From the point of view of structural mechanics, for most cases, the whole point of looking at boundary value problems is for the calculations of stresses in order to determine failure conditions. Typically, the body is considered to be "almost rigid" so that we are not that much interested in the displacements (of course there are problems involving localized deformations involving contact where displacements become very important). But, by and large, calculation of displacements are not generally important and are used primarily as a means for determining the stresses. In view of this, it is philosophically very appealing if we could write everything in terms of stresses, eliminating the need to find the displacements entirely. Hence, stress functions and the use of the compatibility equation are a very natural way to go.

This is not always very successful, however, since the compatibility conditions imply the existence of single valued displacements only for simply connected regions. Nevertheless, the use of stress functions turns out to be a very useful technique since it "reduces" the problem to the determination of a single scalar function that satisfies a certain PDE. Using such approaches a very wide class of problems were solved in the early part of the 20th century and form the basis of many practical solutions for a wide range of technologically important problems (see e.g. [Timoshenko and Goodier (1970)]).

With the advent of finite element methods, however, the focus shifted to methods based on displacements since these were the easiest to understand and program. So much so, that the interest in stress function formulations have almost disappeared. The problem with stress function formulations is that they usually involve high order PDEs (typically fourth order) and consequently, the increased smoothness requirements have made them difficult to use in a FEM setting.

The advent of symbolic and numerical programs have changed the picture again. These programs allow for easy implementation of a variety of specialized numerical schemes for specific problems. This, in turn, means that one can indeed do a stress function formulation and implement it for special geometries in less time than it takes to implement FEM schemes. They serve also as excellent pedagogical tools to illustrate features of solutions.

In view of this, it is worth the effort to develop stress functions for elastoplastic problems and implement them numerically using a finite difference scheme. Consequently, we shall do this here.

Recall that the governing equilibrium equations for plane stress (or plane strain) are given by (7.14), they can be trivially satisfied by using a *stress function* $A(x,y)$ such that

$$\sigma_{11} = A_{,22}, \quad \sigma_{22} = A_{,11}, \quad \sigma_{12} = -A_{,12} \qquad (7.27)$$

You can verify this by substituting (7.27) into (7.14). What the above equation means is that, for plane problems, it is a very simple matter to find a stress field that satisfies the equations of equilibrium: Pick any function $A(x,y)$ then the stress field obtained by using (7.27) will automatically satisfy the equations of equilibrium. The question arises as to whether a compatible strain field can be found that corresponds to the stress field given by (7.27).

For plane stress, the in-plane components of the strain are given by (7.15). If these strains are to be obtained from in-plane displacement fields $\mathbf{u} = u_\alpha(x,y)\mathbf{e}_\alpha$, then

$$\varepsilon_{11} = u_{1,1}, \varepsilon_{22} = u_{2,2}, 2\varepsilon_{12} = u_{1,2} + u_{2,1}. \qquad (7.28)$$

⊙ **7.16 (Exercise:). Strain Compatibility equations**
For the conditions (7.28) to be satisfied, show that the strains must satisfy

$$2\varepsilon_{12,12} - \varepsilon_{11,22} - \varepsilon_{22,11} = 0. \qquad (7.29)$$

Remark: The above conditions are also sufficient for simply connected domains.

⊙ **7.17 (Exercise:). Plane Stress Compatibility equations**
By substituting (7.27) into (7.15) and then substituting the result into (7.29) show that the Airy stress function must satisfy

$$\nabla^4 A = \nabla^2(\nabla^2 A) = A_{,1111} + A_{,2222} + 2A_{,1122} = g(x,y) \qquad (7.30)$$

where $g(x,y) = E(2\varepsilon^p{}_{12,12} - \varepsilon^p{}_{11,22} - \varepsilon^p{}_{22,11})$.
What are the corresponding equations for plane strain? (See [Mendelson (1983)] for further details.)

We have thus converted the problem to finding A that satisfies (7.30). Of course this requires consideration of boundary conditions and this is what we will focus on next.

7.4 Boundary conditions for the stress function

The last point that needs to be observed is the nature of the boundary conditions for the Airy stress functions. To understand the required boundary conditions (BCs), consider the case of a simple Bernoulli Euler beam: We know that the resulting differential equation is a fourth order differential equation and we need two boundary conditions at either end. For example, we need to know the displacement and moment or displacement and slope or some such combination. Similarly, for the Airy stress function, we notice that it is a fourth order differential equation and hence needs two boundary conditions along the edge.

How do we get them? We know the traction along the boundary, which can be written in terms of the stress components. We then use (7.27) to find the conditions along the boundary.

We will illustrate this with a simple example. Consider a square domain of side a i.e. $0 \leq x, y \leq a$. Along the side $y = 0$, there is a tensile traction of magnitude p. Along the sides $x = 0$ and $x = a$, there is a shear traction of magnitude q. And, along the top surface $y = a$ there is a tensile load of magnitude $p + 2q$ (see Fig. 7.2).

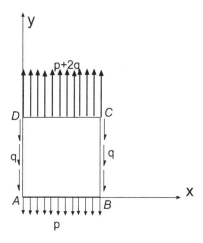

Fig. 7.2 An example problem for a square plate loaded on its edges as shown. There is a normal force of P per unit area on the lower edge. The two lateral edges have shear loads while the top has a normal load of $P + 2q$ per unit length. Thus the BCs are (1) $\sigma_{11} = p, \sigma_{12} = 0$ along the lower edge, (2) $\sigma_{22} = 0, \sigma_{12} = -q$ along the right edge (3) $\sigma_{11} = P + 2q, \sigma_{12} = 0$ along the top edge and (3) $\sigma_{22} = 0, \sigma_{12} = q$ along the bottom.

We need to find the boundary conditions for the Airy stress function $A(x, y)$. We now do this as follows:

(1) Start at the corner $A(x, y = 0)$ and set $A(0, 0) = 0$. This is acceptable since the stresses depend only upon the second derivatives of A.
(2) Along the side AB $(y = 0)$ note that $\sigma_{22} = A_{,11} = p$. So integrate twice with respect to x and set the initial values to zero to get $A(x, 0) = px^2/2, 0 < x < a$. Thus we know the value of A along this boundary.

(3) Along the side AB ($y = 0$), note that $\sigma_{12} = -A_{,12} = 0$. Integrate once with respect to x and set the initial condition to zero and get $\partial A/\partial n(x,0) = \partial A/\partial y(x,0) = 0, 0 < x < a$. Thus we know the value of the normal derivative of A along the boundary.

(4) Along the side BC ($x = a$) note that $\sigma_{11} = A_{,22} = 0$. Again integrate twice with respect to y. Now, you can't choose the value of A or $\partial A/\partial y$. We have to use the end values from the previous step. We see that $A(a,0) = pa^2/2$, $\partial A/\partial y(a,0) = 0$. We use these and get $A(a,y) = pa^2/2, 0 < y < a$.

(5) Along the side BC ($x = a$) note that $\sigma_{12} = -A_{,12} = -q$. Again integrate once with y. Now, you can't choose the value of $A_{,y}$, we have to use the end value from the previous step. We use these to get $\partial A/\partial x(a,y) = +qy, 0 < y < a$.

> ⊙ **7.18 (Exercise:). BCs for the Airy stress function**
> *Repeat the above procedure for the remaining boundaries (CD and DA) and obtain the values of A and $\partial A/\partial n$ along the whole boundary.*
> *Remark: Since we are going around a closed loop, what should be the value of A at the end of the process? What happens if the boundary tractions are not in equilibrium?*

The above procedure can be generalized for a curved boundary in the following way: Given the boundary tractions $\mathbf{t}_n$, let its x and y components be t_x and t_y. Then, we know that

$$\begin{aligned} t_x &= \sigma_{11} n_x + \sigma_{12} n_y = A_{,22}\, n_x - A_{,12}\, n_y, \\ t_y &= \sigma_{12} n_x + \sigma_{22} n_y = -A_{,12}\, n_x + A_{,11}\, n_y. \end{aligned} \quad (7.31)$$

Now if the arc length along the edge is denoted by l, From Fig. 7.1(c), we see that the components of the normal to the boundary are given by

$$n_x = \frac{dy}{dl}, \quad n_y = -\frac{dx}{dl}. \quad (7.32)$$

Substituting (7.32) into (7.31), the latter can be written as

$$\begin{aligned} t_x &= A_{,22} \frac{dy}{dl} + A_{,12}\, dxdl = \frac{d}{dl}(A_{,2}), \\ t_y &= -A_{,12} \frac{dy}{dl} - A_{,11}\, dxdl = -\frac{d}{dl}(A_{,1}). \end{aligned} \quad (7.33)$$

Integrating with respect to the arc length l, we get

$$A_{,2} = \int_0^l t_x\, dl \quad A_{,1} = -\int_0^l t_y\, dl, \quad (7.34)$$

we can combine this into a single vector equations and write it as

$$\nabla A(l) = \mathbf{i}_3 \times \left(\int_0^l \mathbf{t}_n(\xi)\, d\xi \right) = \mathbf{i}_3 \times \mathbf{F}, \quad (7.35)$$

where ∇ is the two-dimensional gradient. We can state it in words as follows:

> **THE GRADIENT OF THE AIRY FUNCTION**
>
> In words, equation (7.35) means the following: *in order to find the gradient of A at some point on the boundary, find the cross product of the unit vector in the z direction with the net force* **F** *acting on the boundary starting from some fixed point.*

⊙ **7.19 (Exercise:). The normal derivative of A**
From the equation (7.35), derive that

$$\partial A/\partial n = \nabla A \cdot \mathbf{n} = \mathbf{n} \cdot \mathbf{i}_3 \times \mathbf{F} = -\mathbf{e}_t \cdot \mathbf{F}. \qquad (7.36)$$

Hint: Use the properties of the scalar triple product and the fact that $\mathbf{n} \times \mathbf{i}_3 = -\mathbf{e}_t$ *where* $\mathbf{e}_t$ *is the tangent to the curve (see Fig. 7.1c).*

> **NORMAL DERIVATIVE OF THE AIRY FUNCTION**
>
> The equation (7.36) can be stated in words as follows: The normal derivative of the Airy stress function at a point on the boundary is equal to the component of the net force along the tangent to the curve at the point.

⊙ **7.20 (Exercise:). The value of A along the boundary**
Using the fact that $\frac{dA}{dl} = \nabla A \cdot \mathbf{i}_t$ and the fact that $\mathbf{e}_t = \frac{dx}{dl}\mathbf{i}_1 + \frac{dy}{dl}\mathbf{i}_2$, take the inner product of (7.35) with $\mathbf{e}_t$, use integration by parts and show that

$$A(l) = \left(\int_0^l (\mathbf{r}(\xi) \times \mathbf{t}_n(\xi))\, d\xi - \mathbf{r}(l) \times \mathbf{F}(l) \right) \cdot \mathbf{i}_3. \qquad (7.37)$$

Stated in words, the above equation means that the value of A at a point on the boundary is equal to the difference between the net moment of the tractions acting up to that point minus the moment of the net force up to that point.

The above two results then allow us to find the value of A and its normal derivative along the boundary for any plane stress problem. However, for many situations it turns out to be easier to simply integrate the basic equations by hand rather than use (7.36) and (7.37).

The fourth order PDE equation (7.30) on p.227 together with (7.36) and (7.37) are the governing equations for plane problems of isotropic elasto-plastic materials.

7.5 Numerical solution

We can solve the governing PDEs by using a finite difference scheme. Since we intend this to be a project for you to implement as a term project, we are deliberately being rather brief as to the details since we expect you to try it out yourself.
We consider a region that has been discretized by a square grid as shown in Fig. 7.3

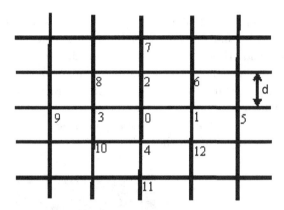

Fig. 7.3 A portion of a plate that has a square finite difference grid inscribed on it, attention is centered on the grid point marked 0.

In terms of this grid, the LHS and the RHS of (7.30) are given by

$$\nabla^4 A = \frac{1}{d^4} [20 A_0 - 8(A_1 + A_2 + A_3 + A_4)$$
$$+ 2(A_6 + A_8 + A_{10} + A_{12} + A_5 + A_7 + A_9 + A_{11}];$$
$$g(x,y) = \frac{E}{d^2} [0.5((\varepsilon^p{}_{12})_6 + (\varepsilon^p{}_{12})_{10} - (\varepsilon^p{}_{12})_8 - (\varepsilon^p{}_{12})_{12}) \qquad (7.38)$$
$$- ((\varepsilon^p{}_{11})_2 + (\varepsilon^p{}_{11})_4 - 2(\varepsilon^p{}_{11})_0)$$
$$- ((\varepsilon^p{}_{22})_3 + (\varepsilon^p{}_{22})_1 - 2(\varepsilon^p{}_{22})_0)],$$

where d is the grid size.

This is then repeated at every grid point (and the BCs are used to find these at grid points close to the boundary). We then have a set of equations of the form

$$\sum_{j=1}^{N} K_{ij} A_j = g_i, \qquad (7.39)$$

where K_{ij} is the stiffness matrix which can be generated and stored once and for all. g_i is the right-hand side which will change from time step to time step due to evolving plastic strains and boundary conditions. This will allow us to find the value of A and hence the stress when the plastic strain field is known. We can thus find the change in the strain $\triangle \varepsilon$ from one iteration to the next.

It only remains to obtain the equations for plastic flow knowing σ and $\triangle \varepsilon$: we need to solve (7.24) and (7.25.). For now, we will propose the following very explicit, forward Euler method (which will work only if you take very small increments):

$$\phi = \frac{\kappa^2 H[f(\sigma,\kappa)]\langle [\mathbf{N}]^T(\mathbb{C}_s)[\triangle\varepsilon]'\rangle}{\kappa^2[\mathbf{N}]^T(\mathbb{C}_s)[\mathbf{N}] + dk/ds} \quad (7.40)$$
$$\triangle\varepsilon_p = \frac{\phi \mathbb{A}[\sigma]'}{\kappa^2}.$$

In the above equation $H(x)$ is the Heaviside step function which is zero if $x < 0$ and 1 if $x \geq 0$. Also $<x> = 0.5(x + \|x\|)$. These two functions will ensure that ϕ is nonzero only if the yield criterion is met and the loading criteria are met. We can update the arc length and other parameters in a similar manner. Be warned, however, this numerical procedure tends to give poor accuracy. A much better scheme is discussed in Chapter 10 on numerical procedures.

If you apply this to a plate with a square hole (square since it is easier to grid), you will get the plastic flow distribution shown in Fig. 7.4.

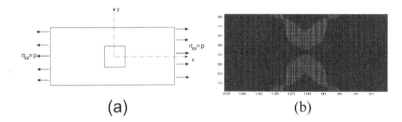

Fig. 7.4 (a) A $10ft \times 4ft$ plate with a $0.3ft \times 0.3ft$ square hole under uniaxial extension as shown. (b) Distribution of the "damage" parameter s representing the total amount of plastic flow. Notice the characteristic "wing-like" pattern emerging from the corners at $45°$ angle. This is characteristic of ductile materials.

We hope you got enough confidence to be able to attempt your own elastoplasticity problems with this method. There are a number of problems that you can solve with these methods without too much trouble.

7.5.1 *Modifying the MATLAB PDE toolbox to solve elastoplasticity problems*

The previous method is based on using the Airy stress functions to solve the plasticity problem. An alternative way to do this is to modify MATLAB's PDE solver (PDETool) to solve elastoplasticity problem. The newest version of the solver is capable of solving plane stress and plane strain elasticity problems with body forces. We can modify it to solve elastoplasticiy problems and display the results. The trick is the following: We use MATLAB to take care of all the discretization, element assembly etc. and we focus only on the plastic flow solution. This is made particularly easy by the very nice GUI that is built in. The MATLAB program to do this is available online (see authors' websites listed in the preface). The starting point of this method is the displacement formulation of the elastoplasticity problem, (7.17). The steps are the following

(1) We discretize the domain into a triangular finite element mesh using the PDE tool GUI. Thus, we will be interested in the nodal displacements $\mathbf{u}^i (i = 1, \ldots, N)$ and the nodal plastic strains $\varepsilon_p^i, (i = 1, \ldots, N)$.
(2) We pretend that we are solving an elasticity problem (say plane stress) and set up the displacement, traction or mixed boundary conditions using the MATLAB GUI.
(3) We export these to the workspace so that we have now obtained the global stiffness matrix, $\mathcal{K}$ and the global boundary conditions, .
(4) To this the effect of the plastic strain is to just add a right-hand side.
(5) For each iteration, (a) we solve an elasticity problem, (b) obtain the displacement, (c) calculate the strains and stresses at the centroids, (d) interpolate and move it to the nodes, (e) check the yield condition and update the plastic strains if the yield criterion is satisfied (f) calculate the right-hand side of (7.17), (g) update the body forces using the calculated plastic strains (g) recalculate new strains until convergence is reached. The computational cycle is shown in Fig. 7.5.

For illustration purposes, this method was applied to solve the problem of a plate with a circular hole and the resulting "damage zone" is shown in Fig. 7.6. Note that there is extensive plastic flow from the hole toward the boundary of the plate roughly at a forty five degree angle.

7.6 Concluding remarks and summary

(1) The governing displacement equations of equilibrium for an elastoplastic material is identical to that of an elastic material with body force terms that are determined by the gradients of the plastic strain.
(2) It should be noted that the plane stress case differs significantly from the plane strain case both in the equations of equilibrium as well as the reduced yield functions and flow rule.
(3) It is possible to solve the plane problems using an Airy stress function leading to an inhomogeneous biharmonic equation. The boundary conditions on the Airy stress function are the same as that for an elastic case.

7.7 Projects

(1) Plate with a square hole: Develop a MATLAB program to solve the plate with a square hole where the load p is gradually increased from zero. At what load does the damage reach the outer boundary? Increase the size of the hole and see

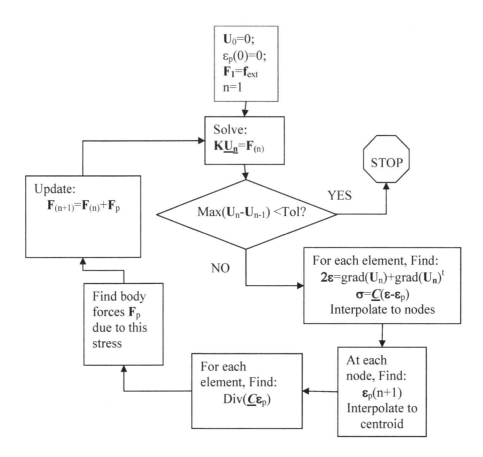

Fig. 7.5 A flowchart showing the implementation of elastoplasticity using MATLAB's PDETool solver. See authors' websites listed in the preface for the listing of the program.

what happens. Look at Chapter 10 on numerical solutions and use the Convex Cutting Plane algorithm for updating the plastic strains.

(2) Consult Mendelson's book on plasticity [Mendelson (1983)] and solve the problem of the elastoplastic torsion of a bar with a square cross-section using the Prandtl Stress function and the iterative scheme described in this chapter.

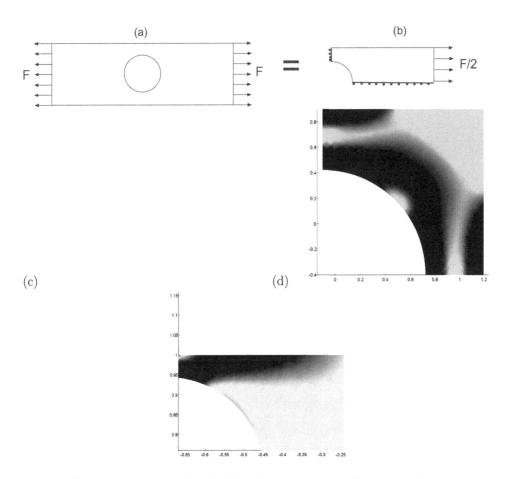

Fig. 7.6 (a) A thin plate with a hole loaded with uniaxial tension; the lateral surface is traction free. (b) Using symmetry to consider 1/4 of the body. The two cut sides have zero normal displacements and zero shear tractions (i.e. they are equivalent to frictionless roller joints). (c) and (d) the plastic strain field is shown. Note the spread of the plastic flow to the boundary from the sides of the hole. (c) For the square plate, there is extensive plasticity since the hole dimensions are large compared to the plate. The thin sections of the plate fail by bending (the "neutral axis" is visible). (d) For the longer rectangular plate, we see that the spread is more toward the sides.

Chapter 8

Examples of Other Yield Surfaces: Associative and Non-associative Plasticity

Learning Objectives

After learning the material in this chapter you should be able to do the following:

(1) Describe the meaning of the terms "isotropic yield surface", "isotropic hardening" and "kinematic hardening" and explain the different ways in which the word "isotropic" is used.

(2) Describe the meaning of the terms "associated flow rule" and "non-associated flow rule".

(3) Write down expressions for the yield surface and flow rules for general isotropic yield surface with kinematic and isotropic hardening.

(4) Derive the expressions for the flow rules for an anisotropic, quadratic yield surface.

(5) Write down the expressions for pressure dependent yield surfaces.

(6) Derive the flow rules from thermodynamical considerations.

CHAPTER SYNOPSIS

In this chapter we introduce generalizations of the J2 plasticity model and show examples of other yield surfaces and other flow rules. We introduce both associated and non-associated flow rules and show examples of material models where they are employed. The central result is that most inelastic material models are specified by two scalar functions— a yield function f which separates elastic from inelastic response and a plastic potential H whose gradient provides the direction of plastic flow. We also provide the MATLAB code for the calculation of the plastic strain for a general quadratic yield surface.

Chapter roadmap

This chapter contains a summary of the various common ways in which J2 plasticity has been extended beginning with a general description of isotropic and kinematic hardening models. We then follow it up with a description of a variety of different classes of yield functions including some anisotropic ones. Now you may wonder why we want to generate so many yield surfaces. Why can't we approximately use the J2 yield surface for everything. The reason is that once we get past small deformations and begin to look at metal forming operations, a precise description of yield surfaces becomes critical. One key property which depends critically on the exact yield surface is the rate of thinning of a sheet metal (related to the so-called "r" value (see e.g., [Marciniak and Kuczinsky (1967); Kocks *et al.* (1998)]])). It is for this reason that we develop increasingly sophisticated yield surfaces.

Next we generalize the constitutive equation for the plastic flow direction by introducing the notion of a plastic potential. We then follow it up with a description of the Bauschinger effect and the Armstrong Frederick model [Armstrong and Frederick (1966)] for this effect. We implement this in a MATLAB code and show how various effects associated with cyclic loading can be simulated.

8.1 The general, small strain elastoplastic model

We have now dealt with the small strain J2 plasticity at length and have seen the development of the theory as well as the numerical solutions for this case. We are now going to generalize this procedure for a much broader class of yield surfaces, dealing both with a wider class of problems as well as a wider class of materials. We begin our developments with the generalization of hardening to account for cyclic loading conditions.

8.1.1 *General characteristics of yield surfaces*

Recall that the constitutive equations for the most widely used plasticity model (referred to as J2 plasticity) are given by[1] (cf. (6.60))

$$\text{Stress Response: } \boldsymbol{\tau} = 2\mu(\boldsymbol{\gamma}_e), \ p = 3Ke,$$

$$\text{Elastic domain: } f(\sigma, \kappa) = \frac{\sqrt{(\boldsymbol{\tau} \cdot \boldsymbol{\tau})}}{\kappa} - 1 \le 0,$$

$$\text{Yield Condition: } \dot{\varepsilon}_p = \dot{\varepsilon} - \dot{\varepsilon}_e; \ \dot{\varepsilon}_p = 0 \Rightarrow f \le 0, \dot{\varepsilon}_p \ne 0 \Rightarrow f = 0, \quad (8.1)$$

$$\text{Flow rule: } = \dot{\varepsilon}_p = \phi \mathbf{N}, \ \mathbf{N} := \frac{\partial f}{\partial \boldsymbol{\tau}} = \frac{\boldsymbol{\tau}}{\kappa^2},$$

$$\text{Non-negativity of dissipation: } = \boldsymbol{\sigma} \cdot \dot{\varepsilon}_p = \phi \ge 0.$$

Note that we have modified the form of the yield function slightly and also have not included the plastic arc length and the hardening parameter (since we will discuss this in more detail later). As usual, we will again write out the Kuhn-Tucker conditions so that you can become intimately familiar with it.

⊙ **8.1 (Exercise:). Kuhn-Tucker Conditions**
(1) Show that the rate of dissipation $\boldsymbol{\sigma} \cdot \boldsymbol{\varepsilon}_p$ happens to be equal to the plastic multiplier ϕ in this case. (Hint: Take the inner product of (8.1)d with σ and use the definition of the yield surface and the yield conditions.) Thus, the non-negativity of the plastic multiplier is just a statement of the fact that plastic flow is dissipative.
(2) Show that the combination of (8.1c) and (8.1e) can be written as

$$\dot{\varepsilon}_p = \phi \mathbf{N}, \ \phi \ge 0, f \le 0, \phi f = 0. \quad (8.2)$$

It is this form that is widely used in the computational literature and is referred to as the "Kuhn Tucker" conditions.

---

[1] We have deliberately written these equations in a different form than in Chapter 6 since, different books represent the same equation in different ways and we want you, the reader, not be fazed by the multiplicity of representations of the same equations. In the subsequent exercises, we will show that all these representations are equivalent and may be written in a unified form.

While the above is the most widely used and most common constitutive law for the plastic strain, it is by no means the only one. Nor is it a very good fit for the actual response of materials such as Aluminum, Titanium etc. although it appears to be reasonable for steel.

From a modeling point of view, what kind of generalizations can we consider? We will list some characteristics of elastoplastic materials and see some possibilities of generalization:

(1) The elastic response is linear, with constant moduli that represent an isotropic elastic response. We could generalize this to be nonlinear and write

$$\boldsymbol{\sigma} = \mathbf{f}(\boldsymbol{\varepsilon} - \boldsymbol{\varepsilon}_p), \tag{8.3}$$

where $\mathbf{f}$ is a nonlinear function. We would like to recover hyperelastic response when $\dot{\boldsymbol{\varepsilon}}_p = 0$ so that we want

$$\boldsymbol{\sigma} = \frac{\partial \psi}{\partial \boldsymbol{\varepsilon}}, \tag{8.4}$$

where ψ is the Helmholtz potential. For large strains, a substantial re-examination of all the issues are needed and we will defer it to the third part of the book.

(2) The yield surface is quadratic and the plastic flow is in the direction of the deviatoric stress. In Chapter 6, we saw that we can visualize the yield surface as enclosing the region where no plastic flow occurs. We can generalize these by saying that *the yield function can be* any *function of $\boldsymbol{\sigma}$ such that the equation $f(\boldsymbol{\sigma}) = \kappa$ represents a smooth closed surface enclosing a region satisfying $f(\boldsymbol{\sigma}) < \kappa$ in stress space where no plastic flow occurs. The boundary of the region represents that set of points where plastic flow can be initiated.*

Having generalized the notion of the yield function and yield surface, we now turn to the plastic flow rule and see how to generalize this.

The normality flow rule

In J2 Plasticity, we assume that the plastic strain rate is in the direction of the deviatoric stress. There are various levels of generalization that are possible.

> ### ASSOCIATIVE PLASTICITY: THE NORMALITY FLOW RULE
>
> The most widely used general flow rule for the plastic strain is that the plastic strain rate be in a direction of the outward normal to the yield surface at that point, i.e.
>
> $$\dot{\varepsilon}_p \propto \frac{\partial f}{\partial \boldsymbol{\sigma}} \Rightarrow \dot{\varepsilon}_p = \phi \frac{\partial f}{\partial \boldsymbol{\sigma}} = \phi \mathbf{N}. \tag{8.5}$$
>
> Models of elasto-plastic behavior that utilize this flow rule are called *associative plasticity models*. We will use the symbol **N** to represent the gradient of the yield function with respect to the stress. The parameter of proportionality is called by various names (dependent upon the exact way in which it is introduced) such as "equivalent strain rate", the "Kuhn Tucker parameter", "consistency parameter", etc.

A generalization such as this is called the **normality flow rule** and there are very compelling reasons (which we will discuss at length in Chapter 9 on thermodynamics) to assume such a form. So much so that this is the MOST popular generalization and it is commonly extended to ALL inelastic variables irrespective of whether they represent plastic strain or not.

However, there are many materials (such as porous materials, rocks, concrete, clay etc.) for which such an assumption is too restrictive. Instead, a much more general assumption is usually made. A broad generalization is that *the direction of the plastic strain rate is determined by the current state of the material*, i.e.

$$\dot{\varepsilon}_p \propto \mathbf{P}(\sigma, \varepsilon_p) \Rightarrow \dot{\varepsilon}_p = \phi \mathbf{P} \tag{8.6}$$

where we have to specify direction of plastic flow **P**. Note that we can always choose to define the sense of **P** such that $(\boldsymbol{\xi}) \cdot \mathbf{P} \geq 0$.

This level of generality is usually a bit too much since we have to find more and more constitutive equations, and again, based on compelling reasons, it is usual to assume the following:

> ### THE PLASTIC POTENTIAL
>
> We will assume that there exists a scalar function $H(\boldsymbol{\sigma}, \varepsilon_p, \mathbf{Q})$ called the *plastic potential* such that
>
> $$\dot{\varepsilon}_p = \phi \frac{\partial H}{\partial \boldsymbol{\sigma}}. \tag{8.7}$$
>
> The "normality" or "associated" flow rule, is then a special case of (8.7) with $H = f$. Otherwise the flow rule is called a non-associative flow rule.

In the next section, we will consider a variety of examples of yield surfaces and flow rules all of which fall into the category (8.7).

8.2 Examples of general yield surfaces

8.2.1 *Pressure dependent yielding and the strength-differential or S-D effect*

Many porous materials (such as sintered or powder compacted materials) show a marked dependence of the yield strength on the mean normal stress. For example consider two samples of Inconel718, a common aerospace material. One subject to pure tension and the other to pure compression. It has been found that the yield strength in compression is higher than that in tension. Note that this is NOT due to pre-yielding unlike the case of the Bauschinger effect (such as that discussed in Chapter 3). Rather, there is an inherent difference in the yield strength in tension and compression. This is called the *Strength-Differential* or the *S-D effect*.

A simple way to model this is to introduce the notion that the scale parameter κ depends upon the current mean normal strain. Thus, to account for the S-D effect, one could consider yield functions of the form

$$f(\boldsymbol{\sigma}) = \frac{\sqrt{\boldsymbol{\tau}\bullet\boldsymbol{\tau}}}{\kappa(s,p)} - 1, \tag{8.8}$$

where $\kappa(s,p)$ is the scale parameter which now depends upon the negative of the mean normal stress $p = -\sigma_{ii}/3$. We can see, in this case that the size of the yield surface will change depending upon whether it is tension or compression.

A specific example of such a yield surface is that of the Drucker Prager yield surface for which

$$\frac{\sqrt{\boldsymbol{\tau}\bullet\boldsymbol{\tau}}}{\kappa(s,p)} - 1 = 0, \quad \kappa(s,p) = \kappa_0(s) + a(s)\,p. \tag{8.9}$$

This yield surface is determined by two constants κ_0 and a. This model is the prototype for a wide range of models used in the granular media literature.

Another such model of a yield surface is the Gurson model which is widely used in the study of ductile void growth, damage and a host of other failure related models. In this model, the yield function is given by

$$f(\boldsymbol{\sigma},\kappa,\alpha) = \boldsymbol{\tau}\bullet\boldsymbol{\tau} + \kappa^2 \left\{ 2\alpha q_1 \cosh(\frac{q_2 p}{\kappa}) - 1 - q_3^2 \alpha^2 \right\}, \tag{8.10}$$

where α is a measure of the volume fraction of the voids or other damage in the material, and q_1, q_2 and q_3 are constants. This model has been widely used for slightly porous materials and for characterization of damage or fatigue in some materials.

8.2.2 *Three-invariant or isotropic yield functions*

So far, we have dealt with isotropic yield functions that depend upon one invariant (J2 plasticity) and two invariants (pressure dependent yielding materials) of the stress. We now consider yield functions that involve all three invariants of the stress. Such models are called *isotropic yield surfaces*.

In order to develop such yield functions, it is useful to introduce some notations for different invariants. Typically, plasticity models do not use the standard invariants in the modeling of the material. Rather, the three most common invariants are

$$p = -\frac{1}{3}\sigma_{ii}, \quad q = \left\{\frac{3}{2}\boldsymbol{\tau}\cdot\boldsymbol{\tau}\right\}^{\frac{1}{2}} = \left\{\frac{3}{2}\tau_{ij}\tau_{ij}\right\}^{\frac{1}{2}}, \quad r = \left\{\frac{9}{2}\boldsymbol{\tau}^2\cdot\boldsymbol{\tau}\right\}^{\frac{1}{3}} = \left\{\frac{9}{2}\tau_{ij}\tau_{jk}\tau_{ik}\right\}^{\frac{1}{3}}. \tag{8.11}$$

Thus a general yield surface may be written as

$$f(\boldsymbol{\sigma}, \kappa) = f(p, q, r, \kappa) = 0. \tag{8.12}$$

In most cases, it is assumed that the trace or projection of the yield surface onto deviatoric (or Π) plane (which will be described shortly) for different values of the pressure p are self similar, so that generally, the yield surface is written in the form

$$\frac{t(q,r)}{\kappa(p)} - 1 = 0, \tag{8.13}$$

where $t(q,r)$ is an "equivalent deviatoric stress magnitude". For example, for the von Mises yield surface, $t(q,r) = q$.

A more general equivalent deviatoric stress magnitude and one that is somewhat commonly used is given by

$$t(q,r) = \frac{1}{2}\left[1 + \frac{1}{K} - \left(1 - \frac{1}{K}\right)\left(\frac{r}{q}\right)^3\right]. \tag{8.14}$$

This rather odd looking form is based on the need to account for the fact that for many rocks and granular materials, the triaxial tension and triaxial compression tests lead to different yield strengths. The parameter, K, is the ratio of triaxial tension to triaxial compression strengths. By this we mean tension or compression testing with a superimposed mean normal tension or mean normal compression. For example, a simple tension test with the lateral surfaces free of traction corresponds to a mean normal tension (since the mean normal stress is positive).

Visualizing the yield surface in the Π plane

One of the major difficulties in visualizing yield surfaces is that they "live" in a six-dimensional space. We don't know about you, but we have difficulty visualizing things in 3-D let alone 6-D. However, yield functions that are functions of the invariants σ are functions of only three variables so they can be visualized in 3-D.

In order to do this, we need the fact that *the invariants of any symmetric second order tensor can be written purely in terms of the eigenvalues.* Thus for example, if $\sigma_1, \sigma_2, \sigma_3$ are the eigenvalues of the stress, then

$$p = \frac{1}{3}\sigma_{ii} = \frac{1}{3}(\sigma_1 + \sigma_2 + \sigma_3). \tag{8.15}$$

⊙ **8.2 (Exercise:). Invariants in terms of eigenvalues**
Obtain expressions for the invariants q and r in (8.11) in terms of $\sigma_1, \sigma_2, \sigma_3$. Using this or otherwise, show that the von Mises yield function is of the form

$$\frac{\sqrt{\tau \cdot \tau}}{\kappa} - 1 = 0 \Rightarrow \frac{\{(\sigma_1 - \sigma_2)^2 + (\sigma_2 - \sigma_3)^2 + (\sigma_3 - \sigma_1)^2\}^{\frac{1}{2}}}{\kappa} = const. \quad (8.16)$$

Hint: Pretend that σ is diagonal and then find the invariants.

Now what did we gain by introducing the eigenvalues? We can actually visualize these yield surfaces. We imagine a 3-D coordinate system with $x = \sigma_1$, $y = \sigma_2$, $z = \sigma_3$. Then we can show that the von Mises yield surface is in the form of a "tube" with axis along the vectors $\mathbf{n} = \sigma_1 \mathbf{i}_1 + \sigma_2 \mathbf{i}_2 + \sigma_3 \mathbf{i}_3$ as shown in Fig. 8.1.

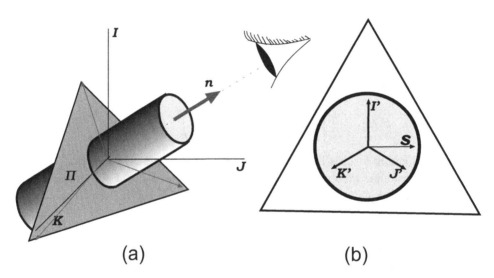

Fig. 8.1 (a) A schematic figure of the von Mises yield surface in the principal stress plane. The plane perpendicular to the axis $\mathbf{n}$ is the *deviatoric stress plane* and is referred to as the Π plane. (b) A plan view of the Π plane showing the trace of the von Mises yield surface as a circle. $\mathbf{S}$ is the vector representing simple shear stress.

In terms of the Π plane, q, r can be considered as coordinates of points on the Π plane, while p represents the distance along the normal $\mathbf{n}$ to the Π plane.

In fact *any isotropic yield function that depends only upon the deviatoric stress will then depend only upon q and r and the yield surface will be a prism whose axis is along the $\mathbf{n}$. It is characterized by its trace on the Π plane.*

Thus, in view of this visualization, a different approach to defining yield functions than using the invariants is the use of the eigenvalues $\sigma_1, \sigma_2, \sigma_3$ of the stress and writing yield surfaces in terms of these.

> ⊙ **8.3 (Exercise:). Visualizing the Tresca Yield surface in the Π plane**
> The Tresca Yield surface is defined by the requirement that the maximum shear stress reaches a critical value. Since the maximum shear stress is the difference between the principal stresses, the Tresca surface takes the form
>
> $$\max[\{|\sigma_1 - \sigma_2|, |\sigma_2 - \sigma_3|, |\sigma_3 - \sigma_1|\}] = \kappa \qquad (8.17)$$
>
> Show that in the Π plane the trace of this yield surface is in the form of a hexagon. (See e.g., [Mendelson (1983)] for an excellent discussion of the Π plane.)

Such yield functions can be easily generalized and we can generate families of yield surfaces of the form

$$\frac{\{|\sigma_1 - \sigma_2|^m + |\sigma_2 - \sigma_3|^m + |\sigma_3 - \sigma_1|^m\}^{\frac{1}{m}}}{\kappa} - 1 = 0. \qquad (8.18)$$

This family of yield surfaces is called the Hosford family [Hosford Jr (1972)]. As m increases from a value of $m = 2$ (which is the von-Mises yield surface), the yield function more and more closely approximates the Tresca yield function. Indeed the Tresca yield function is obtained as the limit of the above yield function as m tends to infinity. Another interesting isotropic yield function was proposed by Karafillis and Boyce [Karafillis and Boyce (1993)]:

$$\begin{aligned}(1-c)[(S_1 - S_2)^{2k} + (S_2 - S_3)^{2k} + (S_3 - S_1)^{2k}] \\ + c\frac{3^{2k}}{2^{2k-1}+1}[S_1^{2k} + S_2^{2k} + S_3^{2k}] = 2Y^{2k},\end{aligned} \qquad (8.19)$$

where S_i are the eigenvalues of the stress deviator, Y is the yield stress in tension, and k and c are material constants. This equation describes a yield surface that lies between the Tresca surface and the so-called "upper bound" surface, and also includes the von Mises surface as a special case.

Normality flow rules for isotropic yield functions

Recall that the normality flow rule states that $\dot{\varepsilon}_p = \phi \partial f / \partial \boldsymbol{\sigma}$. For isotropic yield functions that are written in terms of the invariants, p, q and r, calculating the derivatives is really straightforward.

⊙ **8.4 (Exercise:). Derivatives of the invariants**
Using index notation or otherwise, show that

$$\frac{\partial p}{\partial \boldsymbol{\sigma}} = -\frac{1}{3}\mathbf{I}, \frac{\partial q}{\partial \boldsymbol{\sigma}} = \frac{3\boldsymbol{\tau}}{q}, \text{ and } \frac{\partial r}{\partial \boldsymbol{\sigma}} = \frac{9}{2}\boldsymbol{\tau}^2 - q^2\mathbf{I}. \qquad (8.20)$$

Hence, obtain an explicit expression for $\partial f/\partial \boldsymbol{\sigma}$ in terms of $f_{,p}, f_{,q}$ and $f_{,r}$. What would be the explicit form of the normality flow rule for the yield functions (8.9), (8.10), and (8.13)-(8.14)?.

On the other hand, for yield functions written in terms of the principal stresses, obtaining direct tensorial representations of the normality flow rule is very cumbersome and not feasible for computational purposes. However, it is much easier to work in the principal stress space. Specifically, we use the following result:

> If a function is an isotropic scalar function of a tensor (i.e. it is only a function of the invariants) then its derivative with respect to the tensor has the same eigenvectors as the tensor itself, although the eigenvalues can be completely different.

In other words, if the yield function $f(\boldsymbol{\sigma})$ depends only upon the invariants (or only upon the eigenvalues with the order of the eigenvalues being immaterial) then $\partial f/\partial \boldsymbol{\sigma}$ has the same eigenvectors as $\boldsymbol{\sigma}$. Thus, if the plastic strain rate is given by

$$\dot{\boldsymbol{\varepsilon}}_p = \phi \frac{\partial f}{\partial \boldsymbol{\sigma}}, \qquad (8.21)$$

then since $\dot{\boldsymbol{\varepsilon}}_p$ and $\boldsymbol{\sigma}$ have the same eigenvectors, the eigenvalues of $\dot{\boldsymbol{\varepsilon}}_p$ are given by

$$\dot{e}_i^p = \phi \frac{\partial f}{\partial \sigma_i}, i = 1, 2, 3. \qquad (8.22)$$

⊙ **8.5 (Exercise:). Flow rule in terms of principal stresses**
For the yield function defined by (8.18), use (8.22) and obtain the eigenvalues of $\dot{\boldsymbol{\varepsilon}}_p$.

Note that all of the yield surfaces have been written with no shift tensor $\boldsymbol{\alpha}$. It is a simple matter to include this (if necessary) by simply replacing $\boldsymbol{\sigma}$ in the yield functions with $\boldsymbol{\xi}$.

While working in the three-dimensional space of principal stresses is considerably easier from an experimental point of view, it is cumbersome to use in numerical codes due to the need for frequent calculations of eigenvalues and eigenvectors. But in most commercial codes, this is automatically done and in many cases, if you want to write your own user subroutine, the eigenvalues and eigenvectors of the stress are available as parameters.

Up until now we have focussed attention only on isotropic yield functions. We now turn to a class of anisotropic yield functions that have been popular in the literature.

8.2.3 Anisotropic yield functions

While there are a number of possible examples of isotropic yield surfaces, examples of anisotropic yield surfaces are less common. When do we need such yield functions? Typically, processed sheets, wires and rods where the grains have been aligned in some preferential directions (so-called Textured materials) show strong anisotropy in their yield surfaces, i.e. the tensile yield strength of samples cut in different orientations will be quite different. So if you are doing structural analysis of rolled sheets it is important for you to pick a reasonable yield function that reflects the anisotropy present.

The simplest choice is a generalization of the von Mises criterion, i.e. an anisotropic yield function that is quadratic. The most general anisotropic, quadratic yield surface is of the form

$$\frac{\sqrt{\sigma \cdot \mathcal{A} \sigma}}{\kappa} - 1 = 0 \tag{8.23}$$

where $\mathcal{A}$ is a symmetric positive semi-definite fourth order tensor (i.e. its eigenvalues are all non-negative). The symmetry of the yield surface is then determined by the structure of $\mathcal{A}$ (similar to that of the elastic moduli).

Normally rolled sheet commonly exhibit orthotropic symmetry; they have three perpendicular mirror planes coinciding with the direction of rolling **r**, the direction perpendicular to direction of rolling in the plane of the rolled sheet **n** and the thickness direction **t** as shown in Fig. 8.2.

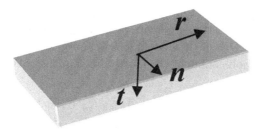

Fig. 8.2 A piece of sheet metal illustrating the rolling, normal and thickness directions. It is customary to choose axes such that the **r** is along the x-axis, **n** is along the y axis and **t** is the z direction.

Yield surfaces that have orthotropic symmetry can be easily constructed by fixing the coordinate axes along the directions of symmetry. A specific and very widely used example of such a yield surface for rolled sheet is Hill's anisotropic yield surface ([Hill (1948)]) which is of the form

$$f(\boldsymbol{\sigma}) = \frac{\sqrt{F(\sigma_{22}-\sigma_{33})^2 + G(\sigma_{33}-\sigma_{11})^2 + H(\sigma_{11}-\sigma_{22})^2 + 2L\sigma_{23}^2 + 2M\sigma_{31}^2 + 2N\sigma_{12}^2}}{\kappa}$$
$$- 1 = 0, \qquad (8.24)$$

where F, G, H, L, M, and N are the six material constants. A common simplification of (6.9) is to assume that the material is isotropic in the plane of the sheet (planar isotropy). Such a simplification reduces the number of material constants from six to three. The relative simplicity of Hill's surface makes it the most widely used anisotropic yield function nowadays, despite its well recognized shortcomings, for example, the surface is quadratic, thus the yield locus will always appear as an ellipse, no matter which section is taken, while the experiments show that the real yield locus is not an ellipse, at least in the plane-stress plane.

Nevertheless, Hill's surface has some good properties, besides its simplicity. For example, it includes in its formulation *all* stress components, that is, also shear stresses. However, it appears not to be a good fit for aluminum sheet. So Hill has suggested several modified yield functions (see [Hill (1990, 1993)]). A variety of other yield surfaces have been developed for anisotropic sheet metals along with various experimental protocols to determine the material parameters (see e.g., [Barlat (1987); Barlat et al. (1997); Bohlke et al. (2008); Choi et al. (1999)]) an excellent discussion of yield surfaces, their role and their determination is provided in [Banabic et al. (2000)].

Most of the other commonly used functions generate only "yield loci", i.e. *sections of the yield surface under the assumption that certain components of the stress are zero*. For example if we assume that the sheet is subject to only plane stress, then we can get a yield locus in terms of just three variables— σ_{xx}, σ_{yy} and σ_{xy} — which might be useful for plane stress problems even though full three-dimensional problems cannot be solved. One of these yield loci was proposed by Gotoh [Gotoh (1977)], who considered a polynomial of the fourth order in the components of the stress:

$$A_1\sigma_{xx}^4 + A_2\sigma_{xx}^3\sigma_{yy} + A_3\sigma_{xx}^2\sigma_{yy}^2 + A_4\sigma_{xx}\sigma_{yy}^3 + A_5\sigma_{yy}^4$$
$$+(A_6\sigma_{xx}^2 + A_7\sigma_{xx}\sigma_{yy} + A_8\sigma_{yy}^2)\sigma_{xy}^2 + A_9\sigma_{xy}^4 = 1. \qquad (8.25)$$

This function is general enough to include shear stress terms in its formulation and is useful for plane stress problems. The shape is more general than an ellipse, but still it cannot approximate a real yield locus. In the '90s Barlat and co-workers ([Barlat (1987); Barlat et al. (1997); Choi et al. (1999)]) proposed several yield surfaces. The basic idea of these surfaces is explained in [Karafillis and Boyce (1993)]. The method adopted in this approach is to use one of the well-established *isotropic* yield functions, say $f = f_{iso}(\sigma)$, in conjunction with an "isotropic plasticity equivalent (IPE) stress transformation", described by the fourth order tensor $\mathcal{L}$:

$$f = f_{iso}(\tilde{\sigma}); \quad \tilde{\sigma} = \mathcal{L}\sigma, \tag{8.26}$$

where σ is the actual stress and $\tilde{\sigma}$ is the "isotropic equivalent stress tensor". The IPE transformation is a linear transformation of the stress space. The anisotropic yield function f is obtained by composing the IPE transformation $\mathcal{L}$ with an isotropic yield function f_{iso}. The resultant yield function will have the same symmetry as $\mathcal{L}$, so, by choosing an appropriate IPE transformation, one can easily assign specific symmetries to the yield surface. In the case of orthotropic symmetry, typical of rolled sheets, the tensor $\mathcal{L}$ will have six independent components.

We then assume any isotropic yield function (say the Hosford family) we like for f_{iso}. Due to the anisotropic linear transformation $\mathcal{L}$ applied to the stress, we will get an anisotropic yield surface. A full discussion of this method is beyond the scope of this book but interested readers may want to read [Banabic *et al.* (2000)] and the papers by Barlat, Hosford, Karafillis and Boyce that were cited earlier to learn more on this and other methods. A survey of these and a vast number of other yield surfaces for sheet metal, along with a new method for generating yield surfaces that match experimental data very well are given in Chapter 6 of [Mollica (2001)] as well as [Mollica and Srinivasa (2002); Weng and Phillips (1977a,b)].

8.3 Examples of plastic potentials and non-associative flow rules

Until now we haven't yet considered non-associative plastic flow rules nor have we considered hardening parameters. It is to these issues that we turn next.

There are many cases, where the flow rule may be changed from the normality flow rule to accommodate behavior wherein the plastic strain rate is *not normal to the yield surface*. Materials that do not obey the normality rule are usually referred to as frictional materials or non-associative materials and they include such materials as granular materials, rocks, metallic foams, certain polymers etc.

The primary way in which this effect becomes manifest is when the mean normal stress plays a role in the yield surface. For materials where there is permanent volume change, the processes that account for permanent plastic volume change (dilatancy in the case of granular materials or consolidation in the case of powders, foams etc.) are pressure dependent and are physically very different from that which causes permanent distortions (deviatoric strains). Thus the rules for changing volume are quite different that the flow rules for the distortion of the body, the former being attributed to void growth, coalescence or collapse, while the latter is primarily attributed to slip or sliding type processes. In view of this, it is generally assumed that there are *two* functions that describe the elastoplastic response (1) the yield function f that determines the onset of plastic deformation and (2) a separate function H called the plastic potential whose derivative gives us the plastic flow rule.

For non-associative materials, the constitutive equations are given by

$$\text{Elastic constitutive equations: } [\boldsymbol{\sigma}] = \mathbb{C}[\boldsymbol{\varepsilon} - \boldsymbol{\varepsilon}_p]$$
$$\text{Yield Condition } f(\boldsymbol{\xi}/\kappa) - 1 = 0, \; \boldsymbol{\xi} = \boldsymbol{\sigma} - \boldsymbol{\alpha}.$$
$$\text{Flow Rule: } [\dot{\boldsymbol{\varepsilon}}_p] = \phi[\mathbf{P}],$$
$$[\mathbf{P}] = \left[\frac{\partial H}{\partial \boldsymbol{\sigma}}\right], H = H(\boldsymbol{\sigma}, \boldsymbol{\alpha}, \boldsymbol{\varepsilon}_p, \kappa) \quad (8.27)$$
$$\text{Kuhn-Tucker Condition: } \phi \geq 0, f \leq 0, \phi f = 0.$$

⊙ **8.6 (Exercise:). Finding ϕ for the general flow rule (8.27)**
Recall the procedure described in Chapter 5 for finding ϕ. We will now use the same procedure to find ϕ here, assuming that $\boldsymbol{\alpha} = 0$ and $\kappa = \text{const}$.
Step 1: Assume $\phi > 0$ and differentiate f with respect to time to get

$$df/dt = (\partial f/\partial \boldsymbol{\sigma})\bullet\dot{\boldsymbol{\sigma}} = 0. \quad (8.28)$$

Step 2: Substitute for $\dot{\boldsymbol{\sigma}}$ by differentiating the elastic constitutive equation (8.27)a and the flow rule (8.27)c to obtain (using the fact that $\mathbb{C}$ is a symmetric matrix)

$$\mathbb{C}\frac{\partial f}{\partial \boldsymbol{\sigma}}\bullet\dot{\boldsymbol{\varepsilon}} - \phi\mathbb{C}\frac{\partial f}{\partial \boldsymbol{\sigma}}\bullet\mathbf{P} = 0. \quad (8.29)$$

Step 3: Show that the first term on the left-hand side of the above equation is nothing but $\hat{g} := df/dt|_{\phi=0}$, so that we finally have

$$\phi > 0 \Rightarrow \phi = h\hat{g}, \; h = \frac{1}{\mathbb{C}\frac{\partial f}{\partial \boldsymbol{\sigma}}\bullet\mathbf{P}}, \; \hat{g} = \mathbb{C}\frac{\partial f}{\partial \boldsymbol{\sigma}}\bullet\dot{\boldsymbol{\varepsilon}}. \quad (8.30)$$

Step 4: Using this value of ϕ show that the tangent modulus is of the form (switching to Matrix notation)

$$\dot{\boldsymbol{\sigma}} = \mathcal{K}\dot{\boldsymbol{\varepsilon}}, \; \mathcal{K} = \mathbb{C}\{\mathbb{I} - h\mathbf{P}\otimes(\mathbb{C}\mathbf{N})\}. \quad (8.31)$$

For illustrative purposes, we will now consider some examples of materials whose elastoplastic response is given by non-associative flow rules.

(1) **Drucker-Prager Model for granular materials that show pressure dependent yield:**

The yield function is a generalization of the von Mises yield function and is of the form

$$f(\boldsymbol{\sigma}) = t(q,r) - p\tan\beta - d = 0, \quad (8.32)$$

where p, q and r are the stress invariants introduced in (8.11) and $t(q,r)$ is a positively homogeneous function of q and r, β is called the *friction angle* and d is a positive constant.

The plastic potential is assumed to be of the form

$$H(\boldsymbol{\sigma}) = \sqrt{a^2 + q^2} - p\tan\psi, \quad (8.33)$$

where a and ψ are constants.

With this assumption, the plastic flow rule becomes (using the chain rule for differentiation, and (8.20))

$$\dot{\boldsymbol{\varepsilon}}_p = \dot{\boldsymbol{\gamma}}_p + \frac{1}{3}\mathrm{tr}\dot{\boldsymbol{\varepsilon}}_p \mathbf{I},$$

$$\dot{\boldsymbol{\gamma}}_p = \phi\frac{\partial H}{\partial \boldsymbol{\tau}} = \phi\frac{3\boldsymbol{\tau}}{2\sqrt{a^2 + q^2}}, \quad (8.34)$$

$$\mathrm{tr}\dot{\boldsymbol{\varepsilon}}_p = -\phi\frac{\partial H}{\partial p} = \phi\tan\psi.$$

(2) **Plasticity Model for crushable foam materials used in impact protection:**

For these materials, the yield surface is of the form

$$\sqrt{q^2 + \alpha^2(p - p_0)^2} - B = 0, \quad (8.35)$$

where α and B are hardening parameters, one of which represents the "size" of the trace of the yield surface with respect to the deviatoric plane and the other representing the "shape" of the yield surface.

The plastic potential is assumed to be of the form

$$H(p, q, r) = \sqrt{q^2 + \frac{9p^2}{2}}, \quad (8.36)$$

so that the flow rule is of the form

$$\dot{\boldsymbol{\varepsilon}}_p = \dot{\boldsymbol{\gamma}}_p + \frac{1}{3}\mathrm{tr}\dot{\boldsymbol{\varepsilon}}_p \mathbf{I},$$

$$\dot{\boldsymbol{\gamma}}_p = \phi\frac{\partial H}{\partial \boldsymbol{\tau}} = \phi\frac{3\boldsymbol{\tau}}{2\sqrt{\frac{9p^2}{2} + q^2}}, \quad (8.37)$$

$$\mathrm{tr}\dot{\boldsymbol{\varepsilon}}_p = -\phi\frac{\partial H}{\partial p} = -\phi\frac{9p}{2\sqrt{\frac{9p^2}{2} + q^2}}.$$

⊙ **8.7 (Exercise:). Derivation of Flow Rules:**
Derive (8.34) and (8.37) from their corresponding plastic potentials.

⊙ **8.8 (Exercise:). Visualization of the yield surfaces**
Rewrite the yield functions (8.32) and (8.35) in terms of principal stresses. Use a software such as Mathematica or MATLAB that is capable of surface plotting and obtain the shape of the yield surfaces. (See the ABAQUS materials manual for some visualizations of these surfaces as well as the plastic potentials.)

There is a vast literature on plasticity of granular media and foams and you can consult those to learn more about non-associative flow rules (see e.g., [Runesson and Mroz (1989)]–general flow rules for non-associative plasticity, [Collins and Houlsby (1997)]–thermodynamics of geomaterials, [Dorris and Nematnasser (1982)]–granular materials, [Klisinski and Mroz (1988)]–flow and damage in concrete, [Lambrecht and Miehe (1999)]–soil mechanics, and [Wang and McDowell (2005)]–metallic honeycombs.)

8.4 Changes in the size, location and shape of yield surfaces and methods to quantify them

Up until now, we have considered only "fixed" yield surfaces, i.e. they cannot yet expand or change shape. We have already seen in Chapter 3 on trusses that if you want to simulate cyclic loading, you have to include hardening effects and in Chapter 6 we saw how isotropic hardening can be included in J2 plasticity. We are now in a position to generalize these notions and assume that the elastic domain is given by a function of the form $f(\boldsymbol{\sigma}, \mathbf{Q})$, where $\mathbf{Q}$ represents a list of parameters (usually called internal variables) that change the size and shape of the yield surface. This is quite a generalization. A typical assumption is that

- the shape of the yield surface is fixed, while its size can change, and
- its "center" can change.

These two conditions can be accommodated by writing the yield surface as a function of a scaled and shifted stress in the form

$$f\left(\frac{\boldsymbol{\sigma} - \boldsymbol{\alpha}}{k}\right) = 1. \tag{8.38}$$

For future convenience, in the rest of the chapter we will write $\boldsymbol{\xi} = \boldsymbol{\sigma} - \boldsymbol{\alpha}$.

In the above equation, $\boldsymbol{\alpha}$ is usually referred to as the *back stress* and is the *shift parameter*, representing the way in which the yield surface moves around, and κ is a *scale parameter* related to the size of the yield surface. The new parameter ξ is the "effective stress" or *the plastic driving force* whose value determines whether plastic flow occurs or not.

The effect of these parameters are shown schematically in a two-dimensional stress space in Fig. 8.3. We can see that increasing κ makes the yield surface

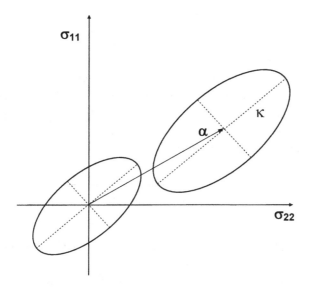

Fig. 8.3 Schematic diagram of a moving and expanding yield surface. α and κ are the parameters that describe the motion of the yield surface. (1) the translation parameter (α) representing the motion of the point initially at $\sigma = 0$ and (2) a scaling parameter (κ) representing the expansion or contraction of the yield surface. In the current figure, the yield surface has expanded by a factor of 1.2. Note however that the major and minor axes are parallel, i.e. the yield surface does not "rotate". In reality, the motion and change of shape is much more complicated than this.

bigger whereas changing α moves the yield surface around. *But the shape and the orientation of the yield surface are unaltered.* We need additional parameters to represent distortions of the yield surface beyond just these two simple modes.

> ⊙ **8.9 (Exercise:). Homogeneity of the yield function**
> Show that the yield function $f((\sigma - \alpha)/\kappa)$ defined by (8.38) satisfies the condition that $f(\lambda\sigma, \lambda\alpha, \lambda\kappa) = f(\sigma, \alpha, \kappa)$ for any constant λ. Using this or otherwise show that
>
> $$\frac{\partial f}{\partial \sigma} \cdot (\xi) = -\frac{\partial f}{\partial \kappa} \kappa. \tag{8.39}$$

Apart from this scaled form, in most cases of practical interest, it turns out that the yield function f can be written as a *positively homogeneous function of degree*

1 in $\boldsymbol{\xi}$, i.e. f satisfies

$$\text{for all } \lambda > 0,\ f(\lambda(\boldsymbol{\xi}), \kappa) = \lambda f(\boldsymbol{\xi}, k). \tag{8.40}$$

This is an important assumption since then we can use (8.40) with $\lambda = 1/\kappa$ and, when (8.40) is true, we can rewrite (8.3) as

$$f(\boldsymbol{\xi}) = \kappa \tag{8.41}$$

which is a common form for yield surfaces in the literature.

⊙ **8.10 (Exercise:). The shape of the yield surfaces**
Show that all the yield surfaces discussed in this chapter, namely the Drucker-Prager Yield surface (8.9), the Gurson Model (8.10), Hill's quadratic yield surface (8.21) and the power law yield functions (8.18) can be rewritten in a form that satisfies the positive homogeneity properties.

Thus the core elements of most elasto-plastic models are given by the following conditions:

- A list of state variables: $\mathbb{S} = \{\boldsymbol{\sigma}, \boldsymbol{\varepsilon}, \boldsymbol{\varepsilon}_p, \kappa, \boldsymbol{\alpha}, ...\}$. It is usual to denote the variables $(\kappa, \boldsymbol{\alpha}, ...)$ as *the inelastic variables* $\mathbf{Q}$. Note that the most common ones, the shift and the scaling parameter are listed here, and recall that κ is the scale parameter and $\boldsymbol{\alpha}$ is the shift parameter.
- Three scalar functions: the strain energy function $\psi = \psi(\boldsymbol{\varepsilon}, \boldsymbol{\varepsilon}_p)$, the yield function $f((\boldsymbol{\xi})/\kappa)$ and a plastic potential $H(\boldsymbol{\xi}, \mathbf{Q})$, which is defined such that $\partial H/\partial \xi \bullet(\boldsymbol{\xi}) > 0$.
- Then the stress response and the kinetic equation for $\boldsymbol{\varepsilon}_p$ are given by

$$\boldsymbol{\sigma} = \frac{\partial \psi}{\partial \boldsymbol{\varepsilon}},$$

$$\dot{\boldsymbol{\varepsilon}}_p = \phi \mathbf{P} = \phi \frac{\partial H}{\partial \boldsymbol{\xi}},\ \boldsymbol{\xi} = \boldsymbol{\sigma} - \boldsymbol{\alpha}, \tag{8.42}$$

$$\phi \geq 0, f((\boldsymbol{\xi})/\kappa) \leq 0, \phi f = 0.$$

- These equations should be augmented with equations that show how the inelastic variables $\mathbf{Q}$ change with time. This is a major aspect of ongoing research and discussion.
- We introduce two definitions that are commonly used to characterize the changes $\mathbf{Q}$. (1) A material is said to undergo *pure isotropic hardening* if $\boldsymbol{\alpha} = 0$ and only κ changes, i.e. only the scale parameter changes so that yield surface expands or contracts with the point $\boldsymbol{\sigma} = 0$ remaining fixed. (ii) A material is said to undergo *kinematic hardening* if $\kappa = const.$ and $\boldsymbol{\alpha}$ changes, i.e. the yield surface moves around in stress space without changing size.

Until now, we have paid scant attention to the parameters $\mathbf{Q}$ which generally represent how the size and shape of the yield surface changes with time. We now consider this in detail. In general, the scale parameter κ and the shift parameter $\boldsymbol{\alpha}$ depend upon time in an indirect way. They do not depend on how long the material has been exposed to a particular sequence of stresses but roughly on "how much plastic deformation has occurred". In other words, on how much internal structural changes have accumulated. In order to do this in a very approximate manner, we introduce the notion of an "accumulated plastic strain". This concept was already introduced in Chapter 5. We revisit this here for completeness and for the purposes of generalization. The idea behind this is that as the material is deformed, the plastic strain traces a path in a six-dimensional space[2]. We try to find a way to calculate the distance travelled along this path.

The plastic arc length, equivalent stress and equivalent plastic strain rate

> ⊙ **8.11 (Exercise:). Equivalent plastic strain for J2 Plasticity**
> *For a J2 plasticity model, recall that the plastic arc length is defined through $s = \int_0^t \|\dot{\boldsymbol{\varepsilon}}_p\| \, dt$. Using this and the flow rule in the form (8.1), show that*
>
> $$\kappa \dot{s} = \phi = \boldsymbol{\sigma} \cdot \dot{\boldsymbol{\varepsilon}}_p. \tag{8.43}$$
>
> *In other words, if we think of κ as the "equivalent stress" the "equivalent plastic strain rate" satisfies the condition that the product of the two is equal to the rate of plastic working.*

What do we do for a general, anisotropic material? It should be apparent that plastic flow in different directions are not equivalent. So what can we use to define some sort of an equivalent measure? Note that because of our rather special form for the yield function, we can use the value of the scale parameter $\kappa > 0$ as the "equivalent stress" at any instant. We also know that the rate of work dissipated is $(\boldsymbol{\xi}) \cdot \dot{\boldsymbol{\varepsilon}}_p \geq 0$ so we can define the "equivalent strain rate" as the rate of plastic working divided by κ. Thus, we can define the plastic arc length through

$$\dot{s} = \frac{\boldsymbol{\xi} \cdot \dot{\boldsymbol{\varepsilon}}_p}{\kappa} \;\Rightarrow\; s = \int_0^t \frac{(\boldsymbol{\xi})}{\kappa} \cdot \dot{\boldsymbol{\varepsilon}}_p \, dt. \tag{8.44}$$

---

[2] For our purposes, 6-D is just like 3-D, only there is more of it:-))

⊙ **8.12 (Exercise:). Plastic arc length**
If the plastic strain rate satisfies the normality flow rule (try to recall what this means without looking at the equations for normality flow rule) and the yield surface is of the form (8.38), show by using (8.44) that

$$\dot{s} = -\phi \frac{\partial f}{\partial \kappa}. \tag{8.45}$$

Further, if the yield surface is positively homogeneous of degree one, i.e. it satisfies (8.40), then show that it can be simplified as

$$\dot{s} = \phi/\kappa. \tag{8.46}$$

⊙ **8.13 (Exercise:). Equivalent plastic strain rate for Hill's yield surface**
Write the anisotropic yield surface (8.21) in the form (8.38), and then use (8.45) to find a form for the rate of accumulated plastic strain.

We emphasize that, other than for J2 plasticity, the definition for equivalent plastic strain rate given in (8.46) is not the same as norm of the plastic strain rate. Nevertheless, (8.46) is easily generalizable to finite strains etc., and is physically motivated in terms of the rate of plastic working.

8.4.1 *Isotropic hardening*

Recall that the parameter κ is a scale parameter that represents the "size" of the yield surface. In view of the fact that generally, the size of the yield surface expands with increasing deformation, we will stipulate that

$$\dot{\kappa} = \Theta(s, \kappa)\dot{s} = -\phi \Theta(s, \kappa) \frac{\partial f}{\partial \kappa}. \tag{8.47}$$

If you recall, the form above is identical to that obtained in Chapter 3. The definition of s is different, accounting for a general three-dimensional motion. The kinds of isotropic hardening laws listed there are reproduced here for convenience:

Linear hardening: $\Theta = C_1 \Rightarrow \kappa = C_1 s.$ (8.48)

Swift Law: $\Theta = mC_2(s_0 + s)^{m-1} \Rightarrow \kappa = C_0 + C_2(s_0 + s)^m.$ (8.49)

Voce Law: $\Theta = -C_4 C_5 e^{-C_5 s} \Rightarrow \kappa = C_3 + C_4 e^{-C_5 s}.$ (8.50)

In all these cases, it is seen that we have explicit expressions for κ as functions of s, although it need not be the case always.

8.4.2 The Bauschinger effect, the back stress α and certain predictable characteristics of cyclic loading

Cyclic loading poses certain demanding challenges on models because of the fact that very small microscopic changes in the internal structure accumulates over a number of cycles and causes eventual cracking. It is a technologically important issue however and most current models use a fairly conservative definition of failure and have models based on the notion of fatigue limit and ways of counting cycles. These models have limited predictive capability and are based on the notion of infinite life, i.e. the predictable phenomenon is that if the conditions for the applicability of the model are satisfied (stresses much smaller than the 0.2% offset yield strength and periodic loading) then one can design the structural member such that there is no likelihood of failure.

On the other hand, materials subjected to large loads (compared to the yield strength) and somewhat more random loads such as that experienced during ground motions etc. cannot be modeled this way. Observations have shown that stress strain curve is extremely complex for a general loading even in one dimension (see [Khan and Huang (1995)], section 6.3.1) but certain general features can be observed:

More generally, if we plot the magnitude of the stress versus the arc length of the strain path, the resulting curve is not the same for different strain paths although the arc length is the same. This path dependence of the stress is called the Bauschinger effect.

(1) **Shakedown:** If the material is subjected to cyclic tension and compression, of equal magnitude, the stress gradually increases and reaches an asymptotic limit. This is called Shakedown (see Fig. 8.4a).

(2) **Ratchetting:** If the sample is subject to cyclic loading between two fixed stress values (see Fig. 8.4c), there is a progressive increase in the strain from one cycle to the next. This phenomenon is called Ratchetting.

(3) **Mean stress relaxation:** If the sample is subject to cyclic strain between two values of the strain, the maximum stress slowly reduces. This is called mean stress relaxation (see Fig. 8.5d).

(4) **Bauschinger effect:** Most of the behavior of materials undergoing cyclic loading are describable in terms of the Bauschinger effect [Bauschinger (1881); Bate and Wilson (1986)]. The basic idea behind this phenomenon is explained by means of the following experiments. Consider three identical samples (A, B, C) of a metal. Sample A is subject to pure monotonic tension, sample B to pure monotonic compression and Sample C is subject to an initial monotonic tension followed by compression. The three stress strain curves are shown in Fig. 3.9. If we now plot the magnitude of the stress versus the magnitude of the strain, we see that the curves for A and B now coincide. But the curve for C is different from the original curve. This effect is called the Bauschinger effect.

In order to model such phenomena, we have to reconsider some of our basic assumptions. The first notion is that our definition of yield will be "departure from linearity" and not 0.2% offset. When this definition of yield is used, all the repeatable observations can then be attributed to the changes in the *hardening behavior* and *no changes need to be made in the elastic response*.

If we now consider the way in which the yield surface changes when we use this more stringent definition, the yield surface changes its shape substantially. Moreover, the centroid of the yield surface moves by large amounts. Thus, the role of the *shift parameter* α emerges clearly. On the other hand, if we assume the yield is defined through backward extrapolation from the plastic region, the role of the shift parameter is much diminished. Thus we see that there is some leeway in how we define these quantities (see [Khan and Huang (1995)] Chapter 6 for a much detailed discussion of the role of the shift tensor).

8.4.3 *Quadratic yield surface plasticity with kinematic hardening*

The simplest extension of the J2 plasticity that also accounts for the motion of the yield surface is by the introduction of a yield function whose centroid *translates* with changes in plastic strain. Thus the basic constitutive equation (8.1) is modified as follows:

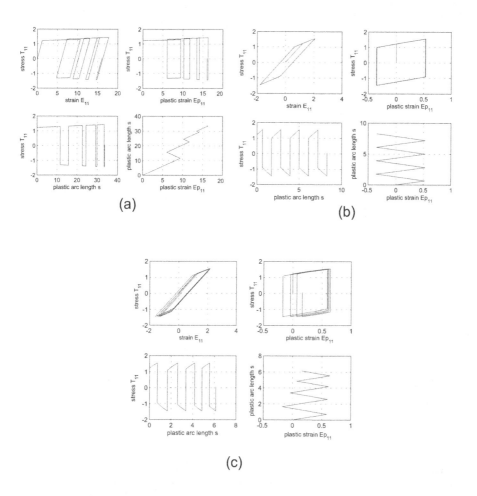

Fig. 8.4 Cycling around a point with nonzero average strain. (a) pure isotropic hardening showing the "shakedown phenomenon (b) pure kinematic hardening and (c) combined effect showing ratchetting phenomenon.

Elastic constitutive equations: $[\boldsymbol{\sigma}] = \mathbb{C}[\boldsymbol{\varepsilon} - \boldsymbol{\varepsilon}_p]$

Yield Condition: $\dfrac{\sqrt{[\boldsymbol{\xi}]^T (\mathcal{Y})[\boldsymbol{\xi}]}}{\kappa^2} - 1 = 0,\ \boldsymbol{\xi} = \boldsymbol{\sigma} - \boldsymbol{\alpha}.$

Flow Rule: $[\dot{\boldsymbol{\varepsilon}}_p] = \phi[\mathbf{N}],$

$$[\mathbf{N}] = \left[\dfrac{\partial f}{\partial \boldsymbol{\sigma}}\right] = \dfrac{1}{\kappa^2}(\mathcal{Y})[\boldsymbol{\xi}] \qquad (8.51)$$

Plastic Arc Length : $\dot{s} = \|\dot{\boldsymbol{\varepsilon}}_p\| = \dfrac{\phi}{\kappa}$

Hardening function : $\dot{\kappa} = \Theta(s, \kappa)\dot{s},\ \dot{\boldsymbol{\alpha}} = \phi\boldsymbol{\Theta}$

Kuhn-Tucker Condition: $\phi \geq 0,\ f \leq 0,\ \phi f = 0.$

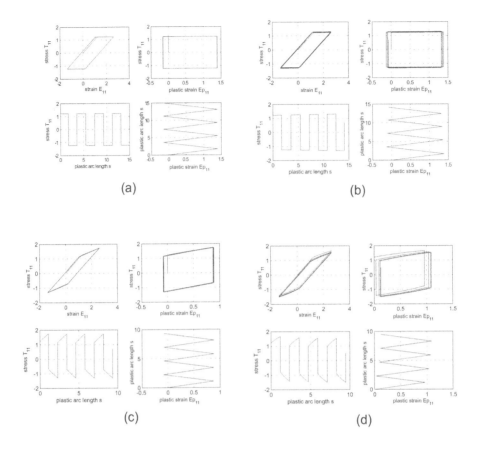

Fig. 8.5 Uniaxial cyclic response of different material models. The material is subject to a uniaxial strain cycle between two fixed strain values with zero average strain. T_{11} reference to σ_{11} and E_{11} refers to ε_{11}. T_{11} reference to σ_{11} and E_{11} refers to ε_{11}. (a) Response of a perfectly plastic material. Notice the immediate appearance of a limiting curve (b) Response of material with a small amount of isotropic hardening. Notice that the stress-strain curve is a spiral. (c) Response of the material with pure kinematic hardening. Compare (b) and (c) to see the difference between isotropic and kinematic hardening curves and (d) response of a material with combined effects showing mean stress relaxation.

Notice that we have now included a constitutive equation for the back stress $\boldsymbol{\alpha}$ in the formulation. Also note that the *projection operator* $\mathcal{Y}_p$ for plane stress introduced in equation (7.23) on p.224 is a special case of the general second order tensor $\mathcal{Y}$ in (8.51b) so that the plane stress case can be accommodated easily as a special case of the above formulation.

The value of $\boldsymbol{\alpha}$ denotes the location of the centroid of the yield surface. Thus $\boldsymbol{\alpha} = 0$ represents the initial yield surface which is only a function of the stress.

Now what constitutive equation should one use for $\boldsymbol{\alpha}$? Since $\boldsymbol{\alpha}$ represents the translation of the yield surface, we can list some of its characteristics as follows:

- If we plot $\boldsymbol{\alpha}$ versus $\boldsymbol{\varepsilon}_p$ in one dimension, we find that $\boldsymbol{\alpha}$ should increase gradually and saturate to a finite value as the plastic strain increases.
- If we consider a 3-D monotonic loading problem, $\boldsymbol{\alpha}$ generally increase in the direction of the pre-stress.

Based on these two criteria, a simple constitutive equation for the evolution of the back stress was suggested by Armstrong and Frederick [Armstrong and Frederick (1966)] and is of the form

$$\dot{\boldsymbol{\alpha}} = c_1 \dot{\boldsymbol{\varepsilon}}_p - c_2 \boldsymbol{\alpha} \dot{s}. \tag{8.52}$$

It can be shown that this model will have a few of the features that we listed for cyclic loading (especially for constant amplitude cycles). However, for complex loading paths, the response is not quite satisfactory. Much more elaborate kinematic hardening laws have been proposed in the literature (see e.g., [Chaboche (1989b,a, 1991, 1993a,b)] and new proposals are being made based on the results of simulations of discrete dislocations ([Guruprasad and Benzerga (2008)]). The central idea behind these kinematic hardening laws is to account for more features of the plastic strain path.

For ease of computation, we can write this as

$$\dot{\boldsymbol{\alpha}} = \phi \boldsymbol{\Theta} \tag{8.53}$$

where, by using (8.51), we have defined $\boldsymbol{\Theta} = c_1 \mathbf{N} - c_2 \boldsymbol{\alpha}/\kappa$. In this form, it is easier to set up the differential equation for ALL the inelastic parameters since then all of them are of the form $\dot{\mathbf{x}} = \phi f(\sigma, \mathbf{x})$.

8.4.4 *Homogeneous motions and governing differential equations*

We are now at a point where we can consider a body subject to homogeneous deformation where some combination of stress and strain are controlled, i.e. we have the familiar control equation

$$\mathbb{A}[\dot{\sigma}] + \mathbb{B}[\dot{\varepsilon}] = [\mathbf{f}(t)], \tag{8.54}$$

where, as usual ($\mathbb{A}$) and ($\mathbb{B}$) reflect the combination of stress and strain that is under our control. The remaining equations are to be obtained from (8.51) and (8.52). Since, by now you should have carried out this derivation many times, we will leave the derivation of the equations as an exercise, merely providing the results with hints as necessary.

⊙ **8.14 (Exercise:). Quadratic plasticity with hardening**
Show that for a material whose constitutive equations are given by (8.51) and (8.52), and where the prescribed loading is given by (8.54),

(1) The conditions for plastic flow are determined as follows: Plastic flow does not occur if

$$[\boldsymbol{\xi}]'(\mathcal{Y})[\boldsymbol{\xi}]/\kappa^2 < 1 \text{ or } [\mathbb{C}[\mathbf{N}]'\{\mathbb{AC}+\mathbb{B}\}^{-1}[\mathbf{f}(t)] < 0. \qquad (8.55)$$

(2) If plastic flow occurs, then ϕ is given by

$$\phi = h\hat{g}, \ \hat{g} = [\mathbf{N}]'(\mathbb{C})[\dot{\varepsilon}], \ h = \frac{1}{[\mathbf{N}]'\mathbb{C}[\mathbf{N}] + [\mathbf{N}]'[\Theta] + \Theta(s,\kappa)/\kappa^2}. \qquad (8.56)$$

(Hint: Differentiate the yield function, use the rate form of the elastic response and the evolution laws for κ and α.)

(3) If plastic flow occurs, then

$$[\dot{\boldsymbol{\sigma}}] = \mathbb{C}\{(\mathcal{I}) - h[\mathbf{N}][\mathbf{N}]'\mathbb{C}\}[\dot{\varepsilon}]. \qquad (8.57)$$

(Hint: Use the rate form of the elastic law together with the equation for $\dot{\varepsilon}_p$ and ϕ also revisit section 6.5.1 on p.207.)

These together with (8.54), (8.52) and (8.51e,f) result in 20 differential equations for the 20 unknowns $\{\boldsymbol{\sigma}, \boldsymbol{\varepsilon}, \boldsymbol{\alpha}, \kappa, s\}$. Notice that, as we have done in the previous chapters, we completely eliminate the plastic strain.

In order to see the influence of the back stress, a number of simulations of cyclic uniaxial tension were carried out for different types of hardening characteristic (ranging from ideal plasticity to nonlinear kinematic and isotropic hardening). The MATLAB programs that are used to generate these graphs are available online (see authors' websites listed in the preface.

The results are shown in Figs. 8.5 and 8.4. The captions are self explanatory. We urge you to run the program with different constants and different stress and strain cycles and observe the response.

8.5 Summary

(1) There are a number of generalizations that are possible based on the J2 plasticity model. They are (1) replace the simple yield function with a general form, (2) introduce expansion and shift parameters, and (3) replace the form for $\dot{\varepsilon}_p$

with a more general form derived from a plastic potential.
(2) The common features of these models are (1) The elastic constitutive equation is the same (reflecting the fact that the yield and flow behavior can be altered easily but not the elastic response), (2) The plastic strain rate is proportional to the gradient of some scalar function of the stress and possibly other inelastic variables, (3) the use of the Kuhn Tucker conditions to represent the elastic and plastic behaviors.
(3) If the plastic strain rate is proportional to the gradient of the yield function with respect to the stress, the corresponding flow rule is called the normality flow rule. Theories of plasticity based on this notion are usually called associative plasticity.
(4) Many different yield functions can be obtained by using functions of either the principal invariants or the eigenvalues of the stress. In general, the yield functions are chosen to be convex functions of the stress. Some anisotropic yield functions have been developed for rolled sheets etc. but this is an area of considerable active research.
(5) The notion of plastic arc length can be generalized to anisotropic media by introducing the rate of change of arc length as the ratio of the current rate of mechanical power dissipation and the current yield strength.
(6) The Bauschinger effect is commonly modeled by introducing the notion of a "back-stress" $\boldsymbol{\alpha}$ which has its own evolution equation.
(7) It is quite straightforward to find out if plastic flow takes place; our criteria is based on the notion that plastic flow takes place if the change in the stress and strain under the assumption of no plastic flow violates the Kuhn-Tucker conditions.
(8) If plastic flow occurs, the value of ϕ (and hence the tangent modulus) can be found by differentiating the yield function and substituting the evolution laws as well as the elastic law in rate form. This invariably leads to an expression of the form $\phi = h(df/dt|_{\phi=0}) = h\hat{g}$ where $\hat{g}$ is the so-called *strain space loading function* and h is a scalar function of the state variables which is less than 1.
(9) The resulting differential equations can be implemented in MATLAB using either ODE23 or ODE45 even though tangent modulus has discontinuities.

8.6 Projects

(1) **Cyclic Hardening response:** By closely observing the Figs. 8.5 and 8.4, write a brief note on the influence of isotropic and kinematic hardening on the cyclic response of materials. What other simulations would you perform to gain greater understanding? Run these simulations either by writing your own code or by modifying the MATLAB code available online (See authors' websites in the preface).
(2) Develop and write a MATLAB program to simulate the response of a crushable

foam (see equations (8.35)–(8.37)) (Assume that $\kappa = constant$) when subject to cyclic combined tension, shear and internal pressure, (i.e. the loading is prescribed by $\varepsilon_{11} = a\sin(wt), \varepsilon_{12} = b\sin(2wt)$ and $\sigma_{33} = c\sin(3wt)$, all other components of stress being zero).

(3) Repeat the process for the Drucker Prager Model (see equations (8.32)–(8.34)). Here, consider a cyclic compression, i.e. $\varepsilon_{11} = 1 - \cos^2(wt)$ with all other components of stress equal to zero.

(4) Show that, in general, given a free energy function $\psi(\varepsilon, \varepsilon_p)$ from which the stress is given by

$$\boldsymbol{\sigma} = \frac{\partial \psi}{\partial \boldsymbol{\varepsilon}} \Rightarrow \dot{\boldsymbol{\sigma}} = (\mathcal{C}_e)[\dot{\boldsymbol{\varepsilon}}] - (\mathcal{C}_p)[\dot{\boldsymbol{\varepsilon}}_p], \tag{8.58}$$

$$\text{where } (\mathcal{C}_e) = \frac{\partial^2 \psi}{\partial \varepsilon^2}, (\mathcal{C}_p) = \frac{\partial^2 \psi}{\partial \varepsilon \partial \varepsilon_p}, \tag{8.59}$$

if $\dot{\boldsymbol{\varepsilon}}_p = \phi \mathbf{P}(\boldsymbol{\sigma}, \mathbf{Q})$ and the set of all inelastic variables $[\mathbf{Q}]$ (including the plastic strain, the hardening parameters etc.) obey evolution equations of the form $\dot{\mathbf{Q}} = \phi \Theta(\boldsymbol{\sigma}, \mathbf{Q})$ then if plastic flow occurs we have

$$\phi = h\hat{g}, \hat{g} = df/dt|_{\phi=0} = [\mathbf{N}]^t(\mathcal{C}_e)[\dot{\boldsymbol{\varepsilon}}], h = -\frac{1}{[\mathbf{N}]^t(\mathcal{C}_p)[\mathbf{P}] + [\partial f/\partial \mathbf{Q}]^t[\Theta]}, \tag{8.60}$$

where $[\mathbf{N}] = [\partial f/\partial \boldsymbol{\sigma}]$. What is the tangent modulus in this case?
(In this way, some of the most general inelasticity models (including certain types of Damage models etc.) can be brought into a common framework)

(5) Carry out a literature search on the different kinematic hardening models developed and write a short paper on their comparative merits and advantages.

8.7 Exercises

(1) Show that for the case of a material with the normality flow rule, irrespective of the exact form of the yield function and the hardening parameters, the tangent modulus matrix is symmetric and is of the from $(\mathcal{K}) = \mathbb{C} - h\mathbb{C}[\mathbf{N}][\mathbf{N}]^t(t\mathbb{C})$ where $[\mathbf{N}] = [\partial f/\partial \boldsymbol{\sigma}]$.

(2) Show that, for non-associative plasticity with the stress given by $\boldsymbol{\sigma} = \mathcal{C}(\boldsymbol{\varepsilon} - \boldsymbol{\varepsilon}_p)$, and with a yield function f and a plastic potential H, the tangent modulus is of the form $(\mathcal{K}) = (\mathbb{C}) - h\mathbb{C}[\mathbf{P}][\mathbf{N}]^t$ where $[\mathbf{P}] = [\partial H/\partial \boldsymbol{\sigma}]$ and $[\mathbf{N}] = [\partial f/\partial \boldsymbol{\sigma}]$. In other words it is not symmetric.

Chapter 9

Thermodynamics of Elasto-plastic Materials: The Central Role of Dissipation

Learning Objectives

After reading this chapter you should be able to do the following

---

(1) Enumerate the list of constitutive relations for a general rate independent elasto-plastic material that has a yield function and a plastic flow potential.

(2) Define the "entropic equation of state" and list the assumptions regarding it that leads to the definition of absolute temperature;

(3) Carry out the manipulations that lead to the definition of the Helmholtz and Gibbs potentials;

(4) Explain the physical meaning of the Helmholtz potential and the Gibbs potential and explain why the former is called the *isothermal work function* while the latter is called the *isothermal complementary function*;

(5) Derive the equations for the entropy and the internal energy in terms of these two potentials;

(6) Derive the work dissipation equation in terms of the Helmholtz and Gibbs potentials and their derivatives;

(7) List the various assumptions that lead to the normality flow rules and compare the subtle differences between them;

(8) Describe the Maximum Rate of Dissipation Hypothesis (MRDH).

(9) Demonstrate how MRDH allows us to define both the yield function and the plastic potential;

(10) Describe how history dependence and the Bauschinger effect are introduced in this formulation and derive the form for the history dependence.

> **CHAPTER SYNOPSIS**
>
> In this chapter, we justify the various ad-hoc assumptions that we introduced in Chapters 6-8 from a thermodynamical point of view. Specifically, we show that *if the rate of mechanical dissipation is a homogeneous function of the plastic strain rate and independent of the elastic strain, then the assumption that processes occur in such a way as to maximize the rate of dissipation implies the existence of a yield function that is convex and that the plastic strain rate is along the normal to the yield surface.*
>
> We also show how to incorporate the backstress (or shift) tensor and the isotropic hardening parameter by introducing a specific kind of history dependence of the stored energy.

Chapter roadmap

This chapter is fundamentally different from the others in the book since it is more philosophical in nature and presents a point of view as to how one could unify the constitutive relations for different dissipative materials using a common thermodynamical framework. Thus, the approach presented here is more explanatory than procedural, and may be skipped on first reading without loss of continuity. Indeed *you do not need to know anything in this chapter in order to implement elastoplasticity models. On the other hand, we urge you to explore this approach and hope that it will provide a deeper understanding of elastoplasticity and stimulate you to work on the thermodynamics of dissipative phenomena, irrespective of whether they agree or disagree with the approach presented here.*

The chapter begins with a brief discussion of the balance laws in "conservation form" and then goes on to introduce macroscopic state variable and a statistical notion of entropy. Then the entropic equation of state is introduced, followed by a discussion of heating and working. In many ways this discussion is a continuum version of the discussion in Chapter 2. We then introduce the notion of a Legendre transformation and hence introduce the Helmholtz and Gibbs potentials. We derive the equation for the rate of mechanical dissipation next, followed by the *maximum rate of dissipation hypothesis* MRDH. We show, by means of an example, how the maximization of the rate of dissipation together with the assumption of positive homogeneity of the rate of dissipation delivers both the yield function and the normality rule. We also show an example of a non associative plasticity model that results from the MRDH. We then show how to model the Bauschinger effect and isotropic hardening by means of a history dependent Helmholtz potential and how to model the history dependence using differential equations.

9.1 Generalization of J2 plasticity: the rationale

In order to understand the nature of the constitutive equations of plasticity and to see its connection with other constitutive equations (say for viscoelasticity etc.), it is vital to examine the thermodynamic underpinnings of both and see that there is a large amount of overlap. The vital difference between the different constitutive equations is in the way in which mechanical power is dissipated, intuitively, viscoelastic materials have dissipative mechanisms that are counterparts of viscous dampers, while elastoplastic materials have dissipative mechanisms that are the counterparts of dry friction. As with particle dynamics with dry friction, problems involving dry friction are quite hard to formulate and solve and require special attention.

In Chapter 8 on general response functions of rate independent plasticity, the general structure of elastoplastic response functions were presented by starting from J2 plasticity and generalizing various concepts in a piece-meal manner. Thus, by this process of generalization, we introduced the "history" variables, $\mathbf{Q}$, which included the plastic arc length s, the yield stress κ, and the back stress $\boldsymbol{\alpha}$, as well as other variables that we might need.

We then stipulated the following set of constitutive relations:

MECHANICAL CONSTITUTIVE LAWS: SMALL STRAIN

List of state variables: $\mathbb{S} = \{\boldsymbol{\sigma}, \boldsymbol{\varepsilon}, \boldsymbol{\varepsilon}_p, \mathbf{Q}\}, \mathbf{Q} = \{s, \kappa, \boldsymbol{\alpha}\}$,

The strain energy function: $\psi = \psi(\boldsymbol{\varepsilon}_e), \boldsymbol{\varepsilon}_e = \boldsymbol{\varepsilon} - \boldsymbol{\varepsilon}_p$,

The elastic response function: $\boldsymbol{\sigma} = \dfrac{\partial \psi}{\partial \boldsymbol{\varepsilon}_e}$,

The yield function: $f(\boldsymbol{\xi}, \mathbf{Q}), \boldsymbol{\xi} := \boldsymbol{\sigma} - \boldsymbol{\alpha}$,

The plastic potential: $H(\boldsymbol{\xi}, \mathbf{Q})$,

The flow rule: $\dot{\boldsymbol{\varepsilon}}_p := \dot{\boldsymbol{\varepsilon}} - \dot{\boldsymbol{\varepsilon}}_e = \phi \dfrac{\partial H}{\partial \boldsymbol{\xi}}$,

The plastic flow conditions: $\phi \geq 0, f \leq 0, \phi f = 0$,

The hardening Laws: $\dot{\mathbf{Q}} = \phi \boldsymbol{\Theta}(\mathbb{S})$.

(9.1)

The *normality flow rule* (also called the *associated flow rule*) is given by the condition that $H = f$. We note that we need constitutive equations for three scalar functions ψ, f and H, and one list or (column vector) function $\boldsymbol{\Theta}$ which includes the evolution laws for the list of all the variables which are collectively referred to as $\mathbf{Q}$.

At first sight, the laws of plasticity appear very different from those of elasticity, non-Newtonian fluid mechanics or those of viscoelastic bodies. Some essential aspects of this difference are

(1) The equation is *implicit* in nature, i.e. we don't specify the stress as a function of the history of the kinematical parameters. Rather, we specify that the stress depends upon ε_e which is not a purely kinematical quantity. Rather its rate of change is prescribed by a constitutive equation that depends upon the stress.
(2) Unlike the other theories, much of the theory of plasticity employs a very "geometric" approach. We speak of yield "surfaces", "normality", "plastic arc length" etc.
(3) There are discontinuities in the response functions.

Of these, the most important difference is the use of *geometrical notions*. For example, in classical elasticity, the stress is the gradient of the strain energy function; we don't speak of it as being "normal" to the "strain energy surface". We routinely do so for the plastic strain rate $\dot{\varepsilon}_p$ in plasticity.

While such a geometrization allows for a great deal of clarity, it hides some of the similarities between models for other materials and that for plasticity. It is for this reason, that while we have used some geometrical concepts in deference to the usual way in which the subject is taught, we have deliberately stayed away from an explicit geometrization of the constitutive equations and have abstained from drawing many "yield surfaces".

We believe that such geometrization actually obscures some aspects of the theory. Especially because *it is not necessary and in some cases even detrimental to always insist on a geometrical viewpoint and hinders some extensions to materials without yield surfaces etc.*

Extremum principles

You may be aware that, in classical elasticity, the fact that $\boldsymbol{\sigma} = \partial \psi / \partial \boldsymbol{\varepsilon}$ is the result of a statement of the second law of thermodynamics (which you might have seen in the form of Castigliano's first theorem, i.e. that a equilibrium state of a body is that which minimizes the strain energy among all the allowable configurations). An alternative way of stating this is that "for an elastic body, the work done during any deformation is independent of the path" or equivalently "The work done in any closed cycle of deformation is zero". This sort of derivation of the constitutive equation from a global statement of the response is very appealing to some people and we do not hesitate to admit that we belong to this group. Variational principles such as Castigliano's theorems and other such have the appearance of "magic", since they seemingly enable us to deduce so much from so little—in spite of the fact that we are aware that we cannot get out more than we have already put in. Perhaps the attraction of such variational principles is due to the fact that they suggest new directions for research and exploration.

In any case, the question arises whether similar "variational" statements can be made regarding the "normality" flow rule. The short answer is yes.

In this chapter we will show how the equations of plasticity can be obtained in a way which is consistent with thermodynamical restrictions rather than from an ad-hoc approach that was presented earlier in Chapter 8. In the process, we will also introduce a "variational" statement that will allow us to "derive" the normality flow rule as well as other classes of flow rules. In general, classical variational principles are used to determine entire trajectories (as in the case of Hamilton's principle) or equilibrium configurations (as in the case of the Principle of minimum potential energy). There is another class of variational principles of which the prime example is "Gauss' theory of least constraint" which gives us *constitutive* laws. The principle that we are introducing in this chapter belongs to the second category and will give us the normality flow rule[1].

In the approach to be presented in this chapter, the entire constitutive structure is derived from constitutive assumptions for two functions

- The strain energy or Helmholtz potential function which may be history dependent and
- the rate of dissipation function and a requirement that among all allowable processes, the actual process is that which is the most dissipative.

The first part of this statement is obvious: since elasticity is assumed to be a subset of plasticity and we all accept that the elastic response can be derived from a suitably chosen strain energy function, we ought to be able to define a suitable strain energy function for the elastoplastic materials also. The second part of the previous paragraph needs some quick elaboration:

Since we know that plasticity is a dissipative phenomenon, let us calculate how much mechanical work is dissipated. We know that it should be something like the stress times the plastic strain rate. Thus, let us *introduce* a variable ζ to be the mechanical power per unit mass that is irrecoverably lost[2] and define it as

$$\varrho\zeta := \boldsymbol{\xi} \bullet \dot{\boldsymbol{\varepsilon}}_p = (\boldsymbol{\sigma} - \boldsymbol{\alpha}) \bullet \dot{\boldsymbol{\varepsilon}}_p. \qquad (9.2)$$

Then, from the normality flow rule (9.1f) we note that

$$\varrho\zeta = \boldsymbol{\xi} \bullet (\phi \frac{\partial H}{\partial \boldsymbol{\xi}}) = \phi, f(\xi, \mathbf{Q}) \qquad (9.3)$$

where $f = \boldsymbol{\xi} \bullet \partial h / \partial \boldsymbol{\xi}$ is a function of the state variables. In words we obtain the fact that the rate of dissipation is ϕ times some function of the state and other variables.

If H is a positive homogeneous function[3] of $\boldsymbol{\xi}$ (as is often the case) then the above equation reduces to $\zeta = \phi H$ since $(\partial H/\partial \boldsymbol{\xi}) \bullet \boldsymbol{\xi} = h$. Thus, we see that *the*

---

[1] In the case of Gauss's principle, it too gives us a "normality" law: The constraint forces are perpendicular to the constraint surface.

[2] We will refer to it as the rate of dissipation. Many papers and books use the term "dissipation" or "dissipation function". We have chosen to explicitly introduce "rate" when referring to this quantity.

[3] A positive homogeneous function $f(x)$ is one that satisfies $f(kx) = kf(x)$ for all positive k. If it is differentiable and satisfies the identity $x(df/dx) = f$.

plastic potential and the consistency parameter are intimately related to the rate of dissipation.

We can go a step further. By taking the magnitude of both sides of equation (9.1f), and solving for ϕ we will obtain the following expression:

$$\phi = \frac{\|\dot{\varepsilon}_p\|}{\|\partial h/\partial \boldsymbol{\xi}\|}. \tag{9.4}$$

Now substituting the above expression onto the right-hand side of (9.3b), we will get

$$\zeta = \|\dot{\varepsilon}_p\| \frac{h}{\|\partial h/\partial \boldsymbol{\xi}\|} \Rightarrow \zeta = R(\xi, \mathbf{Q})\,\|\dot{\varepsilon}_p\|,\ R = \frac{h}{\|\partial h/\partial \boldsymbol{\xi}\|} \tag{9.5}$$

Thus, for any rate independent model, the rate of mechanical dissipation being a product of the function R and the norm of the plastic strain rate, is *a positive homogeneous function of* $\|\dot{\varepsilon}_p\|$. We will show in the subsequent sections, that this is a rather general result that is true for all rate independent materials.

You might be wondering at this stage as to how did we "suddenly" introduce ζ, and where did we get the equation (9.2). In order to see this, we must enter into some of the elements of thermodynamics that we have been skirting up to now. In order to appreciate this section, it is a good idea to be familiar with thermodynamical notions that were introduced in Chapter 2. So, we would urge you, dear reader, to go back and read Chapter 2 if you have not already done so.

9.2 Foundations of the thermodynamics of continua

Continuum thermodynamics is a combination of conservation laws and the idea of a "local equation of state". There are many subtle assumptions that have to be carefully examined in order to fully appreciate the concepts that are introduced. However, once we accept the concepts, continuum thermodynamics provides a *systematic framework to consistently formulate constitutive equations*.

9.2.1 *Governing balance laws for small deformation*

We will now introduce, in an abbreviated form, the governing thermomechanical balance laws for a continuum. These are the counterparts of similar laws introduced for lumped systems and for the truss and beam systems in Chapters 2, 3 and 4. Thus, we have a body $\mathbb{B}$ whose particles occupy positions $\mathbf{X}$ at time $t = 0$ (see Fig. 9.1). The displacement of a typical particle which was at $\mathbf{X}$ is given by a displacement vector $\mathbf{u}(\mathbf{X}, t)$. We define the velocity ($\mathbf{v}$) and acceleration ($\mathbf{a}$) as the first and second time derivatives of the position $\mathbf{u}$.

Further, corresponding to the definitions of "extensive parameters" in classical thermodynamics, we shall assume that at each instant of time, the internal (or

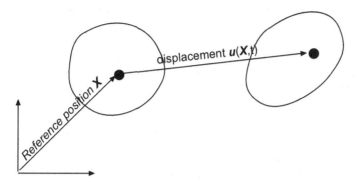

Fig. 9.1 A schematic figure of a continuum undergoing a deformation. The reference position **X** of the particle serves also as its label. The displacement **u**(**X**, t) is the vector that goes from the reference position to the current position.

potential) energy per unit mass is $\exists(\mathbf{x}, t)$ and its total energy density (or the Hamiltonian) $\mathbb{H}(\mathbf{x}, t)$ per unit mass is the sum of the internal (or potential) energy density $\exists$ and the kinetic energy density, i.e.

$$\mathbb{H}(\mathbf{x}, t) = \exists + \frac{1}{2}\mathbf{v}\bullet\mathbf{v}. \tag{9.6}$$

These quantities are governed by the usual balance laws that are assumed to be obeyed by all materials. Specifically, we introduce the following balance laws (assuming small deformations):

$$\begin{aligned} \dot{\mathbf{x}} &= \mathbf{v}, \\ \varrho\dot{\mathbf{v}} &= \operatorname{div}\boldsymbol{\sigma}, \\ \boldsymbol{\sigma}^t &= \boldsymbol{\sigma}, \\ \varrho\dot{\exists} &= -\operatorname{div}\mathbf{q} + \boldsymbol{\sigma}\bullet\dot{\boldsymbol{\varepsilon}} \end{aligned} \tag{9.7}$$

where $\boldsymbol{\sigma}$ is the Cauchy stress tensor (representing the momentum flux density per unit area into a material region), **q** is the heat flux (or energy flux per unit area in thermal form *leaving* the system), and $\dot{\boldsymbol{\varepsilon}}$ is the strain rate, with $\boldsymbol{\varepsilon} := 1/2(\nabla u + \nabla u^t)$. Also, the superposed dot represents the time derivative, the superscript t represents the transpose and $\operatorname{div}(\cdot) = \nabla\bullet(\cdot)$ stands for the divergence operator with respect to the spatial coordinates, and where ∇ stands for the gradient operator.

In order to understand the developments further we need to gain a clear picture of the various terms in equations (9.7); we need to observe that *all conservation laws share a common format*. Thus, if we consider an infinitesimal volume element dV and a scalar, vector or tensorial quantity ϕ that is conserved (say the momentum), then the law of conservation of this quantity, Π may be written as follows:

> **STANDARD FORM OF CONSERVATION LAWS**
>
> The rate of change of a conserved quantity in an infinitesimal volume = flux of the quantity flowing into the infinitesimal volume. This, in equation form, reads
>
> $$\frac{d\phi}{dt}dV = \mathrm{div}(\Pi)dV, \qquad (9.8)$$
>
> where Π is the flux density (influx of the quantity per unit area).
>
> The difference between our use of the word balance law and the words conservation law is the fact that *only the latter will have the left-hand side the rate of change of a quantity and the right-hand side will be in divergence form*

Thus the divergence of a quantity indicates the rate of inflow of the quantity into a region. Applying this understanding to equation (9.7) and noting that ϱ is a constant, we see that equation (9.7b) refers to the conservation of momentum ($\varrho\mathbf{v}$ being the momentum per unit volume). However, (9.7d) does not refer to a conserved quantity. This is because it is the total (kinetic+potential) energy that is conserved and not just the potential energy.

⊙ **9.1 (Exercise:). Conservation of Energy**
Show by using (9.6) and (9.7) that the specific total energy (total energy per unit mass) given by $\mathbb{H} = \beth + 1/2\mathbf{v}\cdot\mathbf{v}$ is conserved, and the time rate of change of the specific total energy can be expressed as

$$\varrho\dot{\mathbb{H}} = \mathrm{div}(-\mathbf{q} + \boldsymbol{\sigma}[\mathbf{v}]) = \mathrm{div}\,\Gamma, \qquad (9.9)$$

where Γ is the specific energy flux that is the sum of the heat flux and the mechanical working ($\Gamma = \boldsymbol{\sigma}[\mathbf{v}] - \mathbf{q}$).
Hint: Start from the middle terms in (9.9) expand out the right-hand side using (9.7b). Then use (9.6) for the left-hand side and show that we will get (9.7c).

9.2.2 The macroscopic state variables

Fundamentally, the starting point of any thermodynamical analysis is the introduction of convenient "state variables" that can be used to approximate the behavior of the material. While there was and continues to be philosophical debates as to the criteria for selection of state variables in general, we subscribe to the notion that it may be more practical to consider each system independently and our needs in question and ask as to what would be a suitable choice, rather than to try and figure out a universal method for deciding on state variables. In this spirit, we will see that our choice (9.10) will suffice quite well for our purpose.

Thermodynamics would be unnecessary if we could easily track the motion (position and velocity) of each atom in the body. The list of positions, momenta and type of each and every atom at any instant of time is called a "microstate" of the system at that time. However, this listing is impossible; there are too many atoms and we don't usually need that much information.

Continuum mechanics gets around this problem by "smearing out" or smoothing out all of this information and representing this "smoothed" or smeared behavior by means of variables representing the aggregate or "smoothed out" behavior. These variables are called the "macrostate variables" or simply "state variables", and we track these macrostate variables as functions of time. The consequent loss of information due to ignoring the details of the microstates is at the heart of thermodynamics.

The basic local state variables introduced so far are the displacement $\mathbf{u}$, the velocity $\mathbf{v}$, the stress $\boldsymbol{\sigma}$, the heat flux $\mathbf{q}$ and the internal energy $\exists$. The evolution of these variables is determined by equations (9.7)–(9.9). This set of local state variables (and their spatial and time derivatives) is sufficient to characterize materials such as gases, and Newtonian viscous fluids. Thus, the first true 3-D continuum that you were exposed to as an undergraduate (remember fluid mechanics?) was a fluid for which the stress was a function of the gradient of the velocity. Similarly for linear elasticity, the stress is a function of the gradient of the displacement.

However, in order to deal with a wider class of materials, these state variables must be augmented by other local state variables that represent internal microstructural features of the materials. To understand how this occurs, we will consider a concrete example:

Consider the case of a thermoelastic solid. Let us assume that, when the body is thermally and mechanically isolated, it occupies a "natural" or stress-free configuration[4] κ_r. To be precise

---

NATURAL CONFIGURATION

A natural configuration of a material is the configuration it attains when all external stimuli are removed, i.e. "when no one is looking or sensing it". Clearly, this is an abstract idea and for practical purposes it means that when the sensing instruments are not disturbing the body so that it may be considered "isolated" for all practical purposes.

When such a material is heated or deformed, its configuration changes. However, when the external stimulus is reversed, the material reverts to its original or natural configuration. This is the continuum generalization of a spring and is a very suc-

---

[4] A configuration means "relative arrangement of its parts". Here, it means assigning a unique position to each particle in the continuum.

cessful representation of most solids under normal operating conditions. For such a material, its mechanical state is determined by its deviation from some suitably chosen natural configuration. The response of such materials will be called *elastic response*.

On the other hand, for most real materials, when the external stimulus is sufficiently large, the "microstructure", i.e. the positions of the atoms or groups of atoms that are invisible to the naked eye, may be fundamentally altered, and the effect of this alteration will be evident in the fact that there is a persistent change in some *macroscopic property* of the material, say the shape or the hardness or the ductility etc. When such internal structural rearrangements take place, an *irreversible* change occurs—the material will NOT recover to its original shape when you remove the external stimulus, but will typically take on a new shape. For example, imagine a metallic coat hanger on which a very heavy coat is hung. Its shape changes and remains changed even after the coat is removed—in other words, the natural shape of the coat hanger (i.e. its configuration) has changed. The original "natural" configuration of the material is altered by the structural rearrangement and the material is said to have undergone a "permanent" change of shape or size. Thus[5]

CHANGING NATURAL CONFIGURATIONS

The fundamental difference between elastic and inelastic materials is that, for the latter, the "natural configuration" (the configuration that it will revert to when all external stimuli are suddenly removed) can change, i.e. there is a persistent change in shape. We introduce a "moving configuration" to account for this.

Usually, this kind of microstructural change does not happen immediately. There is a definite time scale associated with the manifestation of the microstructural change. So, if you deform the body very quickly, there may not be enough time for the change to occur and the "natural configuration" will not change and for all practical purposes, the material will behave like an elastic material. Materials for which the above considerations are reasonable will be said to possess instantaneous elasticity.

When an external stimulus, in the form of stress is applied to the body, the body changes its shape. *For an inelastic body, this change of shape is also accompanied by a change in its natural configuration as well as changes in its microstructure.* Thus the stress in the body has two effects: it ensures that the current configuration is different from its natural configuration (causes the body to be distorted); it

---

[5]Some of this material is probably repetitive from previous chapters, but we have tried to make each chapter as independent as possible from the previous chapters so bear with us as we rehash some old materials.

also causes changes in the natural configuration itself. How do we measure these changes? The change in the overall shape of the body when compared to its initial shape is measured by the total strain $\varepsilon = 1/2(\nabla \mathbf{u} + \nabla \mathbf{u}^T)$. We will introduce a new variable called the "elastic strain" ε_e which has properties similar to the total strain ε to measure the difference between the natural configuration and the current configuration (see Fig. 9.2).

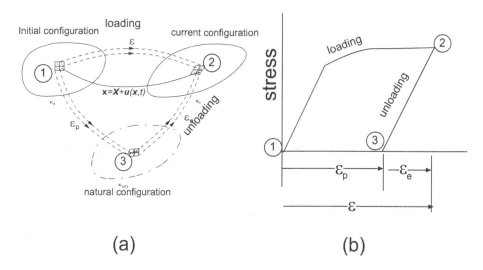

Fig. 9.2 (A) A schematic figure of the notion of a relaxed state. The body starts at state (1) and is deformed to state (2). If it is unloaded rapidly, it will reach a stress-free state (3). The accompanying stress, strain graph is shown. *For small deformations* the total strain ε is the sum of the elastic strain ε_e and plastic strain ε_p.

THE ELASTIC STRAIN

In small deformation inelasticity, the elastic strain ε_e is introduced as a measure of the difference in size and shape between the (changing) natural configuration and the current configuration as shown in Fig. 9.2. In general, it is not a purely kinematical quantity and its value is not equal to that of the total strain. Instead, we prescribe an evolution equation for it (or more precisely for $\dot{\varepsilon}_p := \dot{\varepsilon} - \dot{\varepsilon}_e$).

Thus, in view of the introduction of the "elastic strain", the set of dynamical state variables now becomes

$$\mathbb{S}(\mathbf{X}, t) = \{\exists, \varepsilon_e, \mathbf{u}, \mathbf{v}\}. \tag{9.10}$$

Each of the above variables is a function of $(\mathbf{X}, t)$. Note that the strain ε is not explicitly in the list since it is obtained from the spatial gradient of $\mathbf{u}$ and hence is a *derived* or *dependent* state variable. Similarly, the plastic strain is not in the list since it is the difference between ε and ε_e.

For future reference, most of our models will include only ε_e and $\exists$ or its equivalent as the ONLY state variables—not even the total strain. Thus our state variables will be

$$\mathbb{S}' = \{\varepsilon_e, \exists\}. \tag{9.11}$$

In most applications it is more convenient to use the temperature instead of the energy as a state variable. Similarly, it may be more convenient to use the stress rather than the displacement or its gradient as a state variable. These considerations will be delayed until the appropriate time. For now, (9.11) is more than sufficient.

9.3 The entropy and the equation of state

Since we have "smeared out" the positions, velocities and types of the individual atoms, we have "lost" information. This loss of information can be quantified as follows:

THE ENTROPY

A large number of different microstates (list of positions, velocities and types, of the atoms) will all give the same value for the "smeared out" variables. One measure of information "lost" is the logarithm of the total number of different microstates. They all give the same macrostate. This number is called the "entropy" of the system.

In general there are three distinct contributions to the entropy (or uncertainty) of the system: uncertainty in the positions of the atoms (called configurational entropy) whose changes depends upon the strain in the material), uncertainty in the velocity of the atoms which is called the "thermal entropy" and depends upon the temperature, and the uncertainty in the "type" of atom which is called the "mixing entropy" or the "entropy of mixing" and depends upon the concentration of different constituents in the body.

Of course, these are not independent of each other but it is a convenient way to think about the connection between entropy and uncertainty and allows us to carry out qualitative reasoning about the nature of the materials. Thus we could approximately write that

Total entropy = Configurational + Thermal+Mixing + Coupling terms.

Thus, what the entropy measures is the degree of "uncertainty" or "redundancy" regarding the actual microscopic state of the system given the macroscopic state variables. For example, if the entropy is 100, then we could roughly say that there are 10^{100} (remember the "log" in the definition of entropy?) microstates that will all have the same value for the macroscopic state variables. In physical applications, since the number of microstates are very large and for historical reasons, we will multiply the log of the microstates by the Boltzmann constant which is about 1.38×10^{-23}. This will convert it into smaller numbers.

> ⊙ **9.2 (Exercise:). Changing the entropy of a system**
> *Consider the following examples and find out whether the main change in the entropy is "configurational", "thermal" or "mixing" entropy.*
> *a) Melting a block of ice.*
> *b) Heating a block of steel until it becomes red hot*
> *c) Condensing steam to make water.*
> *d) Rolling an ingot of steel into a flat sheet.*
> *e) Adding milk and stirring coffee.*
> *Construct three examples where you are changing either the configurational, thermal or mixing entropy.*

One way of looking at the second law of thermodynamics is to say that

The total entropy of an *ISOLATED SYSTEM*[6] cannot decrease.

Since "isolated systems" are, by definition, unobservable, we cannot really know them. We get around this by imagining that the entire universe (or at least the system + "surroundings"+ observers) is isolated and then apply the rule to conclude that if we track the entropy exchange between the system and surroundings, then during any change of state, the entropy of the system + surroundings cannot decrease, i.e. entropy can be "produced" but cannot be destroyed.

9.3.1 Equations of state

The entropic equation of state

The behavior of a system subjected to external stimuli is determined in a large part by the changes in its entropy. Thus we need to connect up the changes in the

---

[6]We have highlighted the words "isolated system" because students forget this and think "entropy always increases" and then arrive at spurious paradoxes such as how can a metal crystallize from a melt? Doesn't its entropy decrease? How does a biological system grow? etc. Students are fascinated by such questions and this could serve as a great way to initiate a dialogue on isolated systems.

entropy to the changes in the macroscopic state variables of the system. We will assume that the entropy is a function of the macrostate variables that have been selected for our smeared out description[7].

The starting point in the thermodynamical development of the constitutive response is the introduction of the "entropic equation of state", i.e. assuming an expression for the entropy as a function of the state variables of the form

$$\eta = \eta(\beth, \mathbb{S}_m), \qquad (9.12)$$

where $\mathbb{S}_m$ are all the "mechanical state variables other than the internal energy".

While the physical underpinnings of the notion of entropy and the entropic equation of state has been extensively investigated both in classical macroscopic thermodynamics (in terms of Carnot cycles as well as more abstract notions of inaccessibility due to Caratheodory, or the notions of information or statistical uncertainty due to Shannon or notions of complexity due to Kolmogorov), the focus of this book is more practical—for the kinds of material models that we are interested in all these different approaches speak with one voice. Thus, we shall formally assume (without further justification here) the existence of a specific entropy density η (per unit mass) which is a function of the state variables $\mathbb{S} = (\beth, \mathbb{S}_m)$ of the material, and has the following attributes ([Callen (1985)])

PROPERTIES OF ENTROPIC EQUATION OF STATE

(1) It is an increasing function of the internal energy (i.e. more microstates become available as the potential energy increases), and the absolute temperature T is defined by

$$T = \frac{1}{\partial \eta / \partial \beth} > 0. \qquad (9.13)$$

(2) The maximization of the total entropy functional $S = \int_\Omega \varrho \eta \, dv$ over the set all allowable states, consistent with the constraints imposed and the processes allowed, delivers an equilibrium state of the body when it is thermally and mechanically isolated (i.e. no heating or working or exchange of mass allowed).

(3) The entropy tends to zero as the absolute temperature tends to zero.

(4) The entropy is a constant in those adiabatic processes during which the material responds elastically.

---

[7]This is not an obvious or trivial assumption. It usually goes under the "hypothesis of local equilibrium" and has far reaching consequences. Interested readers should consult [Groot and Mazur (1962)].

The energetic equation of state

In view of the assumption (9.13) above, it is easy to see that the entropic form of the equation of state can be inverted to give the internal energy as a function of the entropy and the other state variables. This latter form is more commonly used and is referred to as the energetic form of the equation of state.

Having introduced the notion of an equation of state, we will consider a special equation of state wherein the entropy depends only upon one mechanical and one nonmechanical variable: the elastic strain ε_e and the internal energy $\beth$. We thus have $\eta = \hat{\eta}(\mathbb{S}')$. We shall now introduce the energetic form of the equation of state by (in principle) solving $\eta = \hat{\eta}(\mathbb{S}')$ for $\beth$ and writing

$$\beth = \hat{\beth}(\varepsilon_e, \eta). \tag{9.14}$$

MECHANICAL VARIABLES

For future reference, we will call everything other than the temperature, the internal energy, the entropy and the concentration of species (if we have different atoms mixed in) as "mechanical" or configurational variables, and we will assign the symbol $\mathbb{E}$ to them. Specifically, ε_e, ε and ε_p are mechanical variables. The mechanical variables themselves may consist of "elastic variables" like ε_e and inelastic variables like the hardening parameter κ, the back stress α and other such terms which are collectively labeled as $\mathbf{Q}$ as in equation (9.1a).

9.3.2 Heating and working

Just from the equation of state (9.14), we can begin deducing a lot about the system and gaining some insight.

First we note from (9.13), that we can write the absolute temperature as

$$T = \partial \beth / \partial \eta. \tag{9.15}$$

Next by taking the time derivative of $\beth$ in (9.14) we get

$$\dot{\beth} = \frac{\partial \beth}{\partial \eta}\Big|_{\varepsilon_e \text{ const.}} \dot{\eta} + \frac{\partial \beth}{\partial \varepsilon_e}\Big|_{\text{ const.}} \dot{\varepsilon}_e. \tag{9.16}$$

The above equation is a very "revealing equation" and is one of the keys to understanding dissipative processes. We will take each term on the right-hand side of (9.16) and examine it more closely.

- The first term on the right-hand side of (9.16), namely $(\partial \beth / \partial \eta|_{(\varepsilon_e \text{ const.})})\dot{\eta}$ represents *rate of increase of energy due to changes in the entropy when the mechanical variables are fixed.* Note that by using (9.13), it can be written as $T\dot{\eta}$.

We are now going to claim that the first term on the right-hand side of (9.16) is a quantification of the intuitive meaning of "heating" .

Reflect on this for a moment—what do we mean when we say that we are "heating" a body (say a pot of water) from a microscopic point of view? We are supplying energy, of course, but is it evident to you that you are increasing the "randomness" or "possible microstates" of the system? The body will get "hotter" if you increase the randomness in the velocities, i.e. if you increase the thermal entropy. On the other hand, if you increase the configurational entropy alone (boil the water to make steam so that the number of possible positions increases dramatically) then is it obvious to you that you might not raise its temperature even though you are "heating" it? Thus "heating" (i.e. increasing the entropy by external means without mechanical energy) is not the same thing as "making something hotter" (raising the temperature of the body). Please reflect on this very carefully.

We hope we have convinced you that we have a way of identifying what "heating/cooling a body" means: it is changing the energy of the system by changing the entropy but not the mechanical variables, i.e. without supplying mechanical power. If the entropy is increased we have "heated" the body. If it is decreased, we have "cooled" the body.

We can heat or cool the body in two ways: we can "supply heat" or more precisely we can heat the body from outside (the div**q** in the energy equation (9.7)) or the body can " heat itself" by internal friction: this is what happens in viscoelastic and other dissipative systems—the body heats itself by internal friction. The key point is that *while there are TWO ways to "heat" a body i.e. increase its entropy, without changing its configuration*, by "supplying heat" or by "internal friction", there is only ONE way to cool it (decrease its entropy without changing its configuration) —from the outside, by keeping it in an environment that will cool the body. The body cannot cool itself by "undoing friction".

This is the origin of irreversibility and the second law of thermodynamics .

- The second term on the right-hand side of (9.16), i.e. $\partial \exists /\partial \varepsilon_e|_\eta \text{ const.} \dot{\varepsilon}_e$ represents "changing the energy of the body by changing its configuration, but NOT its entropy". This is a quantification of the notion of changing the "mechanical energy" or "strain energy" and represents how much of the mechanical power that we have put in is "recoverable". If we think of the body as a "mechanical energy bank", this term represents your "bank balance" that is withdrawable by you. The first term represents the "transaction costs" for doing business with the bank. As with banks, so with continua, you want to do business with bodies that have "low transactional costs", i.e. low internal friction.

Why and how this second term represents recoverable work is a rather complex topic. To make it more transparent, we will have to introduce a mathematical trick called a Legendre transformation, and we turn to this topic next.

9.4 Equivalent forms of the equation of state: Legendre transformations

While it is possible to gain insight into the behavior of systems by considering the entropic or energetic equations of state, these are not convenient to formulate or manipulate. So we perform a mathematical trick called Legendre transformation to obtain formulations that are more useful. One of the most useful Legendre transformations is the one that exchanges the roles of entropy and temperature. This leads us to the Helmoltz potential. A further Legendre transform exchanges the role of stress and elastic strain. This leads to the Gibbs potential. Each of them are useful in specific applications. We will make heavy use of the Helmholtz form

From a physical point of view, while the entropic equation of state in the form (9.14) is the one that arises naturally, it is far from being the most convenient one, since neither the internal energy nor the entropy are easily measurable. Moreover, from the point of view of engineering, what we are interested in are questions like "how much work can we get from the system?" or "how much work should we put in to the system to get what we want?".

The two most convenient forms are the ones involving the Helmholtz potential and the Gibbs potential, both of which deal with work.

Now, Recall from Chapter 2, that

When a body goes from one equilibrium state to another state which is at the same temperature, the *maximum amount of work that you can possibly extract* from the system, (or the minimum amount of work that you need to supply to the system) is equal to the difference in the Helmholtz potential between the two states. This is what we colloquially call the "strain energy function" in mechanics. Perhaps a more accurate name would be "isothermal work function" or "isothermal work potential".

The Gibbs potential is the negative of the complementary energy function (as used in Castigliano's theorem) and is the *minimum amount of work lost* when a dead load (i.e. a suddenly applied constant load) is applied to a dissipative system and the system goes from one equilibrium state to another at the same temperature.

The proofs of the above statements are based on the second law of thermodynamics (see [Callen (1985)] for an explanation) and are beyond the scope of this book[8]. Mathematically speaking these are just clever manipulations of the two functions (9.13) and (9.14). These potentials are obtained by "swapping one variable for another". For example, the internal energy $\exists$ is a function of the strain, the entropy and other variables. If we swap the roles of the entropy and the temperature then we get the Helmholtz potential or the isothermal work potential ψ which is a function of the strain, temperature and the other variables. If we further swap the roles of the stress and the strain, we get the Gibbs potential or the isothermal complementary work potential. We can, in principle, go on swapping the roles of different variables to get more and more different potentials (also called the generalized Massieu functions see [Callen (1985)]) but there must be some compelling physical reason why you want to use a particular potential.

9.4.1 The Helmholtz potential

The Helmholtz potential is useful because it "swaps" the role of entropy and temperature. So everything is written in terms of temperature rather than entropy. This makes it MUCH easier to gain insight into the behavior of systems. While we show you how to get the Helmholtz potential from the equation of state (9.14) and the definition (9.13) of temperature, we NEVER do this process. Rather, we assume a form DIRECTLY for the Helmholtz potential rather than for the internal energy and proceed with our calculations.

The mathematical steps for obtaining the Helmholtz potential from the equation of state (9.14) and the definition of the temperature (9.13) are the following:

(1) Solve (9.13) for η in terms of T, i.e. obtain $\eta = \hat{\eta}(\varepsilon_e, T)$.
(2) Now, substitute this into the right-hand side of the equation (9.14) and get

$$\exists = \hat{\exists}(\varepsilon_e, \hat{\eta}(\varepsilon_e, T)) = \tilde{\exists}(\varepsilon_e, T). \tag{9.17}$$

(3) Then, DEFINE the Helmholtz potential as

$$\psi = \psi(\varepsilon_e, T) = \tilde{\exists}(\varepsilon_e, T) - T\hat{\eta}(\varepsilon_e, T). \tag{9.18}$$

---
[8]That is a formal way of saying, we don't want to do it:-))

This process is called a "Legendre transformation" of (9.13).

⊙ **9.3 (Exercise:). Finding Helmholtz potentials**
Given an entropic equation of state of the form

$$\eta(\beth, \varepsilon_e) = C \ln \left\{ \frac{\beth - 1/2\mathbb{C}(\varepsilon_e)\bullet(\varepsilon_e)}{C} \right\}, \qquad (9.19)$$

where C is a scalar constant and $\mathbb{C}$ is the fourth order tensor representing the elastic constants,
1. *Show (using (9.13)) that*

$$T = \frac{\beth - 1/2\mathbb{C}(\varepsilon_e)\bullet(\varepsilon_e)}{C}. \qquad (9.20)$$

2. *Substituting (9.20) into (9.19). Show that $\eta = C \ln T$.*
3. *Finally show that the Helmholtz potential ψ is given by (use (9.18))*

$$\psi = 1/2\mathbb{C}(\varepsilon_e)\bullet(\varepsilon_e) + CT(1 - \ln T). \qquad (9.21)$$

(Hint: First find $\beth$ in terms of temperature by solving (9.20) for $\beth$.)

Note that in the above equation of state, the entropy depends only upon the temperature. In such cases, the material is said to possess only "thermal" entropy. It indicates that there is not much uncertainty in the positions of the atoms but a large uncertainty in the velocity of the atoms. This is characteristic of crystalline solids where the positions are very precisely fixed by the lattice structure but the atoms are jiggling around their mean positions very rapidly, so that the velocities are very uncertain.

⊙ **9.4 (Exercise:). The Maxwell relations**
Continuing from the previous example,
1. *Show that*

$$-\partial \psi / \partial T = C \ln T = \eta, \quad \text{and that,} \qquad (9.22)$$
$$\psi + T \partial \psi / \partial T = CT + 1/2\mathbb{C}(\varepsilon_e)\bullet(\varepsilon_e) = \hat{\beth}(\varepsilon_e, T) \qquad (9.23)$$

Hint: Start from (9.21) and use (9.20).
2. *Finally show that*

$$\left. \frac{\partial \psi}{\partial \varepsilon_e} \right|_{T \text{ const.}} = \left. \frac{\partial \beth}{\partial \varepsilon_e} \right|_{\eta \text{ const.}} \qquad (9.24)$$

Consider each of these above results and state in words what they reveal about the response of the system.

We hope that the five results that given as an exercise gives a feel for how to go from the entropic equation of state to the definition of the Helmholtz potential.

One of the key results that we obtained in the process is that $-\partial\psi/\partial T = \eta$.

THE ENTROPY AND THE HELMHOLTZ POTENTIAL

The fact that the entropy is the negative of the gradient of the Helmholtz potential with respect to the temperature, and that the derivative of ψ with respect to ε_e is the same as the derivative of $\exists$ with respect to ε_e, i.e. are results of general validity and is a direct result of the definition of the temperature and the definition of the of the Helmholtz potential as a Legendre transformation of the energy. Thus

$$\eta = -\frac{\partial \psi}{\partial T}, \quad \frac{\partial \exists}{\partial \varepsilon_e}\Big|_{\eta=const.} = \frac{\partial \psi}{\partial \varepsilon_e}\Big|_{T=const.} \qquad (9.25)$$

Thus, *the change in energy at fixed entropy is equal to the change in the Helmholtz potential at fixed temperature; since the latter is the change in the recoverable work, we can conclude that the change in energy at fixed entropy is equal to the amount of mechanical energy that can be recovered.*

It must be evident to you by now that finding the Helmholtz potential from the energetic equation of state is not a simple task. However, instead of starting with the entropic or energetic equation of state and carrying out the tedious manipulations to obtain the Helmholtz potential, we usually just start with a constitutive relation for the Helmholtz potential and obtain all results directly. This was the approach taken in chapter 2. Here we went through the entropic equation of state to gain insight into heating and working and provide motivation for the derivation of the Helmholtz potential.

The equation (9.18) is an example of one such equation, i.e. rather than starting with (9.19) we could have started with (9.21) and then obtained both the entropy and energy through (9.22) and (9.23). This is what is more convenient and we will follow this process whenever possible.

⊙ **9.5 (Exercise:). Helmholtz potential for thermoelastoplasticity**
Consider a Helmholtz potential function of the form

$$\psi = 1/2\mathbb{C}(\varepsilon_e - \alpha T)\bullet(\varepsilon_e - \alpha T) + CT(1 - \ln T) \qquad (9.26)$$

Find the entropy and internal energy associated with this Helmholtz potential. (Hint: Use (9.22) and (9.23).)
What is the physical significance of α? What is C? Is the entropy still a purely thermal entropy or does it depend upon the elastic strain? If the entropy depends upon both temperature and elastic strain then there is uncertainty in both position and velocity of the atoms. Can you rationalize why this is so, for this model?

9.4.2 Further transformations: The Gibbs potential

In the previous subsection, we showed a mathematical process to convert the entropic equation of state into a constitutive equation for the Helmholtz potential. In the process, we were able to eliminate the entropy in favor of the temperature. We can do this process repeatedly with any state variable, provided some invertibility conditions are met. We will carry out this process by eliminating the elastic strain from the list of state variables, introducing the stress instead. We will start now with the Helmholtz potential which is given by

$$\psi = \psi(\varepsilon_e, T) = \psi(\varepsilon_e, T). \tag{9.27}$$

Note that this is a fairly minimalist assumption since no inelastic variables **Q** are included. We will show later how to extend this to include these other terms.

For future use, we define :

$$\mathbf{S} = \frac{\partial \psi}{\partial \varepsilon_e}. \tag{9.28}$$

We will show later that **S** is equal to σ/ϱ in the body and is referred to as the *Kirchhoff stress*.

We follow the same steps as before:

(1) Solve (9.28) for ε_e in terms of **S**, to get $\varepsilon_e = \varepsilon_e(\mathbf{S}, T)$.
(2) Substitute into the equation (9.27) and get $\psi = \psi(\mathbf{S}, T)$.
(3) Define a new quantity Φ through

$$\Phi = \Phi(\hat{\mathbf{T}}, T) = \psi(\mathbf{S}, T) - \mathbf{S} \bullet (\varepsilon_e(\mathbf{S}, T)). \tag{9.29}$$

We have now completed the transformation from the Helmholtz potential ψ into the "Gibbs Potential" Φ. Note that the process is the same as with swapping the temperature for the entropy. Before we discuss what we gained from this process, we need to make sure that you have understood the process. So here is an exercise

⊙ **9.6 (Exercise:). Obtaining Φ from ψ**
Given ψ in the form (9.26) show that

$$\mathbf{S} = \mathbb{C}(\varepsilon_e - \alpha T \mathbf{I}). \tag{9.30}$$

Invert this expression and you will get $\varepsilon_e = (\mathbb{C})^{-1}\mathbf{S} + \alpha T \mathbf{I}$, where $(\mathbb{C})^{-1}$ is the compliance matrix. Now substitute into (9.29) and obtain

$$\Phi = -1/2(\mathbb{C})^{-1}\mathbf{S} \bullet \mathbf{S} - \alpha T \mathbf{I} \bullet \mathbf{S} + CT(1 - \ln T). \tag{9.31}$$

Thus we obtain a general quadratic form for Φ in terms of **S**. In order to gain insight into the terms that are present in Φ we will have to differentiate it with respect to $\mathbf{S}$ and T.

⊙ **9.7 (Exercise:). Derivatives of Φ and their physical significance**
Show that we can obtain the elastic strain, the entropy and the internal energy from Φ in (9.29) i.e. show that

$$\frac{\partial \Phi}{\partial \mathbf{S}} = -(\varepsilon_e + \alpha T \mathbf{I}),$$
$$\frac{\partial \Phi}{\partial T} = -C \ln T - \alpha \mathbf{S} \cdot \mathbf{I} = -\frac{\partial \psi}{\partial T} = \eta, \qquad (9.32)$$
$$\exists = \Phi + \frac{\partial \Phi}{\partial \mathbf{S}} \cdot \mathbf{S} + \frac{\partial \Phi}{\partial T} T = \psi + \frac{\partial \psi}{\partial T} T.$$

By using (9.32c) obtain a specific form for $\exists$ in terms of the Kirchhoff stress and the temperature.

Again, although these results are obtained for a specific model, the result that the derivatives of Φ with respect to the state variables gives us the physical quantities listed in (9.32) is valid in general. Thus, instead of starting from the entropic equation of state, we can start either with the Helmholtz potential or with the Gibbs potential.

We have obtained several different forms the thermal and mechanical variables so for. We can summarize the major results in the table shown in Fig. 9.3.

9.5 The "heat" or entropy equation and dissipative processes

The evolution equations for mechanical state variables of a body should be such that the rate of dissipation of mechanical power should be non-negative. By finding an expression for the mechanical dissipation in terms of the derivatives of the Helmholtz potential, we are able to identify that "energy release rate" and the driving forces. Based on this identification we may state that the body changes its state, if its energy release rate is sufficient to overcome frictional resistance. If we quantify these, then we are on our way to establishing evolution equations that have a sound thermodynamical basis.

In the previous section, we used the state variables to gain insight into the system, but it left unanswered as to the process by which the state of a system is changed.

	Energy	**Helmholtz potential**	**Gibbs Potential**
Eqn. of state	$\exists(\varepsilon_e, \eta)$	$\psi(\varepsilon_e, T)$	$G(\mathbf{S}, T)$
Thermal relations	$T = \dfrac{\partial \exists}{\partial \eta}\bigg|_{\varepsilon_e = const.}$	$\eta = -\dfrac{\partial \psi}{\partial T}\bigg|_{\varepsilon_e = const.}$	$\eta = -\dfrac{\partial G}{\partial T}\bigg|_{S = const.}$
Mechanical Relations	$\mathbf{S} = \dfrac{\partial \exists}{\partial \varepsilon_e}\bigg|_{\eta = const.}$	$\mathbf{S} = \dfrac{\partial \psi}{\partial \varepsilon_e}\bigg|_{T = const.}$	$\varepsilon_e + (\alpha T)\mathbf{I} = -\dfrac{\partial G}{\partial \mathbf{S}}\bigg|_{T = const.}$
Relationships		$\exists(\varepsilon_e, T) = \psi + T\dfrac{\partial \psi}{\partial T}$	$\psi(\mathbf{S}, T) = G + \mathbf{S} \bullet \dfrac{\partial G}{\partial \mathbf{S}}$

Fig. 9.3 A summary of the relationships between the Energy, Helmholtz potential and Gibbs Potential and the various thermal and mechanical quantities. Here T is the temperature, η is the entropy, ε_e is the elastic strain and $\mathbf{S}$ is the Kirchhoff stress defined in this book as $\boldsymbol{\sigma}/\varrho$.

In this section, we will now consider how the material changes its state, i.e. what is the process undergone by the material. The main point regarding allowable processes is that only those processes that "produce" entropy will be allowed, according to the second law of thermodynamics. In order to meet this fundamental restriction, we need to obtain an expression for the rate of production of entropy. We begin with the equation of state in the form $\exists = \exists(\eta, \varepsilon_e)$ (cf. (9.14)) and substitute it into the energy equation (9.7d). By using the chain rule, we will get

$$\varrho \frac{\partial \exists}{\partial \eta} \dot{\eta} = -\text{div}\mathbf{q} + \boldsymbol{\sigma} \bullet \dot{\boldsymbol{\varepsilon}} - \varrho \frac{\partial \exists}{\partial \varepsilon_e} \bullet \dot{\varepsilon}_e. \qquad (9.33)$$

This equation is the starting point of the application of the second law of thermodynamics to continua. In order to understand the significance of the terms that appear in the above equation, we need to look at each group of terms in the above equation and understand their physical significance. We will look at each term in the above equation in turn:

- The term on the left-hand side of equation (9.33) is the rate of "heating" of the body as discussed earlier. The terms on the right-hand side can be represented as different ways of "heating" or cooling the body.
- The term $-\text{div}\mathbf{q}$ represents the rate of energy entering (or leaving) the body as heat from external sources. This is the only means to "cool" the body, i.e. decrease its entropy without changing the mechanical variables.
- The term $\boldsymbol{\sigma} \bullet \dot{\boldsymbol{\varepsilon}}$ is the total internal mechanical power supplied to the body due to deformation. Using the "bank" analogy introduced earlier, this is the amount of "money" that you plan to "deposit" in the "bank".
- The term $\varrho(\partial \exists/\partial \varepsilon_e) \bullet \dot{\varepsilon}_e$ is the rate of increase of the energy of the system *without increasing the entropy of the system*; this is what we mean by "recoverable

work". This is the "money" that will actually be "credited" to your account, i.e. that you have actually managed to save. The rest will be lost due to "spending" or internal friction including the transaction cost of the bank.

Thus, in words, equation (9.37) can be interpreted as

HEATING AND WORKING

The rate of heating of the body ($\varrho T \dot{\eta}$)—i.e. rate of increase in energy of the body due to changes in entropy— is equal to the sum of the rate of energy entering the body as heat ($-\text{div}\mathbf{q}$) and the rate of internal power supplied minus the rate of increase of recoverable work, i.e. it is the sum of the rate of external heating (or cooling) and the rate of internal heating (you can't cool a body with internal friction) due to internal damping or friction in the body.

It therefore, stands to reason that the *mechanical dissipation per unit mass*, ζ, i.e. the rate at which the mechanical power supplied is irrevocably lost per unit mass is

$$\varrho\zeta = \boldsymbol{\sigma}\cdot\dot{\boldsymbol{\varepsilon}} - \varrho\frac{\partial \exists}{\partial \varepsilon_e}\cdot\boldsymbol{\varepsilon}_e \mid_{\eta=const.} = \boldsymbol{\sigma}\cdot\dot{\boldsymbol{\varepsilon}} - \varrho\dot{\exists}\mid_{\eta=const.} \tag{9.34}$$

Thus ζ is the "transaction cost" of depositing "money" in your "mechanical power bank". Different materials will use this "transaction cost" differently. An investigation of this phenomenon is what is at the heart of this book. Thus ζ is the internal friction in the system. The right-hand side of the above equation (9.34) is the "energy release rate" and represents the mechanical power that is available to overcome the friction.

9.5.1 Results in terms of the Helmholtz potential

While the use of the energetic equation of state provides a direct means of identifying the rate of power dissipation, it is the Helmholtz potential that allows us to write the energy release rate in terms of the temperature rather than the entropy. This simplifies the study of realistic problems (where temperature changes are not large) more efficiently.

Imagine that, instead of starting with the entropic equation (9.14) you have chosen the equation of state in terms of the Helmholtz potential, i.e. $\psi(\varepsilon_e, T)$. If you use the result (9.24), then after some manipulation, you would get

$$\varrho\zeta = \boldsymbol{\sigma}\cdot\dot{\boldsymbol{\varepsilon}} - \varrho\frac{\partial \psi}{\partial \varepsilon_e}\cdot\boldsymbol{\varepsilon}_e = \boldsymbol{\sigma}\cdot\dot{\boldsymbol{\varepsilon}} - \varrho\dot{\psi}\mid_{T=const.} \tag{9.35}$$

You might think "wait a minute, the only difference between (9.34) and (9.35) seems to be replacing $\exists$ with ψ". But there is a subtle but vital difference between the two versions; note that when we do $\partial\exists/\partial\varepsilon_e$, we are holding the entropy fixed, whereas when we do $\partial\psi/\partial\varepsilon_e$, we are holding the *temperature* fixed. If you feel that the result (9.35) was obtained by some sleight of hand, then do it the hard way—the next exercise is the hard way to obtain the same result.

⊙ **9.8 (Exercise:). The Heat equation: Helmholtz potential form**
Substitute $\exists = \psi + T\partial\psi/\partial T$ into the left-hand side of the fourth of ((9.7)), expand the time derivative terms and get, on simplification,

$$-\varrho T \frac{d}{dt}\left\{\frac{\partial\psi}{\partial T}\right\} = \boldsymbol{\sigma}\boldsymbol{\cdot}\dot{\boldsymbol{\varepsilon}} - \varrho\frac{\partial\psi}{\partial\varepsilon_e}\bigg|_{T=const.}\boldsymbol{\cdot}\dot{\boldsymbol{\varepsilon}}_e - div\mathbf{q}. \qquad (9.36)$$

(Hint: substitute $\psi = \psi(\varepsilon_e, T)$, use the chain rule and product rule to find $d/dt\{\psi + \partial\psi/\partial T\}$.)
Now, use the fact that the entropy η is related to the Helmholtz potential through (see equation (9.22)) $\eta = -\partial\psi/\partial T$ and end up with the following "heat equation" in terms of the Helmholtz potential:

$$\varrho T\dot\eta = div\mathbf{q} + \boldsymbol{\sigma}\boldsymbol{\cdot}\dot{\boldsymbol{\varepsilon}} - \varrho\frac{\partial\psi}{\partial\varepsilon_e}\boldsymbol{\cdot}\dot{\boldsymbol{\varepsilon}}_e. \qquad (9.37)$$

We hope you can recognize that the last two terms on the right-hand side of the above equation are identical to the right-hand side of (9.35). We have obtained the same result by a slightly round about way.

When the Helmholtz potential ψ is given by (9.26), we can obtain a specific form for ζ and this is the next exercise:

⊙ **9.9 (Exercise:). Mechanical dissipation equation**
Substitute the expression (9.26) into (9.35) and hence obtain the result that

$$\begin{aligned}\varrho\zeta &= (\boldsymbol{\sigma} - \varrho\partial\psi/\partial\varepsilon_e)\boldsymbol{\cdot}\dot{\boldsymbol{\varepsilon}} + \varrho\partial\psi/\partial\varepsilon_e\boldsymbol{\cdot}(\dot{\boldsymbol{\varepsilon}}_e - \dot{\boldsymbol{\varepsilon}}) \\ \Rightarrow \zeta &= (\mathbf{S} - \mathbb{C}(\varepsilon_e - \alpha T))\boldsymbol{\cdot}\dot{\boldsymbol{\varepsilon}} + \mathbb{C}(\varepsilon_e - \alpha T)\boldsymbol{\cdot}(\dot{\boldsymbol{\varepsilon}}_e - \dot{\boldsymbol{\varepsilon}}).\end{aligned} \qquad (9.38)$$

Now based on elasticity, if we now require that $\boldsymbol{\sigma} = \varrho\mathbb{C}(\varepsilon_e - \alpha T)$ and introducing $\dot{\boldsymbol{\varepsilon}}_p = \dot{\boldsymbol{\varepsilon}} - \dot{\boldsymbol{\varepsilon}}_e$, show that the second law of thermodynamics requires that

$$\mathbf{S}\boldsymbol{\cdot}\dot{\boldsymbol{\varepsilon}}_p \geq 0, \; \mathbf{S} = \boldsymbol{\sigma}/\varrho. \qquad (9.39)$$

Note that it is only when we introduce the notion of dissipation that we begin to use the plastic strain rate. Until then we can do all the calculations using only the elastic strain and its derivatives.

We have thus obtained the dissipation inequality, which we had proposed in an ad-hoc fashion earlier, from thermodynamical considerations. Note that in general, the form of this inequality depends upon (1) the form of ψ and (2) the assumption made with regard to the stress $\boldsymbol{\sigma}$. Thus we find that in this case, the dissipation inequality does not include any "backstress" terms since it was not included in ψ.

9.5.2 Generalization of the state variables and obtaining small-strain, continuum versions of spring-dashpot models

We can generalize the results of the previous subsections in two steps: first to include a simple form of back-stress, and next to include generalizations of spring mass models that we introduced earlier.

We have hitherto included only the elastic strain $\boldsymbol{\varepsilon}_e$ and the temperature, as the state variables. As an example of a model that includes the backstress, let us first generalize the Helmholtz potential by adding one more variable, the total strain $\boldsymbol{\varepsilon}$. If you add this variable, then we have $\psi = \psi(\boldsymbol{\varepsilon}, \boldsymbol{\varepsilon}_e, T)$. And, the mechanical power dissipation equation will now become

$$\boldsymbol{\sigma} \cdot \dot{\boldsymbol{\varepsilon}} - \varrho\left(\frac{\partial \psi}{\partial \boldsymbol{\varepsilon}} \cdot \dot{\boldsymbol{\varepsilon}} + \frac{\partial \psi}{\partial \boldsymbol{\varepsilon}_e} \cdot \dot{\boldsymbol{\varepsilon}}_e\right) = \varrho \zeta \geq 0. \tag{9.40}$$

⊙ **9.10 (Exercise:). General expression for mechanical dissipation**
Show that if $\psi = \psi(\boldsymbol{\varepsilon}, \boldsymbol{\varepsilon}_e, T)$ then the procedure introduced in Exercise (9.5.1), will deliver equation (9.40).
In general, show that if we have a Helmholtz potential of the form $\psi = \psi(\mathbb{E}, T)$, where $\mathbb{E}$ is any set of mechanical state variables written in list (or column vector) form OTHER than the temperature, then the mechanical power dissipation equation becomes

$$\boldsymbol{\sigma} \cdot \dot{\boldsymbol{\varepsilon}} - \varrho\left(\frac{\partial \psi}{\partial \mathbb{E}} \cdot \dot{\mathbb{E}}\right) = \varrho \zeta \geq 0. \tag{9.41}$$

Returning to (9.40), if we identify the plastic strain rate as $\dot{\boldsymbol{\varepsilon}}_p = \dot{\boldsymbol{\varepsilon}} - \dot{\boldsymbol{\varepsilon}}_e$, then we can group terms together and obtain

$$\boldsymbol{\sigma} - \varrho\left(\frac{\partial \psi}{\partial \boldsymbol{\varepsilon}} + \frac{\partial \psi}{\partial \boldsymbol{\varepsilon}_e}\right) \cdot \dot{\boldsymbol{\varepsilon}} + \varrho \frac{\partial \psi}{\partial \boldsymbol{\varepsilon}_e} \cdot \dot{\boldsymbol{\varepsilon}}_p = \varrho \zeta \geq 0. \tag{9.42}$$

Next we will assume that the stress is given by

$$\boldsymbol{\sigma} = \varrho\left(\frac{\partial \psi}{\partial \boldsymbol{\varepsilon}} + \frac{\partial \psi}{\partial \boldsymbol{\varepsilon}_e}\right), \tag{9.43}$$

and we will define the *backstress* $\boldsymbol{\alpha}$ through

$$\boldsymbol{\alpha} := \frac{\partial \psi}{\partial \boldsymbol{\varepsilon}}. \tag{9.44}$$

Then, the rate of dissipation becomes

$$\frac{\partial \psi}{\partial \varepsilon_e} \cdot \dot{\varepsilon}_p = (\mathbf{S} - \frac{\partial \psi}{\partial \varepsilon}) \cdot \dot{\varepsilon}_p = (\mathbf{S} - \boldsymbol{\alpha}) \cdot \dot{\varepsilon}_p = \zeta \geq 0. \tag{9.45}$$

Thus you can see that we obtain an expression with *back-stress* by choosing an appropriate Helmholtz potential.

Furthermore, notice that equation (9.43) is composed of two terms that are added together—two "springs" as it were—one with extension ε and another with extension ε_e. But it is not necessary that the springs be uncoupled! This is equivalent to Fig. 2.1c on p.33.

⊙ **9.11 (Exercise:). Obtaining the stress response and rate of dissipation equation from ψ**
Re-examine the worked out example that lead to the stress response (9.43) and the rate of dissipation (9.45) above and in words describe the procedure to obtain the stress response and the rate of dissipation starting with any Helmholtz potential ψ.

Were you able to write down the procedure? If not, consider the next example; We will get more ambitious here and try out a model motivated purely from a spring dashpot analogy (see Fig. 9.4)

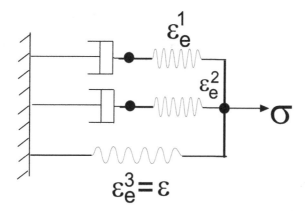

Fig. 9.4 A mechanical model of an inelastic material composed of three springs and two dashpots. The extension of the spring are the elastic strains $\varepsilon_e^i, i = 1, 2, 3$.

(1) The model in Fig. 9.4 has three springs. We need three mechanical state variables to represent the extension of each spring. So we will have three "elastic

strain" variables[9] $\varepsilon_e^1, \varepsilon_e^2$ and $\varepsilon_e^3 = \varepsilon$.

(2) We will now assume a Helmholtz potential of the form $\psi(\varepsilon_e^1, \varepsilon_e^2, \varepsilon, T)$ for the model. You have considerable freedom in choosing the form of this function. It is not necessary for the springs to be "uncoupled" even though it is not possible to represent it in the mechanical analogy! We thus use the mechanical analogy only for guidance.

(3) Now our mechanical power dissipation equation is given by

$$\varrho\zeta = \boldsymbol{\sigma}\bullet\dot{\boldsymbol{\varepsilon}} - \varrho\dot{\psi}|_{T=const.} = \boldsymbol{\sigma}\bullet\dot{\boldsymbol{\varepsilon}} - \varrho\sum_{i=1}^{3}\frac{\partial\psi}{\partial\varepsilon_e^i}\dot{\varepsilon}_e^i. \tag{9.46}$$

(4) First note that the velocities of the two dashpots are each equal to the velocity of the end minus the rate of extension of the springs. So we have two plastic strain rate variables

$$\dot{\varepsilon}_p^i = \dot{\varepsilon} - \dot{\varepsilon}_e^i \Rightarrow \dot{\varepsilon}_e^i = \dot{\varepsilon} - \dot{\varepsilon}_p^i \ (i=1,2). \tag{9.47}$$

Note that, by our very definition, $\dot{\varepsilon}_p^3 = 0$. By using (9.47) in (9.46) and grouping terms, we get

$$\left(\boldsymbol{\sigma} - \varrho\sum_{i=1}^{3}\frac{\partial\psi}{\partial\varepsilon_e^i}\right)\bullet\dot{\boldsymbol{\varepsilon}} + \varrho\sum_{i=1}^{2}\left\{\frac{\partial\psi}{\partial\varepsilon_e^i}\bullet\dot{\varepsilon}_p^i\right\} = \varrho\zeta. \tag{9.48}$$

(5) Finally, note that the total force on the dashpot is the sum of the forces on each spring, and you can see that this is precisely the form of the first term on the LHS of the equation (9.48). So we gracefully give in to the temptation and set

$$\boldsymbol{\sigma} = \varrho\sum_{i=1}^{3}\frac{\partial\psi}{\partial\varepsilon_e^i}. \tag{9.49}$$

And our final mechanical power dissipation equation becomes

$$\sum_{i=1}^{2}\left\{\frac{\partial\psi}{\partial\varepsilon_e^i}\bullet\dot{\varepsilon}_p^i\right\} = \zeta. \tag{9.50}$$

⊙ **9.12 (Exercise:). Converting Spring Dashpot models**
Obtain the stress response and the mechanical power dissipation equation for each of the spring dashpot models in Fig.2.2 on p.34. Ignore the masses.

---

[9] Why is the "elastic strain" for the last spring the same as the total strain? Can you think of an answer?

We have now exhausted all we can get out of the Helmholtz formulation that is generally valid irrespective of the form of the evolution equation for the plastic strain rate chosen.

The key result is that, no matter what the form of ψ, we will always assume the following *state laws*:

$$\eta = -\frac{\partial \psi}{\partial T}, \quad \boldsymbol{\sigma} = \sum \frac{\partial \psi}{\partial \boldsymbol{\varepsilon}_e^i}, \quad \beth = \psi + T\frac{\partial \psi}{\partial T}, \quad (9.51)$$

and the following dissipation restriction,

$$\mathbf{S} \cdot \dot{\boldsymbol{\varepsilon}} - \varrho \dot{\psi}|_{T=const.} = \zeta \geq 0. \quad (9.52)$$

In this formulation, we do not introduce the "plastic strain" as a variable at all, we simply introduce a "short-hand notation" for $(\dot{\boldsymbol{\varepsilon}} - \dot{\boldsymbol{\varepsilon}}_e)$, i.e. we have thus introduced the plastic strain rate *indirectly as the difference between the total strain rate and the elastic strain rate*.

Of course for small deformation, this is not a big deal, but, we will see that this allows us to introduce finite strain plasticity in a somewhat *direct* manner, in a way that is closely analogous to small strain plasticity.

THE PLASTIC DRIVING FORCE

In general, if the rate of dissipation ζ can be written in the form $\boldsymbol{\xi} \cdot \dot{\boldsymbol{\varepsilon}}_p$, where $\boldsymbol{\xi}$ is a suitably defined second order tensor that is obtained from the derivative of the Helmholtz potential and $\dot{\boldsymbol{\varepsilon}}_p := \dot{\boldsymbol{\varepsilon}} - \dot{\boldsymbol{\varepsilon}}_e$, then, $\boldsymbol{\xi}$ will be referred to as *the driving force for the plastic deformation or simply plastic driving force*.

The form of $\boldsymbol{\xi}$ depends upon the form for the Helmholtz potential and the assumptions regarding stress and the plastic strain rate. Thus, for the models that we have considered,

$$\begin{aligned}
\psi = \psi(\boldsymbol{\varepsilon}_e, T) &\Rightarrow \boldsymbol{\sigma} = \varrho \frac{\partial \psi}{\partial \boldsymbol{\varepsilon}_e}, \boldsymbol{\xi} = \frac{\partial \psi}{\partial \boldsymbol{\varepsilon}_e} = \boldsymbol{\sigma}/\varrho = \mathbf{S}. \\
\psi = \psi(\boldsymbol{\varepsilon}, \boldsymbol{\varepsilon}_e, T) &\Rightarrow \boldsymbol{\sigma} = \varrho \frac{\partial \psi}{\partial \boldsymbol{\varepsilon}_e} + \varrho \frac{\partial \psi}{\partial \boldsymbol{\varepsilon}}, \boldsymbol{\xi} = \frac{\partial \psi}{\partial \boldsymbol{\varepsilon}} = \mathbf{S} - \frac{\partial \psi}{\partial \boldsymbol{\varepsilon}_e}.
\end{aligned} \quad (9.53)$$

Furthermore, the mechanical power dissipation equation can be thought of as a balance between the "driving forces" and the friction. Specifically, the left side of the equation, $\boldsymbol{\xi} \cdot \dot{\boldsymbol{\varepsilon}}_p$ is the "energy release rate", i.e. the rate at which mechanical energy is available for wasting. The right-hand side is the "internal frictional resistance". So we can imagine the mechanical power dissipation equation as a statement that "plastic flow occurs if sufficient energy is available to compensate for frictional losses".

A parallel development may be undertaken using the Gibbs potential and this will be the content of the next guided exercise.

⊙ **9.13 (Exercise:). Plasticity using the Gibbs potential**
Substitute the expression (9.32d) into the equation (9.13c) and use the equations (9.32a,b,c) to obtain the heat equation in terms of the Gibbs potential,

$$\varrho T \dot{\eta} = div\mathbf{q} + \boldsymbol{\sigma} \bullet \dot{\boldsymbol{\varepsilon}} - \mathbf{S} \bullet \frac{d}{dt} \{\partial \Phi / \partial \mathbf{S}|_{T=const.}\}, \quad \mathbf{S} := \boldsymbol{\sigma}/\varrho. \tag{9.54}$$

(Hint: Make use of the chain rule and product rule of differentiation to find the time derivative of $\exists$ in terms of Φ (use (9.32)) and its partial derivatives.)

$$\zeta = \mathbf{S} \bullet (\dot{\boldsymbol{\varepsilon}} - \frac{d}{dt} \partial \Phi / \partial \mathbf{S}|_{T=const.}) \geq 0. \tag{9.55}$$

Show further that if we define the plastic strain rate $\dot{\boldsymbol{\varepsilon}}_p$ through

$$\dot{\boldsymbol{\varepsilon}}_p := \dot{\boldsymbol{\varepsilon}} - \frac{d}{dt} \partial \Phi / \partial \mathbf{S}|_{T=const.}. \tag{9.56}$$

then we get

$$\mathbf{S} \bullet \dot{\boldsymbol{\varepsilon}}_p = \zeta. \tag{9.57}$$

Note that as with the Helmholtz potential, here too, we obtain an expression (9.56) for the plastic strain rate *with no need to introduce the plastic strain a priori*. Again, as with the case of the Helmholtz potential, we see that the equation (9.56) is written in rate form and thus $\dot{\boldsymbol{\varepsilon}}_p$ becomes just a convenient short-hand for the terms on the right-hand side which involve the $\boldsymbol{\sigma}, T$ and their rates without any mention of any unloading process, whether fast or slow. This way of obtaining the plastic flow rate is quite useful for geomaterials (see e.g., [Collins and Houlsby (1997)] for an excellent review of this approach).

9.6 Constitutive Laws for $\dot{\boldsymbol{\varepsilon}}_p$ and the satisfaction of the second law

We note that as far as the second law of thermodynamics is concerned, *any constitutive law for $\dot{\boldsymbol{\varepsilon}}_p$ is acceptable so long as the condition*

$$\boldsymbol{\xi} \bullet (\dot{\boldsymbol{\varepsilon}} - \dot{\boldsymbol{\varepsilon}}_e) = \boldsymbol{\xi} \bullet \dot{\boldsymbol{\varepsilon}}_p \geq 0 \tag{9.58}$$

is met. Note that, for example, the trivial assumption that $\dot{\boldsymbol{\varepsilon}}_p = \phi \boldsymbol{\xi}$ is an obvious choice so long as $\phi \geq 0$. *This is a thermodynamic justification for the J2 plasticity* laws that we introduced earlier where $\boldsymbol{\xi} = \boldsymbol{\tau}$.

THE PLASTIC WORK

Note that if $\psi = \psi(\boldsymbol{\varepsilon}_e, T)$ then $\boldsymbol{\xi} = \boldsymbol{\sigma}/\varrho = \mathbf{S}$ and the non-negativity of the rate of dissipation reduces to

$$\boldsymbol{\xi} \bullet \dot{\boldsymbol{\varepsilon}}_p = \mathbf{S} \bullet \dot{\boldsymbol{\varepsilon}}_p = \zeta \geq 0. \tag{9.59}$$

In words, if the Helmholtz potential depends only upon the elastic strain and the temperature, then the driving force for plastic flow $\boldsymbol{\xi}$ is just the Kirchhoff stress, and the rate of dissipation is identical to the "plastic working" per unit mass.

By assuming different forms for $\dot{\boldsymbol{\varepsilon}}_p$, we can obtain a wide range of models ranging from viscoelasticity, viscoplasticity and even certain kinds of rate independent elastoplastic models. In other words, while $\boldsymbol{\varepsilon}_e$ plays the role of the *displacement of an elastic spring*, $\dot{\boldsymbol{\varepsilon}}_p$ plays the role of the *velocity of a dashpot*. So, we can extend any of the lumped models introduced in Chapter 2, so long as we satisfy (9.58). We will explore some of these models first.

9.6.1 Viscoelastic and viscoplastic models

We will begin by stipulating that the Helmholtz potential is of the form $\psi(\boldsymbol{\varepsilon}_e, T)$ so that any constitutive equation for $\dot{\boldsymbol{\varepsilon}}_p$ must satisfy (9.59).

Let us try the following simple form for $\dot{\boldsymbol{\varepsilon}}_p$:

$$\dot{\boldsymbol{\varepsilon}}_p = \dot{\boldsymbol{\varepsilon}} - \dot{\boldsymbol{\varepsilon}}_e = \mu \boldsymbol{\sigma}, \ \mu > 0 \tag{9.60}$$

Does this satisfy the non-negativity of the rate of dissipation? Let us check by computing the rate of dissipation:

$$\mathbf{S} \bullet \dot{\boldsymbol{\varepsilon}}_p = \mu \mathbf{S} \bullet \boldsymbol{\sigma} = \mu \frac{\|\boldsymbol{\sigma}\|^2}{\varrho} \geq 0. \tag{9.61}$$

Clearly it does satisfy the conditions—in fact, if you assumed that $\dot{\boldsymbol{\varepsilon}}_p = \mu \boldsymbol{\tau}$ then not only will it satisfy the non-negativity criterion, but will also enforce the requirement that there be no permanent volume change.

We can eliminate $\boldsymbol{\varepsilon}_e$ and $\boldsymbol{\varepsilon}_p$ from the above equation by using the constitutive equation $\boldsymbol{\varepsilon}_e = \mathbb{C}^{-1}\boldsymbol{\sigma}$, obtained by assuming that ψ is quadratic in $\boldsymbol{\varepsilon}_e$ so that we finally obtain the following differential equation:

$$\dot{\boldsymbol{\varepsilon}} = \mathbb{C}^{-1}\dot{\boldsymbol{\sigma}} + \mu \boldsymbol{\sigma} \tag{9.62}$$

Do you recognize this as being similar to the equation for a simple spring and dashpot in series? This is the continuum analog.

The next exercise shows that other generalizations are possible that give rise to *viscoplastic* behavior:

⊙ **9.14 (Exercise:). A simple constitutive law for viscoplasticity**
Show that the constitutive law

$$\dot{\varepsilon}_p = \mu(\langle \|\boldsymbol{\sigma}\| - \kappa \rangle \mathbf{N}), \quad \mathbf{N} := \frac{\boldsymbol{\sigma}}{\|\boldsymbol{\sigma}\|}, \tag{9.63}$$

(where, as usual, the symbol $\langle x \rangle$ stands for $1/2(x+|x|)$) satisfies the dissipation inequality (9.59). What happens if $\|\boldsymbol{\sigma}\| < \kappa$? What happens if $\|\boldsymbol{\sigma}\| > \kappa$?

You can generate a huge variety of different constitutive models by choosing different forms for $\dot{\varepsilon}_p$ as long as the rate of dissipation is positive.

⊙ **9.15 (Exercise:). A multiple spring-dashpot model**
Consider the continuum analog that we developed for the 3 spring 2 dashpot model shown in Fig. 9.4 on p.291. The stress was given by (9.49) and the rate of dissipation by (9.50). Note that, in this case the driving force was NOT the stress. Assume that

$$\dot{\varepsilon}_p^i = \mu^i \frac{\partial \psi}{\partial \varepsilon_e^i}. \tag{9.64}$$

Is this an allowable form? What kind of response will result? Will it be like a viscoelastic solid or a fluid or will it be an elastoplastic material? Why?

9.6.2 Rate independent models without yield criteria

The previous examples were all that of *rate dependent materials*, i.e. the response of the material would be faster or slower depending on the value of the driving forces. An indicator of the rate dependence of the response is that the constitutive law for $\dot{\varepsilon}_p$ depends only upon the state variables ε_e and T but is independent of their rates. Thus since the LHS has a rate term but the RHS doesn't, the overall constitutive law becomes rate dependent.

How would you develop a constitutive law that is *rate independent*? One possible way is to consider a constitutive law that has rate terms on both the LHS and RHS in such a way that they cancel out, i.e. they are homogeneous in the rate terms. At the same time, we need to meet the conditions of the second law—namely (9.59). How do we do it? Consider the constitutive law below:

$$\dot{\varepsilon}_p = \langle \mathbf{A} \cdot \dot{\varepsilon} + b\dot{T} \rangle \mathbf{N} \tag{9.65}$$

where $\mathbf{A}$ and $\mathbf{N}$ could be any second order tensor functions of the state variables $\mathbf{S}$. The condition (9.59) will then demand that $\mathbf{S} \cdot \mathbf{N} > 0$. Any form for $\mathbf{N}$ that meets this criterion will be acceptable, for example a form such as $\mathbf{N} = \mathcal{K}\mathbf{S}$ where $\mathcal{K}$ is a fourth order positive definite tensor will satisfy the requirements. The parameter b can be any scalar function of $\mathbb{S}$.

This form, while being quite general involves both mechanical and thermal loading and needs the specification of many constitutive functions. We can assume, for example that $\mathbf{A} = \mathcal{C}\mathbf{N}$ where $\mathcal{C}$ is the elastic modulus, i.e. $\mathcal{C} = \varrho \partial^2 \psi / \partial \varepsilon_e{}^2$. This will mimic a sort of *normality rule*, for this model.

The structure of this theory allows for the modeling of a large class of soils and materials that do not have a well-defined yield functions but whose response is rate independent. A variant of this model is described by [Pastor *et al.* (1990)] and is periodically reported in the literature, but it is not very commonly used.

Notice that *none* of the models in this section have included kinematic hardening. This is because the Helmholtz potentials were assumed to be functions of only ε_e and T. If we want to include back stress and other hardening effects, more variables with additional evolution laws have to be incorporated in the Helmholtz or Gibbs potentials. Since further analysis is required for these kinds of problems, we will delay their introduction until later.

9.7 The maximum rate of dissipation criterion

At this stage, we see that the second law of thermodynamics provides scant guidance with regard to the choice of kinetic laws. This is in marked contrast to the laws governing the state variables such as stress, entropy and energy: each of these are readily derivable from the Helmholtz potential, while we have only a mild restriction on the rate of dissipation. How do we get guidance for the form of the kinetic laws?

For rate independent plasticity, the normality flow rule and the consistency condition can be motivated from a number of different starting points; three of the most common approaches are mentioned below [10]:

MAXIMUM PLASTIC WORK CRITERION

This is sometimes referred to as the maximum plastic dissipation condition and is stated as follows:

> For a given plastic strain rate $\dot{\varepsilon}_p$ among all the stresses $\boldsymbol{\sigma}$ that satisfy $f \leq 0$, i.e. among all the elastic states, the actual stress is the one that results in the largest possible value of $\boldsymbol{\sigma} \bullet \dot{\varepsilon}_p$, (usually referred to as the plastic work or the plastic dissipation).

This is by far the most common statement adopted in many books. It has been

---
[10] We list a statement of each of these assumptions here because of the fact that each of these assumptions go under a myriad of different names causing confusion as to which assumption is being referred.

established that the above condition gives rise to an associative plasticity model (i.e. a model where the plastic strain rate is normal to the yield surface), and further that the yield surface is convex (see e.g. [Simo and Hughes (1998)] p. 99). However, it is limited in scope—it is applicable only to rate independent plasticity, does not allow for the presence of the *backstress* α and, as stated, is applicable only to small deformations.

An alternative formulation that is somewhat more restrictive (being applicable to materials that harden) is referred to as (see [Kachanov (2004)] p. 86, para 5).

DRUCKER'S STABILITY POSTULATE

Consider an element of a hardening medium in which there is an original stress σ^0; to this element we apply additional stresses (in general of arbitrary magnitude), and then remove them. We assume that changes take place sufficiently slowly for the process to be regarded as isothermal. Then it is postulated that
(1) in the process of loading additional stresses produce positive work.
(2) For a complete cycle of additional loading and unloading, additional stresses do positive work if plastic deformation takes place.

It has been shown that the above postulate is more restrictive than the maximum plastic work assumption and leads, for a certain class of materials, to an associative plasticity model with a yield surface that is convex, again with no allowance for a backstress.

There are two other formulations that are closely related, Il'iushin's postulate [Il'iushin (1961)] and the work inequality [Naghdi and Trapp (1975b)], the former is postulated for small deformations while the latter is applicable to large deformations also. The statement of the work inequality is as follows:

THE WORK INEQUALITY

The work done during a closed cycle of homogeneous deformation of an elasto-plastic material is non-negative.

It has been shown that, the above assumption together with conditions on rate independent plasticity, leads to (1) the existence of a strain energy function ψ, the derivative of which gives the stress, (2) the existence of a convex yield surface in the space of "driving forces for plastic deformation" (defined as $-\partial\psi/\partial\varepsilon_p$) and (3) an associative flow rule in the space of driving forces. If ψ is a function of $\varepsilon - \varepsilon_p$ then it can be shown that the driving force is the same as the stress. (see [Naghdi and Trapp (1975a); Lin and Naghdi (1989); Srinivasa (1997)] for extensive discussions).

In every case, the general developments follow the same pattern:

- The existence of a yield function is assumed *a-priori*.
- Some general assumptions are made regarding the plastic strain rate (such as being homogeneous and first order in the total strain rate).
- One of the postulates above is then used to show that the resulting model is such that the yield surface is convex and the flow rule is associative. Specifically, non-associative plasticity is completely ruled out by these assumptions.

In other words, the main use of these statements is to show that for a particular class of elasto-plastic materials (with constitutive equations given by (9.1)), it is possible to show that there are restrictions on the forms of the yield surface $f = 0$ (it has to enclose a convex body) and on the plastic potential H (it has to be same as f). But, these are not general enough for our purposes. Because, as stated, they do not provide detailed guidance for the evolution laws for viscoplasticity, viscoelasticity and indeed any kind of dissipative behavior other than rate independent plasticity with a well-defined yield surface.

We will not say much more about these assumptions since they are all specific to rate independent elastoplastic materials. Rather we will start from a much stronger statement that is presumed to be valid for a variety of dissipative materials irrespective of whether they are elastoplastic or not.

In this section, we propose to show how the entire structure of associative plasticity (including the presence of a yield function etc.) can be derived from thermodynamical principles together with the hypothesis that the *rate of dissipation is maximized*. We hasten to add that, just as with any other variational principle, we do not get what we have not already introduced in a different form. Nevertheless, the approach taken here allows us to provide a perspective on the constitutive equations for elastoplastic material in the context of other dissipative behavior and to identify the similarities as well as the essential differences among different dissipative materials. Moreover, this assumption allows us to gain insight into all kinds of dissipative behavior other than plasticity.

The starting point of our subsequent analysis is the following rather simplistic statement[11].

---

[11] A somewhat frivolous analogy to the behavior of people may be instructive and fun. Imagine that you have been given (or have earned) some money (i.e. mechanical power). You may choose to deposit some of it in a bank (so that it can be recovered later) and "spend" some of it—exchange it for goods and services that cannot be converted back into the same amount of money. Clearly, your behavior and mentality is characterized not only by how much money you save but also what goods and services you choose to spend it on. The former, your investment habits, is described by the work potential functions ψ and Φ and the latter, your spending habits, by the rate of dissipation function.

> Just as the constitutive relations for the Helmholtz and Gibbs potentials describe how much work is recoverable along any given process for any particular material, we will introduce a constitutive relation for the rate of dissipation ζ which describes how much work is lost irrevocably along any process.

Unlike the Helmholtz and Gibbs potentials which are functions of the state, the relation for ζ depends upon the *process*, i.e. it is necessarily a function of the path taken by the body in state-space and will be referred to as the *dissipation function*. Then the allowable processes or paths in state space are those for which (9.35) is met along the entire path. There will be many processes that will be allowed from a given state. Each with a different amount of work lost.

Thus, two materials which have the same Helmholtz and Gibbs potentials will undergo very different processes, due to the fact that their *dissipative behavior may be markedly different* so that the set of allowable processes may be different. We will see such an example in the subsequent sections.

We now postulate a form for the rate of dissipation for elasto-plastic materials; we will assume that the rate of dissipation along any path is given by

$$\zeta = \zeta(\mathbb{S}, \dot{\varepsilon}_p) \geq 0. \tag{9.66}$$

(1) Since any change in $\dot{\varepsilon}_p$ should be a dissipative process, we will assume that $\zeta > 0$ if $\dot{\varepsilon}_p \neq 0$. Furthermore (2) since we want to recover elastic response, we will demand that $\zeta(\mathbb{S}, 0) = 0$, i.e. processes involving no permanent deformation are non-dissipative.

Some Examples: Consider the following two very simple examples of the rate of dissipation functions

$$\zeta = k\dot{\varepsilon}_p \cdot \dot{\varepsilon}_p \text{ and } \zeta = k \|\dot{\varepsilon}_p\|. \tag{9.67}$$

It is obvious that both these functions satisfy the criteria laid out above.

Now, for example, assuming that the state variables are $\mathbb{S} = \{T, \varepsilon_e\}$, the equation (9.52) reduces to

$$-\frac{\partial \psi}{\partial \varepsilon_e} \cdot \dot{\varepsilon}_p = \mathbf{S} \cdot \dot{\varepsilon}_p = \zeta(\mathbb{S}, \dot{\varepsilon}_p). \tag{9.68}$$

We note that $\dot{\varepsilon}_p = 0$ always satisfies the above equation (since $\zeta(\mathbb{S}, 0) = 0$). Also, for a given state $\mathbb{S}$, any value of $\dot{\varepsilon}_p$ that satisfies (9.68) will be an allowable process according to the second law of thermodynamics.

⊙ **9.16 (Exercise:). Irreversibility of plastic flow**
Show that given a state $\mathbb{S}$, if $\dot{\varepsilon}_p$ satisfies (9.68), then $-\dot{\varepsilon}_p$ will not be allowable. In other words, the equation (9.68) automatically enforces irreversibility.

Since ANY evolution equation of the form $\dot{\varepsilon}_p = f(\mathbb{S})$ is acceptable so long as (9.68) is satisfied. The question is how do we choose a suitable evolution equation?

One requirement that will pick out a specific evolution law is the assumption that the direction of plastic flow is one which maximizes the rate of dissipation. We will formally state this criterion as follows:

THE MAXIMUM RATE OF DISSIPATION HYPOTHESIS

We consider dissipative materials that are characterized by two fundamental scalar functions—the Helmholtz potential $\psi(\mathbb{S})$ and the rate of dissipation $\zeta(\mathbb{S},\dot{\mathbb{S}})$. The constitutive equation for any dissipative process (i.e. the evolution equation for $\dot{\mathbb{S}}$), is then determined by the condition that, among all allowable processes (subject to whatever external constraints are imposed on the set of possible processes), the actual process proceeds in such a way as to *maximize* the rate at which mechanical work is dissipated.

Notice that, in this form, the statement can apply to any continuum model that can be described by the two functions and hence we can include any continuum analog of a spring-dashpot system. As with all good ideas, this idea too has been discovered (or invented) and reinterpreted with continuously broadening scope until it is now considered as a fairly general idea[12].

One of the principal differences between this criterion and the other criteria stated earlier is that here we are keeping the state variables such as the stress, temperature etc., fixed and are trying to find the kinetic equation, i.e. we vary the value of the $\dot{\mathbb{S}}$ keeping the state $\mathbb{S}$ fixed. On the other hand, with the maximum plastic work criterion, the value of $\dot{\varepsilon}_p$ is fixed and we vary the *state* of the material. From a philosophical point of view, don't you think that varying the stress is inappropriate since at any instant of time, we *already know the state* and we are trying to *find* the plastic strain rate?

We will refer to the above criterion as "MRDH" (Maximum Rate of Dissipation Hypothesis) since we don't want to write out its full form each time. Notice that the MRDH criterion does NOT assume that the material is a rate independent elastoplastic material and hence, may be used for any kinetic equation including viscoelasticity, viscoplasticity etc. (see [Collins and Houlsby (1997); Rajagopal and Srinivasa (2004a,b)]).

---

[12] A version of this idea dates back to the Rayleigh dissipation function in classical mechanics. However, it appears that Ziegler ([Ziegler (1963, 1983); Ziegler and Wehrli (1987)]) appears to have been the first proponent of this idea as a general principle in plasticity and rheology. Ziegler seems to have "backed off" from his original more sweeping and general maximization principle, perhaps as a result of the initial withering criticisms of his ideas. Nevertheless, other authors, notably, [Collins and Houlsby (1997)], [Houlsby and Puzrin (2000)] have developed these ideas for a wide class of elastoplastic materials. Recently [Rajagopal and Srinivasa (2004c)] have expanded and clarified the scope of this idea and have shown its relationship with Onsager's ([Onsager (1931)]) reciprocity principle. In this book, we make heavy use of the ideas developed by [Rajagopal and Srinivasa (2004c)]. [Rice (1971)] in his paper on inelasticity has a statement regarding the existence of a plastic potential. However, a close reading of the result reveals that, as stated, it is applicable only to *scalar* internal variables that only depend on their own scalar driving forces thus making it too restrictive to apply even to classical viscoplasticity.

One way to think about such a criterion and a "heuristic" justification of the criterion is based on the following observation[13]:

The second law of thermodynamics implies that an isolated system will go to a state of maximum entropy subject to whatever external constraints are imposed on its state. It is silent as to how exactly this process occurs. If we think of entropy as the height of a hill, the second law states that an isolated system goes to the highest point of the hill possible, but the route taken is not specified. The MRDH suggests that *route taken is the route of steepest ascent*, i.e. it will follow a process that produces as much entropy as possible given the constraint. Collins, Houlsby and coworkers call those elastoplastic materials that satisfy the "MRDH" as being "Hyperplastic" similar to the use of the word "hyperelastic" for materials that satisfy $\boldsymbol{\sigma} = \partial \psi / \partial \boldsymbol{\varepsilon}$.

Consider a simple model where

$$\psi = \psi(\boldsymbol{\varepsilon}_e, T), \quad \zeta = \zeta(\boldsymbol{\varepsilon}_e, T, \dot{\boldsymbol{\varepsilon}}_p); \quad \dot{\boldsymbol{\varepsilon}}_p = \dot{\boldsymbol{\varepsilon}} - \dot{\boldsymbol{\varepsilon}}_e. \tag{9.69}$$

The precise form of these two functions are immaterial at this time. We will further assume that

$$\mathbb{S} = \partial \psi / \partial \boldsymbol{\varepsilon}_e. \tag{9.70}$$

How do you find the constitutive equation for $\dot{\boldsymbol{\varepsilon}}_p$ from the maximum rate of dissipation criterion?

The simple way is to do calculus— we maximize ζ over all possible values of $\dot{\boldsymbol{\varepsilon}}_p$—but wait, we have a constraint. All allowable values of $\dot{\boldsymbol{\varepsilon}}_p$ must meet equation (9.59) on p.295! Thus, we have to carry out a constrained maximization. Since this is such an important point, we have devoted an appendix to the method of Lagrange multipliers since it has been our experience that this beautiful, elegant and remarkable trick is not well known among many graduate students and practicing engineers. We urge you to refresh your memory regarding this.

We carry out a constrained maximization by using a "Lagrange multiplier", Λ, to enforce the constraint. The trick apparently due to Lagrange, is to maximize an *augmented function*

$$\Gamma(\mathbb{S}, \dot{\boldsymbol{\varepsilon}}_p, \Lambda) := \zeta(\mathbb{S}, \dot{\boldsymbol{\varepsilon}}_p) - \Lambda(\boldsymbol{\xi} \cdot \dot{\boldsymbol{\varepsilon}}_p - \zeta(\mathbb{S}, \dot{\boldsymbol{\varepsilon}}_p)), \tag{9.71}$$

where Λ is an additional new variable that has to be found. We now maximize the function Γ with respect to $\dot{\boldsymbol{\varepsilon}}_p$ and Λ by methods of calculus and obtain

$$\begin{aligned}\frac{\partial \Gamma}{\partial \dot{\boldsymbol{\varepsilon}}_p} &= 0 \Rightarrow \mathbb{S} = \frac{\Lambda - 1}{\Lambda} \frac{\partial \zeta}{\partial \dot{\boldsymbol{\varepsilon}}_p}, \\ \frac{\partial \Gamma}{\partial \Lambda} &= 0 \Rightarrow \mathbb{S} \cdot \dot{\boldsymbol{\varepsilon}}_p = \zeta(\mathbb{S}, \dot{\boldsymbol{\varepsilon}}_p). \quad \leftarrow \text{(the constraint)}\end{aligned} \tag{9.72}$$

We thus get an *implicit constitutive equation for* $\dot{\boldsymbol{\varepsilon}}_p$ *in terms of the Kirchhoff stress* $\mathbb{S}$. In words, equation (9.72)a implies the following.

---

[13] Continuing with the analogy to human nature, our estimation of your behavior is very low indeed; we assume that given a certain amount of money to spend, you will spend it at the highest possible rate given your spending patterns (i.e. the form for ζ). A sort of a principle of "maximum laziness".

> **GENERALIZED NORMALITY CONDITION**
> The MRDH implies that the plastic driving force, $\mathbf{S}$, is directed along the normal to the surface of constant rate of dissipation function, i.e. it is proportional to $\partial \zeta / \partial \dot{\varepsilon}_p$ which is the gradient of the rate of dissipation function (see Fig. 9.5 for a pictorial view of this criterion). Thus,
>
> $$\mathbf{S} = \phi \partial \zeta / \partial \dot{\varepsilon}_p, \qquad (9.73)$$
>
> with ϕ determined by the satisfaction of the constraint (9.59).

Finding Λ:
We have yet to eliminate the Lagrange multiplier, Λ, from (9.72a). We observe that we don't really need to find Λ but only the proportionality parameter, $\phi = (\Lambda - 1)/\Lambda$.

⊙ **9.17 (Exercise:). Finding ϕ**
Take the inner product of (9.72a) with $\dot{\varepsilon}_p$ and use the definition of ϕ as well as (9.72b) to obtain

$$\xi = \phi \frac{\partial \zeta}{\partial \dot{\varepsilon}_p} \quad \text{where} \quad \phi = \frac{\zeta}{\frac{\partial \zeta}{\partial \dot{\varepsilon}_p} \cdot \dot{\varepsilon}_p}. \qquad (9.74)$$

Now that we have seen the general result, we can consider a few examples:

⊙ **9.18 (Exercise:). Evolution equations that satisfy MRDH**
Consider a material for which $\psi = \psi(\varepsilon - \varepsilon_p, T)$. Let the rate of dissipation be given by $\zeta = K(\dot{\varepsilon}_p \cdot \dot{\varepsilon}_p)$ (this particular quadratic form for ζ is called the Rayleigh dissipation function). Show that the constitutive equation that satisfies the MRDH is (Hint: Use (9.74))

$$\dot{\varepsilon}_p = \sigma/(\varrho K). \qquad (9.75)$$

Show that, on the other hand, if $\zeta = \kappa \|\dot{\varepsilon}^p\|$ then we get $\phi = 1$ and

$$\dot{\varepsilon}_p / \|\dot{\varepsilon}_p\| = \mathbf{S}/\kappa \Rightarrow \dot{\varepsilon}_p = (\zeta/\kappa^2)\mathbf{S}. \qquad (9.76)$$

In other words we get a form similar to that of J2 Plasticity (except for the fact that we have not enforced plastic incompressibility).
Show that according to (9.76), $\|\sigma\| = \kappa$. Hence show that if $\|\mathbf{S}\| < \kappa$ then the solution (9.76) above is NOT valid. (Hint: Just find the norm of σ using (9.76) and you will get the result.)
Show that if $\|\mathbf{S}\| < \kappa$ then the only solution to (9.68) is that $\dot{\varepsilon}_p = 0$.
In other words, the yield function and the normality rule for J2 plasticity (without plastic incompressibility) are obtained from the dissipation function ζ.

9.7.1 Graphical understanding of the MRDH

In order for you to visualize how the MRDH "works" see Fig. 9.5.

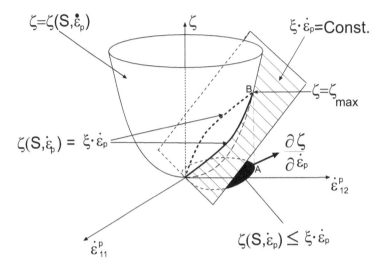

Fig. 9.5 A schematic figure demonstrating the notion of the maximum rate of dissipation hypothesis. The cup like figure is the dissipation function $\zeta(\mathbb{S}, \dot{\varepsilon}_p)$ (note that it is automatically non-negative). The vertical axis is the rate of dissipation and the horizontal plane represents two components of the six-dimensional $\dot{\varepsilon}_p$ space. The slanted plane is the set of points that satisfy the linear relation $\mathbf{S} \bullet \dot{\varepsilon}_p = \zeta$. Note that it changes with $\boldsymbol{\xi}$. The partially dotted curve is the set of the points were the dissipation relation (9.58) is satisfied. The projection of this curve onto the horizontal plane represents the *allowable values of* $\dot{\varepsilon}_p$. The *highest point* on this curve is the point B where the rate of dissipation is the maximum possible. Its projection is the point A. This is the point chosen by the MRDH. At this point $\mathbf{S}$ is along the direction $\partial \zeta / \partial \dot{\varepsilon}_p$ which is normal to the set of allowable values of $\dot{\varepsilon}_p$. This is the generalized normality rule. For some cases, its "dual" is the usual normality rule in plasticity.

To understand the figure, assume that $\dot{\varepsilon}_p$ has only two components and ζ is a nice positive function of $\dot{\varepsilon}_p$ and is drawn in the form of a cup. If you imagine the cup to be full of water up to any height, the surface of the water is called the *level set*. The vertical axis represents the rate of dissipation.

Now, recall that the equation (9.68) states that "Only those processes for which the rate of decrease of recoverable work is equal to the rate of dissipation are allowed". The rate of decrease of recoverable work is given by the function $f(\dot{\varepsilon}_p) = \mathbf{S} \bullet \dot{\varepsilon}_p$. If you expand out the terms you will realize that it is a *linear function of* $\dot{\varepsilon}_p$ *since the state of the system and hence* $\mathbf{S}$ *is fixed*. So, its graph will be a plane in 2-D. This plane is also shown in the figure. Notice that the plane "slices" though the cup along a curve that passes through the origin.

The intersection of the plane and the cup corresponds to the set of allowable values of $\dot{\varepsilon}_p$ since the intersection points lie on the cup and on the plane; so they satisfy $\mathbf{S}\dot{\varepsilon}_p = \zeta(\mathbf{S}, \dot{\varepsilon}_p)$.

The MRDH implies that *among all the values on the intersection curve, the actual value of $\dot{\varepsilon}_p$ corresponds to the HIGHEST point on the intersection curve.* At this point, it turns out that $\mathbf{S}$ is NORMAL to the shadow of the intersecting curve! It is also normal to the "level set". This is the generalized normality condition.

9.8 Rate-independent plasticity: How to get the yield function and the flow rule by using MRDH

> ⊙ **9.19 (Exercise:). Non-hardening J2 plasticity**
> *Show that if we have a dissipation function of the form $\zeta = \kappa \|\dot{\varepsilon}_p\|$ and we add the additional constraint of plastic incompressibility, i.e. $tr(\dot{\varepsilon}_p) = \mathbf{I}\bullet\dot{\varepsilon}_p = 0$, then we will obtain $\dot{\varepsilon}_p = \tau\zeta/\kappa$, which is the constitutive equation for J2 plasticity, with an explicit form for the Kuhn-Tucker parameter, i.e. $\phi = \frac{\zeta}{\kappa}$. (Hint: You have to use two Lagrange multipliers Λ and Γ and maximize $\Gamma = \zeta - \Lambda(\sigma\bullet\dot{\varepsilon}_p - \zeta) - \Gamma(\mathbf{I}\bullet\dot{\varepsilon}_p)$.)*

It may seem somewhat astonishing to you that if we start with a dissipation function of the form $\zeta = \kappa \|\dot{\varepsilon}_p\|$ and use MRDH, we get the entire evolution equation (form of the yield function, normality rule, as well as the Kuhn-Tucker condition) by just the maximization process. This is NOT an accident. We will now show that this is a general property of certain types of dissipation functions which are *positive homogeneous functions*.

POSITIVE HOMOGENEOUS DISSIPATION FUNCTIONS

The rate of dissipation $\zeta(\mathbf{S}, \dot{\varepsilon}_p)$ is said to be positive homogeneous if $\zeta(\mathbf{S}, \lambda\dot{\varepsilon}_p) = \lambda\zeta(\mathbf{S}, \dot{\varepsilon}_p)$ for all non-negative λ. Such a function, if it is differentiable, satisfies *Euler's theorem*, i.e.

$$\zeta(\mathbf{S}, \lambda\dot{\varepsilon}_p) = \lambda\zeta(\mathbf{S}, \dot{\varepsilon}_p) \Leftrightarrow \dot{\varepsilon}_p \bullet \partial\zeta/\partial\dot{\varepsilon}_p = \zeta. \quad (9.77)$$

We will now assume that ζ is a positive homogeneous function of $\dot{\varepsilon}_p$. In other words, let us assume that

$$\zeta(\mathbb{S}, k\dot{\varepsilon}_p) = k\zeta(\mathbb{S}, \dot{\varepsilon}_p). \; k > 0 \tag{9.78}$$

⊙ **9.20 (Exercise:). Positive homogeneous does not mean linear**
Show that $\zeta = \kappa \|\dot{\varepsilon}_p\|$ is a positive homogeneous function but that it is not linear. In general, the requirement that $\zeta \geq 0$ for all $\dot{\varepsilon}_p$ precludes ζ from being linear. Such positive homogeneous functions are called **gauge functions**.
Show that

$$\zeta = k_1 \|\dot{\gamma}_p\| + k_2 \frac{\dot{e}_p^2}{\|\dot{\gamma}_p\|}, \dot{e}_p = tr(\dot{\varepsilon}_p), \; \dot{\gamma}_p = \dot{\varepsilon}_p - \frac{1}{3} tr(\dot{\varepsilon}_p)\mathbf{I} \tag{9.79}$$

is also a positive homogeneous function of $\dot{\varepsilon}_p$.

The graph ζ with respect to $\dot{\varepsilon}_p$ of the functions above looks like a "cone" with its sharp point at the origin.[14] The wall of the cone is made up of straight lines.

9.8.1 The yield function

We will now state the major result regarding homogeneous rates of dissipation functions:

EXISTENCE OF A YIELD FUNCTION

Consider a dissipative material whose state variables are $\mathbb{S}' = \{\varepsilon_e, T\}$. If the rate of dissipation function ζ is positive homogeneous in $\dot{\varepsilon}_p$ and is independent of ε_e then, the maximum rate of dissipation assumption implies that there exists a convex yield function $f(\mathbf{S})$ such that for any Kirchhoff stress state $\mathbf{S}$ such that $f(\mathbf{S}) < 1$ there will be no plastic flow. Further plastic flow occurs only if $f(\mathbf{S}) = 1$ and the direction of the plastic strain rate will be normal to the yield surface.

This is a very crucial result and comes out of the theory of gauge functions. We will not prove all the results; we will just show how the yield function $f(\mathbf{S})$ arises. The construction is simple: given a positive homogeneous rate of dissipation function ζ, we will define $f(\mathbb{S})$ through

$$f(\mathbf{S}) = \max_{\dot{\varepsilon}_p \neq 0} \frac{\mathbf{S} \bullet \dot{\varepsilon}_p}{\zeta(\mathbb{S}, \dot{\varepsilon}_p)}, \tag{9.80}$$

---

[14] Before the advent of the ubiquitous plastic covers, food items in India were wrapped in old newspapers which were cleverly rolled into a cone. Positively homogeneous functions ζ look like those conical wrappers.

The function $f(\mathbf{S})$ is the *maximum of the ratio of the energy release rate* $\mathbf{S}\bullet\dot{\varepsilon}_p$ *and the rate of dissipation* ζ. If this maximum is less than 1, then for any nonzero $\dot{\varepsilon}_p$, sufficient energy is not available to overcome friction and no plastic flow takes place. When it becomes equal to one, then sufficient energy is available and the plastic flow occurs.

This function is called the "polar" of the gauge function ζ and the fact that it is convex follows from results in convex analysis.

Certain conditions regarding boundedness have to be satisfied by ζ in order for this maximum to be well-defined [15]. But these are not critical for our understanding, rest assured that for the relatively simple cases where we use this construction, the maximum can be found very easily.

⊙ **9.21 (Exercise:). Classical normality Rule**
Show that if $\zeta = \kappa\,\|\dot{\varepsilon}_p\|$ then $f(\mathbf{S}) = \|\mathbf{S}\|/\kappa$. (Hint: show that for this case the maximum is achieved when $\dot{\varepsilon}_p = \phi\boldsymbol{\xi}$. Use the Cauchy-Schwartz inequality which says that $\mathbf{a}\bullet\mathbf{b} \leq \|\mathbf{a}\|\,\|\mathbf{b}\|$ with equality only if $\mathbf{a}$ and $\mathbf{b}$ are parallel.)
Also show that if $\zeta = K\,\|\dot{\varepsilon}_p\|^2$ then the maximum in (9.80) can be made arbitrarily large by choosing $\dot{\varepsilon}_p$ to have an arbitrarily small magnitude.

In general, the function $f(\mathbf{S})$ is not well-defined unless ζ is positive homogeneous. Said differently, a properly defined and useful yield function $f(\mathbf{S})$ is obtained only as a result of assuming that the rate of dissipation is a positive homogeneous function of $\dot{\varepsilon}_p$.

From the definition (9.80) it is possible to deduce that

(1) $\mathbf{S}\bullet\dot{\varepsilon}_p \leq f(\mathbf{S})\zeta$ for all $\dot{\varepsilon}_p \neq 0$. (This is a generalization of the Cauchy-Schwartz inequality.)
(2) if $f(\mathbf{S}) < 1$ then $\mathbf{S}\bullet\dot{\varepsilon}_p < \zeta$, i.e. there is NO nonzero value of $\dot{\varepsilon}_p$ that is allowable, in other words if the state $\mathbb{S}$ is such that $f(\mathbf{S}) < 1$ then only elastic response is possible.
(3) if $f(\mathbf{S}) = 1$ then by the definition of $\eta(\mathbf{S})$, there is at least one, nonzero value of $\dot{\varepsilon}_p$ for which $\boldsymbol{\xi}\bullet\dot{\varepsilon}_p = \zeta$ and plastic flow becomes possible.

Thus, the function $f(\mathbb{S})$ is the *yield function* for this material.

---

[15] see [Rockafellar (1970)] Section 15 starting on p. 128. Although all the results in that section are proved by assuming that ζ itself is convex, it turns out for our limited purposes, (i.e. proving normality and convexity conditions) we need to assume only that *the level sets of ζ are bounded*.

The normality flow rule

It can further be shown[16] that the generalized normality rule (9.74) becomes

$$\mathbf{S} = \frac{\partial \zeta}{\partial \dot{\varepsilon}_p} \text{ which then implies that } \dot{\varepsilon}_p = \phi \frac{\partial f}{\partial \boldsymbol{\xi}}. \quad (9.81)$$

We state these results without proof, since the proof involves some rather tedious aspects of complex analysis. The interested reader is directed to read the paper by [Eve et al. (1990)] as well as the chapter on gauge functions by [Rockafellar (1970)].

9.8.1.1 Summary of conclusions drawn by applying MRDH

Thus the entire structure of the constitutive equations for rate independent plasticity can be obtained from the following considerations:

- The state variables are $\mathbb{S} = \{\varepsilon_e, T\}$. We may later augment it with inelastic variables $\mathbf{Q} = \{\kappa, \beta\}$.
- The Helmholtz potential is $\psi = \psi(\mathbb{S})$.
- The rate of dissipation function is $\zeta = \zeta(\mathbf{Q}, T, \dot{\varepsilon}_p)$ and is a positive homogeneous function of $\dot{\varepsilon}_p$. Notice that all we require is that ζ does not depend upon ε_e.
- The stress is given by $\boldsymbol{\sigma} = \varrho \partial \psi / \partial \varepsilon_e$.
- The dissipation equation that must be satisfied is $\mathbf{S} \bullet \dot{\varepsilon}_p = \zeta$, where $\mathbf{S} = \partial \psi / \partial \varepsilon_e$.
- By using the MRDH, the plastic strain rate is given *implicitly* by the relation $\mathbf{S} = \partial \zeta / \partial \dot{\varepsilon}_p$
- Alternatively, we can replace ζ by the yield function, defined through $f(\mathbf{S}) = \max_{\dot{\varepsilon}_p \neq 0} \mathbf{S} \bullet \dot{\varepsilon}_p / \zeta(\mathbb{S}, \dot{\varepsilon}_p)$, and no plastic flow occurs if $f(\mathbb{S}) < 1$.
- It is possible to show that the yield function is of the form $f(\mathbf{S}, T; \mathbf{Q})$ and, is possible to obtain the normality flow rule in the form $\dot{\varepsilon}_p = \phi \partial f / \partial \mathbf{S}$ where the parameter ϕ satisfies the Kuhn Tucker conditions. This last point was not proved in this chapter but was simply stated.

Up to now, we have considered rates of dissipation that are independent of ε_e. What happens if ζ does depend upon ε_e? It can also be shown (The proof is omitted) that *if the Helmholtz potential is $\psi(\varepsilon_e, T)$ and the dissipation function is of the form $\zeta = h(\varepsilon, \varepsilon_p, T) g(\mathbf{Q}, T\dot{\varepsilon}_p)$ then the yield surface will be of the form $f(\mathbf{S}, \mathbf{Q}, T) = h(\varepsilon, \varepsilon_p, T)$ where $f(\mathbf{S}, \mathbf{Q}, T) = \max_{\dot{\varepsilon}_p \neq 0} \mathbf{S} \bullet \dot{\varepsilon}_p / g(\mathbf{Q}, \dot{\varepsilon}_p)$ and that the flow rule is of the form $\dot{\varepsilon}_p = \phi \partial f / \partial \mathbf{S}$.* In other words, specific forms for ζ provide us with plastic potentials also.

---

[16]the proof is omitted since it is somewhat hard and technical. But see [Eve et al. (1990)] for details.

The next example shows a sampling of the results.

> ⊙ 9.22 (Exercise:). Non-associative flow rule
> Consider a dissipation function given by $\zeta = k(tr\varepsilon_e)\|\dot{\varepsilon}_p\|$. Clearly, it is explicitly dependent on ε_e—it depends upon the mean normal strain. Show that the function $f(\mathbb{S})$ defined by (9.80) now becomes $f(\mathbb{S}) = \|\mathbb{S}\|/(k(tr\varepsilon_e))$, i.e. we will now get pressure dependent yield surfaces.
> Show further that the generalized normality condition (9.74) when inverted, gives $\dot{\varepsilon}_p = \phi\sigma$.
> Show that this flow rule does NOT satisfy the normality condition. In other words, non-associative flow rules can be obtained by choosing ζ to explicitly depend upon the strain.

9.9 The Bauschinger effect and history dependence

The previous sections dealt with models where the free energy ψ depended only on current values of the state variables $\mathbb{S}(\varepsilon_e, T)$ and we invariably found that *from a thermodynamical perspective* this will not allow us to include back stresses or Bauschinger effects, although isotropic hardening behavior can be accommodated by introducing inelastic variables $\mathbf{Q}$ and allowing ζ to depend upon them.

However, these are macroscopic variables that do not adequately reflect the microstructural aspects of the response of the material. For example, in materials like aluminum, copper etc. the primary mode of plastic deformation is the movement and accumulation of dislocations through the crystal lattice. As these dislocations move, they get entangled with one another creating cells and networks. A material with an extensive network of entangled dislocations is said to be "heavily cold worked". This network can be destroyed and reformed if the mode of deformation is switched: (say from extension to shearing). The presence of the network has multiple effects. For one, the dislocations impede the motion of other dislocations and thus increase the hardness (this is the main reason for the strain hardening of materials). Also, the dislocation networks and cells have "directionality", once they are formed, it is easier for mobile dislocations to move forward than backward leading to some aspects of the Bauschinger effect. We cannot do full justice to the motion of dislocations and their modeling in this "macroscopic simple minded" approach that is used throughout the book[17].

It is clear that the tangle of dislocations must be accounted for in some way to model hardening behavior. From an energy and entropic point of view, the dislocations increase the level of disorder in the crystal by disturbing the lattice

---

[17] For more details consult any materials science book such as [Hertzberg (1976)], Chapters 2–4 or [Dieter (1986)]. If you want to simulate dislocations and their interactions see in detail, see e.g., [Benzerga and Shaver (2006); Guruprasad and Benzerga (2008)].

structure and hence contribute to the (configurational) entropy. Moreover, the dislocated regions are regions of higher energy so that there is an energy increase due to the dislocations also. Thus the energy and entropy of the system must be altered to account for these effects. G. I. Taylor seems to have been the first to have recognized this (see [Taylor and Quinney (1934)]) and made statements to the effect that only about 80% of the work done is dissipated and that the remaining 20% is stored. Subsequently a number of investigations of this feature have been undertaken (see e.g., [Bever et al. (1973)]), including recent computational work using discrete dislocations (see [Benzerga et al. (2005)]). This observation has led to a number of researchers to include a "Taylor factor" of about 0.8 to multiply the stress power term in the energy equation. Such ad-hoc extensions, while they may be expedient for carrying out a quick engineering task have no place in the development of a theory. Rather, one should go back to the fundamentals and see how to incorporate them in a proper manner.

One way to account for this explicitly is to introduce additional state variables (called microstructural variables) to describe the dislocated structure in detail. However, this is a very complex phenomenon and as yet there seems to be no clear and accepted way to explain the effects by adding just a few additional variables.

An alternative way of approaching this phenomenon, which is simple but crude, is to use an approach that has been very successful in viscoelasticity and simply say that the *energy of the system depends upon the history of the plastic deformation*. This is a phenomenological approach that, surprisingly seems to work reasonably well (see e.g., [Valanis (1980); Haupt and Kamlah (1995); Kamlah and Haupt (1997); Makosey and Rajagopal (2001); Mollica (2001)] to see just a few papers that have dealt with this issue).

One way to achieve this is to consider a Helmholtz free energy of the form

$$\psi = \psi_1(\varepsilon_e(t), T(t)) + T\psi_2 + \psi_3, \qquad (9.82)$$

where the first term depends upon the current values of the state variables $\varepsilon_e(t)$ and T, ψ_2 and ψ_3 in the second and third terms are *functionals* that depend upon the history of the accumulated plastic strain and temperature *but not the elastic strain or temperature*. This idea appears to have been advocated by [Kratochvil and Dillon (1969)] in a pioneering work on work hardening. The material here is based on the ideas that can be found in their work.

Two of the most popular ways to introduce the history dependence are:

- Write out an explicit integral of the history of the accumulated plastic strain, typically with some sort of "fading dependence" on past history, i.e. the material remembers only the recent history but not the distant past. This is the approach commonly followed for viscoelastic materials. Such models are called *integral models*. There is an extensive discussion of models like these (usually called endochronic models, see [Valanis (1980)]).

- Write a differential equation for the variable in question. Integration of the differential equation will give rise to history dependence. Such models are called *rate type models*.

These two approaches are not equivalent[18]. However, there are some constitutive relations that are amenable to be written either way. The problem with directly writing down the history dependence is that the resulting equations turn out to be EXTREMELY complex! From a practical point of view, rate type models are much more amenable to solution since all they require is to keep track of a finite set of additional variables which have to be integrated. On the other hand, for integral type models, the entire past history needs to be kept in memory. Since this is impossible, either the integrals are approximately computed (by keeping only a finite number of past points) or they are approximated by a rate type model in a roundabout fashion (using certain series representations using exponentials (so-called Prony series) etc.). Rather than take this roundabout way, we will start with a rate form directly.

You may be wondering how a rate type model can be used to introduce history dependence. We will demonstrate this by means of a one-dimensional example:

Consider a history integral for the free energy of the form

$$\psi(t) = \int_{-\infty}^{t} \exp^{a(\tau-t)} f(\tau)\, d\tau. \tag{9.83}$$

It is a trivial matter to see that this is the same as a solution to the differential equation

$$\dot{\psi} = f(t) - a\psi. \tag{9.84}$$

Thus, from a practical point of view, rather than propose a history integral of the form (9.83) we might as well treat ψ itself as a variable whose evolution equation is given by (9.84). We will follow this principle: rather than give explicit history integrals for ψ_2 and ψ_3 we will treat them as additional variables and provide differential equations for them. What is the justification for this idea. We are expecting ψ_2 and ψ_3 to be related to the entropy and internal energy associated with the dislocations. Thus we expect them to be proportional to the dislocation density. In this case, all we are saying is that rather than assume that ψ_i is some constant times the dislocation density and then give an evolution equation for the dislocation density (as was done by [Kratochvil and Dillon (1969)]), we will directly give an evolution equation for ψ_i.

Given the form (9.82) for the free energy, we will define the stress, the entropy

---

[18] See [Truesdell and Noll (2004)] as well as [Green and Naghdi (1973)] for an extensive discussion of the similarities and differences between these approaches.

and the internal energy as

$$\sigma = \frac{\partial \psi}{\partial \varepsilon_e} = -\frac{\partial \psi_1}{\partial \varepsilon_e},$$
$$\eta := -\frac{\partial \psi}{\partial T} = -\frac{\partial \psi_1}{\partial T} - \psi_2, \quad (9.85)$$
$$\exists := \psi + T\eta = \psi_1 + \psi_3 - T\frac{\partial \psi_1}{\partial T}.$$

Note that the form (9.82) has been chosen in such a way that the entropy and the internal energy have terms that depend upon the history of the plastic strain but the form for the stress is *independent* of the history of the accumulated plastic strain since it does not depend upon ψ_2 or ψ_3. From this form, we see that ψ_2 is related to the configurational entropy of the dislocations while ψ_3 is related to the internal energy of the dislocation network. Admittedly, we might expect the configurational entropy contribution to be quite small but that the internal energy contribution to be quite large.

⊙ **9.23 (Exercise:). The heat equation and the rate of dissipation function**

Substitute (9.85) into the energy equation (9.13) and simplify the result and show that the heat equation reduces to

$$\varrho T \dot{\eta} = div\mathbf{q}\varrho(\frac{\partial \psi_1}{\partial \varepsilon_e} \cdot \dot{\varepsilon}_p - T\dot{\psi}_2 - \dot{\psi}_3), \quad (9.86)$$

and hence identify the rate of mechanical dissipation, ζ as

$$\zeta = (\frac{\partial \psi_1}{\partial \varepsilon_e} \cdot \dot{\varepsilon}_p - T\dot{\psi}_2 - \dot{\psi}_3). \quad (9.87)$$

Thus the rate of change of the history dependent terms ψ_2 and ψ_3 that we have added to the Helmholtz potential appear in the terms representing the loss of recoverable work. As you will see shortly, these additional terms ψ_1 and ψ_2 will eventually lead to the Bauschinger effect and other hardening effects.

9.9.1 A simple model for the Bauschinger effect

In order to show how the shift tensor and the isotropic hardening terms appear in the equations when the history dependent terms ψ_2 and ψ_3 are included, consider the following constitutive specific assumption for ψ_1:

$$\psi_1 = 1/2\mathbb{C}(\varepsilon_e) \cdot (\varepsilon_e) + CT(1 - \ln T), \quad (9.88)$$

so that

$$\sigma = \varrho \frac{\partial \psi_1}{\partial \varepsilon_e} = \varrho \mathbb{C}(\varepsilon_e). \quad (9.89)$$

We will also assume for the sake of this example that $\zeta = \kappa_0 \|\dot{\varepsilon}_p\|$, where κ_0 is a constant.

We will further assume that $\psi_2 = 0$, i.e. there is a negligible entropy change due to the accumulation of dislocations and that the history dependent term ψ_3 is given by an evolution equation for ψ_3 which is of the form[19]

$$\dot{\psi}_3 = A \|\dot{\varepsilon}_p\| + \boldsymbol{\alpha} \cdot \dot{\boldsymbol{\varepsilon}}_p, \tag{9.90}$$

where A and $\boldsymbol{\alpha}$ are themselves given by a differential equation. Substituting these in (9.87) and grouping terms, we get (using the fact that $A\|\dot{\varepsilon}_p\| = A\zeta/\kappa^o$),

$$(\mathbf{S} - \boldsymbol{\alpha}) \cdot \dot{\boldsymbol{\varepsilon}}_p = \boldsymbol{\xi} \cdot \dot{\boldsymbol{\varepsilon}}_p = (1 + A/\kappa_0)\zeta. \tag{9.91}$$

Note that in keeping with our definition for "driving force for plastic deformation" we have $\boldsymbol{\xi} = \mathbf{S} - \boldsymbol{\alpha}$ and we now see that a generalized "back stress" appears in the dissipation inequality, from thermodynamical considerations. The main point to observe here is that *we are free to choose any constitutive equation we want for $\boldsymbol{\alpha}$ without violating any thermodynamical requirement.*

It is a simple matter to invoke the MRDH now:

⊙ **9.24 (Exercise:). Evolution law with kinematic and isotropic hardening**
Maximize $\zeta = \kappa_o \|\dot{\varepsilon}_p\|$ under the constraint that (9.91) be met and show that we will get

$$\boldsymbol{\xi} = (\kappa_0 + A) \frac{\dot{\boldsymbol{\varepsilon}}_p}{\|\dot{\varepsilon}_p\|}. \tag{9.92}$$

Show also that the yield condition is given by

$$f(\boldsymbol{\xi}) = \frac{\|\boldsymbol{\sigma} - \varrho\boldsymbol{\alpha}\|}{\varrho(\kappa_0 + A)} = \frac{\|\boldsymbol{\sigma} - \varrho\boldsymbol{\alpha}\|}{\kappa} = 1, \tag{9.93}$$

where $\kappa = \kappa_0 + A$ is the "equivalent hardening parameter", whose evolution equation is given by

$$\dot{\kappa} = \dot{A} = f(s)\dot{s}, \quad \dot{s} := \|\dot{\varepsilon}_p\| \tag{9.94}$$

Thus the isotropic hardening parameter is given by $\varrho(\kappa_0 + A)$ and *any type of isotropic hardening can be represented by a suitable choice for the rate of change of A.*

We can now generalize this result to different kinds of hardening behavior as

---

[19]This is the crucial step for incorporation of hardening in a thermodynamically consistent way so please read this portion carefully.

you will see in the following exercise.

⊙ **9.25 (Exercise:). General evolution Law**
Let ζ be a positive homogeneous function of $\dot{\varepsilon}_p$. Let κ_0 be the "intrinsic frictional resistance" of the material and let the plastic arc length be defined by

$$\dot{s} = \frac{\zeta}{\kappa_0}. \tag{9.95}$$

Let the energy "locked up" in the dislocation structure be given by

$$\dot{\psi}_3 = A\dot{s} + \boldsymbol{\alpha} \bullet \dot{\varepsilon}_p. \tag{9.96}$$

Then using the MRDH, show that maximizing the dissipation will result in

$$\boldsymbol{\xi} = \mathbf{S} - \boldsymbol{\alpha} = (\kappa_0 + A)\frac{\partial \dot{s}}{\partial \dot{\varepsilon}_p}. \tag{9.97}$$

Then, by using standard results of convex analysis[20], we can show that there exists a yield criterion of the form

$$f\left(\frac{\boldsymbol{\xi}}{\kappa_0 + A}\right) = 1, \tag{9.98}$$

and that the normality rule will become

$$\dot{\varepsilon}_p = \phi \frac{\partial f}{\partial \boldsymbol{\xi}}. \tag{9.99}$$

Furthermore, we are free to specify evolution equations for κ_0, A and α anyway we want and still maintain full thermodynamical compatibility. The definition of the arc length parameter is closely related to that introduced in an ad hoc manner in Chapter 8 equation (8.44) on p.255. Note that s is non-dimensional and $\dot{s}$ has the dimensions of strain rate. This parameter is closely related to the "internal clock" or "endochronic" theory of plasticity (see e.g., Section 6.5 of [Khan and Huang (1995)], esp. Eqns (6.130) and (6.131)).

Thus, in this model, both isotropic and kinematic hardening comes from the history dependence of ψ. Note, however, that the rate of dissipation, ζ, involves only κ_0 and DOES NOT involve A. *In this model, the hardening of the material is NOT due to the increase in the rate of loss of mechanical power but actually due to the accumulation of dislocations in the material. The same is true of the Bauschinger effect.*

This is a key point and is in keeping with the general discussions in the materials science literature that most of the hardening is not due to the conversion of mechanical work into heat but because of the mechanical energy being "locked up" in the newly formed dislocation structures. This is the effect that is modeled by the terms ψ_2 and ψ_3.

---
[20] The proofs are omitted here since they are quite technical and do not provide insight.

We have yet to show the explicit history dependence of ψ_2. We will now make this explicit by choosing special evolution laws for A and $\boldsymbol{\alpha}$ as follows:

$$\dot{s} = \|\dot{\boldsymbol{\varepsilon}}_p\|$$
$$\dot{\psi}_3 = A\dot{s} + \boldsymbol{\alpha}\bullet\dot{\boldsymbol{\varepsilon}}_p \qquad (9.100)$$
$$\dot{A} = \Theta(s, A)\dot{s}, \quad \dot{\boldsymbol{\alpha}} = c_1\dot{\boldsymbol{\varepsilon}}_p + c_2\boldsymbol{\alpha}\dot{s}.$$

The first equation represents the plastic arc length parameter, and the third defines the isotropic and kinematic hardening functions respectively. The middle equation shows how these two functions contribute to the stored energy.

We can collect all the results that we have obtained in this section and summarize the results as follows:

THERMODYNAMICAL CONSTITUTIVE LAWS: SMALL STRAIN

List of state variables: $\mathbb{S} = \{\boldsymbol{\sigma}, \boldsymbol{\varepsilon}, \boldsymbol{\varepsilon}_p, s, \mathbf{Q}\}, \mathbf{Q} = \{A, \kappa_0, \boldsymbol{\alpha}\}$,

The Helmholtz function: $\psi = \psi_1(\boldsymbol{\varepsilon}_e, T) + \psi_3, \ \boldsymbol{\varepsilon}_e = \boldsymbol{\varepsilon} - \boldsymbol{\varepsilon}_p$,

The rate equation for ψ_3 : $\dot{\psi}_3 = A\dot{s} + \boldsymbol{\alpha}\bullet\dot{\boldsymbol{\varepsilon}}_p$

The elastic response function: $\boldsymbol{\sigma} = \dfrac{\partial \psi_1}{\partial \boldsymbol{\varepsilon}_e}$,

The entropy: $\eta = -\dfrac{\partial \psi_1}{\partial T}$,

The yield function: $f\left(\dfrac{\boldsymbol{\xi}}{\kappa_0 + A}\right) - 1, \ \boldsymbol{\xi} := \boldsymbol{\sigma} - \boldsymbol{\alpha}$,

$$f\left(\lambda\dfrac{\boldsymbol{\xi}}{\kappa_0 + A}\right) = \lambda f\left(\dfrac{\boldsymbol{\xi}}{\kappa_0 + A}\right) \forall \lambda > 0 \qquad (9.101)$$

The flow rule: $\dot{\boldsymbol{\varepsilon}}_p := \dot{\boldsymbol{\varepsilon}} - \dot{\boldsymbol{\varepsilon}}_e = \phi\dfrac{\partial f}{\partial \boldsymbol{\xi}}$,

The plastic flow conditions: $\phi \geq 0, \ f \leq 0, \ \phi f = 0$,

The plastic arc length: $\dot{s} = \dfrac{\zeta}{\kappa_0} = \dfrac{(\mathbf{S} - \boldsymbol{\alpha})\bullet\dot{\boldsymbol{\varepsilon}}_p}{\kappa_0}$,

The hardening Laws: $\dot{\mathbf{Q}} = \phi\boldsymbol{\Theta}(\mathbb{S})$,

The Heat equation: $\varrho T\dot{\eta} = -\text{div}\mathbf{q} + \varrho(\mathbf{S} - \boldsymbol{\alpha})\bullet\dot{\boldsymbol{\varepsilon}}_p$.

where $\boldsymbol{\Theta}$ is a short form for the collection of evolution equations for A, κ_0 and $\boldsymbol{\alpha}$.

Finally, note the since the stress is independent of ψ_3, we see that the *mechanical equations of evolution are actually independent of the specific form for ψ_3*.

Integral representations for the stored energy

It appears that in order to model history dependence, we have introduced a whole host of additional variables and functions. This is not a bad thing—these additional variables give us some insight into what is actually happening. However, they mask the exact nature of the history dependence and seem to be very different from viscoelasticity models. To make the history dependence more explicit let us assume that $\Theta(s, A) = (K - A)$ where K is a constant and further that $c_2 = 0$. Then it is possible to write (9.100) as an integral equation for ψ_2. This history integral is of the form:

$$s = \int_0^t \|\dot{\varepsilon}_p\| \, dt,$$
$$\psi_3 = K(1 - e^{-s}) + c_1 \int_{-\infty}^{t} \int_{-\infty}^{t'} e^{-c_2(s(t') - s(\tau))} \dot{\varepsilon}_p(t') \bullet \dot{\varepsilon}_p(\tau) \, d\tau \, dt'. \qquad (9.102)$$

In other words, the simple set of ODEs given in (9.100) represents a very complicated integral form for ψ_3 in terms of the history, involving 3 time intergals. Note that it would not have been possible to directly intuit the form of the history integral for ψ_3. Even if we did, it would be almost impossible to directly evaluate this at every time step. Rather, the use of the additional variables allows us to simply solve the differential equations given by (9.100) and hence capture the entire past history.

⊙ **9.26 (Exercise:). Integral to rate form**
Show by differentiation that (9.102) leads to (9.100).

As you can see, we can express certain very complicated history dependencies in a very compact form as ODEs. However not all history dependencies can be converted into such ODES. Nevertheless, this approach is very suitable since these lead to a finite state-space representation that is numerically integrable without much effort.

9.10 Summary

If you find this chapter and particularly this section rather hard, this is to be expected. This chapter introduces a host of rather complex concepts that need to be understood and cogitated over. There is much to be done regarding the thermodynamics of these materials and we haven't even touched on finite deformations yet! We hope that you will use this as a starting point to gain a deeper understanding of the thermodynamics of inelastic materials and will help the community of researchers do a better job in explaining these phenomena.

In this chapter we have tried to justify the assumptions that we made in plasticity from the point of view of thermodynamics. From a practical point of view, one could view such justifications as no more than a restatement of the assumed laws in a more complicated format. However, the main point here is that there is a desire for consistency— we want to find out to what extent are our ad-hoc generalizations consistent with the foundations of thermodynamics and what further assumptions are needed. Furthermore such an outlook allows us to explore extensions of these laws to more complicated situations. Here, we would like to provide a warning: There is a general tendency in the literature on continuum thermodynamics to use it as a "crank"—a constitutive equation mill, as it were, where we make some assumptions regarding the Helmholtz potential and the dissipation function (or a kinetic Law) with scant justification and simply obtain expansions and expressions in terms of the invariants etc. Anyone who has had even the briefest encounter with experimentation knows that such an exercise is a fool's errand because we have limited capability to obtain 3 constants not to mention whole functions (which are an uncountable infinity of constants). We urge you to make sure that you use your common sense and physical intuition to choose a suitable simplistic model.

Finally a note on MRDH: This appears to be a deceptively simple hypothesis and has provided excellent results in a variety of contexts; however it should be used with caution, if at all, since it is not justified by any microscopic argument that we know of. Rather it appears to summarize a wide range of empirical observations on various evolution laws, both linear and nonlinear. Linear evolution laws (corresponding to quadratic dissipation functions) have however been well studied and the MRDH is equivalent to the Onsager reciprocity theorems (see [Groot and Mazur (1962)] and [Rajagopal and Srinivasa (2004c)] for an extensive discussion) for this case. In that sense, it seems to be well grounded. But we have boldly applied it to nonlinear and non-differentiable functions with great benefit. This area of thermodynamics of dissipative processes is comparatively new and we hope that, whether you agree with the methodology or not, this chapter will inspire you to work in the area and help us understand the scientific underpinnings better. In this regard, plasticity (and its rigid body dynamics counterpart—friction) is quite a challenge due to its discontinuous nature and serves as an interesting counterexample to a variety of assumptions in classical physics while at the same time being practically very useful.

Salient points:

(1) The foundation of a thermodynamic framework for the constitutive assumptions are given by either a Helmholtz potential ψ or by the Gibbs potential Φ.
(2) In the Helmholtz formulation, the equations of state are:

$$\eta = -\frac{\partial \psi}{\partial T}, \quad \boldsymbol{\sigma} = \frac{\partial \psi}{\partial \varepsilon_e}, \quad \exists = \psi + T\eta. \tag{9.103}$$

(3) The mechanical dissipation equation in the Helmholtz formulation is
$$(\boldsymbol{\sigma}/\varrho - \boldsymbol{\alpha}) \cdot \dot{\boldsymbol{\varepsilon}}_p = \zeta. \qquad (9.104)$$
(4) It may be convenient to write the fundamental laws in the Gibbs formulation and it is a relatively simple matter to transform Φ to ψ.
(5) If the energy release rate can be written as $\boldsymbol{\xi} \cdot \dot{\boldsymbol{\varepsilon}}_p$ for some suitably chosen $\boldsymbol{\xi}$, then $\boldsymbol{\xi}$ will be called the "driving force for plastic deformation".
(6) If we assume that ζ is a positive homogeneous function of $\dot{\boldsymbol{\varepsilon}}_p$ and if we use the MRDH, then we obtain a convex yield surface in $\boldsymbol{\xi}$ space and the fact that $\boldsymbol{\xi} = \partial \zeta / \partial \dot{\boldsymbol{\varepsilon}}_p$.
(7) If we assume that ψ contains terms that depend upon the past history of the plastic strain and temperature *but not on the past history of the strain*, then we can model the Bauschinger effect as well as certain hardening effects due to the formation of dislocation cells.
(8) Besides these rate independent models there are a number of models that are suitable for rate dependent materials. See, e.g., [Freed et al. (1991)], to cite just a few.

9.11 Homework

(1) By using the methods of Chapter 6, obtain an expression for Kuhn Tucker parameter ϕ in equation 9.101 (assuming T=const.). Also obtain an expression for the Tangent Modulus $\mathcal{K}$. Ans:
$$\phi = \frac{\mathbf{N} \cdot \mathbb{C} \dot{\boldsymbol{\varepsilon}}}{\mathbf{N} \cdot \mathbb{C} \mathbf{N} + \mathbf{N} \cdot \boldsymbol{\Theta}_\alpha + \Theta_A} \qquad (9.105)$$
where $\mathbf{N} = \partial f / \partial \boldsymbol{\xi}$ is the normal to the yield surface, $\mathbb{C} = \partial^2 \psi / \partial \boldsymbol{\varepsilon}_e^2$ is the elastic modulus and the evolution equations for the backstress and the hardening parameter are given by $\dot{\boldsymbol{\alpha}} = \phi \boldsymbol{\Theta}_\alpha$ and $\dot{\kappa}_0 + \dot{A} = \phi \Theta_A$. In general, these latter forms themselves will involve the plastic arc length parameter s.)

This is a very useful exercise to test your own understanding of the constitutive formulation.

(2) Consider a material whose free energy function ψ is given by $\psi = 1/2 \mathbb{C} \boldsymbol{\varepsilon}_e \cdot \boldsymbol{\varepsilon}_e$ and whose rate of dissipation function ζ is given by $\mu x \ln(x + \sqrt{1+x^2})$ where $x = \|\dot{\boldsymbol{\varepsilon}}_p\|$, $\dot{\boldsymbol{\varepsilon}}_p = \dot{\boldsymbol{\varepsilon}} - \dot{\boldsymbol{\varepsilon}}_e$. By using the MRDH, find the evolution equation for $\dot{\boldsymbol{\varepsilon}}_p$. (Hint: You need to simplify intermediate results very carefully and obtain by the maximization procedure, that
$$\mathbf{S} = \boldsymbol{\sigma}/\varrho = \mu \ln\left(x + \sqrt{1+x^2}\right) \dot{\boldsymbol{\varepsilon}}_p / x. \quad x = \|\dot{\boldsymbol{\varepsilon}}_p\| \qquad (9.106)$$
Next, take the magnitude of both sides and simplify the resulting equations and get
$$\dot{\boldsymbol{\varepsilon}}_p = (1/\mu) \sinh(\|\mathbf{S}\|) \frac{\mathbf{S}}{\|\mathbf{S}\|}. \qquad (9.107)$$
This form of viscoelastic response is quite well known in polymer mechanics.

Chapter 10

Numerical Solutions of Boundary Value Problems

Learning Objectives

By learning the material in this chapter, you should be able to do the following:

(1) Idealize, simplify and model the problem on inelastic deformations at hand.

(2) Write down the general governing equations for a J2 elastoplastic body.

(3) Choose an appropriate technique to solve the problem numerically.

(4) Approximate the primary variable field (such as the displacement field) using the moving least squares approximation and impose the essential boundary conditions.

(5) Convert the governing differential equations into a set of nonlinear algebraic equations by making use of the approximation and the weak form (weighted residuals) of the governing differential equations.

(6) Solve for the primary variable field (displacement field) using the Newton-Raphson algorithm.

(7) Describe the complete numerical method for solving a general boundary value problem involving inelastic deformations.

(8) Integrate the constitutive equations using the convex cutting plane algorithm using MATLAB.

CHAPTER SYNOPSIS

This chapter is focused on describing how one can carry out a numerical solution procedure for boundary value problems for small deformations, especially, for those problems that are out of reach for analytical solution procedures. The steps related to converting the governing differential equations to a solvable set of algebraic equations are listed. A mesh-free method that utilizes moving least squares approximation for the displacement function is used in solving the governing differential equations. The variational procedure and the Newton-Raphson algorithm are used for this purpose. An important step in a typical numerical procedure involving elastic-plastic deformations is the integration of the constitutive equations. A convex cutting plane algorithm to integrate the constitutive equations that were developed in the earlier chapters is described in detail with a couple of examples. Some example problems are solved using the entire numerical procedure described in this chapter.

Chapter roadmap

Before one embarks on solving a boundary value problem related to an application involving inelasticity, there are idealization exercises that one has to carry out to simplify the geometry, loading conditions and the constitutive model. These that involve important decisions are discussed in section 10.2. With so many numerical procedures available, the choice of the right numerical technique gains importance. This is also discussed in section 10.2 briefly. In a typical numerical solution process, there are standard steps involved which are elaborated in the subsequent chapters.

Section 10.3 discusses mesh-free approach to solving the governing differential equations numerically. This involves generation of the moving least squares approximation for the trial function, choosing the weight function, and formulation and evaluation of weak form integral of the balance law. Finally, a set of non-linear algebraic equations are obtained that are solved using the Newton-Raphson technique.

Integration of the plastic flow equations is an important step in the above procedure and is dealt with separately in section 10.4. Those readers who are only interested in adopting a simple and suitable procedure for integrating the plastic flow equations for their own numerical method of solving the governing differential equations can skip the previous sections and come directly to this section. The elegance of the convex cutting plane algorithm is described and shown using simple homogeneous examples.

In section 10.5, a couple of boundary values problems are solved using the method described in the chapter. One is a one-dimensional example of a rod with varying cross-section while the other is a standard plate with a circular hole problem.

10.1 Background

Imagine a situation in which you have to finalize on a design pertaining to an engineering problem that involves inelastic deformations. For example, the application could be the design of the casing of a turbine. The design requirement may be such that a particular design life is specified for which the component or the structure last. Testing the casing for different loading cases with different materials is prohibitively expensive and therefore, one needs to rigorously analyze for appropriate stresses and inelastic deformations that may occur in the casing under typical load combinations acting on it during its life period that may be critical to design.

Analysis for stresses and deformation gives an insight into the safety against critical loads in the design of the casing. A verification by a standards organization involves: looking into these calculations to certify for safety and design life. Such a certification goes a long way in establishing customer confidence in the product.

We first seek to find out if such simplifications are possible by which one may not overly lose accuracy in the estimation of the actual stresses and inelastic deformations that the casing may undergo. These can reduce the complexity of the problem that may arise due to the geometry, loading, material or boundary conditions.

A crude approximation of the casing could be that of a cylindrical shell subjected an internal pressure and other loads that may arise due to the internal conditions of the casing with simple boundary conditions. While this may be alright for a first cut preliminary estimate, it may not be convincing enough an estimate for a final design since the stresses and strains computed by this approximation may be off from the actual conditions by more than that is acceptable to be representative of the real situation in the casing. Any forceful simplification such as the above that is done to suit the analytical technique for solution will hurt badly like a big foot trying to fit into the small shoe. Therefore, the only option left is to turn to the numerical procedures or techniques that allow for more variety of conditions of the geometry, loading, material and boundary conditions.

In this chapter, we will walk you through the process of numerically solving the governing equations of an initial boundary value problem (IBVP) for the simulation of a general small deformation inelastic application. After discussing the various steps and tasks involving the numerical solution process, a few examples are presented for illustration.

Depending on the nature of the problem, in most cases, reasonable decisions are made on idealizing the geometry, the dimension, the loading, the material response and the boundary conditions. Upon carrying out these, we will end up with a set of governing differential equations (GDE) of a IBVP with appropriate initial and boundary conditions (ICs & BCs). Now, we are ready to apply one of the available numerical techniques to solve this set of equations to obtain the time variation of displacements, stresses and strains apart from the variation of the state variables for the entire structure or component solved for.

10.2 The modeling exercise

As discussed in the elasto-plastic beam chapter, the tasks involved in this are: idealization of the geometry, determining typical design loading cases, a reasonable constitutive model, and selection of an appropriate numerical procedure to solve the set of governing differential equations that arise out of the modeling exercise.

The important decisions on simplifications will have to be made on the geometry, the external loading / displacement conditions and the constitutive model.
Simplifications in the geometry: This is an important step considering the difference in the computational costs that we may incur depending on the simplification made. The question to ask is: Do we lose a lot of information on results when the geometry can be simplified into a one-dimensional or a two-dimensional geometry with the other dimensions playing a secondary role? Of course, in a case where there is no other go, one would go for a full three-dimensional analysis. The size of the problem scales to the power of the dimension while solving numerically and therefore, it is very important to find out possibilities of reducing the geometry to cut down on computational costs.

In addition, certain simplifications in the geometrical shape are preferred that may not affect the design decision or finalization drastically. This will make life easier when it comes to modeling the geometry of the domain of the problem.
Simplifications in the loading: In such cases where the temperature changes that may occur do not contribute to much of change in the analysis results can be ignored. For example, if one is designing the casing for temperatures that may be well within one third of the melting temperature of the metal used, a few tens of degrees of temperature change is not going to affect the mechanical properties of the material. Moreover, unless there are hard displacement constraints in the structure which may not be very difficult to detect, the thermal stresses developed may not be of much significance.

Therefore, to start with, for example, we will consider only mechanical loading. This helps in illustrating the numerical solution process in a simple way, though the scope of the method developed is not limited to this assumption. All the temperature effects on the behavior will be ignored in the formulation developed here. Also, it is possible to decouple the thermal analysis and mechanical analysis given that the mechanical properties don't change significantly within the temperature ranges anticipated.
Boundary conditions: As discussed in the elasto-plastic beams chapter, there is usually less clarity on what kind of boundary conditions to apply since it is easier to apply idealized boundary conditions than the boundary conditions that may have a separate constitutive law to offer! We will go with the simplest of boundary conditions which will let us err, if at all, on the safer side in relation to the design decision to be made.
Choice of the constitutive model: Modeling the material of the structure or its

components is a very important step since we are interested in finding the precise inelastic nature of the component that will answer the design question we have. The choice heavily depends on the nature of such a question and it needs one to be experienced enough to make such a choice. Unless we are interested in the small increase in hardening for small plastic deformations, a perfectly plastic assumption or at best a simple hardening model would suffice. A more detailed model may be necessary if we are interested in low cycle fatigue type of application. Structural collapse load calculations, often, may not need detailed (or complex) models to predict the factor of safety accurately.

Development of the governing differential equations: Given that only mechanical forces are assumed to be acting, effectively and the heat produced due to plastic dissipation is neglected, it boils down to satisfying the equations of mechanical equilibrium. The density of the material is assumed to be a constant.

As discussed in the chapter on boundary value problems for J2 elastoplasticity, a general boundary value problem consists of satisfying the conditions of equilibrium, the strain displacement relations and the constitutive equations apart from the appropriate set of boundary conditions.

Take for example, a body $\mathbb{B}$ with a boundary $\partial \mathbb{B}$, one part of which the displacements are prescribed while on the remaining part, tractions are specified. Let the displacement field in the body be given by $\mathbf{u}(\mathbf{x}) = u_i(x,y,z)\mathbf{i}_i$, and the stresses are given by $\boldsymbol{\sigma} = \sigma_{ij}\mathbf{i}_i \otimes \mathbf{i}_j$.

The local forms of the balance of linear and angular momentum are given by (including the effects due to inertia and body forces),

$$\sigma_{ij,j} + \rho b_i = \rho \ddot{u}_i, \quad \sigma_{ij} = \sigma_{ji}. \tag{10.1}$$

Also, the strains are related to the displacements through,

$$\varepsilon_{ij} = \frac{1}{2}(u_{i,j} + u_{j,i}). \tag{10.2}$$

Note that the strains are assumed to be small in this exercise.

The governing set of equations is complete once the equations related to stress-strain relations are added.

As in the last few chapters on plastic flow, we will use general J2 plasticity type of equations for plastic flow given by equation (8.51) on p.259:

$$\text{Elastic constitutive equations: } [\boldsymbol{\sigma}] = (\mathcal{C})[\boldsymbol{\varepsilon} - \boldsymbol{\varepsilon}_p],$$

$$\text{Yield Condition: } : \frac{\sqrt{[\boldsymbol{\sigma}]^t(\mathcal{Y})[\boldsymbol{\sigma}]}}{\kappa^2} - 1 = 0$$

$$\text{Flow Rule: } [\dot{\boldsymbol{\varepsilon}}_p] = \phi[\mathbf{N}],$$

$$[\mathbf{N}] = \left[\frac{\partial f}{\partial \boldsymbol{\sigma}}\right] = \frac{1}{\kappa^2}(\mathcal{Y})[\boldsymbol{\sigma}], \tag{10.3}$$

$$\text{Plastic Arc Length: } \dot{s} = \frac{\phi}{\kappa},$$

$$\text{hardening function: } \kappa = \hat{\kappa}(s),$$

$$\text{Kuhn-Tucker Condition: } \phi \geq 0, f \leq 0, \phi f = 0.$$

Now, it is a good idea to list out the variables of interest, the input and the output variables pertaining to the problem. In general, the idealization of the geometry and the material properties are pre-defined while the actual input variables include traction on the boundary, the body forces and the displacement boundary conditions. With these input variables available, we wish to find out the variation of displacements, stresses, strains, extent of plastic deformation (or plastic strain) and other internal parameters as a function of input variables and the spatial location.

Choice of the numerical technique: We are now ready to use an appropriate numerical method to solve the above set of governing differential equations. In most numerical methods, a set of primary field variables are approximated spatially at discrete times. In the problem we are formulating, the primary variable is the displacement field and therefore, an approximation is made to this displacement variable. Alternatively, some methods seek to approximate the underlying partial differential equations. A number of methods have been developed to accomplish these tasks. While finite difference methods (FDM) seek to approximate the governing differential equations, the other methods assume a primary variable field that has a finite number of unknowns. Then, based on the principles of weighted residuals on the governing differential equations, the unknowns are solved.

At no point in the above discussion we have referred to inelastic deformation since we have not stated what we are going to do with the state variables for each material point of the body that define the inelastic state. These variables are also fields but there is usually no necessity to take the spatial derivative of these fields. This helps in locally solving for these state variables at a set of pre-determined points consistent with the displacement field values.

Various numerical techniques can be used to solve an inelastic boundary value problem. While some techniques such as *finite difference methods* (FDM) for regular grids and *finite point methods* (FPM) for irregular grids use finite differential expansions to solve the governing differential equations of the problem directly in its strong form, several other methods such as finite element methods (FEM), meshless methods and *boundary element methods* (BEM) use a weak form or a variational form of the governing differential equations to solve the problem. There are scores of weak form methods available currently with their name arising out of the type of approximation function used for the primary variables (in the mechanical case, the displacements), the weight function used and the way the problem spatial domain is subdivided. *Finite element methods*, in general, subdivide the domain into several *non-overlapping* sub-domains within which displacements are *INTERPOLATED* while the controlling boundary nodes of the sub-domains participate in satisfying the continuity conditions of the displacement function. In contrast, mesh-free methods use displacement approximations as a best fit for a given set of control nodal points. The sub-domains that cover the entire problem domain are *not necessarily non-overlapping* with each nodal point controlling a sub-domain. We will discuss this method in detail in the next section.

The choice of the numerical technique depends on the requirement and the problem at hand. Finite difference techniques are useful in standard shapes that are amenable to discretization in this technique. But, these standard shapes are quite limited and the extensions are usually cumbersome. Finite element techniques which approximate the displacement field, in general, within each sub domain called the element are quite robust in handling complex geometry of the problem we are solving. Simple elements in this method approximate the displacement field into a piecewise linear function. In elastic analysis where we do not have state variables that arise from the constitutive law, the stresses computed turn out to be discontinuous. Another limitation that arises is the field inconsistency in that the stress fields computed do not pertain to the displacement differential fields. Problems such as numerical locking can occur in which the computed values may unreasonably be large for some specific values of material properties. In most metals, it is assumed that there is no volume change in plastic deformations, such numerical locking problems are plentifully possible. Unless one is willing to accept such a limitations of the result, a higher element needs to be used. But, this increases the size of the problem.

No such problems exist in a numerical procedure such as the element-free Galerkin method (*the mesh-free methods*) where the displacements are approximated using a particular functional variation with the prescription of the node supported values. The values of these support values, once prescribed for a finite number of nodes, the displacement field in the domain is known. All other quantities are obtained from the displacement field solution. Since the nodal support values of displacements have to be consistent with the equations of equilibrium, these equations are used to find the values supported by each of the nodes. Because of the smoothness of the approximate function, the displacement continuity and differentiability are automatically satisfied. A larger benefit in terms of no element discretization needs to be done and that the nodes can move without having to maintain a neighborhood that is forced in a finite element technique. This is anticipated to help track plastic flow better. The essential boundary conditions have to be enforced by one of the methods amenable to the element-free method and is not simple as in the finite element techniques. We will discuss this method given that it is relatively new and seem to offer scope for better solutions for a class of problems such as the inelastic deformation problems in which smoothness of the displacement may help and spatial updates may be required.

Steps involved in a typical numerical solution process: Before describing the mesh-free methods applied to inelastic analysis, we will list out the major steps involved in this kind of numerical solution process.

(1) **Approximation of the displacement field:** The first step involves approximation of the primary variable - in this chapter, we will take this as the displacement field: $\mathbf{u}^h := \mathbf{u}^h(\mathbf{x}) \approx \mathbf{u}$.
(2) **Time discretization:** The time duration of the application of the external load is divided into finite number of time steps so that the load is applied incrementally. The following steps are performed for each of the time steps, $t_i \in (0, t_1, t_2, ..., t_i, ..., t_N)$.
(3) **Setting up algebraic equations for the solution:** A weak form of the balance laws in conjunction with the constitutive equations is setup up to obtain the algebraic equations to be solved for in obtaining the solution for the current time.
(4) **Solving the nonlinear algebraic equations:** Typically, in the inelastic analysis, we end up with a nonlinear set of algebraic equations in displacements. A method such as the Newton-Raphson algorithm is used to solve for displacements iteratively.
(5) **Integrating the constitutive equations:** In the above solution process, unlike in the elastic analysis, the state variables that define the state of the material at point may evolve and will lead to a different behavior. This has to be updated each time a displacement iteration is performed to make sure the displacement solution is meaningful. Thus, the constitutive equations have to be integrated till the end of the current time step for each displacement iterative solution obtained and the state variables are updated for use in the next iteration of the above Newton-Raphson algorithm.

10.3 The mesh-free method

In this section, we will visit a simple treatment of one of such class of numerical methods called the *mesh-free method*. A Galerkin weighted residual method is adopted along with the mesh-free approximation. One of the main advantages of the mesh-free approximation is the use of moving least squares approximation of the primary field (in this case, the displacement field, $\mathbf{u}(\mathbf{x})$) that automatically satisfies the smoothness condition required of the strains in most of the problems involving homogeneous media.

Essentially, mesh-free methods involve three important steps:

(1) The generation of admissible point sets (called the *nodes*),
(2) Then, the construction of the trial and test approximations,
(3) Lastly, the numerical evaluation of the weak form (Galerkin or Rayleigh-Ritz procedure) integrals which lead to a system of algebraic equations to be solved [Belytschko *et al.* (1994)].

The generation of a point (or, a nodal) set is the first step in a mesh-free methods.

The idea is to achieve a density distribution that has neither voids or holes nor unnecessary clustering of nodes in some regions of the domain. Algorithms based on a centroidal Voronoi tessellation are available in literature for accomplishing this (the reader may refer to works such as [Qiang et al. (2002)] and [Du et al. (1999)] for further reading).

10.3.1 Moving least squares approximation

Owing to its meshless and local characteristics, moving least squares (MLS) [Lancaster and Salkauskas (1981)] methods have been used extensively to approximate the solution of partial differential equations arising from the initial boundary value problems. MLS based shape functions tend to provide a higher order of continuity compared to the interpolations obtained in the finite element method.

The MLS method requires construction of compact supports of nodes over which its corresponding weight is nonzero. The commonly used shape of compact support is circular or rectangular in shape with a node at its center in two-dimensional cases. The optimum support size for these circular or rectangular supports is usually fixed after some numerical experiments (refer to [Belytschko et al. (1996)] for more details). Nevertheless, the compact support size should be large enough such that an evaluation point is covered by a minimum number of compact supports to avoid problems with inversion in MLS methods. Too large or too uniform compact supports may undermine the MLS approximation formulating the shape functions that over-smooth the necessary stress variations.

As noted earlier, one of the primary tasks in this numerical procedure is approximating the displacement solution field using a finite number of unknowns that can be obtained by enforcing the weak form of the balance law.

10.3.1.1 Basic functions and generation of points

In an effort to understand this moving least squares approximation, we will start with the simple least squares approximation (LS), then, we will introduce the local or weighted least square approximation (WLS) before extending to the moving least square approximation (MLS).

Least squares approximation:

In the least squares approximation or fitting procedure, let us consider a set of n points (or *nodes*) $\mathbf{x}_I$, $I = 1, 2, \ldots, n$ at which the values of the displacements d_I are specified. We seek to find an approximation $\mathbf{u}_{LS}^h(\mathbf{x})$ that fits the actual displacement function, $\mathbf{u}(\mathbf{x})$, the best. Let us require that the approximate fit to the actual displacement involves a polynomial of degree m with the complete set of basis functions $\mathbf{p}^t(\mathbf{x}) = \{p_1(\mathbf{x}), p_2(\mathbf{x}), \ldots, p_m(\mathbf{x})\}$. With a complete set of m basis functions such as this, it is possible to write any polynomial of degree m. Note that the decision of choosing the degree of the polynomial, m, lies with the user. Then, the approximate function, $\mathbf{u}_{LS}^h(\mathbf{x})$ can be written as,

$$\mathbf{u}_{LS}^h(\mathbf{x}) = \sum_{i=1}^{m} p_i(\mathbf{x})\, a_i(\mathbf{x}) = \mathbf{p}^t(\mathbf{x})\mathbf{a}, \tag{10.4}$$

where $\mathbf{a} = \{a_1, a_2, \ldots, a_m\}^t$ is a vector of unknown coefficients. The coefficient vector $\mathbf{a}$ in equation (10.4) can be determined by minimizing an L_2 norm (sum of the squares) of the difference between the specified values d_I and value of the approximating function at that corresponding node (the least squares error), i.e.

$$J = \sum_{I=1}^{n} \left[\mathbf{p}^t(\mathbf{x}_I)\,\mathbf{a} - d_I\right]^2. \tag{10.5}$$

In a matrix form, it looks like,

$$J = [\mathbf{P}\,\mathbf{a} - \mathbf{d}]^t\,[\mathbf{P}\,\mathbf{a} - \mathbf{d}], \quad \mathbf{P} = \{\mathbf{p}^t(\mathbf{x}_1), \mathbf{p}^t(\mathbf{x}_2), \ldots, \mathbf{p}^t(\mathbf{x}_n)\}^t. \tag{10.6}$$

⊙ **10.1 (Exercise:). Coefficients for a least squares fit**
Show that, minimization of J in (10.6) with respect to each of the coefficients, a_i yields, in a matrix form, the following equation

$$\mathbf{A}\mathbf{a} = \mathbf{P}^t\mathbf{d} = \mathbf{C}\mathbf{d} \quad say, \tag{10.7}$$

where $\mathbf{A}$ is called the moment matrix with,

$$\mathbf{A} = \sum_{I=1}^{n} \mathbf{p}(\mathbf{x}_I)\mathbf{p}^t(\mathbf{x}_I) = \mathbf{P}^t\mathbf{P}. \tag{10.8}$$

Solving $\mathbf{a}$ from (10.7) and then substituting in (10.4) gives

$$\mathbf{u}_{LS}^h(\mathbf{x}) = \sum_{I=1}^{n} \Phi_I(\mathbf{x}) d_I = \boldsymbol{\Phi}_{LS}^t(\mathbf{x})\mathbf{d}, \tag{10.9}$$

where $\boldsymbol{\Phi}_{LS}^t(\mathbf{x}) = \{\Phi_1(\mathbf{x}), \Phi_2(\mathbf{x}) \ldots \Phi_n(\mathbf{x})\} = \mathbf{p}^t(\mathbf{x})\mathbf{A}^{-1}\mathbf{C}$.

In this, any change in one of the specified values of displacements, d_I will affect the values of the approximation function $\mathbf{u}_{LS}^h(\mathbf{x})$ thus obtained over the entire domain that is covered by $\mathbf{x}$. This *non-local* dependence poses problems in proper control of the solution obtained.

If one wants to find a suitable approximation function that better approximated at a particular point, say, $\bar{\mathbf{x}}$, then, attention should be paid to the values specified nearer to the point, $\bar{\mathbf{x}}$, compared to the values provided for points far away. This can simply be accomplished by multiplying a weight function that decays away from the point $\bar{\mathbf{x}}$ to each of the error values. Since weight function is mainly dependent on the distance from $\bar{\mathbf{x}}$, we have,

$$w_I = w_I(\|\mathbf{x}_I - \bar{\mathbf{x}}\|), \tag{10.10}$$

thus obtaining, for the least squares norm,

$$J_{\bar{\mathbf{x}}} = \sum_{I=1}^{n} w_I \left[\mathbf{p}^t(\mathbf{x}_I) \, \mathbf{a}(\bar{\mathbf{x}}) - d_I \right]^2, \tag{10.11}$$

or in matrix form,

$$J_{\bar{\mathbf{x}}} = [\mathbf{P}\,\mathbf{a}(\bar{\mathbf{x}}) - \mathbf{d}]^t \, [\mathbf{w}(\bar{\mathbf{x}}, \mathbf{x}_I)] \, [\mathbf{P}\,\mathbf{a}(\bar{\mathbf{x}}) - \mathbf{d}], \tag{10.12}$$

in which, $\mathbf{w}$ is a diagonal matrix with each diagonal consisting of each of the weights, w_I. Note that the coefficients $a_i(\bar{\mathbf{x}})$ are now unique to each of $\bar{\mathbf{x}}$ chosen. The rest of the exercise of finding the appropriate shape functions, $\mathbf{\Phi}_{wLS}(\mathbf{x})$ is similar to the derivations done for least squares as in equations (10.7) to (10.9). This procedure offers better scope for approximating the displacement function better at and near $\bar{\mathbf{x}}$. This approximation is generally termed *weighted least squares approximation*. For this method, the approximating function is of the form

$$u_{WLS}(\mathbf{x}, \bar{\mathbf{x}}) := \sum_{i=1}^{m} p_i(\mathbf{x}) \, a_i(\bar{\mathbf{x}}). \tag{10.13}$$

This approximation is good only for those values of $\mathbf{x}$ which are near $\bar{\mathbf{x}}$. Specifically, we note that

$$u_{WLS}(\bar{\mathbf{x}}) := u_{WLS}(\bar{\mathbf{x}}, \bar{\mathbf{x}}) = \sum_{i=1}^{m} p_i(\bar{\mathbf{x}}) \, a_i(\bar{\mathbf{x}}). \tag{10.14}$$

Moving least squares approximation

Now, if one has to obtain good approximations at all points, one has to *MOVE THIS POINT* $\bar{\mathbf{x}}$ over the entire domain and obtain locally good approximations as we move. Note that the approximation varies with the point $\bar{\mathbf{x}}$ taken! Such an approximation is called *the moving least squares approximation*.

In other words, (1) we pick a point $\bar{\mathbf{x}}$ in the domain (2) Find the coefficients $a_I(\bar{\mathbf{x}})$ that forms the best local fit to the given data (3) evaluate the approximating function u_{WLS} *only at the point* $\bar{\mathbf{x}}$ and (4) move to a new point and repeat the process. In this way we obtain a sort of "moving fit" or "moving average" of the data. This is the moving least squares fit or MLS fit. We now don't have to distinguish between the point $\mathbf{x}$ and $\bar{\mathbf{x}}$ and so we now seek to find the coefficients associated with each point $\mathbf{x}$ in the domain, i.e. $a_i(\mathbf{x})$. Changing the set of equations (10.4) to (10.9) to reflect this change in $a_i(\mathbf{x})$ and in $\mathbf{w}_I$, we obtain the following equations:

$$\mathbf{u}_{MLS}^h(\mathbf{x}) = \sum_{i=1}^{m} p_i(\mathbf{x}) a_i(\mathbf{x}) = \mathbf{p}^T(\mathbf{x})\mathbf{a}(\mathbf{x}), \tag{10.15}$$

where $\mathbf{p}^t(\mathbf{x}) = \{p_1(\mathbf{x}), p_2(\mathbf{x}), ..., p_m(\mathbf{x})\}$ is a column vector of complete basis functions of order m and $\mathbf{a}(\mathbf{x}) = \{a_1(\mathbf{x}), a_2(\mathbf{x}), ..., a_m(\mathbf{x})\}^t$ is a column vector of unknown parameters that depend on $\mathbf{x}$.

The coefficient vector $\mathbf{a}(\mathbf{x})$ in equation (10.15) is determined by minimizing a weighted square error as before,

$$J(\mathbf{x}) = \sum_{I=1}^{n} w_I \left[\mathbf{p}^t(\mathbf{x}_I)\mathbf{a}(\mathbf{x}) - d_I\right]^2. \qquad (10.16)$$

Note that in the above weighted error, only the basis functions $p_i(\mathbf{x})$ are evaluated at $\mathbf{x}_I$, not the coefficients $a_i(\mathbf{x})$. written in matrix form,

$$J(\mathbf{x}) = [\mathbf{Pa}(\mathbf{x}) - \mathbf{d}]^t \, \mathbf{w} \, [\mathbf{Pa}(\mathbf{x}) - \mathbf{d}]^t. \qquad (10.17)$$

w_I are the weight functions. Note that, the weight function, $\mathbf{w}$, is defined differently here. It is supposed to be a function of the distance of the chosen point, $\mathbf{x}$, to all the other points in the domain. Therefore, it now becomes a function of $\mathbf{x}$ directly. Ideally, a dirac delta measure for the weight function with the distance metric $\|\mathbf{x} - \mathbf{x}_I\|$ will render the approximation $\mathbf{u}^h_{MLS}(\mathbf{x})$ to equal the nodal value instead of fitting. So, we see a shift from *fitting* to *interpolation* through a change in the sharpness of the weight function. This is important in choosing the type of weight function.

The stationarity of $J(x)$ with respect to $\mathbf{a}(\mathbf{x})$ yields

$$\mathbf{A}(\mathbf{x})\mathbf{a}(\mathbf{x}) = \mathbf{C}(\mathbf{x})\mathbf{d}, \qquad (10.18)$$

where $\mathbf{A}$ is called *the moment matrix*.

$$\mathbf{A}(\mathbf{x}) = \sum_{I=1}^{n} w_I(x) p(\mathbf{x}_I) p^t(\mathbf{x}_I) = \mathbf{P^t w P}, \qquad (10.19)$$

$$\mathbf{C}(\mathbf{x}) = [w_1 p(\mathbf{x}_1), \ldots, w_n p(\mathbf{x}_n)] = \mathbf{P^t w}. \qquad (10.20)$$

Solving $\mathbf{a}(\mathbf{x})$ from equation (10.18) and then substituting in equation (10.15) gives

$$\mathbf{u}^h_{MLS}(\mathbf{x}) = \sum_{I=1}^{n} \Phi_I(\mathbf{x}) d_I = \mathbf{\Phi}^t_{MLS}(\mathbf{x})\mathbf{d}, \qquad (10.21)$$

where $\mathbf{\Phi}^t(\mathbf{x}) = \{\mathbf{\Phi_1}(\mathbf{x}), \mathbf{\Phi_2}(\mathbf{x}), ..., \mathbf{\Phi_n}(\mathbf{x})\} = \mathbf{p}^t(\mathbf{x})\mathbf{A}^{-1}(\mathbf{x})\mathbf{C}(\mathbf{x})$ is a vector with the Ith component given by

$$\Phi_I(\mathbf{x}) = \sum_{j=1}^{m} p_j(\mathbf{x}) \left[\mathbf{A}^{-1}\mathbf{C}(\mathbf{x})\right]_{jI}, \qquad (10.22)$$

representing the shape function of the MLS approximation corresponding to node I. The method in one dimension is shown in Fig. 10.1. One of the important observations is to note that the approximating function $\mathbf{u}^h_{MLS}(\mathbf{x})$ does not necessarily pass through d_I. Thus, the approach is a curve "fit" and not an "interpolation".

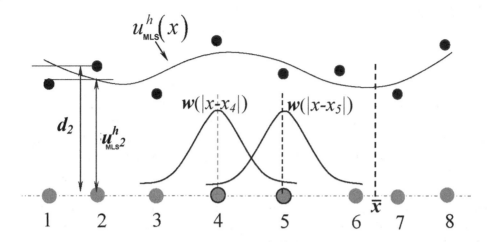

Fig. 10.1 Curve fitting using moving least squares method. The compact support is shown. Unlike the Finite element method, the values at the nodes are not the same as nodal parameters. Also, it is a best fit instead of an interpolation fit. Notice that the weight functions are shown against each of the nodes with the maximum weight at the nodes.

It has been shown in [Belytschko et al. (1996)] that any function which appears in the basis can be reproduced exactly. Thus the order of consistency of an approximation depends on the order of the polynomial which is represented exactly and is complete if it reproduces the function with an arbitrary order of accuracy. In this method, it should be made sure that the moment matrix is non-singular to avoid any problems with consistency of the approximation. Singularity can be avoided by making sure that the weight functions of each node overlaps with weight functions of the neighboring nodes. This has a bearing on the sharpness of the weight functions.

10.3.1.2 *Local / compact support*

The local support of each particle plays a crucial role with respect to the accuracy and stability of the solution, similar as the element size in FEM does. As seen in Fig. 10.1, the weight function w not only ensures the locality of the approximation but also global smoothness of the approximant in the domain Ω. The smoothness of the shape functions $\Phi_I(\mathbf{x})$ is governed by the smoothness of the basis function and the weight function. In general, the weight functions, $w(\mathbf{x})$ are chosen or normalized so that they have a unit value at the center and zero along the boundaries of their support.

One example of weight function that is often used is based on the student's

t-distribution given by,

$$w_I(\mathbf{x}) = \begin{cases} \left(1 + \beta^2 \frac{z_I^2}{z_{mI}^2}\right)^{\left(\frac{1+\beta}{2}\right)} - \left(1 + \beta^2\right)^{\left(\frac{1+\beta}{2}\right)}, & z_I \leq z_{mI} \\ 0, & z_I > z_{mI} \end{cases}, \qquad (10.23)$$

where β is a parameter controlling the shape of the weight function, and the constants $z_I = \|\mathbf{x} - \mathbf{x}_I\|$ is the distance from the sampling point, $\mathbf{x}$ to a node $\mathbf{x}_I$. Figure 10.2 shows the variation of the weight function for different values of β. The parameters z_{mI} represent the domain of influence of node I so that

$$z_{mI} = z_{max} \cdot z_{cI}, \qquad (10.24)$$

in which z_{cI} is the characteristic nodal spacing distance which is chosen such that the node I has enough number of neighbors sufficient for regularity of $A(\mathbf{x})$ in equation (10.18) and z_{max} is a scaling parameter.

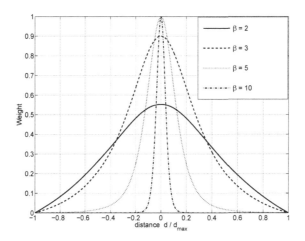

Fig. 10.2 Student t-distribution weight function. The variation of the function with different values of β are shown. Notice that the weight function becomes sharper with the increase in the magnitude of β

10.3.1.3 *Enforcement of essential boundary conditions*

Quite unlike the finite element methods in which the essential boundary conditions (such as specified displacements or temperatures) can be enforced directly without any problem, the shape functions of the moving least squares approximation does not satisfy the crucial Kroenecker delta property[1]. Therefore, the essential

---

[1] this property guarantees that value of the shape function is unity at the node it is associated to and zero elsewhere. This helps in blending the shape functions using the values at the corresponding nodes to generate the displacement approximation

boundary conditions have to be enforced indirectly through a penalty or a Lagrange approach. There are also simpler ways that can be adopted to take care of this problem at the cost of loss of accuracy in the approximation near the node at which this condition is enforced. One such method, is called the *transformation method* and is described below.

The MLS approximation is $\mathbf{u}^h(\mathbf{x}) = \mathbf{\Phi}^t(\mathbf{x})\mathbf{d}$ where $\mathbf{d}$ is the nodal parameter. Since the MLS shape function does not have the Kroenecker delta property, at the essential boundary nodes, $\mathbf{x}_J$, we have $\mathbf{u}^h(\mathbf{x}_J) = \mathbf{\Phi}^t(\mathbf{x}_J)\mathbf{d} \neq \mathbf{d}$. Note that the quantities that we solve for, namely $\mathbf{d}$ are not the values of the approximate function $\mathbf{u}^h$ at the nodes, i.e. $\mathbf{d}$ is a list of nodal parameters and not nodal values of displacements. To understand this, see Fig. 10.1; in this figure, let us assume that node 2 is a node at which the displacement is specified as u_0. Note that d_2 and $u_{MLS}^h(\mathbf{x}_2)$ do not conicide and so it is not easily possible to enforce the condition that $u_{MLS}^h(\mathbf{x}_2) = u_0$.

In the transformation method, the nodal parameters of the nodes belonging to the essential boundary condition are transformed so that these nodal parameters are directly related with the corresponding physical nodal values through an appropriate transformation matrix.

For example, if the essential boundary condition $u^h(\mathbf{x}_J) = u_J$ is to be satisfied at a point $\mathbf{x}_J$ in a one-dimensional problem, then we write the MLS approximation, $u^h(\mathbf{x}_J)$ in terms of the shape functions such that,

$$u^h(\mathbf{x}_J) = \mathbf{\Phi}(\mathbf{x}_J)^t \mathbf{\Lambda}_J^{-1} \mathbf{d} = u_J, \qquad (10.25)$$

with

$$u^h(x) = \mathbf{\Phi}(\mathbf{x})^t \mathbf{\Lambda}_J^{-1} \mathbf{d} = \bar{\mathbf{\Phi}} \mathbf{d}, \text{ say}, \qquad (10.26)$$

$\mathbf{\Lambda}$ is a scaling matrix such that the above equation produces the value of $\mathbf{u}^h$ at $\mathbf{x}_J$ equal to the given value u_J. This can simply be done by taking an identity matrix to be $\mathbf{\Lambda}$ and replacing the Jth column to be the same as the Jth row of the $\mathbf{\Phi}$ matrix. $\bar{\mathbf{\Phi}}(x)$ is thus the transformed shape function matrix. Using this basically amounts to forcing the displacement parameter pertaining to $\mathbf{x}_J$ to be u_J. This is a simple transformation that modifies the MLS approximation to a certain extent.

To illustrate this, if we wish to fit a curve given by the following values:

	x	$f(x)$
1	0.000	15
2	0.111	5
3	0.222	6
4	0.333	10
5	0.444	8
6	0.556	14
7	0.667	5
8	0.778	14
9	0.889	15
10	1.000	7.8

We also require that the curve passes through the data points, 1, 7 and 10, say (denoted as points a, b and c respectively in Fig. 10.3). Then, we can form a scaling transformation matrix for each of the given conditions as given in (10.25) to obtain the transformed shape function matrix $\bar{\Phi}$ as in (10.26) for each of enforcement successively.

The curves obtained on this enforcement compared to the others with a focus on point 7 (point b) is highlighted in Fig. 10.3. Note that the change enforced by the transformation method is local without affecting smoothness in the fitted curve. Additionally, the real advantage arises from the fact that the Λ_J is not fully populated and is easily invertible.

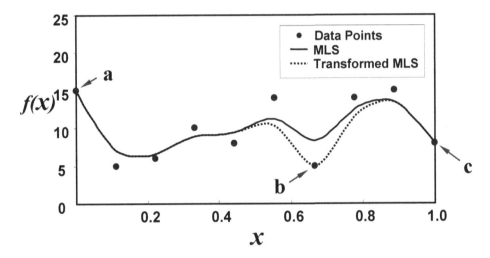

Fig. 10.3 The curve fitting using the variations of MLS approach. As you can see, the scaling transformation method provides the necessary enforcement of an essential displacement condition and local at that.

10.3.2 Weak form of the balance law

A specific weak form of equilibrium equations is the equations of principle of virtual work. In this the virtual variations of the displacements that are kinematically admissible are used. Thus the equilibrium equation in (10.1) can be written in the weak form as,

$$\int_V \sigma_{ij} \delta\varepsilon_{ij} dV + \int_V \rho \delta b_i \delta u_i dV - \int_V \rho \ddot{u}_i \delta u_i dV = \int_{\partial \mathbb{B}} t_i \delta u_i d\partial \mathbb{B}, \quad (10.27)$$

where δ represents the virtual quantities. $\delta\varepsilon_{ij}$ and δu_i are the compatible kinematically admissible virtual strains and the virtual displacements, respectively.

For the sake of convenience and use in the mesh-free formulation and code, we will write the quantities such as displacements, strains and stresses in vector form.

As in chapter 6, we will use the notation $[\cdot]$ to represent column vectors and $(\cdot)$ to represent matrices. Thus, we have, $[u_1 \ u_2 \ u_3]^t$ represents the vector form of the displacements u_i, $[\varepsilon_{11} \ \varepsilon_{22} \ \varepsilon_{33} \ \sqrt{2}\varepsilon_{12} \ \sqrt{2}\varepsilon_{13} \ \sqrt{2}\varepsilon_{23}]^t$ represents the strain tensor, while, $[\sigma_{11} \ \sigma_{22} \ \sigma_{33} \ \sqrt{2}\sigma_{12} \ \sqrt{2}\sigma_{13} \ \sqrt{2}\sigma_{23}]^t$ for the stress tensor.

In the moving least squares approximation, as discussed earlier, the displacement field is approximated using $\mathbf{u}^h(\mathbf{x}) = \mathbf{\Phi}^t(\mathbf{x})\mathbf{\Lambda}^{-1}\mathbf{d} = \bar{\mathbf{\Phi}}^t(\mathbf{x})\mathbf{d}$ which is MLS approximation scaled appropriately as explained above to satisfy the essential boundary condition, $\mathbf{u}^h(\mathbf{x}_J) = \mathbf{d}_J$, $\mathbf{x} \in \partial \mathbb{B}_u$.

In the equation, (10.27) for the virtual work principle, the virtual displacements can now be substituted with the approximate expression for the displacements so that the equation can be written in terms of the nodal parameters. The term left out in the equation (10.27) now is the virtual strain term. For this, we need to write virtual strain in terms of the virtual nodal parameters, so that all the expressions will have the virtual nodal parameters as the common factor. Since, the strain, ε_{ij} is related to the displacement, u_i through, $\varepsilon_{ij} = \frac{1}{2}(u_{i,j} + u_{j,i})$ (note that we are dealing with small deformations case here). Writing the strain in vector form and writing strain components explicitly in terms of derivatives of displacements, we get,

$$[\varepsilon] = (\mathbf{B})[\mathbf{U}] \tag{10.28}$$

where $\mathbf{B}$ for the two-dimensional case is an assemblage of individual matrices $\mathbf{B}_I$ associated with each node, I can be found to be,

$$(\mathbf{B}_I) = \begin{pmatrix} \bar{\Phi}_{I,1} & 0 \\ 0 & \bar{\Phi}_{I,2} \\ \bar{\Phi}_{I,2} & \bar{\Phi}_{I,1} \end{pmatrix} \tag{10.29}$$

so that,

$$\begin{aligned}\varepsilon_I &= \mathbf{B}_I \mathbf{d}_I \\ &= \begin{pmatrix} \Phi_{I,1} & 0 \\ 0 & \Phi_{I,2} \\ \Phi_{I,2} & \Phi_{I,1} \end{pmatrix} \begin{Bmatrix} d_1 \\ d_2 \end{Bmatrix} \end{aligned} \tag{10.30}$$

Notice that all the differential operators found in the matrix are linear operators since linear part of the strain is used in this definition (small strain assumption).

We now have an expression for the displacement field in terms of the nodal displacement parameters, so that the weak form the equilibrium equation (10.1) can now be completed to obtain all expressions in terms of the virtual nodal displacements. The weak form of the equilibrium equation can now be rewritten in vector / matrix form as,

$$\int_\Omega [\delta\varepsilon]^t [\sigma] \, dV - \int_\Omega [\delta\mathbf{u}]^t [\rho\mathbf{b}] \, dV + \int_\Omega [\delta u]^t [\rho\ddot{\mathbf{u}}] \, dV = \int_{\partial\mathbb{B}} [\delta\mathbf{u}]^t [\rho\mathbf{t}] \, d\partial\mathbb{B} \tag{10.31}$$

Inserting the expressions for the displacement field and the strains in terms of nodal displacements, we obtain,

$$\int_\Omega \{\delta(\mathbf{B})\mathbf{U}\}^t \sigma dV - \int_\Omega \{\delta(\mathbf{\Phi})[\mathbf{d}]\}^t [\rho \mathbf{b}] dV + \int_\Omega \{\delta(\mathbf{\Phi})[\mathbf{d}]\}^t \left\{\rho(\mathbf{\Phi})\left[\ddot{\mathbf{d}}\right]\right\} dV = \int_{\partial \mathbb{B}} \{\delta(\mathbf{\Phi})[\mathbf{d}]\}^t [\rho \mathbf{t}] d\partial \mathbb{B} \qquad (10.32)$$

Let us, for now, assume that the process is quasi-static and therefore, neglect inertial and rate effects. Then, the above equation reduces to:

$$\int_\Omega \{\delta \mathbf{Bd}\}^t \sigma dV = \int_\Omega \{\delta \mathbf{\Phi d}\}^t \rho \mathbf{b} dV + \int_{\partial \mathbb{B}} \{\delta \mathbf{\Phi d}\}^t \rho \mathbf{t} d\partial \mathbb{B} \qquad (10.33)$$

Since the virtual nodal displacements are arbitrary, we obtain the governing equation in the integral form to be:

$$\int_\Omega \mathbf{B}^t \sigma dV = \int_\Omega \mathbf{\Phi}^t \rho \mathbf{b} dV + \int_{\partial \mathbb{B}} \mathbf{\Phi}^t \rho \mathbf{t} d\partial \mathbb{B} \qquad (10.34)$$

In the above equation, the right-hand side expressions concern the effect of external forces, namely, the body forces and the tractions. The left-hand side expression is due to stresses in the body and therefore, can be taken to be due to internal balancing forces. Thus, we can write,

$$[F_{int}] = [F_{ext}], \quad \Rightarrow \quad \int_\Omega \mathbf{B}^t \sigma dV = [F_{ext}]. \qquad (10.35)$$

where $[F_{ext}]$ is given by,

$$[F_{ext}] = \int_\Omega \mathbf{\Phi}^t [\rho \mathbf{b}] dV + \int_{\partial \mathbb{B}} \mathbf{\Phi} \rho \mathbf{t} d\partial \mathbb{B} \qquad (10.36)$$

Once the relationship between the stresses and strains is established, the above equation can be completely written in terms of displacement quantities so that it is ready be solved for displacements. If the stress is linear in strain, say,

$$\sigma = \mathbb{C}\varepsilon \qquad (10.37)$$

where $\mathbb{C}$ is a constant, then, the left-hand side expression is a linear function of nodal displacements, i.e.

$$\int_\Omega \mathbf{B}^t \sigma dV = \int_\Omega \mathbf{B}^t \mathbb{C} \varepsilon dV = \int_\Omega \mathbf{B}^t \mathbb{C} \mathbf{B}(\mathbf{d}) dV = \left[\int_\Omega \mathbf{B}^t \mathbb{C} \mathbf{B} dV\right] (\mathbf{d}) \qquad (10.38)$$

Since $\mathbb{C}$ is a constant with respect to $\mathbf{d}$, we obtain a linear algebraic equation in $\mathbf{d}$ which can be solved readily using a standard linear solver. However, such a luxury is possible only in the linear elastic analysis where we encounter a simple constitutive equation such as the one represented by equation (10.37).

10.3.3 Generating and solving the nonlinear algebraic equations

Since the constitutive equation for plastic deformations is nonlinear, the equation (10.35) will yield nonlinear set of algebraic equations in **d** which need to be solved.

One of the popular methods used to solve equation (10.35) the classical Newton-Raphson method or the variations of it. In this method, the solution is obtained iteratively. An incremental analysis is carried out in which the external load in equation (10.36) is applied incrementally to the structure. Let us represent the external forces applied till the end of n incremental steps, to be F_{ext}^n and that the increment for the $(n+1)$th step is taken to be ΔF_{ext} so that,

$$F_{ext}^{n+1} = F_{ext}^n + \Delta F_{ext} \tag{10.39}$$

At the end of each increment, it is made sure that the F_{int}^n is balanced with F_{ext}^n satisfying the equation (10.35). Let us assume that the corresponding nodal parameters obtained are $\mathbf{d}^n$.

Then for the $(n+1)$th increment, satisfying the equation (10.35) will mean that, we satisfy,

$$[F_{int}^{n+1}(\mathbf{d}^{n+1})] = \int_V (\mathbf{B})^t \left[\boldsymbol{\sigma}(\mathbf{d}^{n+1})\right] dV = F_{ext}^n + \Delta F_{ext} = \left[F_{ext}^{n+1}\right]. \tag{10.40}$$

Assuming that we can write a residual function, R in $\mathbf{d}^{n+1}$ that should ideally vanish, we have,

$$R(\mathbf{d}^{n+1}) = F_{ext}^{n+1} - F_{int}^{n+1}(\mathbf{d}^{n+1}). \tag{10.41}$$

The above equation can be solved iteratively for $\mathbf{d}^{n+1}$ using the Newton-Raphson iteration, in which, the $i+1$ approximate to $\mathbf{d}$ can be found using the ith iterate by using the Taylor expansion of the above equation (10.41) about the ith iterate and setting the residual R to be zero, so that,

$$F_{ext}^{n+1} - \frac{\partial F_{int}}{\partial \mathbf{d}}\bigg|_{\mathbf{d}_i^{n+1}} (\mathbf{d}_{i+1}^{n+1} - \mathbf{d}_i^{n+1}) - F_{int}(\mathbf{d}_i^{n+1}) = 0 \tag{10.42}$$

From equation (10.35), we can find,

$$\frac{\partial F_{int}}{\partial \mathbf{d}}\bigg|_{\mathbf{d}_i^{n+1}} = \int_V (\mathbf{B})^t \left[\frac{\partial \boldsymbol{\sigma}}{\partial \mathbf{d}}\bigg|_{\mathbf{d}_i^{n+1}}\right] dV = \int_V (\mathbf{B})^t \left[\frac{\partial \boldsymbol{\sigma}}{\partial \boldsymbol{\varepsilon}} \frac{\partial \boldsymbol{\varepsilon}}{\partial \mathbf{d}}\bigg|_{\mathbf{d}_i^{n+1}}\right] dV \tag{10.43}$$

This can be written in terms of the tangent stiffness matrix $\mathcal{K}$ as given in equation (6.34) in Chapter 6. Thus, we can now write,

$$\frac{\partial F_{int}}{\partial \mathbf{d}}\bigg|_{\mathbf{d}_i^{n+1}} = \int_V (\mathbf{B})^t (\mathcal{K})\bigg|_{\mathbf{d}_i^{n+1}} (\mathbf{B}) \tag{10.44}$$

Thus, equation (10.42) can now be rewritten in the form,

$$F_{ext}^{n+1} - \left(\mathbf{K}^{n+1}\right)_{i+1} (\mathbf{d}_{i+1}^{n+1} - \mathbf{d}_i^{n+1}) - F_{int}(\mathbf{d}_i^{n+1}) = 0$$
$$\Rightarrow \left(\mathbf{K}^{n+1}\right)_{i+1} \delta\mathbf{d}_{i+1} = F_{ext}^{n+1} - F_{int}(\mathbf{d}_i^{n+1}) \tag{10.45}$$

where $(\mathbf{K})$ is called the global stiffness matrix. Notice that the right-hand side expression is known since both the previous iterate of internal balancing force and the external force vectors are known. Also, note that this stiffness matrix changes according to the current displacement and therefore, needs to be updated for each iteration. Computing the stiffness matrix for the entire structure for each iteration is computationally very expensive. Modified Newton Raphson uses less number of updates over a group of iterates to save on this computation. However, this increases the number of iterations to be carried out since the quadratic convergence of Newton-Raphson is lost in doing this. Therefore, there is a trade off involved. If the tangent stiffness matrix directly related to the classical elastic-plastic relation is used, then also, the quadratic convergence advantage of Newton-Raphson scheme will be lost. To preserve the convergence advantage, it is necessary to find the tangent stiffness matrix that is consistent with the integration algorithm used for updating stresses and plastic strains. Computation of such a consistent tangent stiffness matrix will be discussed in a section of this chapter on convex cutting plane algorithm that follows.

Thus,

$$\left(\mathbf{K}^{n+1}\right)_{(i+1)modified} = \left(\mathbf{K}^{n+1}\right)_1 = \int_V (\mathbf{B})^t \left(\mathcal{K}\right)\Big|_{\mathbf{d}_1^{n+1}} (\mathbf{B}) \tag{10.46}$$

At each iteration, the total displacement can be obtained as,

$$\mathbf{d}_{i+1}^{n+1} = \mathbf{d}_i^{n+1} + \delta\mathbf{d}_{i+1} \tag{10.47}$$

and the total strain in the $(n+1)$th increment can be found to be,

$$\varepsilon^{n+1} = (\mathbf{B})^t \left\{\mathbf{d}^{n+1}\right\}_{i+1} \tag{10.48}$$

Once the total strain field $\varepsilon(\mathbf{x},t)$ is obtained, the corresponding plastic strain field $\varepsilon_p(\mathbf{x},t)$ can be obtained by integrating the plastic flow equations. This integration is again not straightforward because of the nonlinearity in the governing equations for the plastic flow. A convex cutting plane algorithm is explained in detail with a few examples in a separate section in this chapter. Using this algorithm, all the state variables are updated at each of the integration points.

In general, it is better to use the tangent stiffness matrix that is consistent with the algorithm used for integration to obtain the global stiffness matrix for each iteration. This will help in preserving the convergence advantage of the Newton-Raphson algorithm used. In the case of convex cutting plane algorithm, the consistent tangent stiffness matrix can be obtained except that the symmetry will be lost. The symmetry lose and even the symmetrization will be a question in terms of convergence. Therefore, the classical tangent stiffness matrix is generally adopted.

Finally, a convergence criterion needs to be introduced to terminate the iteration to indicate that the equilibrium is satisfied within the specified tolerance. Since both displacements and forces play an important role in the iteration, an internal energy criterion as given below can be used.

$$\xi \Rightarrow Tol; \quad \text{where} \quad \xi = \frac{\delta \mathbf{d}_{i+1}^{n+1} \left(F_{ext}^{n+1} - F_{int}(\mathbf{d}_i^{n+1}) \right)}{\delta \mathbf{d}_1^{n+1} \left(F_{ext}^{n+1} - F_{int}(\mathbf{d}^n) \right)}. \tag{10.49}$$

Once the convergence criterion is satisfied, the state variables are updated for the $(n+1)$th step of loading and one proceeds to the next incremental step.

10.4 Integration of the plastic flow equations

The central problem in the constitutive theory of elasto plastic materials is the following: Given the strain field $\varepsilon(\mathbf{x}, t)$ what is the corresponding plastic strain field $\varepsilon_p(\mathbf{x}, t)$? Just as a quick review, the governing equations for plastic flow are:

$$\text{Elastic constitutive equations:} \ [\boldsymbol{\sigma}] = (\mathcal{C})[\varepsilon - \varepsilon_p]$$

$$\text{Yield Condition}: \frac{\sqrt{[\boldsymbol{\sigma}]^t (\mathcal{Y})[\boldsymbol{\sigma}]}}{\kappa^2} - 1 = 0$$

$$\text{Flow Rule}: [\dot{\varepsilon}_p] = \phi[\mathbf{N}],$$

$$[\mathbf{N}] = \left[\frac{\partial f}{\partial \boldsymbol{\sigma}}\right] = \frac{1}{\kappa^2}(\mathcal{Y})[\boldsymbol{\sigma}] \tag{10.50}$$

$$\text{Plastic Arc Length}: \dot{s} = \frac{\phi}{\kappa}$$

$$\text{Hardening function}: \kappa = \hat{\kappa}(s)$$

$$\text{Kuhn-Tucker Condition}: \phi \geq 0, f \leq 0, \phi f = 0.$$

10.4.1 *Non-dimensionalization and its importance*

It is not a good idea to simply code in all these constitutive equations into a computer program in dimensional form. The reasons for this are (1) If you do so, all the geometrical and physical details of the problem being solved will be "hardwired". (2) You will not be able to see which features are negligible and which play a critical role. (3) You will have to provide many more material parameters than needed. (4) You will not be able to compare the response for different materials and at different conditions. (5) Your computer program will have to deal with unnecessary large and small number that may always appear as a product, thus increasing its error.

One of the most important reasons for non-dimensionalization is that it will simplify the constitutive equations considerably. For example if the material is such that its initial yield stress as well as its hardening variables change with temperature, then a blind data fitting will involve a number of temperature dependent parameters. On the other hand it might turn out that the form of the temperature dependence of all the parameters are the same if they are properly institutionalized. It is for this reason that we recommend that you revisit the nondimensionalization procedure taught in your fluid mechanics class (the Buckingham Pi Theorem etc.) and do a systematic nondimensionalization of all the constitutive equations.

In our current case, we need to choose an appropriate stress and strain measure for non-dimensionalization. The basic idea is that the non-dimensional stresses and strains should be of order 1. For example, for steel, the Young's modulus is of the order of $200 GPa$, while the yield strength is about $450 MPa$. This huge difference in the values means that the material will begin yielding when the strain reaches a value of about 0.2%. So if we blindly use these numbers, we will get stresses of the order of 450 MPa and strain values of 0.2%. To make these values much more reasonable, we can nondimensionalize the stresses with the initial yield stress (so that stress values below 1 are below the initial yield stress and hence elastic) and the strains with the initial yield strain given by the ratio of the yield stress to the young's modulus.

⊙ **10.2 (Exercise:). Nondimensionalization of the basic equations**
Systematically non-dimensionalize the constitutive equations (using κ_0 as the nominal stress and κ_0/E as the nominal strain) for an isotropic J2 elastoplastic material and show that the resulting equations take the following form:

$$\text{Stress-Strain Relations}: \bar{\varepsilon}_{ij} = (1+\nu)\bar{\sigma}_{ij} + \nu\bar{\sigma}_{kk}\delta_{ij} + \bar{\varepsilon}^p{}_{ij}$$

$$\text{yield criterion}: \frac{\sqrt{\bar{\tau}_{ij}\bar{\tau}_{ij}}}{\bar{\kappa}} - 1 = 0, \quad , \bar{\kappa} = \frac{\kappa}{\kappa_0}$$

$$\text{flow rule}: \dot{\bar{\varepsilon}}^p = \bar{\phi}\frac{\bar{\tau}_{ij}}{\bar{\kappa}^2}, \quad \bar{\phi} = \frac{\phi E}{\kappa_0^2} \qquad (10.51)$$

$$\text{plastic arc length}: \dot{\bar{s}} = \frac{\bar{\phi}}{\bar{\kappa}}, \quad \bar{s} = \frac{sE}{\kappa_0}$$

$$\text{Voce Hardening}: \bar{\kappa} = 1 + \bar{a}\bar{s} + \bar{b}(1 - e^{-\frac{\bar{s}}{\bar{s}_0}}),$$

$$\bar{a} = a/E, \bar{b} = b/\kappa_0, \bar{s}_0 = s_0 E/\kappa_0$$

Thus, the constitutive equations depend upon 4 non-dimensional parameters, $\nu, \bar{a}, \bar{b}, \bar{s}_0$, while the original dimensional form appeared to depend upon 6 parameters.

In the reminder of this chapter, we shall assume that the equations have been properly non-dimensionalized and so we shall not use the over-bar on the non-dimensional quantities.

In the above equation ($\mathcal{Y}$) is the same as the *projection operator* defined in equation (8.51) on p.259 for the full 3-D case as well as the plane strain case, and $\mathcal{C}$ is the elastic stiffness matrix. For the case of plane stress ($\mathcal{Y}$) is $\mathcal{A}$ defined in (equation (7.23) on p.224) and $\mathcal{C}$ is the plane stress modulus ($\mathcal{C}_s$) defined on (7.19).

We can use any hardening function we want and for our purposes it is not necessary to address this in detail. Clearly this is a nasty set of nonlinear Ordinary differential equations to solve. Our solution strategy is based on using a Backward Euler or a Taylor series expansion scheme to solve this.

The basic Backward Euler scheme for the solution of a system of ODES of the form

$$\frac{d\mathbf{y}}{dt} = f(\mathbf{y}, t) \tag{10.52}$$

is to replace the time derivative by a finite difference operation and calculate the right-hand side at the *end of the time step*, i.e. we solve

$$\frac{\mathbf{y}^{n+1} - \mathbf{y}^n}{\Delta t} = f(\mathbf{y}^{n+1}, t_{n+1}). \tag{10.53}$$

This results in a set of nonlinear algebraic equations that need to be solved by some suitable scheme (such as a Newton-Raphson scheme).

The above scheme is generated by (1) replacing the time derivatives by a simple difference and (2) all the other terms (both known and unknown terms) by their values at time t_{n+1}.[2]

If we apply this to the present plasticity problem we get:
(1) divide the strain path into a set of increments so that we are given $\boldsymbol{\varepsilon}(t_i) = \boldsymbol{\varepsilon}_i, i = 1, \ldots, n$.
(2) for i=0, set the plastic strain and the plastic arc length to their initial values (say 0 if we are starting with a virgin material).

---

[2]If you are wondering why did we make our life complicated by using the unknown values at t_{n+1} rather than the known values at t_n, the reason is that if you do that your integration scheme will be subject to wild oscillations (instability) unless you take many tiny steps. Even then, the solution has a tendency to drift.

(3) For $i = 1, 2, 3, \ldots, N$ solve the following algebraic equations for ε_p^{i+1}:

$$\text{Elastic constitutive equations: } [\boldsymbol{\sigma}] = (\mathcal{C})[\varepsilon^{i+1} - \varepsilon_p^{i+1}]$$

$$\text{Yield Condition : } \frac{\sqrt{[\boldsymbol{\sigma}]^t(\mathcal{Y})[\boldsymbol{\sigma}]}}{\kappa_{i+1}^2} - 1 = 0$$

$$\text{Flow Rule :} [\varepsilon_p^{i+1}] = [\varepsilon_p^i] + \triangle\phi[\mathbf{N}],$$

$$[\mathbf{N}] = \left[\frac{\partial f}{\partial \boldsymbol{\sigma}}\right] = \frac{1}{\kappa^2}(\mathcal{Y})[\boldsymbol{\sigma}] \qquad (10.54)$$

$$\text{Plastic Arc Length :} s_{i+1} = s_i + \frac{\triangle\phi}{\kappa_{i+1}}$$

$$\text{Hardening function :} \kappa_{i+1} = \hat{\kappa}(s_{i+1})$$

$$\text{Kuhn-Tucker Condition :} \triangle\phi \geq 0, f(\boldsymbol{\sigma}, \kappa_{i+1}) \leq 0, \triangle\phi f = 0,$$

where $\triangle\phi = \triangle t \dot\phi$.

So, we have traded a nasty ODE for a nasty set of algebraic equations. We have made progress. There are two ways to go about solving this. The fast, elegant and hard way (e.g. Newton-Raphson) resulting in the so-called "return mapping algorithm" (see [Simo and Hughes (1998)]) or a somewhat slower but easier way. In keeping with the spirit of the course, we will do the easy way.

10.4.2 The convex cutting plane algorithm

The material in this section is based on the work of [Ortiz et al. (1983); Ortiz and Simo (1986)]. The basis of this algorithm is that of a Taylor series approximation.

In order to facilitate the expansion of this algorithm to other problems and also in order to write the algorithm in a concise form, we will introduce the following abbreviations (this is actually a repeat of the variables introduced in Chapter 3, written in matrix form for ease of numerical implementation). As usual square brackets [·] imply that they are written as column vectors while round brackets (·) imply square matrices. Collections of vectors will be written using curly braces {·}. A semicolon after a term means that it is the next element in the column vector (like in MATLAB):

Constitutive Variables : $\mathbb{U} = \{[\boldsymbol{\sigma}], [\boldsymbol{\varepsilon}], [\boldsymbol{\varepsilon}_p], s, \kappa\}$

Independent Variables : $\mathbb{I} = \{[\boldsymbol{\varepsilon}], [\boldsymbol{\varepsilon}_p], \kappa, s\}$

Dependent Variables : $\mathbb{D} = \{[\boldsymbol{\sigma}]\}$

Plastic Variables : $\mathbf{Q} = [[\boldsymbol{\varepsilon}_p]; \kappa; s]$

Initial Values : $\mathbb{I}_0 := \{[\boldsymbol{\varepsilon}]_0, [\boldsymbol{\varepsilon}_p]_0, s_0\}^t, \mathbf{Q}_0 := [[\boldsymbol{\varepsilon}_p]_0; \kappa_0; s_0]$

These are supposed to be known at the beginning of the time step

Iteration Counter : k

Values at iteration k : $\mathbb{U}^k = \{[\boldsymbol{\sigma}^k], [\boldsymbol{\varepsilon}_1], [\boldsymbol{\varepsilon}_p^k], s^k, \kappa^k\}$

Note that the value of $\boldsymbol{\varepsilon}$ is always set to $\boldsymbol{\varepsilon}_1$

Values at the end of the iteration : $\mathbb{U}_1$

Of these $\boldsymbol{\varepsilon}_1$ is known

Flow rule and arc length equations : $[\dot{\mathbf{Q}}] = \phi H(\mathbb{U}) = \phi H(\boldsymbol{\sigma}, s, \boldsymbol{\varepsilon}_p, \kappa)$

For our case : $[[\dot{\boldsymbol{\varepsilon}}_p]; \dot{\kappa}; \dot{s}] = \phi[[\mathbf{N}]; (1/\kappa)(d\kappa/ds); 1/\kappa]$
(10.55)

where $[\mathbf{N}]$ was defined in (10.50). Note that the intermediate values are given by superscripts but the initial and final values are given by subscripts.

With these abbreviations, we are now ready to describe the algorithm: Given the values of $\mathbb{I}_0$ at time $t = t_n$ and the value of the total strain $\boldsymbol{\varepsilon}_1$ at time $t = t_{n+1}$, to find the values of $\mathbb{U}_1$:

The basic trick here is first see if the material yields by not updating the plastic variables, i.e. we see if simply updating the total strain works. In other words, we see if $f((\mathbb{C})[\boldsymbol{\varepsilon}_1 - \boldsymbol{\varepsilon}_0^p], \kappa_0)$ is less than zero. If it is then there is no updating required. If it is not, then we must update it in such a way that $f((\mathbb{C})[\boldsymbol{\varepsilon}_1 - \boldsymbol{\varepsilon}_1^p], \kappa_1) = 0$.

We do this with an iterative process: The first iteration (i.e. $k = 0$) is when we check whether the material has yielded. If yes, then we find the $(k+1)$th iteration by using a Taylor-Series expansion as follows:

$$f((\mathbb{C})[\boldsymbol{\varepsilon}_1 - \boldsymbol{\varepsilon}_1^p], \kappa_1) - f((\mathbb{C})[\boldsymbol{\varepsilon}_1 - \boldsymbol{\varepsilon}_p^k], \kappa^k) \approx \frac{\partial f}{\partial \boldsymbol{\varepsilon}_p} \cdot \triangle \boldsymbol{\varepsilon}_p + \frac{\partial f}{\partial \kappa} \triangle \kappa \bigg|_{\mathbb{U}=\mathbb{U}^k}$$

approximation of the flow rule : $\triangle \boldsymbol{\varepsilon}_p^k = \triangle t \phi \mathbf{N}^k$ (10.56)

approximation of the arc length : $\triangle \kappa^k = \frac{d\kappa}{ds} \triangle s = \frac{d\kappa}{ds} \triangle t \phi / \kappa^k$

Substituting the second and third of (10.56) into the first, and using the fact that $f((\mathbb{C})[\boldsymbol{\varepsilon}_1 - \boldsymbol{\varepsilon}_1^p], \kappa_1) = 0$ we can solve for $\triangle \phi := \triangle t \phi$ to get

$$\triangle \phi = - \frac{f}{\left[\frac{\partial f}{\partial \mathbf{Q}}\big|_{\boldsymbol{\varepsilon} \text{ fixed}}\right]^t [H]} \bigg|_{\mathbb{U}=\mathbb{U}^k}$$
(10.57)

Once we have $\triangle\phi$ we can then use (10.56) to find the increments to ε_p and κ and find ε_p^{k+1} and κ^{k+1}. It should be evident to you that the above approach is very similar to the usual calculation of ϕ and indeed the denominator on the right-hand side of (10.57) is identical to that for the exact calculation of ϕ.

With this motivation, we now present the actual procedure:

(1) Initialize : $k = 0, \mathbb{I}^k = \{\varepsilon_1, \mathbf{Q}_0\} \phi^k = 0$. Note that the total strain alone has been set to the final value but all the other variables have been set to the initial values

(2) Compute the values of the dependent variables :
$$[\sigma]^k = (\mathcal{C})[\varepsilon^k - \varepsilon_p^k]$$
$$\kappa^k = \hat{\kappa}(s^k) \tag{10.58}$$

At this stage we know the value of $\mathbf{U}^k$.

(3) Compute the value of the yield function and check for yielding $f^k = f(\mathbf{U}^k) = f([\sigma]^k, \kappa^k);.$
IF $f^k < TOL$, GO TO EXIT.
ELSE:

(4) Compute the value of the Plastic flow function $H^k(\mathbb{U}^k)$. In our case, this means find $[[\mathbf{N}^k]; 1/\kappa^k]$ using $\mathbb{U}^k$.

(5) Find the consistency parameter:
$$\phi = -\frac{f}{\left[\frac{\partial f}{\partial \mathbf{Q}}|_{\varepsilon \text{ fixed}}\right]^t [H]}\bigg|_{\mathbb{U}=\mathbb{U}^k} = \frac{f}{[\mathbf{N}]^t(\mathcal{C})[\mathbf{N}] + \frac{d\kappa}{(\kappa)^3 ds}}\bigg|_{\mathbb{U}=\mathbb{U}^k} \tag{10.59}$$

The denominator for this comes from using the chain rule of differentiation, the flow rule for plastic strain and the definition of the arc length parameter so that $[\partial f/\partial \mathbf{Q}|_{\varepsilon \text{ fixed}}]^t H$ is the same as

$$[-(\mathcal{C})[\partial f/\partial \sigma], (\frac{\partial f}{\partial \kappa})d\kappa/ds] \begin{bmatrix} [\partial f/\partial \sigma] \\ 1/\kappa^2 \end{bmatrix} = [\mathbf{N}]^t(\mathcal{C})[\mathbf{N}] + \frac{d\kappa}{(\kappa)^3 ds} \tag{10.60}$$

where we have used the fact that $\mathbf{N} = \partial f/\partial \sigma$ and the fact that $\partial f/\partial \kappa = 1/\kappa$.

(6) Update the plastic variables:
$$\mathbf{Q}^{k+1} = \mathbf{Q}^k + \phi H^k \tag{10.61}$$

In our case this becomes

$$[\varepsilon_p^{k+1}] = [\varepsilon_p^k] + \phi[\mathbf{N}^k]; \quad \kappa^{k+1} = \kappa^k + \phi \left(\frac{d\kappa}{\kappa ds}\right)^k. \tag{10.62}$$

(7) $k \leftarrow k+1$ and GO TO 2
END IF
EXIT

A MATLAB code that implements this procedure is available online (see authors' websites listed in the preface).

The convex cutting plane algorithm (CCPA) is by no means the only one that can be used. There are a broad class of algorithms that go under the name of "return mapping algorithms" (see [Simo and Hughes (1998)]) and the CCPA is just one of the simpler ones. Here is a sample of references for what else is out there ([Fotiu and NematNasser (1996); Hartmann et al. (1997); Hofstetter et al. (1993)]).

10.4.3 *Homogeneous deformation examples*

We will now consider a special example to illustrate this general procedure:

Tension-torsion:

The set of experiments that are of great experimental significance are those involving the tension and torsion of the thin walled tube (see Fig. 6.10). If the thickness is much less than the radius of the tube, one can ignore the variation of the shear stress through the tube and so cylindrical sector is subject to a combination of axial tension and shear.

To simulate such an experiment, we simply set $\varepsilon_{11} = h_1(t)$, $\varepsilon_{12} = h_2(t)$ and $\sigma_{22} = 0$. This is a completely strain controlled test and so no equilibrium iterations (i.e. iterations using the equations of equilibrium and boundary conditions) are needed. We simply use these functions together with the algorithm for calculating the plastic strain and we can show the response of the material. Note that here, the strain in the y direction is unknown. This problem is the simplest problem that illustrates the equilibrium iterations since we need to calculate the e_{22} component by enforcing the boundary condition.

The algorithm for the solution is as follows:
(1) Discretize the time interval into N instants t_i, $(i = 1, N)$;
(2) Assume that we know $\boldsymbol{\varepsilon}$ and $\boldsymbol{\varepsilon}_p$ at time $t = 0$;
(3) At time $t = t_1$, we know $(\varepsilon_{11})_1$ and $(\varepsilon_{12})_1$. So, we need to find $(\varepsilon_{22})_1$ and $(\boldsymbol{\varepsilon}_p)_1$;
(4) Set iteration counter $k = 0$. Set $(\varepsilon_{11})_1, (\varepsilon_{12})_1$ set $[\mathbf{Q}]^k = [\mathbf{Q}]_0$;
(5) Compute $\sigma_{11} = E(\varepsilon_{11} - \varepsilon^p{}_{11})$.
(6) Compute $\varepsilon_{22} = -\frac{\nu \sigma_{11}}{E} + \varepsilon^p{}_{22}$.
(7) Using the values of $\boldsymbol{\varepsilon}$ find the corresponding $\boldsymbol{\varepsilon}_p$ by the convex cutting plane algorithm;
(8) Check for convergence of the plastic strain by comparing the new value of the plastic strain with the old value;
(9) Update the plastic strain and the hardening parameter;
(10) if not converged, go to item 5.

Figure 10.4 shows the simulations for a tension-torsion path that is in the form of a figure eight.

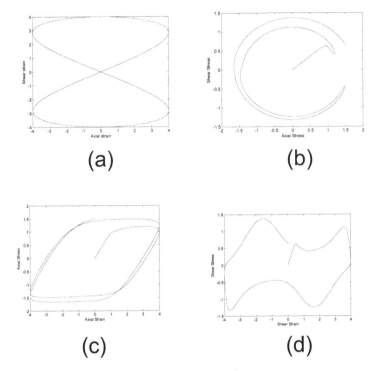

Fig. 10.4 An example of the loading of a thin walled tube in tension and torsion is shown. (a) The loading path is a figure 8, i.e, the material cycles twice as fast in torsion as in tension (b) the resulting stress components showing the elliptical yield surface and its gradual expansion. (c) The axial stress strain curve showing the cycle and (d) the rather complex response in torsion

10.5 Numerical examples of boundary value problems

To demonstrate the method described above, two numerical examples are solved. One is a one-dimensional bar of varying cross-section and the other a plate with a hole subjected to an in-plane remote stress. The program for this meshless method for solving plane stress boundary value problems is available online (see authors' websites listed in the preface).

Material used for numerical examples are treated as linear elastic linear kinematic hardening. Material properties used in the numerical study are as follows: $E = 200 GPa$, $h = 10 GPa$, $\nu = 0.3$, and $\sigma_y = 250 MPa$.

10.5.1 Rod with varying cross-section

A one dimensional rod with linearly varying cross sectional area, as in Fig. 10.5 is considered for the study. Rod has a cross sectional area of $2mm^2$ at $x = 0$ and $1mm^2$ at $x = 1$. A sinusoidal loading is applied with $p = 350N$ as given in Fig. 10.5. The rod is discretized into 11 equally spaced nodes and one point Gauss quadrature rule is used for numerical integration. Load is applied in 50 equal intervals. Figure 10.6 is the load displacement curve at the free end of the rod. Figure 10.7 shows the total strain distribution and the plastic strain distribution obtained for this example at the end of the first and second seconds. The first second is that of the maximum load applied and the second refers to the state upon complete unloading. As can be seen, the end part of rod undergoes plastic deformation which becomes residual strain upon unloading. Since there are no constraints, the stress goes to zero at all points at the end of the loading (at the end of 2 seconds).

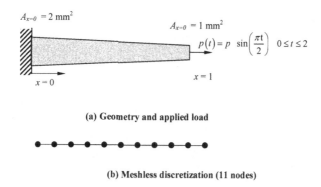

(a) Geometry and applied load

(b) Meshless discretization (11 nodes)

Fig. 10.5 The geometry of the one-dimensional rod with linearly varying cross-section subjected to a sinusoidally time varying load. A single cycle of loading and unloading with $p = 350N$. The rod is discretized into 11 equally spaced nodes in this example.

10.5.2 Plate with a hole subjected to tension

A simple 2-d example in terms of a square plate with a hole is loaded uniformly along one direction as shown in Fig. 10.8(a). The distribution of nodes for this problem is chosen so that density is higher near the hole and tapers down away from the hole since the variation of stresses and strains expected to be higher nearer to the hole. A typical nodal distribution chosen for analysis is shown in Fig. 10.8(b). 153 nodes were used in the analysis. This number was obtained by looking at the convergence of the solution nearer to the hole. Load vs. average displacement plot shows that for the material chosen that exhibits linear hardening, there is a gradual change in stiffness to an asymptotic value as seen in Fig. 10.9. The distribution of stresses at the maximum load stage is shown in Fig. 10.10. The plastic strain distribution in Fig. 10.11 shows an elongated region along the $45°$ angle where the shear is maximum. The total strain distribution at the end of the loading is also shown in Fig. 10.12.

Since this is a plane stress case, it is possible to obtain the stress-elastic strain

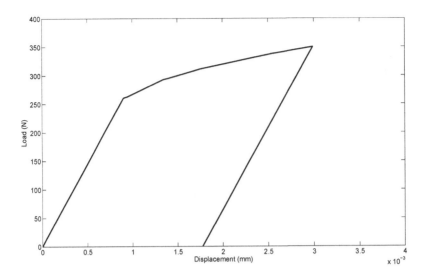

Fig. 10.6 Load to the end displacement is shown here. As can be seen, the end displacement does not proportionally follow the loading time variation.

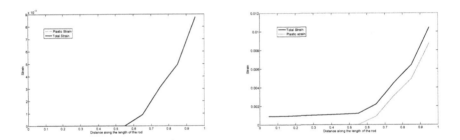

Fig. 10.7 The plot shows the distribution of the total strain and the plastic strain along the rod at the end of the loading.

relation using the condition that the stress is zero in the 3-direction. In effect, this will result in a modification of the elastic constitutive matrix for the plane stress case. In the case of elasto-plastic analysis, instead of deriving the elasto-plastic constitutive stiffness matrix that may involve tedious derivation using the condition that the stress in the 3-direction is zero, the stress can be obtained by using the modified elastic constitutive matrix. As far as using the calculation of global stiffness matrix, this modification was not carried out and the analysis did not affect convergence of the Newton-Raphson iteration greatly. Therefore, this method was adopted in solving this problem.

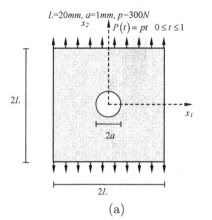 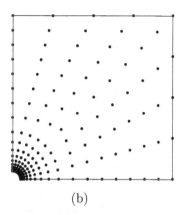

Fig. 10.8 (a) A square plate with a central hole is subjected to pull along one direction. Since it is symmetric about two planes with normals along x_1 and x_2 directions, only one quarter of the plate needs to be analyzed for a quasi-static loading. (b) The nodal distribution of one quarter of the plate is shown. Around 150 nodes is found to be an optimal number for convergence.

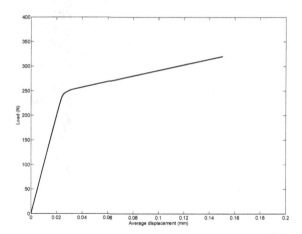

Fig. 10.9 Variation of average displacement of the edge with the load. The transition from elastic to plastic response is smooth because of the gradually expanding plastic zone near the hole.

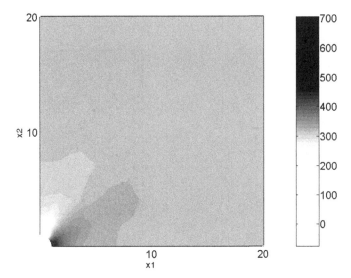

Fig. 10.10 The stress distribution at the end of the loading is shown. Notice that the maximum stress occurs near the point on the hole on the x_1-axis. As expected, the maximum stress is less compared to an elastic case because of the plastic relaxation.

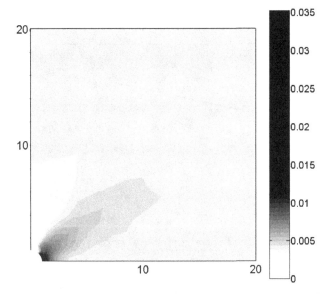

Fig. 10.11 The plastic strain distribution at the end of the loading is shown. The plastic region emanates from the hole, propagates and elongates along the 45° line.

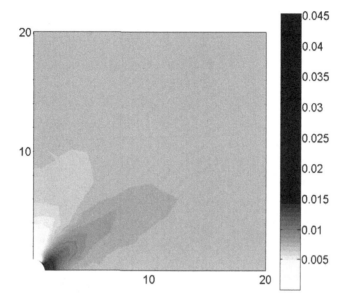

Fig. 10.12 The total strain distribution at the end of the loading.

PART 3
INELASTICITY OF CONTINUA
Finite Deformations

Chapter 11

Summary of Continuum Thermodynamics

Learning Objectives

By reading this chapter, you should be able to

(1) describe the different measures of deformation and strain in a body;

(2) derive the expressions for the time derivatives of some measures of strain in terms of the velocity gradient;

(3) describe the difference between sequential and simultaneous motions;

(4) describe the meaning of an objective tensor and identify whether a tensor is objective or invariant or neither;

(5) describe a method for obtaining an objective rate of a tensor;

(6) use eigenvalue decompositions for a variety of strains to simplify expressions;

(7) list the balance laws for a continuum undergoing finite deformations;

(8) obtain the expression for the rate of dissipation under finite deformations;

(9) derive the expression for the stress for an elastic material;

(10) obtain the simplified form for the stress for a material that is isotropic.

CHAPTER SYNOPSIS

The main point that we are seeking to convey about finite deformations is that (1) the constitutive response is assumed to depend only upon *changes in shape of the body* and NOT upon the orientation of the body in Euclidean space. This idea (or variants of it) is called the principle of objectivity or invariance under superposed rigid body motions or Frame Indifference, depending upon the point of view. We thus conclude that the Helmholtz potential of an elastic body is of the referential stretch tensor $\mathbf{F}^t\mathbf{F}$ and the temperature and the Kirchhoff stress is related to the gradient of the Helmholtz potential with respect to $\mathbf{C}$. If the body is isotropic, then we can define a logarithmic strain tensor $\mathbf{E} = 1/2\ln(\mathbf{F}\mathbf{F}^t)$. Then the kirchhoff stress is equal to $\partial \psi/\partial \mathbf{E}$ and is a form similar to small deformation. Unfortunately, the situation is not so elegant with anisotropic elastic solids and this is an area of active research.

Chapter roadmap

Our original intention was to move this whole chapter to the appendix. However, studies have shown[1] that students and faculty do not pay any attention to appendices other than generally tell them "read the appendix". If you are interested in modeling finite elastoplastic deformation, either because you are responsible for developing models for inelasticity of polymers and/or biomaterials, or because you are interested in materials processing, then you have to be very familiar with the notation and terminology. This chapter is NOT a replacement for a continuum mechanics course[2]. If you have already taken continuum mechanics, then this is a quick overview of the notation and a refresher on some concepts.

---

[1] "Studies have shown that made-up statistics are as convincing as real statistics"—quote from the great philosopher Dilbert.

[2] Since both authors teach continuum mechanics, we will maintain until our last breath that (1) nothing can replace our continuum mechanics course and (2) You can't make judgements about finite deformation models without knowing continuum mechanics. The latter is actually true:-))

11.1 Overview of kinematics for finite deformation

The purpose of this section is to quickly introduce some of the fundamental concepts of finite deformations. It is not possible to do justice to continuum mechanics in just a few pages, so we do not even try. What follows here is neither comprehensive nor detailed. It is just a collection of definitions, facts, interpretations and results that we will find useful. The starting point of any continuum theory is the capability to measure lengths (a ruler), angles (protractor) and time (a clock). If we have these, and a location to start from, then we can measure the location of any point and can develop the whole of kinematics. The formalization of a ruler and protractor is the notion of a "frame".

⊙ **11.1 (DEFINITION). *Frame:*** *A frame $\mathbb{F}$ is a set composed of a point O in space which is the origin, and three mutually orthonormal vectors $\mathbf{i}_i, (i = 1, 2, 3)$ (see Fig. 11.1) which represent three mutually perpendicular directions. We use the three vectors to create a cartesian coordinate system (xyz) associated with the frame[3].*

We don't use $\mathbf{e}_i$ for the unit vectors since the letter $\mathbf{e}$ has been overused for a variety of different things. Let us assume that we have a global frame $\mathbb{F}_r = \{O, \mathbf{i}_1, \mathbf{i}_2, \mathbf{i}_3\}$ and a corresponding cartesian coordinate system (XYZ) attached to our laboratory. This will be called a *laboratory* or *inertial* frame[4] and the coordinate axis is the laboratory axis.

From your dynamics classes, you have perhaps been used to the notation for taking derivatives with respect to a frame (i.e. calculating $[d/dt]|_{(XYZ)}$). Even though this is an abuse of notation (since we are keeping the FRAME fixed and not the axes), we will continue to use this notation. All the kinematics that we have done up until now has been with this laboratory frame since the bodies in question hardly moved at all. We are now considering situations where bodies move and deform by large amounts.

Thus, we have a body $\mathbb{B}$ that starts out at time $t = 0$ occupying a region Ω in 3-D Euclidean space, i.e. the body starts out with a definite size, shape and location. As time progresses, due to the external influences such as forces and heating/cooling of the body, the body changes its configuration[5]—each particle, whose location is

---

[3] Most textbooks confuse a frame with a coordinate system with the X axis along $\mathbf{i}_1$, Y axis along $\mathbf{i}_2$ etc. and use them interchangeably. This creates HUGE problems. So let us be very specific. Imagine a frame to be a "little blob with three unit vectors sticking out of it". We can use it to define any coordinate system that we want. A coordinate system is a grid that is "rigidly attached to the frame". The difference between a frame and a coordinate system is that between the notion of a "window frame" and a "window grille". The same window frame can be used to attach different patterned grills. Similarly, we can use the frame to define a polar coordinate system, a skew coordinate system or some other whimsical curvilinear coordinate system.

[4] A frame in which measurements will satisfy Newton's laws, (i.e. $\mathbf{f} = m(d^2\mathbf{x}/dt^2)|_{(XYZ)}$).

[5] Here, it is probably a good idea for you to review the one-dimensional version of this description in Chapter 3.

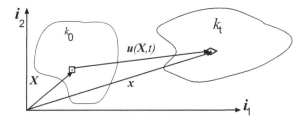

Fig. 11.1 A schematic figure of a body undergoing a large distortion and rotation: Note that the directions of the normals to the surfaces as well as the overall shapes change rather drastically. Observe the change in the shape of the square inscribed around the material point **X** into a parallelogram around **x**. The deformation gradient $\mathbf{F}(\mathbf{X}, t)$ is a mathematical description of this change in shape of the square.

X at time t, suffers a displacement $\mathbf{u}(\mathbf{X}, t)$ and goes to a new location **x**. A picture of this is shown in Fig. 11.1.

⊙ **11.2 (DEFINITION).** *Configuration:* *A fundamental aspect of the description of a body is the notion of a "configuration". The dictionary definition of the word is "relative arrangement of the parts of a system"*[6]. *Here it means the "spatial arrangement of the particles that comprise the body". The configuration of a body at time t is denoted by k_t.*

Thus a configuration is a listing of the labels of every particle and its location. For a finite element program, the configuration of the body at any time t is a listing of the node numbers and their x, y and z coordinates. unlike in a FEM program where nodes are labeled with integers (since there is only a finite number of them), we will label the particles in a continuum by their locations **X** at time $t = 0$. It is impossible to describe the configuration of a continuum (since it has an uncountably infinite number of nodes). *Only changes of configuration can be described by using mathematical functions.* In other words, given a configuration where a particle P is at a position **X**, we can precisely describe how it changes by using a vector function $\mathbf{u}(\mathbf{X}, t)$.

⊙ **11.3 (DEFINITION).** *Material point and its current location:* *At the risk of slightly abusing the notation, we will define the reference position vector of a particle **X** itself as a material point*[7] *and its location at any other time as the position vector $\mathbf{x}(\mathbf{X}, t)$. In other words, we will assume that $\mathbf{X} = \mathbf{x}(\mathbf{X}, 0)$. This will allow us to simplify the description of the motion of the body.*

---

[6]Compare it to the use of the word configuration when describing computer systems: you can see its original usage, "what are the components and how are they arranged?"

[7]It is standard practice to define **X** as the reference location of an "abstract" material point. However, since bodies are always embedded in 3-D space, every material particle of the body, automatically comes with some location associated with it, so it is okay to refer to the particle itself by its location at the first time that the body was encountered.

⊙ **11.4 (DEFINITION). *Referential and Spatial Descriptions:*** *Given the motion* $\mathbf{x} = \mathbf{X} + \mathbf{u}(\mathbf{X}, t)$, *we will assume that, given the current location* $\mathbf{x}$ *at any time* t, *we can solve for its reference location* $\mathbf{X}$. *In other words, if we know the motion and the current location of the material point, we will be able to find the location it used to occupy at time* $t = 0$. *Now, using this, any function* $\hat{f}(\mathbf{X}, t)$ *can be written as a different function* $\tilde{f}(\mathbf{x}, t)$. *The former is called the "referential" or "Lagrangian" description*[8], *while the latter is called the "current" or Eulerian description*[9].

In other words given a property of the system such as the temperature as a function of the material point (i.e. particle "number" or label) we can always rewrite it, in principle, as a function of the current location of the material point in question. The former is very useful for dealing with solid-like behavior, where we track particles, while the latter is useful for studying changes in properties at a given location, typically useful in modeling fluid-like behavior. Not surprisingly, the spatial description is overwhelmingly favored in fluid mechanics, while the referential description is favored in solid mechanics. It is not so clear which one to use when we deal with viscoelastic or elastoplastic materials since the behavior of the material is intermediate between that of a solid and that of a fluid. So here, we must work with both, choosing that which is simple.

11.1.1 Temporal and spatial gradients of the motion

The spatial and time derivatives of the displacement $\mathbf{u}(\mathbf{X}, t)$ have special significance. Thus the first and second time derivatives of $\mathbf{u}$ keeping $\mathbf{X}$ fixed are respectively the velocity and acceleration of the particle in question, i.e.

$$\dot{\mathbf{u}}(\mathbf{X}, t) = \mathbf{v}(\mathbf{X}, t), \; \ddot{\mathbf{u}}(\mathbf{X}, t) = \dot{\mathbf{v}}(\mathbf{X}, t) = \mathbf{a}(\mathbf{X}, t), \quad (11.1)$$

where the superposed dot, $(\dot{\cdot})$, means time derivative (with respect to the laboratory frame) keeping $\mathbf{X}$ fixed. Similarly, the referential gradients (i.e. gradient with respect to $\mathbf{X}$) of $\mathbf{u}$ and $\mathbf{v}$ also have special meaning.

We will define the displacement gradient $\mathbf{H}$ and the deformation gradient $\mathbf{F}$ as

$$\mathbf{H} := \text{Grad}(\mathbf{u}) = \frac{\partial \mathbf{u}}{\partial \mathbf{X}}, \; \mathbf{F} := \text{Grad}(\mathbf{x}) = \mathbf{I} + \mathbf{H} = \frac{\partial \mathbf{x}}{\partial \mathbf{X}}, \quad (11.2)$$

where the symbol Grad is used to define the gradient with respect to $\mathbf{X}$. *The deformation gradient* $\mathbf{F}$ *measures the difference in locations (at a given instant) of two neighboring material points.*

The velocity $\mathbf{v}$ is a great example of a property that is frequently written as a function of the current location. The *velocity gradient* $\mathbf{L}$ which is the "spatial or Eulerian" gradient of $\mathbf{v}$, is defined by

$$\mathbf{L} := \text{grad}(\mathbf{v}) := \frac{\partial \mathbf{v}}{\partial \mathbf{x}}. \quad (11.3)$$

---

[8] Invented by Euler, of course.
[9] Naturally, we would expect it to be invented by someone else, in this case D'Alembert. See [Truesdell (1977)].

where we have used the notation grad to stand for the gradient with respect to **x**. *The velocity gradient measures the difference in the velocities at two neighboring locations in space at a given time.* Note that this definition involves taking derivatives with respect to **x** rather than **X**.

The quantity $\mathbf{L} = \text{grad}\mathbf{v}$ is a very important tensor, so you should become very familiar with it. Most fluid mechanics textbooks (see e.g. [Fox et al. (2006)] Chapter 5) contain a very detailed description of what this tensor means and it is well worth your time to read some of the fluid mechanics textbooks where they describe the physical meaning of each component of **L**.

The connection between these quantities is given by the exercise below:

⊙ **11.5 (Exercise:). Reference and current gradients:**
Show that

$$\text{Grad}(\cdot) = \text{grad}(\cdot)\mathbf{F}, \text{ in other words } \frac{\partial(\cdot)}{\partial \mathbf{X}} = \frac{\partial(\cdot)}{\partial \mathbf{x}}\mathbf{F} \qquad (11.4)$$

(Hint: use the chain rule of differentiation and the definition (11.2)). Hence show that

$$\dot{\mathbf{F}} = \mathbf{LF}. \qquad (11.5)$$

(Hint : note that $\dot{\mathbf{F}} = \text{Grad}(\mathbf{v})$, and then use (11.4).)
*Explain in words what is the difference between **L** and $\dot{\mathbf{F}}$ in terms of what they measure.*

Note that for scalars, we have to be careful since the gradient is written as a column vector and not as a row vector. Hence if ϕ is a scalar then $\text{Grad}(\phi) = \mathbf{F}\text{grad}(\phi)$.

11.1.2 Local motion: Deformation of line, area and volume

For the development of constitutive equations, we are not interested in the motion of the whole body. Rather, we pick a particular particle say $\mathbf{X}_0$ in the reference configuration and consider the motion of a small chunk or elemental volume around this particle (see Fig. 11.2 on p. 361). We will be somewhat casual in defining this elemental volume; suffice to say that, for the purposes of this description, consider a parallelepiped P whose edges are given by three line elements $d\mathbf{X}_i$, $(i = 1, 2, 3)$ (called reference line elements). Due to the deformation, this parallelepiped, *if it is small enough*, will become another parallelepiped p^{10}, whose sides are given by $d\mathbf{x}_i$, $(i = 1, 2, 3)$.

How are these two sets of line elements $d\mathbf{X}_i$ and $d\mathbf{x}_i$ related? By using the chain

---

[10]Well...almost, it will not really become one, but the smaller you make P the closer will p be to a true parallelepiped. But we will not quibble about such nuances here.

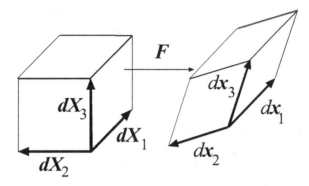

Fig. 11.2 A schematic figure of the deformation of an elemental cube with sides $d\mathbf{X}_i$, $(i = 1, 2, 3)$. The faces of the cube are the area elements. Note that the cube becomes a parallelepiped.

rule, we can see that
$$d\mathbf{x} = \frac{\partial \mathbf{x}}{\partial \mathbf{X}} d\mathbf{X} \Rightarrow d\mathbf{x}_i = \mathbf{F} d\mathbf{X}_i, \ (i = 1, 2, 3) \tag{11.6}$$

⊙ **11.6 (DEFINITION).** *Transformation of line elements:* The deformation gradient is the linear transformation that tells us how line elements in the reference configuration are related to the corresponding line elements in the current configuration. It is defined through the relationship (11.6), see Fig. 11.2.

Similarly if we consider the elemental areas that form the faces of the parallelepiped, the area vector associated with each face can be calculated through
$$d\mathbf{A} = d\mathbf{X}_1 \times d\mathbf{X}_2, \text{ and } d\mathbf{a} = d\mathbf{x}_1 \times d\mathbf{x}_2, \tag{11.7}$$
with other faces being cyclically defined[11].

Then, it can be shown that[12]
$$d\mathbf{a} = \mathbf{F}^* d\mathbf{A}, \ \mathbf{F}^* = \det(\mathbf{F}) \mathbf{F}^{-t}, \tag{11.8}$$
where $\mathbf{F}^*$ is called the adjugate of $\mathbf{F}$, and the symbol $\det(\cdot)$ stands for the determinant.

⊙ **11.7 (DEFINITION).** *Adjugate:* The adjugate of the deformation gradient is the linear transformation that tells us how area elements in the reference configuration become area elements in the current configuration. It is defined through (11.8).

Finally, if we define volume elements through:
$$dV = d\mathbf{X}_1 \times d\mathbf{X}_2 \cdot d\mathbf{X}_3, \text{ and } dv = d\mathbf{x}_1 \times d\mathbf{x}_2 \cdot d\mathbf{x}_3 \tag{11.9}$$
then it can be shown that
$$dv = J dV, \ J = \det \mathbf{F} \tag{11.10}$$

---

[11] you know what we mean—$1 \to 2 \to 3 \to 1$, etc.
[12] Most continuum mechanics text books will have a proof of this and it is usually associated with "Nansens formula" for normals.

⊙ **11.8 (DEFINITION). *Jacobian:*** *The determinant of the deformation gradient is the linear transformation that tells us how volume elements in the reference configuration become volume elements in the current configuration. It is defined through (11.10).*

⊙ **11.9 (Exercise:). Gradient of the Jacobian:**
Show that for any tensor $\mathbf{A}$,
$$\frac{\partial J(\mathbf{A})}{\partial \mathbf{A}} = \mathbf{A}^*, \quad J(\mathbf{A}) = \det(\mathbf{A}), \qquad (11.11)$$
where $\mathbf{A}^*$ is the adjugate.
(Hint: This is a hard problem requiring index notation. Find a good book on linear algebra or continuum mechanics to solve this.)

Now that we have some preliminary ideas on how finite deformations are defined and the way in which line, area and volume elements are transformed by the motion, we want to bring in the notion of strain. We want to generalize the idea of "change in length per unit length." We note that what matters in this idea is the "ratio of lengths." It is this that we define next:

⊙ **11.10 (DEFINITION). *Stretch ratio:*** *The stretch ratio λ associated with a line element $\mathbf{X}$ in the reference configuration, is the ratio of its current length to its original length, i.e.*
$$\lambda = \frac{\|d\mathbf{x}\|}{\|d\mathbf{X}\|} = \sqrt{\frac{\mathbf{F}d\mathbf{X} \bullet \mathbf{F}d\mathbf{X}}{d\mathbf{X} \bullet d\mathbf{X}}} = \sqrt{\mathbf{M} \bullet (\mathbf{F}^t \mathbf{F} \mathbf{M})}, \qquad (11.12)$$
where $\|(\cdot)\|$ is the so-called 2-norm of a vector and is defined as $\|\mathbf{a}\| = \sqrt{\mathbf{a} \bullet \mathbf{a}}$ and $\mathbf{M}$ is the unit vector along $d\mathbf{X}$ and is defined through $\mathbf{M} := d\mathbf{X}/\|d\mathbf{X}\|$.

Just as the displacement gradient $\mathbf{F}$ provides information about changes in line, area and volume elements, the velocity gradient provides information about *the rates of change of line, area and volume elements* as we can see in the next few definitions.

⊙ **11.11 (DEFINITION). *Rate of change of a line element:*** *For our purpose, the most important property of $\mathbf{L}$ is the fact that if $d\mathbf{x}$ is a material line element, then*
$$\dot{d\mathbf{x}} = \mathbf{L}d\mathbf{x}. \qquad (11.13)$$

Thus, $\mathbf{L}$ is the linear transformation that relates the time rate of change of a material line element with its current value.

In other words, the rate of change of a material line element is equal to the velocity gradient times the line element. Thus **L** is the counterpart of **F** for time derivatives.

This is such an important idea that

BEHAVING LIKE LINE ELEMENTS

We will say that a vector behaves like a material line element at any given instant if its time derivative is equal to **L** times the vector at that time instant, i.e. if it satisfies

$$\dot{\mathbf{a}} = \mathbf{L}\mathbf{a} \qquad (11.14)$$

at that time instant.

⊙ **11.12 (Exercise:). Rates of change of area and volume elements**
Show that the rate of change of an elementary area is given by $\dot{\overline{d\mathbf{a}}} = (tr(\mathbf{L})\mathbf{I} - \mathbf{L}^t)d\mathbf{a}$.

Show also that the rate or increase of an elementary volume is $\dot{\overline{dv}} = tr(\mathbf{L})dv$ (Hint: The first one is hard: Do the second one first and then for the first use (11.8).)

HOMOGENEOUS MOTION

We define a **homogeneous motion** as one in which straight lines remain straight. For such a motion, the deformation gradient **F** is only a function of time. The motion is given by $\mathbf{x} = \mathbf{F}(t)\mathbf{X} + c(t)$. In this motion, there is no difference between the behavior of line elements and macroscopic lines in the body.

Thus, in a homogeneous motion, only the first referential and spatial derivatives (i.e. Grad and grad are nonzero but second and higher referential and spatial derivatives are all zero).

> **CLASSICAL CONTINUUM MODEL**
>
> A material model is called a "Classical continuum model" if the response relations of the model, at any point $\mathbf{X}$ can be calculated solely from quantities that can be obtained from suitably chosen homogeneous motions, i.e. only kinematical quantities derived from the first referential or spatial derivatives of the motion are involved. There is no restriction on their histories or material time derivatives.

Thus, referential or spatial derivatives of $\mathbf{F}$ do not figure in the constitutive equations of classical continua. Theories that are not classical include those for which the response depends upon higher gradients (referred to as "higher gradient theories") [13] or those for which the response depends upon values of parameters at other points (referred to as "non-local" constitutive theories) such as theories of self-gravitation.

Thus in the remainder of this chapter, since we are considering only a classical continuum model, we will consider only homogeneous deformations.

⊙ **11.13 (Exercise:). Homogeneous deformation examples**
Consider a square and a circle in the reference configuration. For each of the following deformations in 2 dimensions, calculate and draw the final shape (at time t=1) of the square (which should become a parallelogram of some sort) and the circle (which should become an ellipse). Also calculate $\mathbf{F}$ and $\mathbf{L}$ for these deformations:

$$\text{Uniaxial extension: } \mathbf{x} = (1 + 3t^2)X\mathbf{i}_1 + Y\mathbf{i}_2,$$

$$\text{Isochoric extension: } \mathbf{x} = (1 + 3t^2)X\mathbf{i}_1 + \frac{1}{1 + 3t^2}Y\mathbf{i}_2, \quad (11.15)$$

$$\text{Simple shear: } \mathbf{x} = (X + 3t^2 Y)\mathbf{i}_1 + Y\mathbf{i}_2.$$

11.1.3 *Sequential versus simultaneous action*

Consider the following sequence of deformations: We first elongate the material along the x axis and then shear the material. For small deformations, the order in which these operations are carried out are immaterial: i.e. for small deformations,

---
[13] The difference can be appreciated by considering the constitutive equations for a string versus that for a beam. For a string the tension is determined only by the elongation (first spatial derivative) on the other hand, for a beam the moment depends upon the curvature (second spatial derivative). Thus a string is a classical continuum but a beam is not. Similar considerations hold for a membrane versus a plate.

the above sequence of operations is the same as first shearing the material and then elongating it. *On the other hand, for finite deformations, sequence does matter, the order of operations is important.* Thus there is a fundamental difference between sequential and simultaneous operations. This is illustrated in Fig. 11.3 :

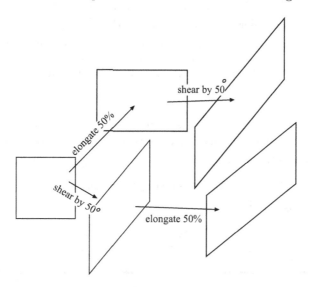

Fig. 11.3 This figure illustrates the importance of sequence of operations. We started with a square, and on the top panel we elongated it horizontally by 1.5 times and then sheared it vertically by 50°. On the bottom panel, we reversed the order, shearing vertically by 50° and then horizontally elongating the resulting configuration by 1.5 times. Notice that the results, although similar, are not the same. The effect in three dimensions is much more drastic.

In general, if two deformations *follow one another* sequentially, then their deformation gradients need to be multiplied, i.e. if

$$\mathbf{x}_1 = \mathbf{f}_1(\mathbf{X}, t), \ \mathbf{x}_2 = \mathbf{f}_2(\mathbf{x}_1, t), \tag{11.16}$$

then

$$\mathbf{F} = \frac{\partial \mathbf{x}_2}{\partial \mathbf{X}} = \frac{\partial \mathbf{x}_2}{\partial \mathbf{x}_1} \frac{\partial \mathbf{x}_1}{\partial \mathbf{X}} = \mathbf{F}_2 \mathbf{F}_1. \tag{11.17}$$

In other words, if the deformation proceeds one after another, then the respective deformation gradients must be multiplied in reverse order (last deformation is the left most), i.e.

$$\mathbf{F} = \mathbf{F}_n \mathbf{F}_{n-1} \mathbf{F}_{n-2} \cdots \mathbf{F}_2 \mathbf{F}_1. \tag{11.18}$$

It should be evident that the matrix multiplication is non-commutative, and the order of multiplication matters.

Now, what happens if for each of the deformations, the displacement gradient $\mathbf{F}_i - \mathbf{I} = \mathbf{H}_i$ is very small? We see that

$$\mathbf{H} = \mathbf{F} - \mathbf{I} = (\mathbf{I} + \mathbf{H}_n)(\mathbf{I} + \mathbf{H}_{n-1})(\mathbf{I} + \mathbf{H}_{n-2}) \cdots (\mathbf{I} + \mathbf{H}_2)(\mathbf{I} + \mathbf{H}_1) - \mathbf{I}. \tag{11.19}$$

Now expanding out the multiplication we see that

$$\mathbf{H} = \sum_{i=1}^{n} \mathbf{H}_i + \text{Higher order terms} .\tag{11.20}$$

In the above expression, all the sequential information is in the higher order terms. If we retain only the lower order terms, all we have is the sum of the $\mathbf{H}_i$ which can be added any way we like. This is the reason for the fact that for small deformations, the sequence of operations does not matter. We tacitly used this fact when we said in the small deformation plasticity chapter (see Chapter 9 Fig. 9.2 p. 275) that

$$\varepsilon = \varepsilon_e + \varepsilon_p \tag{11.21}$$

How do we deal with "simultaneous action" in finite deformations? In other words what if we are "simultaneously stretching and shearing and rotating" an object. The trick is to *add the velocity gradients* i.e.

$$\mathbf{L} = \mathbf{L}_{\text{stretching}} + \mathbf{L}_{\text{shearing}} + \mathbf{L}_{\text{rotating}} \tag{11.22}$$

Hence for simultaneous action

$$\mathbf{L} = \sum_{i=1}^{n} \mathbf{L}_i .\tag{11.23}$$

SEQUENTIAL vs. SIMULTANEOUS ACTION

Thus, sequential actions are represented by multiplication of the deformation gradients keeping the order in mind from right to left, with the first action in the rightmost location and the most recent in the left most. Simultaneous actions are represented by summing the Eulerian or spatial descriptions of the velocities (i.e. $\mathbf{v} = \sum_i \mathbf{v}_i(\mathbf{x}, t)$). If we differentiate this expression with respect to $\mathbf{x}$, the velocity gradients will also be added, i.e. $\mathbf{L} = \sum_i \mathbf{L}(\mathbf{x}, t)$— we can disregard the order here since summation is commutative. The result also holds if the displacement gradients $\mathbf{H}_i$ are small.

Thus, when considering any constitutive equation, it is important to decide whether we view a particular process as a sequential action or a simultaneous action. As we shall see in the next chapter, this point will attain crucial significance in developing constitutive theories.

11.1.4 *Stretch and rotation*

In this subsection we gather some important results regarding some kinematics of finite deformation. You should already know these if you are reading this chapter— we are presenting it only to introduce the notation and as a recapitulation. If you don't know these, then we urge you to learn some continuum mechanics (see e.g. [Chadwick (1976); Spencer (1980)] for introductory material and [Jaunzemis (1967)] or [Truesdell (1977)] for a more comprehensive treatment) before you proceed further.

⊙ **11.14 (DEFINITION).** *Rigid body motion:* A homogeneous deformation $\mathbf{x} = \mathbf{F}(t)\mathbf{X} + c(t)$ where $\mathbf{F}$ is a proper orthogonal tensor, i.e. $\mathbf{F}\mathbf{F}^t = \mathbf{F}^t\mathbf{F} = \mathbf{I}$ is called a rigid body motion. In other words, other than the axis of rotation, any other line in the body will rotate, i.e. change its orientation.

Rigid body motions are what you have learnt in your undergraduate and perhaps graduate dynamics classes. We will find that there is a lot of use for many of the kinematical ideas that we introduced there (angular velocity, rotating frames etc.). But this is only one extreme case (although vitally important) of the kinds of motion that we study. In this kind of motion, note that lengths and relative angles between line elements do not change at all. Thus the "shape" of the material is not changed at all. The only thing that changes is the "orientation" of the body. In other words, *orthogonal tensors represent changes in orientation*.

⊙ **11.15 (Exercise:). Angular velocity**
Show that if $\mathbf{R}(t)$ is any proper orthogonal matrix, then $\dot{\mathbf{R}}\mathbf{R}^t = \mathbf{\Omega}$ is a skew symmetric matrix. Using this or otherwise show that for a rigid body motion $\mathbf{x} = \mathbf{R}(t)\mathbf{X} + \mathbf{c}(t)$, the velocity vector is given by $\mathbf{v} = \mathbf{\Omega}\mathbf{x} + \dot{\mathbf{c}} = \boldsymbol{\omega} \times \mathbf{x} + \dot{\mathbf{c}}$, where $\boldsymbol{\omega}$ is called the axial vector of $\mathbf{\Omega}$.
Hint: differentiate $\mathbf{R}^t\mathbf{R} = \mathbf{I}$ for the first part. Look up a book on index notation for the second.

The opposite extreme case is the one where orientation does not change but lengths and angles change. Such motions are called "pure stretch".

⊙ **11.16 (DEFINITION).** *Pure Stretch:* A homogeneous deformation $\mathbf{x} = \mathbf{F}(t)\mathbf{X} + c(t)$ where $\mathbf{F}$ is a **symmetric positive definite tensor**, is called a pure stretch. In other words, a homogeneous motion is a pure stretch if there are at least three mutually orthogonal directions which do not change their orientation. These three directions are the eigenvectors $\mathbf{u}_i$ of the pure stretch, and the stretch ratios λ_i along these directions are the eigenvalues of the pure stretch. Note that a pure stretch changes the shape of the body. However its orientation is not changed.

Compared to these two extreme cases, it should be evident to you that in a general deformation, both the shape and orientation of the body changes. It stands to reason therefore that a general deformation is some combination of stretch and rotation.

This combination can be considered either as a sequential action of orientation change followed by a stretch or as the simultaneous stretching and rotation. Generally, workers in finite elasticity make heavy use of the former while those in fluid dynamics make heavy use of the latter.

Befitting its role as a model for intermediate behavior (some attributes are solid-like while others are fluid-like) finite deformation plasticity makes use of both.

Now for a general motion, **the polar decomposition theorem**[14] states that the deformation gradient can be decomposed into a pure rigid body rotation and a pure stretch in two different ways, i.e.

$$\mathbf{F} = \mathbf{R}\mathbf{U} = \mathbf{V}\mathbf{R}. \tag{11.24}$$

In the above decomposition, $\mathbf{R}$ is a rotation tensor (proper orthogonal) and $\mathbf{U}$ and $\mathbf{V}$ are pure stretch tensors (symmetric positive definite) that are related to each other by

$$\mathbf{V} = \mathbf{R}\mathbf{U}\mathbf{R}^t \tag{11.25}$$

A MATLAB code for this is available online (see authors' websites listed in the preface).

A very useful way to "visualize" the polar decomposition theorem (see Fig. 11.4 p. 369) is to imagine that we first deform the material by means of a pure stretch $\mathbf{U}$. While this will change the shape, it will leave the orientation (i.e. the three eigenvectors of $\mathbf{U}$) unchanged. Next we rotate the body by $\mathbf{R}$ to change its orientation. We thus get $\mathbf{F} = \mathbf{R}\mathbf{U}$ as a concatenation of $\mathbf{R}$ which follows $\mathbf{U}$.

⊙ **11.17 (Exercise:). The polar decomposition**
Describe the decomposition $\mathbf{F} = \mathbf{V}\mathbf{R}$ along the same lines as was used for $\mathbf{F} = \mathbf{R}\mathbf{U}$. You might find (11.4) helpful.

THE SQUARED STRETCH TENSORS

The Right and Left Cauchy-Green tensors with symbols $\mathbf{C}$ and $\mathbf{B}$ respectively are defined as

$$\mathbf{C} = \mathbf{U}^2 = \mathbf{F}^t\mathbf{F}, \quad \mathbf{B} = \mathbf{V}^2 = \mathbf{F}\mathbf{F}^t. \tag{11.26}$$

If you find the names for these tensors somewhat odd, you are in good company. We are going to call them the squared stretch tensors, with $\mathbf{C}$ being the *referential squared stretch* tensor and $\mathbf{B}$ being the *current squared stretch* tensor.

We will state without proof the following geometrical significance of the two tensors:

> Consider a unit sphere in the reference configuration. If the body is deformed homogeneously, the sphere will become an ellipsoid in the current configuration.

---

[14] This is a special case of the singular value decomposition (SVD) of matrices.

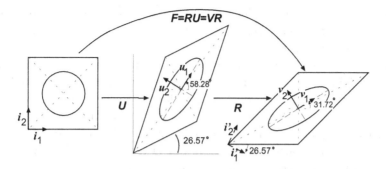

Fig. 11.4 A scale drawing of the simple shear of a unit square with an inscribed circle. The deformation gradient is $\mathbf{F} = \mathbf{I} + \mathbf{i}_1 \otimes \mathbf{i}_2$. The circle becomes an ellipse due to the distortion. The Laboratory frame $\{\mathbf{i}_1, \mathbf{i}_2\}$ and the rotated frame $\{\mathbf{i'}_1, \mathbf{i'}_2\}$ are shown. Note the eigenvectors of the $\mathbf{U}$ tensor and that of the $\mathbf{V}$ tensor. They are not along the diagonals to the parallelogram, although they are close. Note also that the rotation angle of $-26.57°$ is equal to the difference between the angle made by the major axis of the $\mathbf{V}$ ellipse and that of the $\mathbf{U}$ ellipse.

The square of the lengths of the axes of the ellipsoid and their orientations are nothing but the eigenvalues and eigenvectors of $\mathbf{B}$. Similarly, if we draw a sphere in the current configuration and deform the material back to its reference configuration, then the squares of the square of the lengths of the axes of the ellipsoid and their orientations are nothing but the eigenvalues and eigenvectors of $\mathbf{C}^{-1}$. Similarly, if we start with a cube oriented along the reference axes, it will become a parallelepiped in the current configuration. The squares of the length of the sides of the parallelepiped are given by the diagonal components of $\mathbf{C}$, the off-diagonal components contain information regarding the angle between the edges of the parallelepiped. See Fig. 11.4.

Since $\mathbf{U}$ and $\mathbf{V}$ are symmetric, positive definite tensors, they have three orthonormal eigenvectors each. Thus, we can write

$$\mathbf{U} = \sum_{i=1}^{3} \lambda_i \mathbf{u}_i \otimes \mathbf{u}_i, \quad \mathbf{V} = \sum_{i=1}^{3} \lambda_i \mathbf{v}_i \otimes \mathbf{v}_i, \quad \mathbf{v}_i = \mathbf{R}\mathbf{u}_i \qquad (11.27)$$

where λ_i are the eigenvalues of $\mathbf{U}$, and $\mathbf{u}_i$ and $\mathbf{v}_i$ are the eigenvectors of $\mathbf{U}$ and $\mathbf{V}$ respectively.

In words,

EIGENVALUES OF THE STRETCH TENSORS

The referential and current stretch tensors $\mathbf{U}$ and $\mathbf{V}$ have the same eigenvalues. The eigenvectors $\mathbf{v}_i$ of $\mathbf{V}$ are rotated versions of those of $\mathbf{U}$, the rotation matrix being that obtained by the polar decomposition of $\mathbf{F}$, which can then be written as $\mathbf{F} = \sum_{i=1}^{3} \lambda_i \mathbf{v}_i \otimes \mathbf{u}_i$,

It is not a simple matter to find the eigenvalues and eigenvectors of a tensor in closed form (see [Simo and Hughes (1998)] p. 242). On the other hand, if you want to find numerical values, then it is easily computed with MATLAB: To find the eigenvalues and eigenvectors of any symmetric tensor $\mathbf{B}$, use the command "[V,D]=eig(B)" in MATLAB; This will give the matrix of eigenvectors and eigenvalues of $\mathbf{B}$, with the eigenvalues ordered from largest to the smallest.

We can visualize the stretch tensors $\mathbf{U}$ and $\mathbf{V}$ as "stretch ellipsoids" or "lozenges" whose principal axes have "magnitudes" of λ_i in the three directions $\mathbf{u}_i$ and $\mathbf{v}_i$. Thus the ellipsoid of $\mathbf{V}$ is a rotated version of that of $\mathbf{U}$ (see Fig. 11.4 p. 369). This is in analogy with the notion of a strain ellipsoid in small deformation.

The magnitudes of the "stretch ellipsoids" are the three eigenvalues and their orientations are the three orthonormal eigenvectors.

⊙ **11.18 (Exercise:). Visualization of C and B**
Show that the $\mathbf{C}$ and $\mathbf{B}$ have the same eigenvalues. How are their eigenvalues and eigenvectors related to that of $\mathbf{U}$ and $\mathbf{V}$ respectively.

⊙ **11.19 (Exercise:). Interpretation of stretch measures**
Either by your own efforts or by consulting a Continuum mechanics textbook, verify that the geometrical interpretation given above is correct. (Hint: We would urge you to try it out yourself: start with the equation of a circle $\mathbf{X} \cdot \mathbf{X} = 1$ now use the fact that $\mathbf{X} = \mathbf{F}^{-1}\mathbf{x}$ for a homogeneous deformation and show that you will get an equation for an ellipse.)

In fact, for sheet metal forming operations (as well as in a number of imaging applications), this is how strains were measured; you draw "small" circles on the sheet and then deform it, then measure the shape of the approximate ellipse that it forms, or we draw a square grid of lines and then measure the change in the lengths of the sides and the angles between the lines after deformation[15].

⊙ **11.20 (DEFINITION).** *Distortion and reorientation:* A configuration k_2 is said to be "distorted" from its original configuration k_1 if either the volume or shape is different. In other words, the stretch tensor $\mathbf{U}$ between k_1 and k_2 is NOT the identity tensor. A configuration k_2 is said to be "reoriented" from its original configuration k_1 if the rotation tensor $\mathbf{R}$ between k_1 and k_2 is NOT the identity tensor. Thus, in a general motion the body is both continually distorted and reoriented.

---

[15]In modern times, with more sophisticated digital imaging technology, using stereoscopic imaging and following individual material points appears to be the way to go. You might have heard of, or seen such things since they are used for animated sequences in movies (think Terminator 3 and others) that are realistic. If you are interested look up "bulge testing of sheet metals" on the web. See also [Joshi et al. (1992)].

For a simple spring, distortion means that the spring has been stretched while reorientation means that the axis of the spring has been rotated. We know that reorientation of the spring has no effect on its constitutive behavior. In the same way, we expect that for a continuum, it is the distortion (change in size and shape of the continuum that matters and not its reorientation)

Hence, to properly interpret theories, we will need a rotating frame that rotates with the body, i.e. we need a frame that "rotates with the body but still retains three orthonormal vectors". There are many possible ways that one can introduce this rotating frame of reference. Here we will introduce one which is related to a suggestion by Green and Mccinnis [Green and McInnis (1965)] [16].

In order to define this frame, we will use the polar decomposition theorem which says that $\mathbf{F} = \mathbf{R}\mathbf{U} = \mathbf{V}\mathbf{R}$ and use the rotation tensor $\mathbf{R}$ to create a frame that "rotates with the body" in the following way:

⊙ **11.21 (DEFINITION).** *Corotational frame: Associated with each material point $\mathbf{X}$ of the body, we define a corotational frame $\mathbb{F}_c$ through*

$$\mathbb{F}_c = \{\mathbf{x}, \mathbf{i}'_i, (i=1,2,3)\}, \mathbf{i}'_i = \mathbf{R}\mathbf{i}_i, \quad (11.28)$$

where $\mathbf{x}$ is the current location of $\mathbf{X}$, i.e. $\mathbf{x} = \mathbf{X} + \mathbf{u}(\mathbf{X}, t)$ and $\mathbf{i}'_i$ are the rotated versions of the laboratory axes $\mathbf{i}_i$ and where $\mathbf{R}$ is the rotation tensor associated with the polar decomposition theorem.

Thus, at each point in the current configuration of the body, we have a rotating frame whose origin is at $\mathbf{x}$ and whose basis vectors are such that their components (with respect to $\mathbb{F}_r$ are simply the column vectors of $\mathbf{R}$).

⊙ **11.22 (Exercise:).** **Corotational Frame Example**
Consider the simple shear deformation given by

$$\mathbf{x} = (\mathbf{I} + k\mathbf{i}_1 \otimes \mathbf{i}_2)\mathbf{X} \quad (11.29)$$

(a) For $k=1$ show that $\mathbf{R}, \mathbf{U}$ are respectively

$$\mathbf{R} = \begin{pmatrix} 0.8944 & 0.4472 & 0 \\ -0.4472 & 0.8944 & 0 \\ 0 & 0 & 1 \end{pmatrix} \quad \mathbf{U} = \begin{pmatrix} 0.8944 & 0.4472 & 0 \\ 0.4472 & 1.3416 & 0 \\ 0 & 0 & 1 \end{pmatrix}. \quad (11.30)$$

(Hint: use the MATLAB script that was provided earlier.)
(b) Consider a unit square with one vertex at the origin and two sides along the coordinate directions $\mathbf{i}_1$ and $\mathbf{i}_2$. Draw the final shape of the square. (Hint: you should get a parallelogram one of whose sides is on the x-axis.)
(c) Draw the directions of the $\mathbf{i}'_i$ axes: (Hint: the components of the unit vectors are the columns of $\mathbf{R}$.)

---

[16] Commercial programs like ABAQUS report all results in terms of this frame and NOT in terms of the Laboratory frame $\mathbb{F}_r$.

The deformation described in the exercise above is shown in Fig. 11.4 p. 369.

The corotational frame defined above is a special frame with respect to which the material locally undergoes only a pure distortion. No other type of frame has this special property. In commercial programs like ABAQUS, all the stress components are listed in terms of the $\mathbf{i'}_1, \mathbf{i'}_2, \mathbf{i'}_3$ axis such as the one shown in Fig. 11.4.

⊙ **11.23 (Exercise:). Components of U and V**
Show that the components of $\mathbf{U}$ with respect to the laboratory frame $\mathbb{F}_r$ are the same as those of $\mathbf{V}$ with respect to the corotating frame $\mathbb{F}_c$.
(Hint: use the fact that $\mathbf{V} = \mathbf{R}\mathbf{U}\mathbf{R}^t$.)
In other words, $\mathbf{V}$ is the rotated version of $\mathbf{U}$.

Note that the base vectors of the corotational frame are not fixed with respect to the Laboratory frame $\mathbb{F}_r$ but rotate with it. In fact, the rate of rotation of the corotational frame is measured by the tensor Ω which is defined by

$$\Omega := \dot{\mathbf{R}}\mathbf{R}^t = -\Omega^t \;\Rightarrow\; \frac{d\mathbf{i'}_i}{dt} = \Omega\mathbf{i'}_i. \tag{11.31}$$

In the above equation Ω is an "angular velocity tensor" whose three independent components are related to the angular velocity components of the rotating frame.

A counterpart to the polar decomposition is the decomposition of the velocity gradient $\mathbf{L}$. In keeping with the notion that velocity gradients should be added to represent simultaneous action, we will decompose $\mathbf{L}$ into a symmetric and antisymmetric part, i.e. we introduce the idea of stretching and spin tensors through

---

STRETCHING AND SPIN TENSORS

The velocity gradient is composed of a part that represents the "stretching" which acts simultaneously with "rotating" or spinning of the eigenvectors of $\mathbf{D}$. Thus, $\mathbf{D}$ is called the "stretching" tensor and $\mathbf{W}$ is called the spin or vorticity tensor, i.e.

$$\mathbf{L} = \mathbf{D} + \mathbf{W}, \quad \mathbf{D} = 1/2(\mathbf{L} + \mathbf{L}^t), \quad \mathbf{W} = 1/2(\mathbf{L} - \mathbf{L}^t). \tag{11.32}$$

⊙ **11.24 (Exercise:). Vorticity**
Either by your own efforts or by consulting a continuum mechanics textbook, show that $\mathbf{W}$ is the counterpart of the vorticity associated with fluid mechanics dynamics. Hint: Look up a fluid mechanics book.

Note that the Vorticity **W** is NOT directly related to Ω, so the vorticity tensor is NOT a measure of the rate of rotation.

⊙ **11.25 (Exercise:). Rates of the stretch tensors**
Show that

$$\dot{\mathbf{C}} = 2\mathbf{F}^t\mathbf{D}\mathbf{F}, \quad \dot{\mathbf{B}} = \mathbf{L}\mathbf{B} + \mathbf{B}\mathbf{L}^t. \tag{11.33}$$

(Hint: write **C** *and* **B** *in terms of* **F**, *and use* $\dot{\mathbf{F}} = \mathbf{L}\mathbf{F}$.*)*

The above is an important exercise, and we will be using (11.33) frequently.

11.2 Strain measures

Unlike what you have been used to in your undergraduate texts as well as in your linear elasticity class, in finite elasticity, there is no single definition of strain! The pure stretch tensors can be used to define a variety of different strain measures:

(1) The linear strain: $\boldsymbol{\varepsilon} = \mathbf{U} - \mathbf{I}$.
(2) The Lagrangian or referential strain $\mathbf{E} = (\mathbf{U}^2 - \mathbf{I})/2 = (\mathbf{F}^t\mathbf{F} - \mathbf{I})/2 = (\mathbf{C} - \mathbf{I})/2$.
(3) The logarithmic strain: $\mathbf{e} = \ln(\mathbf{V}) = 1/2 \ln \mathbf{B}$.
(4) The Eulerian or current strain: $\boldsymbol{\epsilon} = 1/2(\mathbf{I} - \mathbf{V}^{-2}) = 1/2(\mathbf{I} - \mathbf{B}^{-1})$.

In software packages like ABAQUS etc. when they implement finite deformation, they use logarithmic strain **e**, since it appears to be the most direct counterpart to linear strain measures. We will use this strain measure heavily as we develop constitutive equations for elastic and elastoplastic materials, due to convenience and to be in line with what is being done in most FEM packages.

There is nothing very special about any strain measure. We can define any function of **U** or **V** (within reason) that we want for a strain measure. All the above strain measures become equal to the linear strain measure for small deformations. Generally speaking, the Lagrangian strain, and the Eulerian strain are the ones that are easiest to use since they don't require the use of the eigenvalues and eigenvectors of **U** or **V** for their computation. On the other hand, the logarithmic strain, while being in wide use, is difficult to compute since it requires the eigenvalues of **U** or **V** for computation. Many software programs such as ABAQUS and MARC routinely provide these tensors as part of their customizable subroutines. Thus, if you are planning to implement your own custom constitutive equation in these packages, you need not have to worry about the computational details of these strain measures but can focus on the algorithmic aspects *but you absolutely NEED to know what strain measure they are actually using.*

Coaxial tensors and matrix functions

Having defined the essential kinematical quantities, we will now turn attention to functions of these quantities. Here, one of the important notions is that of coaxiality.

COAXIALITY

Two symmetric tensors are said to be "coaxial" if their eigenvectors are the same (even though their eigenvalues are not). Thus, $\mathbf{C}$ and $\mathbf{U}$ are coaxial and $\mathbf{B}$ and $\mathbf{V}$ are coaxial. Furthermore any two symmetric tensors $\mathbf{A}$ and $\mathbf{B}$ commute, i.e. $\mathbf{AB} = \mathbf{BA}$ if and only if they are coaxial. We will use this notion HEAVILY in the next chapter on finite plasticity.

Loosely speaking, coaxial tensors can be treated somewhat similar to ordinary numbers. We will see later that coaxiality makes calculations quite a bit simpler since we can "cancel out terms" easily.

Matrix functions: Given a function such as $\cos(x)$ or $\tan(x)$, is it possible to extend these for tensors? I.e. can we take the cosine of a tensor? What does it mean? Clearly we can take the square or inverse and other such operations for tensors. Why can't we take log of a tensor? The short answer is that "if a tensor is a real symmetric tensor, then many of the standard scalar operations can be easily extended to them"[17] such extensions will be referred to as matrix functions. Thus the extension of $\cos(x)$ will be called $\cos_m(\mathbf{A})$ etc.

⊙ **11.26 (DEFINITION).** ***Functions of tensors:*** *Given a function $f(x)$, we will define $f_m(\mathbf{A})$ to be*

$$f_m(\mathbf{A}) = \sum_i^3 f(\lambda_i) \mathbf{a}_i \otimes \mathbf{a}_i \tag{11.34}$$

where λ_i are the eigenvalues of $\mathbf{A}$ and $\mathbf{a}_i$ are the three orthonormal eigenvectors of $\mathbf{A}$. This notation is consistent with the usage in MATLAB. Thus the extension of $\ln(x)$ according to the formula (11.34) will be called $\ln_m(x)$.

Many of the common operations that we do on tensors can be considered this way: for example, if $f(x) = x^n$ then $f_m(\mathbf{A})$ is $\mathbf{A}^n$. To carry out these operations: multiply $\mathbf{A}$ by itself n times or do it according to (11.34). For large values of n it is faster to use (11.34). Also, we can now extend it to other operations such as $n = 1/2$ for square roots and $n = -1$ for inverses etc.[18]

---

[17] Hold on, hold on before you put on your mathematicians hat and claim that the definition is much broader (you can do power series in general Banach spaces— see we can also use high brow math terms by looking them up in books:-)). Note that we said "easily" and we can always claim that extension to other types of operators is not "easy".

[18] The result of extending f to a symmetric second order tensors will be another second order tensor that will be coaxial with it. It is NOT the application of the function to the components.

Why is this result useful? It turns out that using the above extension, we can define a variety of strain measures that are coaxial either with $\mathbf{U}$ or $\mathbf{V}$ by means of the following procedure: Find a monotonically increasing function $f(x)$, such that (1) $f(1) = 0$, (2) $f'(1) > 0$ and (3) $f(x) < 0$ if $x < 1$ and $f(x) > 0$ if $x > 1$. Then define tensors

$$\mathbf{E} = \sum_{i=1}^{3} f(\lambda_i)\mathbf{u}_i \otimes \mathbf{u}_i, \quad \mathbf{e} = \sum_{i=1}^{3} f(\lambda_i)\mathbf{v}_i \otimes \mathbf{v}_i, \quad (11.35)$$

then $\mathbf{E}$ will be called a referential strain and $\mathbf{e}$ will be called a current strain. These strain measures are written as $f_m(\mathbf{U})$ or $f_m(\mathbf{V})$ depending on whether we are using $\mathbf{u}_i$ or $\mathbf{v}_i$ as the eigenvectors.

⊙ **11.27 (Exercise:). Common Strain Measures**
Show that (1) $f(x) = 1/2(x^2 - 1)$ defines a strain measure when applied to $\mathbf{U}$. How is it related to $1/2(\mathbf{C} - \mathbf{I})$? (2) Does $f(x) = ln(x)$ define a strain measure? (3) How about $f(x) = x - 1/x$? (4) Show that in each case, given $f(\mathbf{U})$ one can find $\mathbf{U}$, i.e. we can find the stretch given the strain.

⊙ **11.28 (Exercise:). Calculating functions of B**
One of the nice things about modern computational tools is that we can focus on the core idea while relieving us from doing much of the tedious algebra. Eg.: Consider the tensor

$$\mathbf{B} = \begin{pmatrix} 5 & 2 & 0 \\ 2 & 7 & 3 \\ 0 & 3 & 8 \end{pmatrix} \quad (11.36)$$

Using MATLAB, find $\cos(\mathbf{B})$ and $\ln(\mathbf{B})$. Note: This is NOT the same as just taking the cosine of components.

(Hint, Remember that one way to do this is to find the eigenvalues and eigenvectors of $\mathbf{B}$ using the command "[V,D]=eig(B)"; This will give the matrix of eigenvectors and eigenvalues of $\mathbf{B}$. Now find the cosine and log of the eigenvalues. Then convert back, i.e. "cos(B)=V*diag(cos(diag(D)))*V'". What does "diag(cos(diag(D)))" do?[19])
Ans:

$$\cos \mathbf{B} = \begin{pmatrix} 0.0444 & 0.6648 & -0.6934 \\ 0.6648 & -0.3309 & -0.0429 \\ -0.6934 & -0.0429 & 0.1170 \end{pmatrix} \quad (11.37)$$

Also don't confuse it with scalar functions of the components of $\mathbf{A}$. These will NOT have a subscript m.

[19] If you get used to using MATLAB efficiently, you will get quite far. We will have plenty of reasons to get familiar with MATLAB in this book.

In general, computations of functions of tensors is not a simple task and there are special techniques for computing them efficiently and if you are really planning to do repeated calculations, you should look up some of the work that has been done in this regard (see e.g., [Miehe (1993)]).

⊙ **11.29 (DEFINITION).** *Generalized stretch and strain:* In this book we will use the word "stretch" for any matrix function f_m of $\mathbf{U}$ or $\mathbf{V}$ which is such that $f_m(\mathbf{I}) = \mathbf{I}$. On the other hand, we will use the word "strain" for any matrix function f_m which is such that $f_m(\mathbf{I}) = 0$.

Thus, $\mathbf{C}$ and $\mathbf{B}$ are stretches and $\ln \mathbf{C}$ and $\ln \mathbf{B}$ are strains. Thus if we say "the stress depends upon the current stretch", we mean $\boldsymbol{\sigma} = \mathbf{f}(\mathbf{B})$ or a similar equation. Generally $\mathbf{C}$ and its matrix functions will be referred to as the referential variables, while $\mathbf{B}$ and its functions will be referred to as the current strain variables. We will not go any further into the kinematics of deformable bodies. We would urge the reader to learn continuum mechanics before going deeply into finite deformation plasticity since we will assume more than a passing knowledge of the subject in the subsequent sections.

One of the most interesting aspects of the corotational frame $\mathbb{F}_c$ is that the components of $\mathbf{B}$ with respect to $\mathbb{F}_c$ are the same as that of $\mathbf{C}$ with respect to the laboratory frame $\mathbb{F}_r$. The same is true for $\mathbf{U}$ and $\mathbf{V}$. By using this result, we can provide the following picture: imagine two observers one fixed to the laboratory and another rotating with the body. At time $t = 0$, let us assume that they synchronize watches and frames (i.e. at that instant, both their frames have the same origin and the same unit vectors). Subsequently they both measure certain components of the deformation relative to their respective frames and compare answers. It should be clear to you that the rotating observer sees the motion as being a pure stretch (since his rotation coincides with the rotation of the body itself) whereas the laboratory observer sees both stretch and rotation. If the rotating observer measures the components of $\mathbf{B}$, with respect to the rotating frame (to him or her, since the motion is a pure stretch, there will be no difference between $\mathbf{B}$ and $\mathbf{C}$) and compares answers with the fixed observer who is measuring the components of $\mathbf{C}$ their answers will coincide! The same is true for the components of $\mathbf{V}$ and $\mathbf{U}$. In other words, the following equalities are true

$$B'_{ij} = \mathbf{B}\mathbf{i}'_j \cdot \mathbf{i}'_j = C_{ij} = \mathbf{C}\mathbf{i}_i \cdot \mathbf{i}_j \tag{11.38}$$

⊙ **11.30 (Exercise:). Corotational components of B**
Prove the above result.

Now that we have completed a quick overview of kinematics, we will introduce the following definition so that it is simple to describe motions.

11.3 Dynamics and thermodynamics of motion

When we consider small deformations, the assumption is that the bodies do not deform much, i.e. there are essentially "almost rigid". So when we define quantities like stress and heat flux which are forces and energy transfer per unit area, there is no need to worry about how to measure the area since it does not change. On the other hand, for finite deformations, there is substantial change in area. This means that when we talk about quantities per unit area or volume, you have to specify "area or volume at which time instant". Thus, we have a number of different stress tensors and similarly for the heat flux since all of these variables are defined per unit area. This unfortunate state of events is due to the fact that there are a variety of definitions that are all suitable for one purpose or another.

In short, there are three main stress measures that you will find in continuum mechanics. Of these the simplest one is the true or "Cauchy" stress $\boldsymbol{\sigma}$ whose components are defined by the corresponding force components per unit current area. In this definition, the reference configuration plays no role. This is by far the most favored among the different stress measures. Its definition is very simple:

⊙ **11.31 (DEFINITION).** *Cauchy Stress: Consider an area element* $d\mathbf{a}$ *in the current configuration. The Cauchy stress tensor* $\boldsymbol{\sigma}$ *is the linear transformation (or function) which gives the net force vector* $d\mathbf{f}$ *acting on this area element, i.e.*

$$d\mathbf{f} = \boldsymbol{\sigma} d\mathbf{a}, \tag{11.39}$$

i.e. it is the force intensity per unit current area. The component σ_{ij} *of* $\boldsymbol{\sigma}$ *represents the component of the force* $d\mathbf{f}$ *in the direction of* $\mathbf{i}_i$ *per unit area acting on the plane whose normal is* $\mathbf{i}_j$. *Thus the diagonal components are the normal stresses and the off-diagonal components are the shear stresses.*

Apart from this stress, there is also the engineering stress (referred to as the first Piola-Kirchhoff stress or the "nominal stress") $\mathbf{P}$ which represents the force per unit reference area. It is defined in a manner that is similar to $\boldsymbol{\sigma}$, i.e.

⊙ **11.32 (DEFINITION).** *Nominal or Engineering Stress: Consider an area element* $d\mathbf{A}$ *in the reference configuration. The engineering or referential stress tensor* $\mathbf{P}$ *is the linear transformation (or function) which gives the net force vector* $d\mathbf{f}$ *acting on this area element, i.e.*

$$d\mathbf{f} = \mathbf{P} d\mathbf{A} \tag{11.40}$$

The component P_{iA} *of* $\mathbf{P}$ *represents the component of the force* $d\mathbf{f}$ *in the direction of* $\mathbf{i}_i$ *per unit area acting on the plane whose normal* in the reference configuration *is* $\mathbf{i}_A$.

From an experimental perspective, it is much easier to deal with the nominal stress. The reason for this is simple: You have to measure the area once and for all *before* you conduct the experiment. During the experiment, you just keep track of the current force, divide it by the previously measured area and voila! You got the nominal stress! On the other hand, for the true stress, you have to measure or calculate the current cross sectional area of the material: This is not simple at all, since, if the deformation is non-homogeneous, then the area will change from location to location.

If you are going to use some commercial software for modeling large deformations, please be aware if they need input data in terms of true stress or nominal stress.

⊙ **11.33 (Exercise:). Converting from nominal to true stress**
In many software programs such as (ABAQUS), plasticity data in one dimension needs to be given in terms of "true stress" and "true strain" (i.e. Cauchy stress and Logarithmic strain). On the other hand, data for elasticity needs to be in the form of nominal stress (Engineering stress) vs nominal strain (uniaxial displacement gradient H_{11}).
Now consider a simple tension test. Assume that it is isochoric (i.e. volume preserving). Show that, if the motion is an isochoric simple tension, the relationship between them is $e = \ln \lambda = \ln(1 + H_{11})$ and $\sigma_{11} = P_{11}(1 + H_{11})^2$.
Hint: what is the relationship between area change and the length change for an uniaxial extension which is isochoric?

Please note that the above conversion is reasonable ONLY if the deformation is volume preserving. This is a very good approximation for large deformation of most metals, and is seen almost as a "magic formula". But for dilatant materials (porous media), where volume change is fairly significant, please re-examine this formula[20].

At first sight, **P** seems to be quite as simple as **σ**. However, it has a number of peculiarities that become evident only when large rotations are involved. The problems arise from the words *"force per unit area in the reference configuration"*. To understand the subtlety involved, consider the two configurations of a small element shown in the reference and current configurations in Fig. 11.5: It should be clear to you that the force **F** is along the x-direction. According to the definition of the Cauchy stress, since it acts on the side cd whose normal is also only the x axis, we see that $\sigma_{11} = F/cd = df/ab$ all others being zero and it is evident that the material is under uniaxial tension.

But consider **P**: the force is still in the x direction, but the *reference direction of cd (which is the normal to CD) is in the y direction!* Thus according to our definition $F/CD = P_{12}$, all others being zero. Thus, we see that as far as the engineering stress is concerned, its *diagonal components are zero*! And, it is not at all evident

---
[20]There are no magic formulae; there are only hoodwinked customers. So don't go for this kind of magic other than for entertainment purposes.

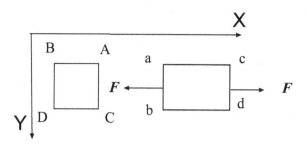

Fig. 11.5 A schematic figure of a body subject to large rotation. The body was originally at ABCD. It was rotated and stretched to the current position abcd. The current force is **f**. Note that for this particular load, the Cauchy stress only has a 11 component, since both the force and the area cd on which it acts are pointed along the x axis. However, the Nominal stress has only 12 components, since the original area CD used to be pointing in the y-direction whereas the force on it points in the x direction.

what kind of force is being applied unless the reference shape is also known. The problem arises because the side CD which was originally pointed in the y-direction got rotated around to the positive x-direction in the course of deformation. Thus, while the Cauchy stress gives us an accurate picture of the nature of the forces, it is not possible to find out the nature of the loading on a body by looking at the engineering stress. So, a word to the wise– " be careful about interpreting the engineering stress in a computer simulation where large rotations are involved such as in sheet metal bending."

⊙ **11.34 (Exercise:). Relationship between the Cauchy and Piola Stresses**
Show that

$$\boldsymbol{\sigma} = 1/J \P \mathbf{F}^t. \tag{11.41}$$

(Hint: Use the fact that $d\mathbf{f} = \boldsymbol{\sigma} d\mathbf{a} = \mathbf{P} d\mathbf{A}$ and then the relationship between $d\mathbf{a}$ and $d\mathbf{A}$).

Once we have the stress, the balance laws that govern the motion are

$$\text{Mass conservation: } \dot{\varrho} = \varrho \,\text{div} \mathbf{v}$$
$$\text{Momentum conservation: } \varrho \dot{\mathbf{v}} = \text{div} \boldsymbol{\sigma}, \; \boldsymbol{\sigma} = \boldsymbol{\sigma}^t \tag{11.42}$$
$$\text{Internal energy balance: } \varrho \dot{\exists} = -\text{div} \mathbf{q} + \boldsymbol{\sigma} \cdot \mathbf{D},$$

where ϱ is the mass per unit current volume, $\exists$ is the internal energy per unit current volume and **q** is the heat flux vector per unit area in the current configuration and div stands for the divergence of the tensor or vector[21]. The form listed above

---

[21] Please refer to a book on continuum mechanics (such as [Chadwick (1976); Spencer (1980); Segel (2007)] for a derivation of these balance laws.

are *identical* to the ones that we used for small deformation (see Chapter 9, equation (9.7) on p.271) except that here (1) all quantities are with respect to current coordinates **x** which change with motion, whereas no such distinction need to be made for small deformations; and (2) here a superposed dot means time derivative holding **X** fixed, which by chain rule corresponds to the following relation

$$(\dot{\cdot}) = \frac{d}{dt}(\cdot)|_{\mathbf{X} \text{ fixed}} = \frac{d}{dt}(\cdot)|_{\mathbf{x} \text{ fixed}} + \text{grad}(\cdot)\mathbf{v}. \tag{11.43}$$

11.3.1 *Thermodynamics or how to avoid creating perpetual motion machines*

A lot of our current ideas of thermodynamics came from our desire to get something for nothing: i.e. how to obtain a device that does all our tasks without "paying" anything. Exploration of this possibility eventually led to the development of two fundamental notions:

(1) *It is impossible for a system to operate in a cycle with no net input but a net output of work*; this is called a perpetual motion machine (pmm) of the first kind. i.e. "it is impossible to extract work in a cyclic process without supplying either work or heat". The principle that makes this inviolate is called the first law of thermodynamics or the law of conservation of energy.

(2) There is another more subtle kind of machine that is also impossible: *it is impossible to create a system that operates in a cycle whose only input is energy in the form of heating and whose only output is energy in the form of mechanical work*, i.e. "it is impossible to completely convert heat into work". The principle that ensures the satisfaction of this is the second law of thermodynamics.

Of course, if you build an actual machine, the laws of nature will guarantee that it will not be a pmm[22]. On the other hand, you can simulate any perpetual motion machine on a computer[23]. Of course, if you want your simulations to reflect reality, then you have to make sure that your simulations do not violate the impossibility of pmm either. The purpose of a thermodynamical framework is to develop a systematic way to make sure that your models incorporate the impossibility of perpetual motion machines of the first and second kinds.

We are now in a position to introduce the equation of state, Helmholtz potentials and the other paraphernalia of thermodynamics. If you are not familiar with these ideas, Chapter 2 has a very approachable introduction to these principles for lumped systems that you really should read and Chapter 9 does the same for continua undergoing small deformations. This section does not contain much in the way of

---

[22]You might end up with a machine that does nothing—it sits in your garage or lab like a mule refusing to do your bidding! But not moving means that it is not a perpetual motion machine and so will satisfy the laws!

[23]and it is done all the time in any video game : look at Mario in the Super Mario brothers happily leaping up to various levels without the need for any external agent.

explanation, but more in the way of the procedure that we will follow.

The main task from a mechanician point of view is to figure out how much work can you get back and how much is lost and the thermodynamical framework that we will present here allows us to carry out this task. The procedure for doing so is listed below

Step 1: Decide on the mechanical state variables

We will quickly take a look at the common structure of thermodynamics (keeping thermoelasticity in mind). We need to begin by introducing the state variables.

The starting point that we will adopt is the following: We have a finite list of "state variables" which are written as $\mathbb{S} = \{\mathbb{S}_m, T\}$ where $\mathbb{S}_m$ are called "mechanical state variables" and stand for "anything but the temperature". Generally it is not easy to decide which state variables to use and it requires a lot of hits and misses.

For the case of thermoelasticity, however the situation is simple. $\mathbb{S} = \{\mathbf{F}, T\}$, i.e. the mechanical state variable is just the deformation gradient tensor.

Step 2: Choose a suitable Helmholtz potential

In keeping with our philosophy of descriptive names, we will call the Helmholtz potential the isothermal work potential since it represents the potential of the body to do work in an isothermal process and is the function that is closest to the notion of a "strain energy function". Throughout this and subsequent chapters, we will use these two words interchangeably so as to reinforce one of the interpretations of the Helmholtz potential. For thermoelasticity, the isothermal work function ψ is a function of $\mathbf{F}$ and T[24].

Step 3: Find the form for the entropy function and the internal energy

This is a simple step and involves differentiating ψ with respect to the temperature. Thus we obtain the entropy function η through

$$-\frac{\partial \psi}{\partial T} = \eta, \ \psi - T\eta = \exists(\mathbb{S}_m, T). \tag{11.44}$$

These results are *identical* to that for small deformations (see Chapter 9 Fig. 9.3 p. 287) and can be taken as the defining characteristics of the Helmholtz equation of state for a body[25].

Step 4: Identify the mechanical dissipation in terms of the rate of change of the isothermal work function

---

[24] Note that elastoplastic materials have strain energy functions too: they are not just functions of $\mathbf{F}$ alone.

[25] This is an assumption based on very flimsy evidence—and is primarily an extension from thermoelasticity. The best we can say for it is that this is the best we can do based on current data and understanding. So don't look for deep physical insight beyond the statement that "if we freeze all other mechanical state variables other than $\mathbf{F}$ we should get thermoelasticity". How to "freeze" all these other variables is not a question that can be answered in general.

If we now substitute (11.44) into (11.42c) and simplify (the steps are the same as for small deformations), we will get the *heat equation for the body* in the form

$$\varrho T \dot{\eta} = -\varrho \frac{\partial \psi}{\partial \mathbb{S}_m} \cdot \dot{\mathbb{S}}_m + \boldsymbol{\sigma} \cdot \mathbf{D} - \text{div} \mathbf{q}. \tag{11.45}$$

As with small deformations (see Chapter 9 Eqn. 9.34 on p.288), we identify the rate of mechanical dissipation (irrecoverable loss of supplied mechanical power) as

$$\varrho \zeta := \boldsymbol{\sigma} \cdot \mathbf{D} - \varrho \frac{\partial \psi}{\partial \mathbb{S}_m} \cdot \dot{\mathbb{S}}_m = \boldsymbol{\sigma} \cdot \mathbf{D} - \varrho \dot{\psi}|_T. \tag{11.46}$$

The right-hand side of the above equation is called the *energy release rate* and is heavily used in the fracture mechanics literature. We will also make use of this notion. It reflects the difference between the mechanical power supply ($\boldsymbol{\sigma} \cdot \mathbf{D}$) and the rate of increase of recoverable work ($\varrho \dot{\psi}|_{T=const.}$) and reflects how much of the power supplied is "lost," i.e. cannot be recovered[26].

THE HEAT EQUATION

The equation (11.46) states that, the difference between the power supplied ($\boldsymbol{\sigma} \cdot \mathbf{D}$) and that which is recoverable ($\varrho \dot{\psi}|_T$) is used to heat (i.e. raise the entropy of) of the solid.

⊙ **11.35 (Exercise:). Dissipation equation for thermoelastic materials**

Show, by using (11.5), (11.32) and the fact that $\boldsymbol{\sigma}$ is symmetric, that, for thermoelastic materials where $\psi = \psi(\mathbf{F}, T)$, we obtain

$$\varrho \zeta = (\boldsymbol{\sigma} - \varrho \frac{\partial \psi}{\partial \mathbf{F}} \mathbf{F}^t) \cdot \mathbf{L}. \tag{11.47}$$

Step 5: Find constitutive equations for the stress and the other mechanical state variables that have been introduced in such a way that the mechanical dissipation ζ is positive

There is no single way to carry out this last step, that can be described without causing a great deal of confusion. Each type of model requires a slightly different procedure depending upon the variables for which constitutive functions are sought. So rather than provide a general recipe for doing so, we will look at some special cases and show how this is done.[27]

---

[26] In terms of the current "home mortgage crisis" sweeping the US, this is the equivalent of a "bad loan". You are not going to get all your money back. So if work is money, you can see the irreversibility in the situation.

[27] A word of caution: In the approach presented here, we are always choosing constitutive equations that are *sufficient* to guarantee that the mechanical dissipation is non-negative. We will make no

11.3.2 Constitutive equations for thermoelasticity

We will demonstrate the application of these steps to a classical thermoelastic solid:
(1) Step 1: The state variables are $\mathbf{F}$ and T.
(2) Step 2: We need to choose some form for ψ. Let us say that we choose a Helmholtz potential that depends only upon the "magnitude" of the deformation (somewhat like a spring), i.e. it is of the form $\psi(\|\mathbf{F}\|, T)$, where $\|F\| := \sqrt{\mathbf{F} \cdot \mathbf{F}}$. We will discuss a few specific forms based on experimental evidence for ψ later.
(3) Step 3: This is the step that reveals quite a lot about the nature of the material. The entropy is $\eta = -\partial \psi / \partial T$ and the internal energy is $\exists = \psi + T\eta$. We can easily compute them for any form for ψ. Note that both the entropy and the internal energy are functions of $\mathbf{F}$. It may happen that the entropy is a function of temperature alone. This is a common idealization for metals and occurs when $\psi = \psi_1(\mathbf{F}) + \psi_2(T)$. Thus isothermal changes in the shape of the body does not change the entropy at all. This type of model is called "energetic" elasticity.
(4) Step 4: For this particular problem, the rate of dissipation has already been calculated as (11.47).
(5) Step 5: In this case, the mechanical state variable is $\mathbf{F}$ and its rate is $\dot{\mathbf{F}}$; hence its conjugate force is $\varrho \partial \psi / \partial \mathbf{F}$ since their product will give us the rate of change of the Helmholtz potential when $\mathbf{F}$ changes. On the other hand, it is more useful to consider $\mathbf{L} = \dot{\mathbf{F}} \mathbf{F}^{-1}$ as a measure of the rate of change of $\mathbf{F}$. Then corresponding driving force is $\varrho \partial \psi / \partial \mathbf{F} \mathbf{F}^t$, as can be seen from the form of (11.47).
(6) Step 6: We expect that a thermoelastic body should have no mechanical dissipation, i.e. in any isothermal process, any work done is completely recoverable. Thus, we want a constitutive equation for $\boldsymbol{\sigma}$ that ensures that the right-hand side of (11.47) vanishes. Let us be guided by linear elasticity in our choice: Recall that for small deformation thermoelasticity we can state the constitutive equation for the stress in the following form:

For linear elasticity, the isothermal work function depends upon the linear strain ε and the temperature; the stress is equal to the derivative of the isothermal work function with respect to the linear strain,

$$\boldsymbol{\sigma} = \partial \psi / \partial \boldsymbol{\varepsilon} \Rightarrow \boldsymbol{\sigma} \cdot \dot{\boldsymbol{\varepsilon}} - \dot{\psi}_{T=const.} = 0. \qquad (11.48)$$

For a thermoelastic material undergoing large deformations, we choose a constitutive equation for $\boldsymbol{\sigma}$ to be such that for all processes that are consistent with the constraints on the body, there should be no mechanical dissipation. A

---

attempt to show that this is the most general possibility. This approach is somewhat different than conventional approaches in continuum mechanics (see [Coleman and Noll (1963)]) that try to prove necessary and sufficient conditions invoking a variety of special assumptions. We will trust in our (and your) common sense in making constitutive choices.

straightforward choice for $\boldsymbol{\sigma}$ related to the derivative of the Helmholtz potentials, and which will satisfy this criterion is[28]

$$\boldsymbol{\sigma} = \varrho(\partial \psi/\partial \mathbf{F})\mathbf{F}^t. \qquad (11.49)$$

Notice that this choice guarantees that (11.47) will be automatically satisfied *provided* $\varrho \partial \psi/\partial \mathbf{F} \mathbf{F}^t$ *is a symmetric tensor*. It can be shown (the proof is not easy) that the symmetry requirement on the derivatives of ψ implies that

$$\varrho \frac{\partial \psi}{\partial \mathbf{F}} \mathbf{F}^t = \left[\frac{\partial \psi}{\partial \mathbf{F}} \mathbf{F}^t\right]^t \Rightarrow \psi = \hat{\psi}(\mathbf{F}^t\mathbf{F}, T) = \hat{\psi}(\mathbf{C}, T) = \hat{\psi}(\mathbf{R}^t\mathbf{B}\mathbf{R}, T). \qquad (11.50)$$

ψ DEPENDS ON **F** ONLY THROUGH **C**

For a thermoelastic body where the stress and isothermal work function depend only upon **F** and T, the symmetry of the stress (or equivalently, the balance of angular momentum) implies that the isothermal work function can depend upon **F** only through the combination $\mathbf{F}^t\mathbf{F} = \mathbf{C}$.

Usually, this result is established by invoking the requirement of objectivity. But for the class of materials that we are investigating, just the symmetry of the stress is enough. Another noteworthy feature of the development here is that we have always insisted on choosing constitutive assumptions that are sufficient to satisfy the second law. In other words, we will just choose whatever constitutive equations we want, based on physical reasoning and check to ensure that the resulting equations are consistent with the fundamental laws, and we will not make any claims that these are the only constitutive equations that will meet our criteria.

What do we mean when we say "the isothermal work function depends upon the tensor **C**"? Recall that **C** can be written in terms of its eigenvalues λ_i^2 which quantify "how much strain", i.e. its magnitudes, and its eigenvectors $\mathbf{E}_i$ which quantify the "orientation" or "direction" of the strain. Furthermore, the eigenvalues of **C**, **B**, **U** and **V** are all related to one another, the relationship being invertible: So as far as the magnitude of the strain goes, the eigenvalues of ANY of these tensors can be used (indeed the eigenvalues of any of the generalized strain measures $f(\mathbf{C})$ or $f(\mathbf{B})$ can be used). Thus, we can say that,

---

For the class of thermoelastic materials considered here, the isothermal work function depends upon the eigenvalues λ_i of the stretch tensor **U**, the orientation of its eigenvectors $\mathbf{u}_i$ (represented by the direction cosines of $\mathbf{u}_i$) relative to the laboratory frame $\mathbb{F}_r$ and the temperature.

---

[28]This is not the only possible choice, however, we can add any tensor **A** that is *perpendicular* to **D**, i.e. a tensor which satisfies $\mathbf{A} \bullet \mathbf{D} = 0$ without affecting the mechanical dissipation. Whether such materials can be called *elastic* is a philosophical issue (see [Rajagopal and Srinivasa (2009)]) in this regard.

⊙ **11.36 (Exercise:). Entropic Elasticity**
Show that if $\psi = T\psi_1(\mathbf{F}) + \psi_2(T)$ then the internal energy is a function of T alone!. In other words, for this model changes in the shape of the body will have no effect on the internal energy at all!

This is called entropic elasticity and is an idealized model for polymeric solids.

⊙ **11.37 (Exercise:). Stress response for an elastic solid**
Show that the stress of an elastic material is given by

$$\boldsymbol{\sigma} = 2\varrho \mathbf{F}\frac{\partial \psi}{\partial \mathbf{C}}\mathbf{F}^t = 2\varrho \mathbf{R}\mathbf{U}\frac{\partial \psi}{\partial \mathbf{C}}\mathbf{U}\mathbf{R}^t. \qquad (11.51)$$

(Hint: Use the chain rule to show that $\partial \psi / \partial \mathbf{F} = 2\mathbf{F}\partial\psi/\partial\mathbf{C}$.)

⊙ **11.38 (Exercise:). Example of a hyperelastic material**
If $\psi = \psi(\|\mathbf{F}\|, T)$ show that (11.47) reduces to

$$\left(\frac{\varrho}{e}\frac{\partial \psi}{\partial e}\mathbf{B} - \mathbf{T}\right) \cdot \mathbf{L} = \zeta, \; e = \|\mathbf{F}\| \qquad (11.52)$$

Hint: use the chain rule and the fact that $\mathbf{F}\mathbf{F}^t = \mathbf{B}$.

Observe that (11.51) implies that $\boldsymbol{\sigma}$ is a function of both $\mathbf{U}$ and $\mathbf{R}$. On the other hand, the following exercise shows again the utility of the rotating frame $\mathbb{F}_c$ introduced earlier:

⊙ **11.39 (Exercise:). Corotating Frame**
Show that the components of $\boldsymbol{\sigma}$ with respect to the corotating frame $\mathbb{F}_c$ are functions of only $\mathbf{U}$: In other words, show that

$$\sigma'_{ij} = 2\mathbf{U}_{ik}(\partial \psi / \partial \mathbf{C}_{kl})\mathbf{U}_{lj} \qquad (11.53)$$

Note that this is written with the LHS in terms of $\mathbb{F}_c$ and the RHS in terms of $\mathbb{F}_r$. Using the fact that the components of $\mathbf{B}$ with respect of $\mathbb{F}_c$ are the same as that of $\mathbf{C}$ with respect to $\mathbb{F}_r$ (i.e. $B'_{ij} = C_{ij}$), we can rewrite this as

$$\sigma'_{ij} = 2\mathbf{V}'_{ik}(\partial \psi / \partial \mathbf{B}'_{kl})\mathbf{V}'_{lj} \qquad (11.54)$$

In this version, everything is written in component form with respect to the corotating frame $\mathbb{F}_c$. (Hint: use the fact that $\sigma'_{ij} = \mathbf{i}'_i \cdot \boldsymbol{\sigma} \mathbf{i}'_j$ and then use the relations (11.51) and the definition (11.28).)

It is for this reason that in implementing your constitutive equations for hyperelastic materials, many commercial programs implement them in terms of σ'_{ij}.

11.3.3 Isotropic elastic materials: Large strain

Intuitively speaking, an isotropic elastic material is one whose response function is not dependent upon the directions $\mathbf{u}_i$ of the stretch tensor but only upon its magnitudes, i.e. the eigenvalues λ_i. Thus, two stretch tenors with the same "amount of stretch" (i.e. the same eigenvalues) in different directions (i.e. with different values for the eigenvectors $\mathbf{u}_i$) will give rise to the same value of ψ. Another way of saying this is that if we are going to perform a specific test, the *initial orientation of the body relative to the Laboratory frame, i.e. testing directions, has no effect on the response.*

Thus, for an isotropic elastic material,

(1) The isothermal work function ψ depends only upon the eigenvalues of $\mathbf{U}$, (or $\mathbf{V}$), i.e.

$$\psi = \psi(\lambda_1, \lambda_2, \lambda_3, T) = \psi(\lambda_2, \lambda_1, \lambda_3, T) = \psi(\lambda_1, \lambda_3, \lambda_2, T) = \cdots \quad (11.55)$$

The "cyclic conditions" listed above is due to the fact that the designations λ_1, λ_2 etc. are arbitrary so that a valid constitutive equation should be one which is invariant (does not change) under permutation of the indices.

For this reason, the constitutive equations are usually described in terms of variables that automatically satisfy the permutation criteria. For example, we can write ψ as a function of $I_1 = \lambda_1^2 + \lambda_2^2 + \lambda_3^2$, $I_2 = \lambda_1^2 \lambda_2^2 + \lambda_1^2 \lambda_3^2 + \lambda_2^2 \lambda_3^3$ and $I_3 = \lambda_1^2 \lambda_2^2 \lambda_3^2$. These are called the principal invariants. However, as shown by Criscione et al., [Criscione et al. (2000)] these make a poor choice of variables due to the fact that they are not experimentally suitable.

A different choice[29] for the invariants would be functions of $\mu_i = \ln \lambda_i$. For example, we can use the invariants $K_1 = \mu_1 + \mu_2 + \mu_3 = \ln I_3$, $K_2 = \sqrt{(\mu_1 - \mu_2)^2 + (\mu_2 - \mu_3)^2 + (\mu_3 - \mu_1)^2}$ and $K_3 = (\mu_1 - \mu_2)(\mu_2 - \mu_3)(\mu_1 - \mu_3)$. But no matter which of these invariants are used, for computational purposes especially with MATLAB, we suggest that you eventually convert to the eigenvalue notation since it is simplest to carry out calculations in terms of them. In other words, it appears that it might just be better to bite the bullet and find the eigenvalues and eigenvectors of $\mathbf{B}$ or $\mathbf{T}$ and do all calculations with

---

[29] See [Criscione et al. (2000)] for a very detailed discussion of invariants and their suitability for experimental purposes.

them. We will follow this procedure when dealing with isotropic elastoplastic materials in this book.

(2) The eigenvectors of $\boldsymbol{\sigma}$ and $\mathbf{V}$ (or $\mathbf{B}$ or $\mathbf{e}$) are the same so that the stress is given by

$$\boldsymbol{\sigma} = \sum_{i=1}^{3} \sigma_i \mathbf{e}_i \otimes \mathbf{e}_i \tag{11.56}$$

where $\mathbf{e}_i$ are the eigenvectors of $\mathbf{V}$ (or $\mathbf{B}$). In other words, *the stress and the current stretch are coaxial.*

(3) The stress is related to the derivative of ψ with respect to $\mathbf{B}$ and the eigenvalues of $\boldsymbol{\sigma}$ depend upon $\mathbf{B}$ only through the eigenvalues of $\mathbf{B}$, i.e.

$$\boldsymbol{\sigma} = 2\varrho \frac{\partial \psi}{\partial \mathbf{B}} \mathbf{B} = 2\varrho \mathbf{B} \frac{\partial \psi}{\partial \mathbf{B}}; \quad \sigma_i = 2\varrho \frac{\partial \psi}{\partial B_i} B_i \tag{11.57}$$

where σ_i and $B_i (= \lambda_i^2)$ are the eigenvalues of $\boldsymbol{\sigma}$ and $\mathbf{B}$ respectively.

(4) For an isotropic material, the stress can also be written in terms of the logarithmic strain $\mathbf{e} = 1/2 \ln \mathbf{B}$ in the form

$$\psi = \psi(\mathbf{e}) \Rightarrow \boldsymbol{\sigma} = \varrho \frac{\partial \psi}{\partial \mathbf{e}} \Rightarrow \sigma_i = 2 \frac{\partial \psi}{\partial \ln B_i} \tag{11.58}$$

(5) Another form for the stress that is computationally convenient is

$$\sigma_i = \varrho \lambda_i \frac{\partial \psi(\lambda_i)}{\partial \lambda_i}, \Rightarrow \boldsymbol{\sigma} = \varrho \mathbf{V} \frac{\partial \psi}{\partial \mathbf{V}} = \varrho \frac{\partial \psi}{\partial \mathbf{V}} \mathbf{V}. \tag{11.59}$$

Thus, in words, we can say that

STRESS RESPONSE IN ISOTROPIC THERMOELASTICITY

For an isotropic thermoelastic material undergoing finite deformation, the stress is equal to ϱ times the derivative of the isothermal work function with respect to the current logarithmic strain $\mathbf{e} := 1/2 \ln \mathbf{B}$. It also happens to be equal to $\varrho \mathbf{V}$ times the derivative of the isothermal work function with respect to the current stretch $\mathbf{V}$.

Comparing this statement with that for linear elasticity, we see that a direct counterpart of the stress in linear elasticity is the Cauchy stress, and a direct counterpart of the linear strain $\boldsymbol{\varepsilon}$ is $\ln \mathbf{V}$. In spite of the great similarity between the linearized strain tensor $\boldsymbol{\varepsilon}$ and $\ln \mathbf{V}$, it is very hard to compute quantities related to $\ln \mathbf{V}$ and so in many cases, it is much better to work directly with $\mathbf{B}$ (unless you are directly working with the eigenvalues and eigenvectors), knowing that any function of $\ln \mathbf{V}$ can be written in terms of $\mathbf{B}$ and vice versa, (although the results are not quite elegant in many cases). For this reason, we will write everything in terms of $\mathbf{B}$ introducing $\ln \mathbf{V}$ only where essential.

⊙ 11.40 (Exercise:). Helmholtz potential for isotropic materials
Consider a Helmholtz potential of the form

$$\psi = \frac{k}{4}(I_3 - \ln I_3 - 1) + \frac{\mu}{2} I_3^{-1/3} I_1, \qquad (11.60)$$

where $I_3 = \det(\mathbf{B})$ and $I_1 = tr(\mathbf{B})$ are the first and third invariants of $\mathbf{B}$. Show that the stress is given by

$$\boldsymbol{\sigma} = \varrho(k/2(I_3 - 1)\mathbf{I} + I_3^{-1/3}\mu(\mathbf{B} - 1/3 tr(\mathbf{B})\mathbf{I})). \qquad (11.61)$$

Show also that $\boldsymbol{\sigma}$ is coaxial with $\mathbf{B}$, i.e. they have the same eigenvectors. Hint: Use (11.57a) and the fact that $\partial I_3/\partial \mathbf{B} = I_3 \mathbf{B}^{-1}$ and $\partial I_1/\partial \mathbf{B} = \mathbf{I}$.

The above constitutive equation is simple and, when $\mathbf{B}$ is very close to $\mathbf{I}$, i.e. for small strains, it will reduce to the equations for linear elasticity.

⊙ 11.41 (Exercise:). A model using $\ln \mathbf{V}$
Consider a material whose response is given by

$$\boldsymbol{\sigma} = \frac{1}{J}(C_1 tr(\ln \mathbf{V})\mathbf{I} + C_2 \ln \mathbf{V}), \; J = \det \mathbf{F} \qquad (11.62)$$

This is called a Hencky material. Is it hyperelastic? What is the isothermal work function associated with it? (Hint: Try a form for ψ that is quadratic in $\ln \mathbf{V}$ and then the fact that $\boldsymbol{\sigma} = \varrho \partial \psi / \partial \ln \mathbf{V}$. Is the $1/J$ necessary?)

Unfortunately, the above "nice" structure for isotropic elastoplastic materials, where the Cauchy stress is a function of a strain like variable, unfortunately does NOT carry over directly for anisotropic elastic materials. This gives rise to a number of difficulties (which have not been satisfactorily resolved in a manner that is easy to understand) when extending to finite deformation plasticity.

⊙ 11.42 (Exercise:). Small strains
Show that if $\mathbf{V} - \mathbf{I}$ is small (i.e. all the entries in its matrix are much less than one, or equivalently, all the eigenvalues of $\mathbf{V}$ are very close to one), then the constitutive equation for $\boldsymbol{\sigma}$ reduces to

$$\boldsymbol{\sigma} \approx \varrho_0(k tr(\mathbf{V} - \mathbf{I}) + \mu(\mathbf{V} - tr(\mathbf{V})\mathbf{I})) \qquad (11.63)$$

where ϱ_0 is the reference density of the material, k is the bulk modulus and μ is the shear modulus.

Hint: This is a hard problem: Assume that the eigenvalues of **V** are $1 + \epsilon_i$ and substitute, systematically keeping only the first powers of ϵ_i. You should refer to some advanced Elasticity text such as [Ogden (1984)]. It is well worth doing so since, for many finite deformation plasticity problems, the "elastic strains" (which will be introduced in the next chapter) will be small.

We have thus concluded a very brief description of the constitutive assumptions for a thermoelastic solid.

Note that we did not have to use any ideas of *Objectivity* or other notions that appear in the elasto-plasticity literature on a constant basis. This is one of the advantages of the thermodynamical approach. We will never have to really invoke all of this paraphernalia for our purposes[30].

But, since enquiring minds want to know, we have added a whole section on invariant and objective tensors.

11.4 Invariant and objective tensors

You can skip this section on first reading because we can build most of the constitutive theory we want without using this section. However, it is important for some discussions in the finite plasticity chapter (Chapter 12).

If you have looked at papers on metal forming or if you have read some of the books and manuals on large deformation software packages like MARC or ABAQUS, you would have seen a lot of talk about "objective rates" or "corotational rates" or "objective updates" etc. Now what is this and why is it important? We first note that the notion of objectivity and objective rates is *not only for continua but is also applicable to lumped parameter systems*. Then, why did we not learn it in our undergraduate classes? Well, we actually learnt it the hard way!

We will give you an example from dynamics that might illustrate the point: Consider a spring: we know that every book on dynamics states a formula of the from $f = -kx$ for the spring. This is an example of a constitutive equation. In your early days in dynamics, you always got this wrong because you were never sure what x meant and which way f should point—the formula is silent in this regard. For example, if the spring was vertical then perhaps the only coordinate was y and there was no x! If the spring was on a rotating platform,... forget it!

---

[30] For many researchers in the field, this is almost tantamount to blasphemy. "You have cleverly (or stupidly) guided the development by making choices that do not invoke this notion of objectivity. We agree and that is precisely our point—You don't need these concepts *for the current models that we have studied*. We are not saying (not yet anyway) that it is not needed for *any* model. Just that for the class of models that we are looking at, namely elasticity here and isotropic elastoplastic materials and later single crystals, we can get everything from the requirement that $\boldsymbol{\sigma}$ be symmetric rather than any notion of frame indifference or Objectivity.

Gradually you began to realize that x did not mean a specific global coordinate axes, rather in reality you realized that $f = -kx$ meant something along the lines of "The algebraic value of the force *along the length of the spring* is equal to $-k$ times the *change in the dimensions of the spring along its length.*"

Notice that the equation $f = -kx$ actually did not say anything about the global direction of the force or that of the displacement—rather the direction of the force and displacements is defined *in terms of the current orientation of the spring.* This is a key idea.

OBJECTIVITY FOR LUMPED SYSTEMS

The constitutive equation of a lumped system (such as a spring or a dashpot) depends only upon how the system is distorted (elongated or compressed) over time and not upon its absolute location, orientation, translation or rotation.

Now let us see another example: that of a dashpot—here we know that $f = -cv$. Let us write it out in words: "The algebraic value of the force *along the length of the dashpot* is equal to $-c$ times *the rate of increase in the dimensions of the dashpot along its length*".

Again, we see that our constitutive equations deal with forces and changes of geometry that are described *without any reference to a global coordinate system at all* but rather *intrinsically with respect to the current orientation of the body itself.* Clearly, taking a dashpot and just running with it in your hand will not generate a damping force; you need to deform the dashpot somehow.

These constitutive equations hide a tacit assumption: changing the orientation of the body (by rotating it, for example) does not affect its response. If we test the spring in different directions in space (i.e. go to outer space and test it "horizontally" or "vertically") then clearly we expect that its response will be the same and will be described in words as we stated before.

As you can see from the above examples, constitutive equations for springs and dashpots are actually written for components of forces and displacements *with respect to an axis that is translating and rotating with the spring or dashpot.*

To make this point even clearer: consider the problem of a spring that moves in a slot cut on a rotating disk (see Fig. 11.6).

The relationship between the force and displacement of the spring is easiest to explain by using an axis (xyz) that rotates with the disk: in terms of this axis, we can write $f_x = -k(x - x_0)$. We can then translate the result to global axes (XYZ) if needed.

This form of the constitutive equation, (written in terms of components associated with a frame that rotates with the body) is very compact and focuses attention only on the essential ingredients. However, it has to be converted into global com-

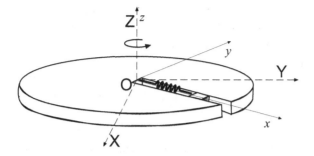

Fig. 11.6 A spring and mass inside a slot on a rotating disk. The global or laboratory frames $\mathbb{F}_r$ is shown as XYZ and the local corotational frame $\mathbb{F}_c$ is shown as xyz. Note that we assume that with respect to $\mathbb{F}_c$ $f_x = -k(x-x_0)$ as the constitutive equation for the spring.

ponents whenever we actually solve a problem according to need.

The situation with continua is quite a bit more complicated. We want our constitutive equations to be written in terms of quantities that are defined *intrinsically with respect to the current orientation of the body itself*. But what do we mean by "current orientation of a deformable body," in general? For the spring and dashpot, it was easy to see since there was only one direction. In a rigid body, we can use the principal axes of inertia to define the (xyz) axes.

How do we do this in a systematic way? This is the reason for introducing all this paraphernalia of "objective tensors". They provide a systematic way (if you understand what is going on) to implement the rather simple requirement of identifying quantities that are defined intrinsically with respect to the current orientation.

In order to develop the notions used to describe objectivity, we will use the corotational frame $\mathbb{F}_c$ defined in equation (11.28) on p.371 to obtain constitutive equations[31]. We will begin with a few definitions and then look at some examples:

⊙ **11.43 (DEFINITION). Equivalent motions:** *Two motions* $\mathbf{u}(\mathbf{X}, t)$ *and* $\mathbf{u}^*(\mathbf{X}, t^*)$ *of a body with the same material points* $\mathbf{X}$ *are said to be "equivalent" if for every material point* $\mathbf{X}$ *and every instant t, the referential squared stretch tensor* $\mathbf{C}$ *of the first motion and that of the second motion ($\mathbf{C}^*$) satisfy the relationship*

$$\mathbf{C}(\mathbf{X}, t) = \mathbf{C}^*(\mathbf{X}, t+a) \qquad (11.64)$$

where a is a constant. In other words the two motions have the same stretch history at every material point.

Roughly speaking, the two motions are *identical* in terms of how the body is distorted. Only their respective orientations are different, i.e. if you calculated their deformation gradients, they would have the same stretch but different rotations.

---

[31] We are taking somewhat of an unorthodox view on the subject. This is based on the fact that students have a very tough time with this concept and we felt that it is better to start from a place where they were somewhat familiar, rigid body mechanics, and extend ideas from there. As we will see shortly, this approach also gives us a rather direct way to consider objective tensors.

You might ask, "why is there a constant a in the definition?" Well, when we talk about the same distortion history, we tacitly know that only time *intervals* matter, not the absolute value of the time. So it doesn't matter if one of the bodies started its motion at a later time; as long as they suffer equal amounts of distortion in equal intervals of time, they are equivalent.

Now why do we use the term "equivalent"? We expect that the constitutive equations depend only upon the "distortion" (remember the discussion of the spring and dashpot) and not upon the orientation. In order to separate it out, we need to be able to say that if two motions are equivalent, then their constitutive equations will be, in some sense, identical. It is not as simple as this but we will show that, appropriately interpreted, this will indeed be true.

⊙ **11.44 (Exercise:). Rotation and shear**
Consider the following two motions[32]

$$\mathbf{x} = (\mathbf{I} - \{(1 - \cos\Theta)\mathbf{i}_1 + \sin\Theta \mathbf{e}2\} \otimes \mathbf{e}3)\mathbf{X} \quad (11.65)$$

and

$$\begin{aligned} x &= X\cos\Theta - Y\sin\Theta + Z(1 - \cos\Theta), \\ y &= X\sin\Theta + Y\cos\Theta - Z\sin\Theta \\ z &= Z \end{aligned} \quad (11.66)$$

where $\Theta = \Omega t$. Show that they are equivalent. Find the rotation matrix that will take the motion (1) to motion (2). (Hint: show that they have the same value of $\mathbf{C}$. TO find the rotation matrix, calculate $\mathbf{F}_2 \mathbf{F}_1^{-1}$. In this particular case, it is a rotation about the Z axis.)

Now that we have the idea of equivalent motions, we can start looking at kinematical quantities that are the same in the two motions. These are the quantities that appear in the constitutive equations for the body. We begin our search with the following definition

⊙ **11.45 (DEFINITION).** *Invariant scalar:* A scalar s, is said to be an "invariant with respect to current orientation", if it has the same value for two motions that are "equivalent" as defined above, i.e. if it satisfies $s(t) = s^*(t+a)$. A vector or tensor is said to be an invariant quantity if its components with respect to the laboratory reference frame $\mathbb{F}_r$ are invariant scalars.

In other words, an invariant vector is of the form $\mathbf{v} = v_i \mathbf{i}_i$ where v_i is an invariant scalar and $\mathbf{i}'_i$ are the basis vectors in the laboratory reference frame $\mathbb{F}_r$. Similarly an invariant tensor is of the form $\mathbf{T} = T_{ij} \mathbf{i}_i \otimes \mathbf{i}_j$ where T_{ij} are invariant scalars. An important example of invariant vectors are the eigenvectors $\mathbf{u}_i$ of $\mathbf{U}$. Thus, invariant vectors and tensors are quantities whose components in the laboratory basis are invariant scalars. Invariant scalars play a crucial role in the development of constitutive equations, since they alone are unaffected by the rotations.

> ⊙ **11.46 (Exercise:). Examples of invariant quantities**
> *(a) Show that any matrix function of $\mathbf{C}$ is invariant as defined above.*
> *(b) Is the mass density an invariant scalar? Is this an assumption or can this be proved mathematically?*
> *(c) Is the kinetic energy an invariant scalar?*

⊙ **11.47 (DEFINITION).** *Objective quantities:* A vector or tensor is said to be **objective** *if its components with respect to the rotating frame $\mathbb{F}_c$ defined in (11.28) are invariant scalars.*

In other words, an objective vector is of the form $\mathbf{v} = v'_i \mathbf{i}'_i$ where v'_i is an invariant scalar. Similarly an objective tensor is of the form $\mathbf{T} = T'_{ij} \mathbf{i}'_i \otimes \mathbf{i}'_j$ where T'_{ij} are invariant scalars.

A very important objective quantity is $\mathbf{B} = \mathbf{F}\mathbf{F}^t$. Similarly, the components of the tensor $\mathbf{V}$ with respect to $\mathbb{F}_c$ are the same as that of $\mathbf{U}$ with respect to $\mathbb{F}_r$ and so $\mathbf{V}$ is an *objective tensor*. On the other hand, the components of $\mathbf{L}$ with respect to $\mathbb{F}_c$ are not invariant scalars, so $\mathbf{L}$ is not an objective tensor. How about $\mathbf{D}$? Can you prove that it is objective?

ASSUMED INVARIANCE OF PHYSICAL QUANTITIES

In the development of classical continuum theories, we will assume that (1) the mass density is an invariant scalar (2) The Cauchy stress is an objective tensor (3) The heat flux vector $\mathbf{q}$ is an objective vector (4) The entropy, energy and temperature are invariant scalars (and hence so are the isothermal work potential ψ and the Gibbs potential).

> ⊙ **11.48 (Exercise:). Objectivity of kinetical quantities**
> *Reflect on the above assumption and rationalize the conclusions. Suggest experiments that could be done to corroborate or invalidate the above assumptions.*

11.4.1 When is a tensor objective?

How do we check if a quantity is objective or not? Fundamentally we need to see if its components are invariant scalars. However we can use a trivial result that shows that we can bootstrap our way to finding objective quantities:

> **PROPERTIES OF INVARIANT QUANTITIES**
> 1. Functions of invariant tensors are invariant.
> 2. Time rates of change of invariant tensors are themselves invariant.
> 3. If $\mathbf{A} = A_{ij}\mathbf{i}_i \otimes \mathbf{i}_j$ is an invariant tensor, then $\mathbf{B} = \mathbf{R}\mathbf{A}\mathbf{R}^t = A_{ij}\mathbf{i}'_i \otimes \mathbf{i}'_j$ is objective, since the components A_{ij} are invariant scalars. Thus, if $\mathbf{A}$ is an invariant tensor then $\mathbf{R}\mathbf{A}\mathbf{R}^t$ is an objective tensor.
> 4. If $\mathbf{A}$ is an objective tensor, then $\mathbf{R}^t\mathbf{A}\mathbf{R}$ is an invariant tensor.

In other words, associated with any objective tensor $\mathbf{A} = A_{ij}\mathbf{i}'_i \otimes \mathbf{i}'_j$ there is an invariant tensor. Simply replace $\mathbf{i}'_i$ with $\mathbf{i}_i$, keep the components the same and voila we have an invariant tensor.

For example $\mathbf{V}$ is an objective tensor because $\mathbf{V} = \mathbf{R}\mathbf{U}\mathbf{R}^t$ and $\mathbf{U}$ is an invariant tensor (being the square root of $\mathbf{C}$).

⊙ 11.49 (Exercise:). Objectivity of D
Show that $\mathbf{D}$ is an objective tensor. (Hint: Use (11.33) and the polar decomposition theorem.)

⊙ 11.50 (Exercise:). Derivatives of invariants
Show that the time rate of change of an invariant quantity with respect to the Laboratory frame $\mathbb{F}_r$ is also an invariant quantity. (Hint: take the time derivative of the condition of invariance $s(t) = s^*(t+a)$.)

⊙ 11.51 (Exercise:). Non-objectivity of rates
Show that the time rate of change of an objective vector with respect to the laboratory inertial frame $\mathbb{F}_r$ is NOT an objective tensor. (Hint: use the fact that, if a vector $\mathbf{v} = v_i\mathbf{e}_i$ is an objective vector, then $\dot{\mathbf{v}}$ has terms that come from the rate of change of $\mathbf{e}_i$ also.)

⊙ 11.52 (Exercise:). Derivative relative to $\mathbb{F}_c$
Show that if $\mathbf{T}$ is an objective tensor with components T_{ij} with respect to $\mathbb{F}_c$ then the tensor

$$\frac{d}{dt}\mathbf{T}\bigg|_{(xyz)} := T'_{ij}\mathbf{i}'_i \otimes \mathbf{i}'_j \qquad (11.67)$$

is an objective tensor. (Hint: T'_{ij} are invariant scalars.)

If you took a course in rigid body dynamics, you might recall that when we discussed particles moving on rotating rigid bodies, we attached a rotating frame (xyz) to the body and wrote down formulas like

$$\frac{d\mathbf{v}}{dt}|_{XYZ} = \frac{d\mathbf{v}}{dt}|_{xyz} + \boldsymbol{\omega} \times \mathbf{v}. \tag{11.68}$$

The term $\frac{d\mathbf{v}}{dt}|_{xyz}$ which represents the rate of change of $\mathbf{v}$ keeping the (xyz) axis fixed is called a *corotational rate*. We can extend this notion to continua in a very simple way:

⊙ **11.53 (DEFINITION).** *Corotational rate:* The derivative of an objective tensor with respect to any frame which rotates with the body is called a corotational rate. The corotational rate defined by $\dot{T}'_{ij}\mathbf{i}'_i \otimes \mathbf{i}'_j$ is called the Green-Naghdi-Mccinnis rate and is written as $\overset{\square}{\mathbf{T}}$ (see e.g., [Simo and Hughes (1998)] p. 255 for a number of other objective rates).

⊙ **11.54 (Exercise:).** **The role of the rate of rotation**
Show that if (xyz) is the corotational frame, then the material time derivative with respect to $\mathbb{F}_r$ is related to that with respect to $\mathbb{F}_c$ by

$$\frac{d}{dt}\mathbf{T}|_{(XYZ)} = \frac{d}{dt}\mathbf{T}|_{(xyz)} + \mathbf{\Omega T} - \mathbf{T\Omega}, \ \mathbf{\Omega} = \dot{\mathbf{R}}\mathbf{R}^t. \tag{11.69}$$

Thus the rate of change of a tensor with respect to the laboratory frame $\mathbb{F}_r$ contains contributions from the rate of rotation of $\mathbb{F}_c$ through the second and third terms on the right-hand side of the above equation.

OBJECTIVE CONSTITUTIVE THEORY

A constitutive equation that relates quantities that are not affected by rotations of the body but only by their distortions will be called objective. Thus a constitutive theory is said to be objective *if its constitutive equations can be shown to be equivalent to a relationship between invariant scalars.*

For example, the constitutive equation (11.49) does not appear to be an objective constitutive equation since it is in terms of $\mathbf{F}$ whose components are not invariant scalars. However, the form (11.54) is an objective constitutive equation since it is written in terms of invariant scalars.

We avoid the use of the phrase "principle of Material Objectivity" because it sounds too formal even though it is merely a working hypothesis or assumption. It does not seem to have the same authority as one of the conservation principles.

11.4.2 Other rotating frames

The rotating frame that we have defined here (using the rotation tensor) is not the only possibility. We can define a new (and different) rotating frame using the antisymmetric part of the velocity gradient tensor $\mathbf{L}$ through

$$\frac{d\mathbf{i}_i''}{dt} := \mathbf{W}\mathbf{i}_i'', \quad \mathbf{W} = 1/2(\mathbf{L} - \mathbf{L}^t) \tag{11.70}$$

Note that $\mathbf{W} \neq \mathbf{\Omega}$ although both are skew symmetric tensors. However, *the difference in the spin rates associated with any two corotational frames is an objective tensor*. We can then define objective tensors using this frame which is frequently referred to as the "Jaumann Frame" and the derivatives that are computed keeping this frame fixed are called "Jaumann derivatives". Thus choosing different rotating frames such as the one defined above and defining objective derivatives through $\dot{\mathbf{T}}|_{(xyz)''} = \dot{T}_{ij}''\mathbf{i}_i'' \otimes \mathbf{i}_j''$ is equivalent to adding objective terms to the corotational derivative (11.67) defined here.

11.4.3 Why do we not need to worry about objective rates?

We will end this admittedly brief introduction to continuum mechanics by addressing some issues about objective rates.

Notice that in our discussion, we barely used anything about objectivity in our discussion of elasticity. How is this possible? Why are there so many research papers on Different Objective rates in the plasticity and viscoelasticity literature? We will try to answer this question here.

The first point to note is that *if you start your constitutive formulations using ONLY invariant scalar kinematical variables to represent the response, you don't have to worry about objectivity. Things will work out just fine*. So if you start by assuming that the isothermal work function depends upon the history of $\mathbf{C}$ (which is an invariant tensor) then everything will work out fine. On the other hand, if you introduce additional tensorial quantities (such as the back stress $\boldsymbol{\alpha}$) for which you want to prescribe evolution equations, you now have to decide which corotational frame you need to keep fixed when you take the time rate of change. In dynamics parlance you have to "choose" a rotating coordinate system.

For example, there are a number of plasticity models that try to set up constitutive theories of the form

$$\dot{\mathbf{T}} = \mathbb{A}\mathbf{D} \tag{11.71}$$

which is referred to as a hypoelasticity theory. By suitably choosing $\mathbb{A}$, we can obtain a number of very interesting response graphs. The structure has amazing flexibility to deal with a huge variety of situations without any difficulty whatsoever. The appealing simplicity of this structure hides a major problem: The left-hand side of the above equation is NOT objective. (Since $\mathbf{T}$ is objective, and the conventional meaning is that $\dot{\mathbf{T}}$ is with respect to the laboratory frame $\mathbb{F}_r$, it is not objective.)

You might say "aha! I know how to fix it, replace $\dot{\mathbf{T}}$ with the derivative with respect to a rotating frame." But which rotating frame? initially, people used the Jaumann rate defined with respect to the rotating frame defined by (11.70). However, there are a number of papers (see [Khan and Huang (1995)] Section 7.8 for an extensive and informative discussion) which show that you get some strange behavior when you consider large rotations. It was later shown that if you use the "Green-Naghdi-Mccinnis" rate $\overset{\square}{\mathbf{T}}$ defined by equation (11.69) on p.395 it appears that these problems disappear. Of course there are a number of other objective rates that are also "well behaved" in this sense. However, it appears that the frame defined through the rotation tensor $\mathbf{R}$ seems to occupy a sweet spot and we suggest that it is a good frame to choose and will give you meaningful results.

Chapter 12

Finite Deformation Plasticity

Learning Objectives

By learning the material in this chapter, you should be able to do the following:

(1) Describe in words, what is meant by an isotropic inelastic material.

(2) Describe the different ways in which inelasticity is being modeled and the core differences between them.

(3) List the two scalar functions that are essential for any model of finite deformation isotropic inelastic behavior.

(4) Given the two functions that describe an inelastic material, develop a procedure for finding governing equations for the deformation gradient and the stress as a function of time.

(5) Decide whether or not a particular model is capable of a thermodynamical interpretation and whether it satisfies the laws of thermodynamics.

(6) Develop your own algorithm or modify the supplied MATLAB program to simulate the response of an inelastic material undergoing homogeneous deformation.

(7) State the limitations of the theory of isotropic inelastic materials as developed here.

> # CHAPTER SYNOPSIS
>
> The *central message* of this chapter is that a relatively straightforward way to generalize finite strain elasticity and obtain an inelastic isotropic material is to replace the strain measure such as $\mathbf{B}$ with an "elastic stretch" and provide an evolution equation for it. Different evolution equations will give rise to different types of plasticity models. The main task is to do this in a thermodynamically consistent way.
>
> The most important equation in this whole chapter is
>
> $$\mathbf{D}_p := \frac{1}{2}\mathbf{V}_e^{-1}(\mathbf{L}\mathbf{B}_e + \mathbf{B}_e\mathbf{L}^t - \dot{\mathbf{B}}_e)\mathbf{V}_e^{-1} = \phi\mathbf{P}(\mathbf{B}_e, \kappa), \qquad (12.1)$$
>
> which relates the plastic flow rate to the velocity gradient $\mathbf{L}$ and the rate of change of the squared elastic stretch $\mathbf{B}_e$.

Chapter roadmap

In the last several chapters, we have studied the fundamental principles of inelasticity under the assumption of small deformations, i.e. inelasticity applied to structural problems. We now embark on the study of inelasticity under large deformations. Different classes of inelastic behavior can be modeled by choosing different "kinetic laws" or "evolution equations" for $\mathbf{D}_p$. This chapter reveals how to choose such laws in a manner that is consistent with thermodynamical principles.

The chapter begins with a discussion of different kinds of plasticity models and when to use them. Then we discuss how to extend small strain plasticity to come up with a finite strain plasticity model in a very simple and direct way. The main idea behind the approach is to introduce a measure of *elastic stretch* $\mathbf{B}_e$ for finite deformation. *We don't need to define a notion of plastic strain at all either to construct the theory or solve boundary value problems.*

The major advantage of this extension is that you can use your understanding of small deformations to model finite deformations. The problem with this purely formal approach is that you may not be sure how to modify the structure for your own purposes. The next few sections deal with the different possibilities for variation and generalization of the model and how to ensure that these meet thermodynamical compatibility. We then discuss the maximum rate of dissipation hypothesis in the context of finite deformation. We next discuss the method for determining the consistency parameter for these models. We next extend the convex cutting plane algorithm to finite deformations. In the final section, we provide a geometrical interpretation of the finite elastic strain using the notion of stress free or natural configurations. Before discussing the extension of these ideas to rate dependent response. Surprisingly, modeling rate dependent response is actually a *trivial exercise* as compared to modeling rate independent response. You might remember that viscous dashpots are trivial to deal with compared to friction. Since plasticity is an extension of friction to continua, it is not surprising that it is very hard to do.

12.1 Introduction to finite deformation inelasticity

Imagine that you are an engineer in a high tech materials company that is exploring the possibility of processing a new alloy. In order to create a suitable die for the processing operation, your team has been asked to simulate the process and arrive at a preliminary evaluation and find out whether the process is feasible and which regions of the part require special attention.

As a group leader, you have to make a number of choices:
What software to use for the simulation?
What material model to adopt? Should it be a custom material model or can you adopt one that is already built-in?
What experiments should be run on the material? What parameters to monitor?
There are a large number of decisions that you have to make before you even start the project. Let us first clarify that reading this chapter will NOT enable you to make these decisions. That requires a detailed knowledge of the various operations of your company and experience with sheet forming technology and software. What this chapter WILL enable you to do, is to get some INSIGHT into the modeling and help you make sensible choices about what is important from among a bewildering array of choices, terminology and papers[1] that is provided in the literature on forming operations. Let's face it, while you are pressed for time, you don't want to simply use some model that is available. You have a desire to understand— to gain some insight, within the time that you have. This is why you chose this career path in engineering in the first place.

Alternatively, you might be a graduate student who has to finish your thesis and your advisor, being the unreasonable person that he or she is(:-)), has categorically stated that she will not give you a Ph.D unless you develop a model for your amazing bio-nano-hysteretic polymer super duper computer shape memory chip that you developed over the last 15 months slaving away in a dark dingy lab with no food or drink[2]. You hunt around on the web, look at several research papers and find that you have to develop a new model—at the very least, you have to modify some other model, but you can't make head or tail of the model that you found—we know, both the authors have been there[3]!

Read this chapter and suddenly you will see all—or at least you will know enough (actually just the hysteretic part) to keep your advisor happy until time or money runs out and he or she HAS to give you the Ph.D:-)).

To cut a long story short, in this chapter we will show you *how to develop a*

---

[1] Additive vs. multiplicative decomposition, Jaumann vs. Upper convected vs. Green-Naghdi McKinnis rates, objective versus non-objective time stepping algorithms ... the choices seem endless and confusing

[2] Notice how we cleverly included all the major buzzwords of the day: bio, nano and info in the same sentence.

[3] Although in our days bio-nano-hysteretic polymers were simply referred to as "vulcanized rubber":-))

model of an isotropic inelastic material based on thermodynamic principles. We will find that a certain objective rate (referred to as the Lie derivative or the Oldroyd Convected Derivative) will be very convenient to use for our purpose. We choose constitutive equations for the Helmholtz potential (or the isothermal work function) and use thermodynamic criteria to guide us on the choice of rates. For isotropic inelastic materials, it provides us with a means to rationalize our choices.

There are many different inelasticity theories: Choose the appropriate one

---

The take home message here is that if you want to quickly estimate macroscopic parameters such as forces and overall shape change with very little experimental data, i.e. if your primary objective is die design and the optimization of bulk processing, then an isotropic material model of the type presented here is sufficient. It will NOT be suitable for many sheet metal applications since plastic anisotropy will play a huge role in the thinning of the sheet, formation of ears during drawing, etc.

Why are there so many restrictions and caveats for the application of inelasticity theories? The reason is that, at its core, the theory of elastoplasticity deals with *microstructural changes in a body.*
What precisely does microstructural change mean?
In a polycrystalline material, other than the gross or macroscopic shape of the body which is called its macrostructure or structure (plate, beam, truss, rod or whatever), the body has other structural elements.

- At the scale of nanometers and Angstroms, the material has a crystal structure called the crystal lattice. This is the atomic scale structure or *crystal structure.*
- Next, there are defects in the long range (of the order of nanometers) arrangement of the lattice. This is referred to as the nanoscale structure or simply *nanostructure.*
- Next, there are numerous crystallites called grains arranged in a random or semi-random order separated by grain boundaries. This is called the microscale structure or *microstructure.*

Any change in the structure at any of these levels would be under the broad purview of "inelasticity". On the other hand, elasticity deals only with change in the macrostructure without any change in any of the microstructural variables. Thus, inelasticity does not refer to one theory but a class of models for various types of changes. Depending upon the scales at which you are interested in, you have to choose one of the many possible approaches.

The situation for polymeric and biological materials is quite different *although in many cases an inelasticity like approach can be quite useful.* Here, the microstructure is composed of long chain molecules that are entangled and crosslinked.

Changes in the entanglement and the formation and disappearance of "physical entanglements" and perhaps the breaking of the crosslinks all give rise to permanent shape and property changes. These responses can also be modeled by inelasticity.

At the macroscopic or "process scale", you might be interested in forces on the die, the change in shape, the places where strain concentration (also-called excessive thinning or thickening or dead zones) occurs and other such features. The material in this chapter is an introduction to some of the methods used for "process modeling" of bulk manufacturing processes.

12.1.1 The objective of a finite inelasticity constitutive model

At its core, for process modeling, the goal of the constitutive theory is simple to state: given the velocity gradient $\mathbf{L}(t) = \nabla \mathbf{v}$ at every point (or maybe only on the boundary), find the deformation $\mathbf{F}(t)$, the stress $\boldsymbol{\sigma}(t)$ and other relevant variables (such as grain size, grain orientation etc.) as a function of time.

Finding $\mathbf{F}(t)$ from $\mathbf{L}$ is not a challenge. Simply solve the differential equation $\dot{\mathbf{F}}(t) = \mathbf{L}(t)\mathbf{F}(t)$. Given $\mathbf{L}$, it is a linear differential equation, MATLAB will make short work of it provided you specify $\mathbf{F}(0)$; the challenge is to find $\boldsymbol{\sigma}$ and the other variables of interest.

Perhaps the simplest theory is to ignore all variables other than the stress and go for a fluid like response for the stress (if you recall, we did this in Chapter 6), i.e. choose a constitutive equation of the form

$$\boldsymbol{\sigma} = -p\mathbf{I} + \mu(\mathbf{D})\mathbf{D}, \tag{12.2}$$

where $\mathbf{D}$ is the symmetric part of the velocity gradient and p is the pressure. Choose a functional form for μ (say, $\mu = const$ or $\mu = \mu_0 \|\mathbf{D}\|$) and that is all—you are off and running. What can be easier than this? A wide range of metal forming software implement this strategy and it is widely used in the industry. There are many very good books and papers on this which show how to choose μ and how to simulate various processes (see e.g. [Banabic et al. (2000); Bohlke et al. (2006); Dixit and Dixit (2008)]).

Then why do we waste so much effort and work on other more complicated models? The reason is that the simple model above—while it has the obvious advantage of simplicity and being practically useful—has some severe limitations.

First, notice that (12.2) is a model for an isotropic fluid. In other words, the model is such that all of its traits (such as resistance to motion etc.) are the same in all directions. This makes it unsuitable for direct application to certain processes. For example rolled sheets are quite anisotropic in their response (i.e. their stress response to the same external load in different directions is different), a fact not taken into account by the model (12.2).

Furthermore, it has no "springiness" or "elasticity" at all. Anyone who has rolled dough[4] knows that as soon as the rolling pin is lifted, the material "recoils", i.e. it "springs back" quite a bit and some clever and repeated rolling is required to obtain a certain shape. Metals are not all that dissimilar although their "springback" is much less pronounced. Nevertheless, since our tolerances are very small (you wouldn't want to buy a car whose body panel appears "warped" would you?) we need great precision in compensating for springback. Other effects of elasticity and nonlinear phenomena are "die swell", "normal stress differences" etc., none of which are accounted for in the simple viscous fluid like model.

If you are interested in forming materials other than metals, such as polymer sheets, then you are really out of luck with the model (12.2); these materials tend to have very large elastic strains—which are completely ignored by the viscous fluid model—that have to be compensated for in making dies for materials processing. Also, if you are interested in studying various other phenomena (diffusion, grain growth etc.) associated with a variety of thermodynamic driving forces, then you have to introduce some aspect of elasticity into the model. All of these will then induce you to consider models other than the simple model (12.2).

The minute you start exploring the literature in order to find more sophisticated approaches, you will be cast out into a vast sea of material models all of which seem to use terminologies such as "objective rates", "co-rotational rates", "multiplicative decomposition", "stress update algorithms" "material spin", "substructure spin"..., it is enough to make your head spin. So a typical approach then would be to say "to heck with all this; let me use whatever built-in model that is available in the software package". In other words, you have now let the programmers at these software companies decide for you (or if you are a programmer in such a company, you will pick whatever you can implement given the structure of your code). As with all of these decisions "caveat emptor" (buyer beware) you have no idea where your model has been:-)), i.e. what has been implemented and whether it is suitable.

Part of the reason for this confusion is the fact that finite deformation inelasticity is still an area of active research and researchers are still exploring how best to model these materials with the very little experimental evidence that is currently available in the public domain. This is the bane of a field where experiments at universities are very expensive and those performed by industry are proprietary. So, the key point is that, if you are working in a company, you probably have access to experimental evidence that is not available to university researchers and so you are in a position to make an informed choice of constitutive formulation.

The aim of this chapter is to introduce you to a systematic way of classifying

---

[4]Such as chapatti, which is a kind of Indian "bread" or wrap.

the different modeling approaches and to help you find your way through some of the jargon and get to the core of the model.

12.2 Classification of different macroscopic inelasticity models

You have to choose between energy (or finite elasticity) based models and "stress-rate based" or "fluid-like" models. If you plan to use the latter approach, then this chapter will NOT be particularly helpful to you. We suggest that you look for some of the recent papers on the subject[5]. Next, you have to choose whether an isotropic model[6] would be sufficient for your purpose or do you need an anisotropic model.

If you want to deal only with texture and anisotropy, then skip to a subsequent chapter. If you want to deal with isotropic materials then stay on.

Fundamentally there are two broad classes of models. The first approach essentially has a more "elasticity-like" feel to it whereas the second approach has a more "fluid-like" feel to it.

- The class of models which use finite elasticity as their base and extend it by defining additional variables (such as elastic and/or plastic strain). In this set of models, there are a number of issues as to how to properly augment the model and how to include anisotropic effects.
- The models are those that utilize a constitutive equation for the *stress rate* in terms of the velocity gradient[7], which is split into an elastic and plastic part. Here, there are a number of issues that have to be dealt with including how to write the stress rate and how to split the velocity gradient.

In general, these two approaches are NOT equivalent. However, [Xiao *et al.* (2007b)] shows how these two approaches can be rationalized and shown to be equivalent for a particular class of constitutive relations. You can see that the first is an extension of the approach used for small deformation inelasticity and it has a certain appeal to it. However, there seems to be no real consensus as to how to define the elastic and plastic strains so as to be meaningful and how to write the various constitutive equations.

In the metal forming industry, most software packages utilize an approach similar to (2) rather than (1) since it is easier to implement.

---

[5]e.g., [Xiao *et al.* (2007b,a)] for a detailed description of a thermodynamically consistent approach that is entirely Eulerian.

[6]By isotropy, we mean that all the constitutive functions are isotropic functions of their respective arguments.

[7]see e.g., [Simo and Hughes (1998)] Chapters 7 and 8 for several examples, as well as a critique of some of these models

Compounding the difficulty in developing a single widely accepted framework for finite inelasticity is the fact that unlike for small deformation, every aspect of finite deformations can be written in several ways that can sometimes be shown to be equivalent. For example, there is no unique measure of strain but a plethora of different measures, each of which has its own peculiar characteristics. Similarly there are at least three different measures of stress, at least two different ways of writing the balance laws—the list goes on.

Nevertheless, we will follow the first type of model rather than the second since this is in keeping with the intent of this book and is consistent with the other chapters, and they are more amenable to a relatively straightforward thermodynamic treatment. Furthermore, the model can be easily modified to account for explicit time dependent behavior. This is important because, you don't want to learn a new kind of model for each type of behavior. The elasticity based models can be easily and gracefully extended to accommodate a wide range of phenomena—it serves as a template for you to use to generate your model.

We also want to encourage programmers in industry to adopt and implement models that are more thermodynamically based, so that many of the arguments about appropriate choices can be settled by resorting to thermodynamical and physical arguments rather than intuition alone. Rather than being difficult to implement, they are quite easy to do (as we will see shortly) and the effort you invest in learning this method of modeling will pay off handsomely in making you more efficient and speeding up the time taken for you to implement a new model by easily and systematically extending pre-existing ones. Your boss/advisor will feel very happy that you were able to implement the model quickly and you will get to go home early to do the things that you REALLY care about:-)).

We will briefly discuss anisotropic effects before developing models of isotropic materials. There are two kinds of anisotropic phenomena that have to be dealt with:

(1) Plasticity of materials whose anisotropy is fixed. For example, they start out as, say, orthotropic materials and remain so. These models are useful when dealing with sheet metal forming operations where the deformations are large but are not so severe as to change the microstructure of the material. This is also the case for plasticity of single crystals. For inelasticity with fixed anisotropy, changes of orientation of the anisotropy axes come into play. This issue has caused many difficulties and we will try to clarify some of them in Chapter 14, where we deal with single crystal plasticity.

(2) Plasticity of materials whose anisotropy changes with deformation, i.e. the material is initially isotropic but due to deformation it becomes anisotropic. This

is typical in materials processing operations, rolling, drawing etc., where the purpose is to induce significant microstructural changes for want of shape and properties (usually referred to as severe plastic deformation). These problems are much harder. There seems to be no satisfactory way to deal with "evolving anisotropy" yet. This is an area of active research and there are many unanswered questions. We will not deal with such models in this book. Luckily for us, we will finish this book before we tackle this difficult topic :-)).

Each of these problems requires a different type of approach, the simplest starting point being isotropic materials. For isotropic inelasticity, the approach can be described as an extension of that for small deformation inelasticity. The main reason why such an approach works is that the orientation of the material does not matter: so the only parameter that is of interest is change of shape and not change of orientation. Now, since small strain inelasticity is written in terms of change of shape (i.e. in terms of strains) it is possible to extend this to finite strains.

With this background, we will plunge in directly to the procedure for the development of the model and then do an extensive discussion.

12.3 Recapitulation of small deformation inelasticity results

Our justification of the steps that we have listed for developing a finite deformation inelastic model, is based on a direct analogy with the small deformation model.

The constitutive equations for an inelastic material under small deformations are listed below along with brief descriptions of the various quantities. In spite of their much perceived complexity, there are only four major steps:

(1) Choose the inelastic state variables $\mathbf{Q}$ to include along with the elastic strain and temperature.

(2) Express $\boldsymbol{\sigma}$ in terms of the state variables through a Helmholtz potential.

(3) Choose a yield function, and

(4) choose evolution equations for the plastic strain rate (which is the difference between the total strain rate and the elastic strain rate) and the inelastic variables for rate independent inelasticity.

State variables and equations

$$\text{State variables}: \mathbb{S} = \{\boldsymbol{\sigma}, \boldsymbol{\varepsilon}, \boldsymbol{\varepsilon}_e, \mathbf{Q}\}, \; \boldsymbol{\varepsilon}_e = \boldsymbol{\varepsilon} - \boldsymbol{\varepsilon}_p, \; \mathbf{Q} = \{\kappa, \boldsymbol{\alpha}\},$$

$$\text{Isothermal work function}: \psi = \psi(\boldsymbol{\varepsilon}_e) = 1/2\mathbb{C}\boldsymbol{\varepsilon}_e \bullet \boldsymbol{\varepsilon}_e,$$

$$\text{Stress response}: \boldsymbol{\sigma} = \frac{\partial \psi}{\partial \boldsymbol{\varepsilon}_e} = \mathbb{C}\boldsymbol{\varepsilon}_e,$$

$$\text{Plastic driving force}: \boldsymbol{\xi} := \boldsymbol{\sigma} - \boldsymbol{\alpha}, \text{ where } \boldsymbol{\alpha} \text{ is the back stress.}$$

Evolution equations

$$\text{Yield function } f \text{ and loading function } \hat{g}: f := f(\boldsymbol{\xi}, \mathbf{Q}), \quad \mathbf{M} := \mathbb{C}\frac{\partial f}{\partial \boldsymbol{\xi}}, \quad \hat{g} := \mathbf{M}\bullet\dot{\boldsymbol{\varepsilon}},$$

$$\text{Plastic potential } H \text{ and its normal } \mathbf{P}: H := H(\boldsymbol{\xi}, \mathbf{Q}), \quad \mathbf{P} := \frac{\partial H}{\partial \boldsymbol{\xi}},$$

$$\text{Evolution equations/kinetic laws}: \dot{\boldsymbol{\varepsilon}}_p = \phi\frac{\partial H}{\partial \boldsymbol{\xi}} = \phi\mathbf{P} \Rightarrow \dot{\boldsymbol{\varepsilon}}_e = \dot{\boldsymbol{\varepsilon}} - \phi\mathbf{P},$$

$$\text{Other inelastic variables}: \dot{\mathbf{Q}} = \phi\tilde{\boldsymbol{\Theta}}(\mathbb{S}),$$

$$\text{Loading criteria and the value of } \phi: \phi \geq 0, \; f \leq 0, \; \phi f = 0$$

$$\Rightarrow \phi = h(f)\frac{\langle \mathbf{M}\bullet\dot{\boldsymbol{\varepsilon}}\rangle}{\mathbf{P}\bullet\mathbf{M} - \frac{\partial f}{\partial \mathbf{Q}}\bullet\tilde{\boldsymbol{\Theta}}},$$

(12.3)

where $h(f)$ is the Heaviside step function which is such that $h(f) = 0$ if $f < 0$ and $h(f) = 1$, if $f = 0$ and $< \cdot >$ is the Macauley bracket which is given by $<x> = 1/2(x + \|x\|)$. These have been listed many times in the previous chapters (see e.g., equation (9.1) on p.267) but we are listing it for easy comparison and look-up.

The model requires the specification of the elastic moduli $\mathbb{C}$, the yield function f and plastic potential H (typically taken to be the same as f) and the hardening functions Θ. A minimalist model for small strain inelasticity would set $\mathbb{C}$ to be that of an isotropic elastic material, $f = H$ and $\Theta = 0$.

12.4 How to develop a minimalist model for isotropic inelastic materials subject to finite deformation

---

We will develop isotropic inelasticity models by starting with an *isotropic finite elasticity model* and extending it using a combination of physical intuition and thermodynamics. The most important and unusual element in this approach is that we *don't* introduce a finite deformation counterpart of plastic strain but we *do* introduce a finite deformation counterpart of elastic strain. Surprisingly, this simplifies the formulation considerably.

There are several ways to introduce an understanding of the behavior of inelastic materials undergoing large deformations. We will follow a somewhat formal procedure (i.e. an approach that uses a mathematical analogy based on small deformation inelasticity guided by finite deformation elasticity) giving more physical insights later.

Analogies and formal extensions are all in the eyes of the beholder however, and there are several ways in which these analogies can be drawn. We will only deal with isotropic materials here since these are the ones where the structure is clearest[8] apart from our own papers on the subject. The interpretation given here is our own, however, based on our experience in teaching these concepts to students and realizing the difficulties faced by students in understanding these concepts.

You might be thinking at this stage "I don't want to see all the gory details, just tell me the procedure to find the constitutive equations". We will indulge your curiosity. Rather than dealing with generalities, we will demonstrate the development of a finite inelasticity model by means of a specific example.

There are only four steps involved in choosing three scalar functions! Choose a constitutive function for σ and another constitutive equation for "rate of elastic stretch", one for the yield function and one for the hardening parameter and you are done! What could be simpler than that?

To set the context, we want you to do the following exercise so that you can remember what isotropic small strain inelasticity looks like.

⊙ **12.1 (Exercise:). Small strain, isotropic Prandtl-Reuss constitutive equations in terms of ε_e:**
Show that the evolution equations (12.3) for an isotropic non-hardening material with a von Mises yield function and the normality rule, can be written as

$$\sigma = \lambda tr(\varepsilon_e)\mathbf{I} + 2\mu\varepsilon_e,$$
$$f(\sigma) = \boldsymbol{\xi}\cdot\boldsymbol{\xi} - \kappa^2 = H, \quad \boldsymbol{\xi} = \sigma_{dev},$$
$$\dot{\varepsilon}_e = \dot{\varepsilon} - \phi\boldsymbol{\xi}, \hspace{4em} (12.4)$$
$$\dot{\kappa} = \phi\Theta(\varepsilon_e, \kappa),$$
$$\phi \geq 0, f \leq 0, \phi f = 0,$$

where the notation $(\cdot)_{dev}$ stands for the deviatoric part of the tensor. What is the most unusual aspect of this set of equations (as compared to equation (12.3) on p.408?

---

[8]The following sections draws heavily from recent papers on finite deformation (most notably from the works of [Onat (1968); Rubin (2001); Rajagopal and Srinivasa (1998)]).

The main thing to observe here is that (1) the constitutive equation for the stress looks exactly like that for linear elasticity with $\boldsymbol{\varepsilon}$ replaced by $\boldsymbol{\varepsilon}_e$, (2) $\dot{\boldsymbol{\varepsilon}}_e = \dot{\boldsymbol{\varepsilon}}$ when $\phi = 0$, i.e. when no plastic flow occurs and (3) *there is no mention of plastic strain!* It is the generalization of this model to finite deformations that we will deal with next.

12.4.1 The modeling procedure

In order to simplify notation and to show the similarity with small strains, we will use the symbol $\mathbf{e}$ for the logarithmic strain $1/2 \ln \mathbf{B}$ and $\boldsymbol{\Gamma}$ for $(\mathbf{e})_{dev}$, the deviatoric part of $1/2 \ln \mathbf{B}$.

(1) **Pick a suitable isotropic finite elasticity candidate model to extend.**
Recall from a previous chapter on continuum mechanics, that the stress response of an isotropic Green-elastic material is given by

$$\boldsymbol{\sigma} = 2\varrho \frac{\partial \psi}{\partial \mathbf{B}} \mathbf{B} = \varrho \frac{\partial \psi}{\partial \mathbf{e}}, \tag{12.5}$$

where ψ is an isotropic function of $\mathbf{B}$.
For example, from the last chapter recall that for a Hencky elastic material

$$\psi = \frac{1}{2}(c_1 (\operatorname{tr} \mathbf{e})^2 + c_2 \|\mathbf{e}\|^2),$$
$$\boldsymbol{\sigma} = \varrho(c_1 (\operatorname{tr} \mathbf{e})\mathbf{I} + c_2 \mathbf{e}). \tag{12.6}$$

Other models can also be chosen as starting points.

⊙ **12.2 (Exercise:). Compressible neo-Hookean model:**
Consider the following form for ψ. Consider a strain energy function of the form $\psi(I_1, I_3)$, where $I_1 = \operatorname{tr} \mathbf{B}$ and $I_3 = \det \mathbf{B}$. Show that the stress is given by

$$\boldsymbol{\sigma} = 2\varrho \left\{ \frac{\partial \psi}{\partial I_3} \mathbf{I} + \frac{\partial \psi}{\partial I_1} \mathbf{B} \right\}. \tag{12.7}$$

Using this or otherwise, obtain the stress for the compressible Neo-Hookean model whose strain energy function is given by

$$\psi = f(I_3) + \mu/2(I_3^{-1/3} \operatorname{tr} \mathbf{B} - 3). \tag{12.8}$$

(Hint: use the fact that $\partial I_3 / \partial \mathbf{B} = I_3 \mathbf{B}^{-1}$.)
Ans:

$$\boldsymbol{\sigma} = 2\varrho(I_3 f'(I_3) \mathbf{I} + \mu I_3^{-1/3} \mathbf{B}_{dev}), \tag{12.9}$$

where $\mathbf{B}_{dev}$ is the deviatoric part of $\mathbf{B}$.

(2) **Extend this model to inelasticity by replacing the squared stretch B with the elastic stretch $\mathbf{B}_e$**

In other words, we simply assume that the elastic response is essentially "unchanged" except that the relationship between the strain and the deformation gradient (i.e. $\mathbf{B} = \mathbf{FF}^t$) has been abandoned. This idea falls under a mathematical structure referred to as an "isomorphism"[9]. Again, in order to simplify notation and to show the similarity with small strains, we will use the symbol $\mathbf{e}_e$ for $1/2 \ln \mathbf{B}_e$ and $\mathbf{\Gamma}_e$ for $(\mathbf{e}_e)_{dev}$, the deviatoric part of $1/2 \ln \mathbf{B}_e$.

With this notation, we will now extend the isotropic elasticity models to inelasticity by introducing an *elastic squared stretch tensor* $\mathbf{B}_e$ to replace the squared stretch tensor $\mathbf{B}$ in the elastic response. This is the crucial conceptual step in the model. This formal extension is based on the idea that "the stress response of inelastic materials is just like that of elastic materials except that the strain is not the total strain". In viscoelastic fluid models the tensor $\mathbf{B}_e$ is usually referred to as the "conformation tensor" (and written as $\mathbf{c}$) and represents the average value of the squared stretch (or conformation) of individual polymeric molecules that were randomly oriented when there was no flow.

Why is this acceptable? We are reasoning by analogy from small deformations: There we replaced $\boldsymbol{\sigma} = \mathcal{C}\boldsymbol{\varepsilon}$ with $\boldsymbol{\sigma} = \mathcal{C}\boldsymbol{\varepsilon}_e$. Here we are replacing $\boldsymbol{\sigma} = \varrho\mathbf{f}(\mathbf{B})$ with $\boldsymbol{\sigma} = \varrho\mathbf{f}(\mathbf{B}_e)$. Another reason is that this replacement "works" in the sense that (1) it produces reasonable results with minimum modifications of existing theories, (2) many other ways of developing finite inelastic models can be shown to be equivalent to this procedure.

At this stage, we will not interpret $\mathbf{B}_e$ as being obtained from some deformation gradient as yet—simply that it has the same attributes as $\mathbf{B}$, i.e. we will assume that it is a symmetric positive definite, objective tensor just like $\mathbf{B}$. This is in line with the introduction of elastic strains in small deformation inelasticity where we did not introduce any notion of deformation gradient for it. We simply said that the stress depends upon an elastic strain.

It is important to also emphasize here what we are NOT doing. *We are NOT going to say that $\mathbf{B} - \mathbf{B}_e$ is the plastic strain.* We are not going to say anything about plastic strain at all! We don't really need it.

Hence, based on our discussion of small deformations and finite deformation elasticity, we will assume that, for the class of inelastic materials that we are going to consider, the equations of state are given by

$$\psi = \psi(\mathbf{B}_e), \quad \boldsymbol{\sigma} = 2\varrho \frac{\partial \psi}{\partial \mathbf{B}_e}\mathbf{B}_e = \varrho \frac{\partial \psi}{\partial \mathbf{e}_e} \tag{12.10}$$

where ψ is an isotropic function of $\mathbf{B}_e$. Thus, the modified forms of the Hencky (12.6) and the neo-Hookean (12.9) models look the same except that $\mathbf{B}$ is replaced by $\mathbf{B}_e$. For example we have the following two extensions of Hencky and

---

[9]This assumption is a special case of a more general statement of "elastic isomorphisms" that have been described in great detail by Bertram [Bertram (2003)]. See also [Rajagopal (1995); Rajagopal and Srinivasa (1998)] for an explanation of this idea for a wider range of inelastic behavior.

neo Hookean materials:

Extension of Hencky Elasticity:

$$\boldsymbol{\sigma} = \varrho(c_1(\text{tr}\mathbf{e}_e)\mathbf{I} + c_2\mathbf{e}_e), \quad \mathbf{e}_e := \frac{1}{2}\ln\mathbf{B}_e.$$

Extension of compressible neo-Hookean Elasticity : (12.11)

$$\boldsymbol{\sigma} = \varrho(2I_{3e}f'(I_{3e})\mathbf{I} + \mu I_{3e}^{-1/3}(\mathbf{B}_e)_{dev}),$$

where $I_{3e} = \det \mathbf{B}_e$.

Note that (12.11a) is the counterpart of (12.4a) where we have replaced $\boldsymbol{\varepsilon}$ by $\boldsymbol{\varepsilon}_e$. However, unlike small deformation inelasticity, where the strain energy function is always quadratic, we can have widely different forms for the strain energy function for finite elasticity; correspondingly there are many different forms for the stress. This is especially true for polymeric materials where many different forms (motivated by various theories of polymer networks) have been proposed. Any of these can be used as a starting point for extension to inelasticity.

As we did for small deformations, it is very convenient to use the so-called "Kirchhoff stress" which is defined as[10] $\mathbf{S} = \boldsymbol{\sigma}/\varrho$. For example, for the extension of Hencky elasticity, the Kirchhoff stress is given by

$$\mathbf{S} = \boldsymbol{\sigma}/\varrho = \frac{\partial \psi}{\partial \mathbf{e}_e} = 2\frac{\partial \psi}{\partial \mathbf{B}_e}\mathbf{B}_e = c_1\text{tr}(\mathbf{e}_e)\mathbf{I} + c_2\mathbf{e}_e. \quad (12.12)$$

Note that $\mathbf{S}$ unlike $\boldsymbol{\sigma}$, is independent of ϱ and is a function only of $\mathbf{B}_e$. Now, since $\mathbf{B}_e$ is not the same as the total strain, we need to find out how it changes with time: In other words, our approach is based on writing an equation for the *elastic strain* rather than one for "plastic strain". You will see that we don't really need to introduce a specific definition of "plastic strain" at all in the theory.

In principle, the next step is really simple: "write down a sensible evolution equation for $\dot{\mathbf{B}}_e$". If you do this, you have completed the theory. But what does "sensible" mean?

To understand that question, let us consider a simpler case:

⊙ **12.3 (Exercise:). Finite elasticity in a new light**
What "evolution equation" for $\dot{\mathbf{B}}_e$ will reproduce finite elasticity? Hint: see equation (11.33) on p.373.

$$\text{Ans: } \dot{\mathbf{B}}_e = \mathbf{L}\mathbf{B}_e + \mathbf{B}_e\mathbf{L}^t \quad (12.13)$$

Note that with this assumption, we have obtained a "kinetic" or Eulerian version of elastic materials. We have an "internal variable" $\mathbf{B}_e$ whose evolution equation

---
[10] We are using a slightly different definition than that used elsewhere. Our definition differs from the conventional one by a constant ϱ_0.

is given by (12.13). It is a trivial matter to show that this evolution equation can be solved exactly with initial condition $\mathbf{B}_e(0) = \mathbf{I}$, to give $\mathbf{B}_e = \mathbf{FF}^t$. In other words the expression $\dot{\mathbf{B}}_e = \mathbf{LB}_e + \mathbf{B}_e \mathbf{L}^t$ is a mathematical way of saying "that the value of $\mathbf{B}_e$ coincides with that of $\mathbf{B}$ for all time".

Clearly, in order to model inelastic response, we don't want (12.13) to be true when inelastic deformations take place. We will stipulate that $\dot{\mathbf{B}}_e$ is something other than what is given in (12.13). We will try a very simple trick: noticing that $\mathbf{LB}_e + \mathbf{B}_e \mathbf{L}^t - \dot{\mathbf{B}}_e$ must be nonzero for inelastic flow, let us define

$$\mathbf{A}_p := \mathbf{LB}_e + \mathbf{B}_e \mathbf{L}^t - \dot{\mathbf{B}}_e. \tag{12.14}$$

Intuitively, it is the difference between the rate of change of $\mathbf{B}_e$ if the response was elastic and the actual rate of change of $\mathbf{B}_e$, and so is a measure of plastic flow. For those of you who are interested in viscoelastic fluids, this is the negative of the Oldroyd Upper convected derivative[11]. For those of you who have a differential geometry background, [Simo and Hughes (1998)] show that it is the same as the negative of the *Lie* derivative of $\mathbf{B}_e$ along the flow and use the notation $L_v \mathbf{B}_e$. Finally, it is connected with the "Rivlin-Ericksen Tensors" that show up in certain kinds of viscoelasticity models (see e.g., [Truesdell (1977)]). All things considered, this is a fairly important result.

(3) Assume a yield function and evolution equations

Based on the previous discussion, our strategy will be to assume a kinetic law for $\mathbf{LB}_e + \mathbf{B}_e \mathbf{L}^t - \dot{\mathbf{B}}_e$ by generalizing from the plastic flow rule for small strain inelasticity $(12.4)_{b,c}$, i.e.

$$f(\mathbf{S}, \kappa) \leq 0, \ \mathbf{A}_p = \phi \tilde{\mathbf{P}}(\mathbb{S}). \tag{12.15}$$

Just as in small strain inelasticity, f is the yield function, ϕ is the Kuhn-Tucker consistency parameter and $\tilde{\mathbf{P}}$ is the plastic flow direction.

Furthermore, the form (12.15) is exactly the same constitutive equation that is used for viscoelastic fluid models (although ϕ and $\tilde{\mathbf{P}}$ do not have the same interpretation) that are based on conformation tensors (see, for e.g., [Grmela and Carreau (1987); Leonov (1987, 1992); Leonov and Padovan (1996)]). We will see later that such models are, in some sense, simpler than the plasticity models considered here since they do not have to worry about yielding behavior etc.

---

[11] The upper convected time derivative or Oldroyd derivative is the rate of change of some tensor property of a small parcel of fluid that is written in the coordinate system "glued" to the fluid particles so that it distorts with the fluid.

(4) **Assume a hardening rule for** κ: Our final step is to specify κ through

$$\dot{\kappa} = \phi\tilde{\Theta}(\mathbf{B}_e, \kappa). \tag{12.16}$$

Generally, this is not a convenient way to specify κ; rather we specify it in an equivalent form using the "plastic arc length" parameter. Since we haven't introduced the finite plasticity generalization of the plastic arc length yet, we will delay writing a specific form for $\tilde{\Theta}$.

(5) If you put this all together, you have now arrived at the full set of constitutive equations. Note that you needed four functions to get this done:

(a) The Helmholtz potential,
(b) A yield function,
(c) A flow rule (i.e. the function $\tilde{\mathbf{P}}$), and
(d) A hardening rule described by the function $\tilde{\Theta}$.

For example, if we choose the Hencky Strain energy function and used a von Mises yield condition like for small deformations (see [Anand (1979)] for a discussion of this energy function), we will get the following set of equations:

INELASTICITY WITH HENCKY POTENTIAL

$$\psi = \frac{1}{2}(c_1(\text{tr}\,\mathbf{e}_e))^2 + c_2\|\mathbf{e}_e\|^2,$$

$$\boldsymbol{\sigma} = \varrho\frac{\partial\psi}{\partial\mathbf{e}_e} = \varrho(c_1(\text{tr}\,\mathbf{e}_e)\mathbf{I} + c_2\mathbf{e}_e),$$

$$f(\boldsymbol{\sigma},\kappa) = \boldsymbol{\tau}\bullet\boldsymbol{\tau} - \kappa^2 \leq 0, \quad \boldsymbol{\tau} = \boldsymbol{\sigma}_{dev}/\varrho, \tag{12.17}$$

$$\mathbf{L}\mathbf{B}_e + \mathbf{B}_e\mathbf{L}^t - \dot{\mathbf{B}}_e = \phi\tilde{\mathbf{P}},$$

$$\dot{\kappa} = \phi\tilde{\Theta}(\mathbf{B}_e,\kappa),$$

$$\phi \geq 0, \quad f(\boldsymbol{\sigma},\kappa) = \boldsymbol{\tau}\bullet\boldsymbol{\tau} - \kappa^2 \leq 0, \quad \phi f = 0.$$

Another example is the following set of constitutive equations based on a compressible neo-Hookean elasticity model (see [Simo and Hughes (1998)] for a discussion of this strain energy function):

> **INELASTICITY WITH COMPRESSIBLE NEO-HOOKEAN POTENTIAL**
>
> $\psi = f(I_{3e}) + \mu/2(I_{3e}^{-1/3}\mathrm{tr}\mathbf{B}_e - 3), I_3 = \det(\mathbf{B}_e),$
> $\boldsymbol{\sigma} = 2\varrho(I_{3e}f'(I_{3e})\mathbf{I} + \mu I_{3e}^{-1/3}(\mathbf{B}_e)_{dev}), \quad \boldsymbol{\tau} = \boldsymbol{\sigma}_{dev}/\varrho,$
> $f(\boldsymbol{\sigma}) = \boldsymbol{\tau} \bullet \boldsymbol{\tau} - \kappa^2 \leq 0, \boldsymbol{\tau} = \mathbf{S}_{dev},$
> $\mathbf{L}\mathbf{B}_e + \mathbf{B}_e \mathbf{L}^t - \dot{\mathbf{B}}_e = \phi \tilde{\mathbf{P}},$ (12.18)
> $\dot{\kappa} = \phi \tilde{\Theta}(\mathbf{B}_e, \kappa),$
> $\phi \geq 0, \quad f \leq 0, \quad \phi f = 0.$

In both these models, we have yet to specify $\tilde{\mathbf{P}}$. This will be done via a finite deformation version of the normality rule. We will now state the normality rule that we are going to derive and use in this chapter.

> **NORMALITY RULE**
>
> For the models considered in this chapter, we will assume that ψ and f are isotropic scalar functions of $\mathbf{B}_e$ and $\mathbf{S}$ respectively, i.e. they depend only on the eigenvalues. In this case, (1) $\mathbf{S}, \mathbf{B}_e$ and $\mathbf{A}_p$ are coaxial, i.e. they have the same eigenvectors, and the normality rule takes the form
>
> $$\mathbf{L}\mathbf{B}_e + \mathbf{B}_e\mathbf{L}^t - \dot{\mathbf{B}}_e = \phi\tilde{\mathbf{P}}, \tilde{\mathbf{P}} = \frac{\partial f}{\partial \mathbf{S}}\mathbf{B}_e, \quad \mathbf{S} = \boldsymbol{\sigma}/\varrho. \quad (12.19)$$

That is it! That is all, nothing more, really! you can generate any number of other models by

- choosing a different constitutive equation for ψ;
- choosing a different yield function;
- choosing a different constitutive equation for $\tilde{\mathbf{P}}$ other than the normality flow rule.

The complete procedure for developing finite inelasticity theories by analogy is listed in Table 12.1. Please peruse this table carefully and convince yourself that the analogies are meaningful.

Of course, we haven't found an explicit expression for ϕ yet since the form of that expression depends upon the numerical scheme that is to be used for integrating the above expressions. Nor have we indicated how to incorporate hardening behavior.

If you want to go directly to the computational aspects, then just skip to a subsequent chapter. If you are experienced in the nuances of finite deformation

Table 12.1 A table illustrating analogous variables in elasticity and inelasticity, both for small deformation and for finite deformation of isotropic materials. The key points observe are that (1) $\mathbf{B}$ plays the role of total strain and (2) for inelasticity we introduce a variable $\mathbf{B}_e$ in analogy to elastic strain.

Concept	Linearized elasticity	Isotropic finite elasticity	Small strain plasticity	Isotropic finite plasticity
Elastic strain	ε	$\mathbf{B}$	ε_e	$\mathbf{B}_e$
Helmholtz potential	$\psi(\varepsilon)$		$\psi(\varepsilon_e)$	
Stress response	$\sigma = \rho_0 \dfrac{\partial \psi}{\partial \varepsilon}$	$\sigma = \rho \dfrac{\partial \psi}{\partial \mathbf{B}} \mathbf{B}$	$\sigma = \rho_0 \dfrac{\partial \psi}{\partial \varepsilon_e}$	$\sigma = \rho \dfrac{\partial \psi}{\partial \mathbf{B}_e} \mathbf{B}_e$
Total strain rate	$\dot{\varepsilon}$	$\dot{\mathbf{B}} = \mathbf{LB} + \mathbf{BL}^t$	$\dot{\varepsilon}$	$\dot{\mathbf{B}} = \mathbf{LB} + \mathbf{BL}^t$
Elastic strain rate	$\dot{\varepsilon}$	$\dot{\mathbf{B}} = \mathbf{LB} + \mathbf{BL}^t$	$\dot{\varepsilon}_e$	$\dot{\mathbf{B}}_e$
Plastic flow rate	0	0	$\dot{\varepsilon}_p := \dot{\varepsilon} - \dot{\varepsilon}_e$	$\mathbf{D}_p := \dfrac{1}{2} \mathbf{V}_e^{-1}(\mathbf{LB}_e + \mathbf{B}_e \mathbf{L}^t - \dot{\mathbf{B}}_e)\mathbf{V}_e^{-1}$
Driving force	0	0	$\xi = \sigma - \alpha$	$\xi = \mathbf{S} = \sigma/\rho$
Normality rule	-	-	$\dot{\varepsilon}_p = \phi \dfrac{\partial f}{\partial \xi}$	$\mathbf{D}_p = \phi \dfrac{\partial f}{\partial \xi}$
Hardening parameters, $\mathbf{Q}=\{\kappa,\alpha\}$	-	-	$\dot{\mathbf{Q}} = \phi \mathbf{\Theta}(\xi, \varepsilon_e, \mathbf{Q})$	$\dot{\kappa} = \phi \Theta(\mathbf{B}_e, \kappa), \quad \alpha = 0$
Loading criteria	-	-	$\phi \geq 0, f \leq 0, \phi f = 0.$	$\phi \geq 0, f \leq 0, \phi f = 0.$

then this is all you will need. But are you happy with the procedure laid out for this? Does it not seem somehow "incomplete" in the sense that you did not get a full understanding of what it "means"? We know that you are not just looking for a procedure, you are also looking for an understanding. Some of the questions for which you may want to find answers are:

- Is there any guidance for choosing $\tilde{\mathbf{P}}$ or f? What is acceptable?
- You may have seen papers where people have used variables other than $\mathbf{B}_e$ (for example, $\mathbf{V}_e$ or $\ln \mathbf{V}_e$) as variables. How do they fit in to the framework?
- How do normality and convexity criteria fit in?
- What if I wanted to develop models that are not "just" plasticity but involve time dependent effects?

We hope to provide you with answers to some of these questions in the following section.

12.5 Dissipative behavior of inelastic materials

The problem with the procedure as outlined in the previous section is that it is a "cookbook" procedure and hides the main aspects of inelasticity. At the very least, inelasticity should incorporate yield behavior, elastic response, dissipative response etc. At the same time, it should appeal to our physical intuition (always considering the possibility that our "intuition" may be wrong) and should avoid introducing inadvertent inconsistencies. For example, you might have chosen a constitutive equation that may result in a model of a perpetual motion machine. Or it might behave in a very strange manner when large rotations are involved. Avoiding these internal conflicts and providing a deeper understanding of finite inelasticity models is at the heart of remaining discussion that follows. We will try to help you gain a deeper understanding of the structure of isotropic inelasticity, through a series of rhetorical questions and answers[12].

Q: How does an inelastic material dissipate mechanical power?

One way to think about the dissipative mechanism is to imagine that the external force deforms the body by $\mathbf{B}$ but the material stores only the energy necessary to recover a portion of the deformation (namely $\mathbf{B}_e$), dissipating the rest of the mechanical power. Thus, the rate of change of $\mathbf{A}_p$ is directly related to dissipative processes.

Finding the expression for the rate of dissipation in terms of the rate of change of the chosen Helmholtz potential ψ will allow you to eliminate perpetual motion machines and suggest simple constitutive forms for $\tilde{\mathbf{P}}$.

---

[12]Somewhat like an FAQ.

Just the fact that we changed $\mathbf{B}$ to $\mathbf{B}_e$ in the isotropic elastic response to the new variable has unexpected consequences. For one thing, the material model is now capable of exhibiting dissipative response whereas the original elastic model was non-dissipative!

Just to jog your memory, recall that the rate of mechanical power supplied per unit volume is $\boldsymbol{\sigma}\cdot\mathbf{L}$ where $\mathbf{L}$ is the velocity gradient tensor. Further, recall that the rate of mechanical dissipation is the difference between the mechanical power supplied and the rate of increase of the isothermal work function, i.e.

$$\boldsymbol{\sigma}\cdot\mathbf{L} - \varrho\dot{\psi} = \varrho\zeta \quad \Rightarrow \quad \mathbf{S}\cdot\mathbf{L} - \dot{\psi} = \zeta, \tag{12.20}$$

where $\mathbf{S}$ is the Kirchhoff stress introduced earlier. Note that the above expression implies that *the mechanical power per unit mass is the inner product between the Kirchhoff stress and the velocity gradient.*

⊙ **12.4 (Exercise:). Dissipation in inelasticity**
Substitute for ψ and $\boldsymbol{\sigma}$ in the above equation (12.20) by using (12.10) and show that

$$\frac{\partial\psi}{\partial\mathbf{B}_e}\cdot[\mathbf{L}\mathbf{B}_e + \mathbf{B}_e\mathbf{L}^t - \dot{\mathbf{B}}_e] = \frac{\partial\psi}{\partial\mathbf{B}_e}\cdot(\mathbf{A}_p) = \zeta \geq 0. \tag{12.21}$$

(Hint: This is somewhat tricky unless you are used to doing these kinds of manipulations. Use the chain rule to expand $\dot{\psi}(\mathbf{B}_e)$. Next substitute for the stress from (12.10) and use the fact that $\mathbf{AB}\cdot\mathbf{C} = \mathbf{A}\cdot(\mathbf{CB})_{sym}$ if $\mathbf{A}$ is symmetric.)

You can see right away that the rate of dissipation depends upon the term $\mathbf{A}_p$. Comparing with the corresponding small strain version (see equation (9.59) on p.295), we see that there, the rate of dissipation (without hardening or backstress) was given by $\boldsymbol{\sigma}\cdot\dot{\boldsymbol{\varepsilon}}_p = \varrho\zeta$.

So, it appears that $\partial\psi/\partial\mathbf{B}_e$ plays the driving force for plastic flow and $\dot{\mathbf{B}}_e - \mathbf{L}\mathbf{B}_e - \mathbf{B}_e\mathbf{L}^t$ appears to play the role of the plastic strain rate $\dot{\boldsymbol{\varepsilon}}_p$. However, we want a closer agreement with small strain plasticity in the sense that we want the Kirchhoff stress to be the driving force, unfortunately in the form that we have right now, it does not appear to be. We will show soon that this form can be reformulated to ensure that $\mathbf{S}$ is the driving force for plastic flow.

The point of the above exercise is that *the rate of dissipation ζ has been written as an inner product between a term that is purely a function of the current state and is dependent upon the gradient of ψ (in this case $\partial\psi/\partial\mathbf{B}_e$) and a term which is related to the time derivative of a state variable and does not include any ψ de-*

pendence (in this case, it is the combination $\mathbf{A}_p$).

From the form (12.21), we see that $\mathbf{A}_p$ is a measure of the rate of plastic flow and $\boldsymbol{\xi} = \partial \psi / \partial \mathbf{B}_e$ is called the plastic driving force, and the rate of dissipation can be written as $\boldsymbol{\xi} \cdot \mathbf{A}_p$.

Note that, this split into $\boldsymbol{\xi}$ and $\mathbf{A}_p$ is NOT unique. We can write it in many different forms (such as $\mathbb{A}\boldsymbol{\xi} \cdot \mathbb{A}^{-1}\mathbf{D}_p$ where $\mathbb{A}$ is any symmetric fourth order tensor) and still obtain the same result, or we can add terms to $\mathbf{A}_p$ that are perpendicular to $\boldsymbol{\xi}$ without changing the rate of dissipation. This has led to much debate in the inelasticity literature and we will try to resolve it in the next subsection. However, from the point of view of the second law of thermodynamics, the following summary is important:

THERMODYNAMICALLY ALLOWABLE EVOLUTION EQUATIONS

The bottom line in introducing thermodynamics is that *any evolution equation (irrespective of whether it is rate dependent or rate independent) for $\mathbf{A}_p$ can be assumed as long as (12.21) is satisfied for all the allowable processes. Said differently, if you assume an evolution equation for $\mathbf{A}_p$, then that evolution equation should satisfy (12.21) allowed by thermodynamical considerations.*

For example, if you assume that $\mathbf{A}_p = \mu \partial \psi / \partial \mathbf{B}_e$ where μ is positive, then ANY process is thermodynamically possible, since (12.21) is always met. On the other hand, if we assume that $\mathbf{A}_p = \mu \mathbf{A}$ where μ is positive and $\mathbf{A}$ is some constant tensor, then only those processes for which $\partial \psi / \partial \mathbf{B}_e \cdot \mathbf{A} \geq 0$ will be possible. Thus depending upon your choice of constitutive equations, processes may or may not be restricted by (12.21).

Returning to (12.21), if we use the constitutive equation $\mathbf{A}_p = \phi \tilde{\mathbf{P}}(\mathbf{B}_e, \kappa)$ and substitute into (12.21), then we would obtain

$$\phi \frac{\partial \psi}{\partial \mathbf{B}_e} \cdot \tilde{\mathbf{P}} = \zeta \geq 0. \tag{12.22}$$

The non-negativity of the rate of dissipation, together with the fact that $\phi \geq 0$ immediately implies the following important condition :

THERMODYNAMICAL RESTRICTION ON $\tilde{\mathbf{P}}$

If the evolution equation $\dot{\mathbf{B}}_e - \mathbf{L}\mathbf{B}_e - \mathbf{B}_e\mathbf{L}^t = \phi\tilde{\mathbf{P}}$, $\phi \geq 0$ satisfies

$$\frac{\partial \psi}{\partial \mathbf{B}_e} \cdot \tilde{\mathbf{P}} > 0, \qquad (12.23)$$

for all processes, then the material response will satisfy the second law irrespective of the process and hence will NOT lead to perpetual motion.

⊙ **12.5 (Exercise:). Satisfying the Second Law**
Do the constitutive equations (12.17) satisfy the second law? How about (12.18)? (Hint: Note that $\frac{\partial \psi}{\partial \mathbf{B}_e}$ and $\tilde{\mathbf{P}}$ are coaxial; so write everything in terms of eigenvalues and eigenvectors and then calculate! This is a hard problem.)

Ans: True. But proving it for the constitutive equations (12.17) is hard. You might notice that the calculations would have been trivial if $\tilde{\mathbf{P}} = \frac{\partial \psi}{\partial \mathbf{B}_e}$! Why?

We hope the above discussion helped you understand how mechanical power is dissipated in an inelastic material and its relationship to the constitutive equation for $\mathbf{A}_p$. A convenient form for the constitutive equation for $\tilde{\mathbf{P}}$ is to make it depend upon $\mathbf{B}_e$ in such a way that checking the satisfaction of the second law should become trivial. We will see in the next section how to do this in a clever way.

Q: Are there other forms for the driving force & plastic flow rate?

Yes, other forms can be obtained, but only some of them are in line with our physical intuition as to what inelasticity should mean.

There isn't ONE correct form, but there are a number of acceptable choices. You might favor one or the other depending upon your CRITERIA of acceptance. *This is one of the primary reasons for the bewildering array of different models of finite deformation inelasticity.* Until the researchers in the area agree that one set of criteria trumps all others so that there is only one UNIQUE choice, we have to live the fact that different people will have different choices. What we are asking you to do, is to list your criteria and to avoid internal inconsistencies in your formulation.

In one way, this is no surprise. Unlike small deformation, where there is just one strain ε we saw that in finite deformation elasticity, there were a number of different but equivalent measures of strain and that the free energy can be written in many equivalent forms depending upon the strain measure used. The situation here is that there are a number of different measures of plastic flow rate and the rate of dissipation can be written in many forms depending upon the measure used. We hope that you will see that using different criteria is equivalent to choosing different constitutive equations—it all depends upon what you think is happening and how best to describe it.

The next exercise reinforces this ambiguity in defining a plastic flow rate.

> ⊙ **12.6 (Exercise:). Other forms for the rate of dissipation**
> Show that ζ can also be written as
>
> $$\mathbf{S} \cdot ((1/2 \dot{\mathbf{B}}_e \mathbf{B}_e^{-1})_{symm} - \mathbf{D}) = \zeta, \qquad (12.24)$$
>
> and
>
> $$\mathbf{S} \cdot \frac{1}{2}(\mathbf{V}_e^{-1}(\dot{\mathbf{B}}_e - \mathbf{L}\mathbf{B}_e - \mathbf{B}_e\mathbf{L}^t)\mathbf{V}_e^{-1}) = \zeta, \qquad (12.25)$$
>
> where $\mathbf{V}_e = \sqrt{\mathbf{B}_e}$. *Find two other forms for ζ. Which (if any) of these forms would you find acceptable or better? Can you list your criteria for choice?*

Thus, once you have an expression for ζ as an inner product between a "driving force" and a term involving $\dot{\mathbf{B}}_e$, then the term involving $\dot{\mathbf{B}}_e$ is a candidate for a plastic flow rate measure.

Q: How does one choose a suitable candidate as a measure of plastic flow?

Let us stop a minute and think about where we are going: we are trying to find a way to make sense of plastic flow and to obtain a way to quantify it that is consistent with our physical intuition. So we will list some of the attributes of plastic flow that we think are important.

(1) **Thermodynamic Consistency:** Plastic flow is dissipative, so any measure of plastic flow should be consistent with and can be made to appear as one of the two terms (the other being the driving force) that appear in the inner product in the expression for the dissipation function.

ACCEPTABLE MEASURES OF PLASTIC FLOW: CRITERION 1

Specifically, this implies that if $\mathbf{D}_p^*$ is a candidate for measure of plastic flow rate, that there should be some driving force $\boldsymbol{\xi}$ which is *linear* in $\partial\psi/\partial\mathbf{B}_e$ such that their inner product is equal to the rate of dissipation, i.e.

$$\mathbf{S}\bullet\mathbf{L} - \dot{\psi}|_{T=const.} = \zeta \;\Rightarrow\; \boldsymbol{\xi}\bullet\mathbf{D}_p^* = \zeta, \text{ where } \boldsymbol{\xi} = \mathbb{A}(\mathbf{B}_e)[\partial\psi/\partial\mathbf{B}_e] \quad (12.26)$$

where $\mathbb{A}$ is a fourth order symmetric tensor which could be a function of $\mathbf{B}_e$. The key point here is that $\boldsymbol{\xi}$ should be linear in $\partial\psi/\partial\mathbf{B}_e$.

---

(2) **Consistency with elasticity:** Plastic flow occurs ONLY when rate of elastic strain is not equal to the rate of total strain. Thus, if we apply the model to a purely elastic material by setting plastic flow rate to zero, we should get a model that is consistent with finite elasticity, i.e.

ACCEPTABLE MEASURES OF PLASTIC FLOW: CRITERION 2

If $\mathbf{D}_p^*$ is a measure of plastic flow then, elastic response should cause it to vanish, i.e.

$$\mathbf{D}_p^* = 0 \Leftrightarrow \dot{\mathbf{B}}_e = \mathbf{L}\mathbf{B}_e + \mathbf{B}_e\mathbf{L}^t. \quad (12.27)$$

---

(3) **Objectivity:** When the body rotates like a rigid body, we would expect no plastic flow to occur, i.e. the measure of plastic flow rate should be an objective or invariant tensor (either is acceptable since we can convert from one to the other by means of a simple transformation).

ACCEPTABLE MEASURES OF PLASTIC FLOW: CRITERION 3

If two motions start from the same configuration and have the same value of the referential stretch $\mathbf{C}(t) = \mathbf{F}^t\mathbf{F}$ for all time, then the value of $\mathbf{D}_p^*$ for the two motions should be invariant apart from orientation, i.e. if two motions 1 and 2 are such that

$$\mathbf{F}_1(t) = \mathbf{Q}\mathbf{F}_2(t) \;\Rightarrow\; (\mathbf{D}_p^*)_2 = \mathbf{Q}(\mathbf{D}_p^*)_1\mathbf{Q}^t \quad (12.28)$$

in other words, we demand that the plastic flow measure be an objective tensor.

The criteria (1), (2) and (3) are the minimum requirements for any sensible measure of plastic flow that is free from inconsistencies. Our approach to satisfying these criteria is to choose a form motivated by criterion (1) and then see if it can be made to satisfy (2) and (3).

To see if $\mathbf{A}_p$ meets the criteria above, note that (12.21) implies that criterion (1) is met with $\boldsymbol{\xi} = \partial \psi / \partial \mathbf{B}_e$. To check criterion (2), remember that for elasticity, the evolution equation for $\mathbf{B}_e$ should be that $\dot{\mathbf{B}}_e = \mathbf{L}\mathbf{B}_e + \mathbf{B}_e \mathbf{L}^t$. Clearly, this is satisfied when $\mathbf{A}_p = 0$. So the form (12.14) for $\mathbf{A}_p$ passes this test.

The next criterion is objectivity; here again, if we assume that $\mathbf{B}_e$ is to be like $\mathbf{B}$ for elastic response, we have to stipulate that $\mathbf{B}_e$ should be an objective tensor like $\mathbf{B}$. But then, $\dot{\mathbf{B}}_e$ (being the time rate of change of $\mathbf{B}_e$, an objective tensor) is NOT objective.

> ⊙ **12.7 (Exercise:). Objectivity of $\mathbf{A}_p$**
> Show that $\mathbf{A}_p$ is an objective tensor if we assume that $\mathbf{B}_e$ is objective.
> Hint: Consider two motions 1 and 2 such that $\mathbf{F}_2 = \mathbf{Q}\mathbf{F}_1$. Then we know that $(\mathbf{B}_e)_2 = \mathbf{Q}(\mathbf{B}_e)_1 \mathbf{Q}^t$. How are $\mathbf{L}_2$ and $\mathbf{L}_1$ related? Using this, show that $(\mathbf{A}_p)_2 = \mathbf{Q}(\mathbf{A}_p)_1 \mathbf{Q}^t$.

Indeed the formula (12.14) is an example of a very popular way to obtain objective tensors from rates of other tensors. The definition $\mathbf{D}_p = \mathbf{A}_p$ motivated from (12.21) thus passes the second test with flying colors.

But we want you to verify for yourself that $1/2(\dot{\mathbf{B}}_e \mathbf{B}_e^{-1})_{symm} - \mathbf{D}$ motivated by (12.24), will not satisfy either the requirement for elasticity nor will it be objective. However, the form

$$\mathbf{D}_p := 1/2(\mathbf{V}_e^{-1}(\dot{\mathbf{B}}_e - \mathbf{L}\mathbf{B}_e - \mathbf{B}_e \mathbf{L}^t)\mathbf{V}_e^{-1}) = 1/2\mathbf{V}_e^{-1}\mathbf{A}_p\mathbf{V}_e^{-1} \qquad (12.29)$$

motivated from (12.25) and being a product of objective tensors, is objective and satisfies all the criteria.

Thus, it appears that the above three criteria alone, while being necessary, are not sufficient to find a unique candidate, since both (12.14a) and (12.29) are acceptable. So, let us add another criterion:

ACCEPTABLE MEASURES OF PLASTIC FLOW: CRITERION 4

The driving force conjugate to any measure of plastic flow should be $\mathbf{S}$—the Kirchhoff stress.

Based on this and upon comparing (12.21) and (12.25), we see that the driving force conjugate to $\mathbf{D}_p$ defined by (12.29) is $\mathbf{S}$ and we have our "best" candidate[13].

At this stage you should be somewhat exasperated at the fact that there appears to be no unique "correct" way to do this unlike classical inelasticity and that you have to use all kinds of subjective criteria to find one that you like. Have no fear, we will try and convince you that the form (12.29) is an excellent way to develop inelasticity models since it meets our other additional criteria (which can also be ignored without any inconsistencies) such as

(i) $\mathbf{D}_p$ should have a convenient geometrical interpretation of its own and[14]
(ii) It should be convenient for numerical algorithms.

Q: Will different choices of the definition for plastic flow rates give different answers?

Definitions of quantities are just that—definitions; and they will not affect the response of the material. What matters is the *constitutive assumptions*. In our case, for the stress and the plastic flow rates. If the constitutive assumptions (based on the different definitions) are not equivalent to each other, then the response will be different. Then, why do we care so much about a "suitable" definition? The reason is that we are seeking a definition that will make our constitutive choices simple and physically appealing.

Q: What is the physical meaning of the form (12.29) for the plastic flow rate. How is it related to small strain inelasticity?

If we consider the evolution of the eigenvectors of $\mathbf{B}_e$, we can show that the form (12.29) can be written as the difference between the rate of stretching and the rate of change of logarithmic strain.

It turns out that the easiest way to see the physical interpretation of plastic flow is through the use of eigenvalues and eigenvectors. So we quickly recapitulate that $\mathbf{B}_e$ can be written as $\mathbf{B}_e = \sum_{i=1}^{3} b_e^i \mathbf{b}_i \otimes \mathbf{b}_i$ where b_e^i are the eigenvalues of $\mathbf{B}_e$ and $\mathbf{b}_i$ are the corresponding eigenvectors. Now consider the following exercise.

---

[13] Our victory is actually illusory since we looked through only a small set of choices in order to find a "best" candidate. There are in reality, infinitely many choices of plastic flow measures all of which satisfy the four criteria. But if you add "ease of computation" as a criterion, then we can show that $\mathbf{D}_p$ is a clear favorite.

[14] We will postpone the geometrical interpretation until the end of this chapter since it can be skipped if you are short of time or if you want to get on with the theory without any further need for explanations.

⊙ **12.8 (Exercise:). Rate of change of the eigenvalues of B_e**
Let the eigenvalues and eigenvectors of B_e be b_e^i and b_i respectively. Show that

$$\mathbf{b}_i \cdot \dot{\mathbf{B}}_e \mathbf{b}_i = \dot{b}_e^i. \tag{12.30}$$

Hint: Start from $\mathbf{B}_e \mathbf{b}_i = b_e^i \mathbf{b}_i$ take the time derivative on both sides and then take inner product with $\mathbf{b}_i$ Why do the terms involving $\dot{\mathbf{b}}_i$ vanish?

⊙ **12.9 (Exercise:). Interpreting D_p**
Show that if we choose (12.29) as the definition of plastic flow rate, i.e.

$$\mathbf{D}_p := \frac{1}{2} \mathbf{V}_e^{-1} (\mathbf{A}_p) \mathbf{V}_e^{-1}, \tag{12.31}$$

then take the inner product on both sides with $\mathbf{b}_i \otimes \mathbf{b}_i$, and obtain

$$D_{ii}^p = D_{ii} - \frac{1}{2}\overline{\ln b_e^i} = D_{ii} - \overline{\ln v_e^i} = D_{ii} - \overline{e_e^i} \tag{12.32}$$

where $D_{ii}^p = \mathbf{b}_i \cdot \mathbf{D}_p \mathbf{b}_i$ and $D_{ii} = \mathbf{b}_i \cdot \mathbf{D} \mathbf{b}_i$ are the normal components of $\mathbf{D}_p$ and $\mathbf{D}$ along the eigenvectors of $\mathbf{B}_e$ and $v_e^i = \sqrt{(b_e^i)}$ are the eigenvalues of the elastic stretch $\mathbf{V}_e$ and e_e^i are the eigenvalues of the logarithmic strain $\mathbf{e}_e$. Discuss the analogy with the small deformation result $\dot{\varepsilon}_p = \dot{\varepsilon} - \dot{\varepsilon}_e$.
(Hint: Use (12.30) and the fact that $\mathbf{B}_e \mathbf{b}_i = b_e^i \mathbf{b}_i$. You will also need to figure out eigenvalues and eigenvectors of $\mathbf{V}_e^{-1}$ in terms of those for $\mathbf{B}_e$.)

The above exercise shows that

PHYSICAL SIGNIFICANCE OF D_p

The normal component of $\mathbf{D}_p$ along with the eigenvectors of $\mathbf{B}_e$ is the difference between the total stretching along that direction and rate of change of logarithmic strain along that direction.

The above result shows how close the definition of $\mathbf{D}_p$ is to $\dot{\varepsilon}^p$ in small strain. Hence, for the remainder of the chapter, we will use the expression (12.31) as the measure of plastic flow rate.

Q: So what is the bottom line? In examples (12.17) and (12.18) you give constitutive equations for $\mathbf{A}_p$. Now you are saying that $\mathbf{D}_p$ is the one. Which is it?

The beauty of expression (12.29) is that writing a constitutive equation for $\mathbf{A}_p$ *is equivalent to writing a different one for* $\mathbf{D}_p$ For example, if you set

$$\mathbf{D}_p = \phi \mathbf{P} \tag{12.33}$$

for ANY function $\mathbf{P}$, this is the same as writing

$$\mathbf{A}_p = 2\mathbf{V}_e \mathbf{D}_p \mathbf{V}_e = 2\phi \mathbf{V}_e \mathbf{P} \mathbf{V}_e = \tilde{\phi} \tilde{\mathbf{P}} \tag{12.34}$$

where $\tilde{\phi} = 2\phi$ and $\tilde{\mathbf{P}} = \mathbf{V}_e \mathbf{P} \mathbf{V}_e$. Furthermore, if $\mathbf{P}$ happens to be coaxial with $\mathbf{B}_e$ then we can simplify even further (by using commutativity) and get

$$\mathbf{D}_p = \phi \mathbf{P} \Leftrightarrow \mathbf{A}_p = 2\phi \mathbf{P} \mathbf{B}_e. \tag{12.35}$$

If they are equivalent, then why do we bother? Well it turns out that $\mathbf{D}_p$ is the one with similarity to small deformation and one for which the normality flow rule works in terms of the Kirchhoff stress whereas $\mathbf{A}_p$ is the one that is easier to compute and implement. So we will obtain the normality rule in terms of $\mathbf{D}_p$ and then transform it into a constitutive equation for $\mathbf{A}_p$ by using (12.35).

Q: What happens to the normality flow rule in finite deformation?

For finite deformations, the normality and convexity conditions reduce to the assumption that we have a convex yield function $f(\mathbf{S})$ and require that the plastic flow rate defined by (12.29) be directed along the gradient to the yield function.

This point has created some confusion in the literature. For example [Simo (1988)] and [Simo and Hughes (1998)] simply assume without proof that normality means that $\mathbf{A}_p$ is directed along the outward normal to the yield surface in stress space (see section 9.2.1.3 of [Simo and Hughes (1998)] p. 309). However, $\mathbf{S} \bullet \mathbf{A}_p$ is NOT the plastic dissipation as we have seen, so, while their answer does not violate any thermodynamic principle, it is not based on "plastic dissipation" as derived here.

We have seen that $\mathbf{D}_p$ defined by (12.29) is a measure of plastic flow rate and that $\mathbf{S}$ (which is the Kirchhoff stress) is the conjugate driving force (see equation (12.25) on p.421).

We are now going to explore the notion of maximum rate of dissipation and the finite deformation generalization of the normality flow rule.

Procedurally, to enforce the normality flow rule,

- Assume a yield function as a convex function of $\boldsymbol{\xi} = \mathbf{S}$.
- Stipulate that $\mathbf{D}_p$ is directed along the outward normal to this yield surface.

For example, we could assume that the yield function is given by $f = \mathbf{S}\cdot\mathbf{S} - \kappa^2$ and then the normality flow rule would be $\mathbf{D}_p = \phi\mathbf{P}$, $\mathbf{P} = \partial f/\partial \mathbf{S} = 2\mathbf{S}$, so that we get

$$\frac{1}{2}\mathbf{V}_e^{-1}(\mathbf{A}_p)\mathbf{V}_e^{-1} = 2\phi\mathbf{S} \Rightarrow \mathbf{A}_p = 4\phi\mathbf{V}_e\mathbf{S}\mathbf{V}_e \tag{12.36}$$

The above flow rule can be further simplified if we notice that $\mathbf{S}$ (which is an isotropic tensor function of $\mathbf{B}_e$) and $\mathbf{V}_e$ are coaxial and hence commute. This means that we can write the normality flow rule as

$$\mathbf{A}_p = \phi\mathbf{S}\mathbf{B}_e. \tag{12.37}$$

We have "swallowed" the constants into the definition of ϕ in the above expression. Now we see that we don't need to find the square root of $\mathbf{B}_e$ or its inverse to implement the flow rule.

NORMALITY RULE

In general, if we have a yield function of the form $f(\mathbf{S}, \kappa)$ *and it is an isotropic function of* $\mathbf{S}$, then with the use of $\mathbf{D}_p$ (defined in (12.29)) as the measure of plastic flow rate and the corresponding conjugate force being the Kirchhoff stress $\mathbf{S}$, the normality flow rule implies that

$$\mathbf{D}_p = \phi\frac{\partial f}{\partial \mathbf{S}} \Leftrightarrow \mathbf{A}_p = \mathbf{L}\mathbf{B}_e + \mathbf{B}_e\mathbf{L}^t - \dot{\mathbf{B}}_e = 2\phi\frac{\partial f}{\partial \mathbf{S}}\mathbf{B}_e. \tag{12.38}$$

It is this form that we will use in the numerical implementation.

Can the normality flow rule be derived from the maximum rate of dissipation criterion?

The normality flow rule can also be considered as a consequence of a maximum rate of dissipation assumption and can be derived from it.[15]

We first recall that

$$\mathbf{S}\cdot\mathbf{D}_p = \zeta \geq 0, \tag{12.39}$$

is the condition that we must satisfy, and *any constitutive equation for* $\mathbf{D}_p$ *that satisfies this condition is acceptable.* As in small strain inelasticity, the normality flow rule can be obtained as that direction of $\mathbf{D}_p$ that guarantees that ζ is maximum.

To see how this happens, as with small strain plasticity, we first stipulate that ζ be a positively homogeneous function of $\mathbf{D}_p$, i.e. assume that

$$\zeta = \zeta(\mathbf{D}_p, \kappa), \quad \zeta(\lambda\mathbf{D}_p, \kappa) = \lambda(\zeta(\mathbf{D}_p, \kappa)), \quad \forall \lambda > 0, \tag{12.40}$$

---

[15] This part is a somewhat technical section involving some elements of convex analysis. You can skip this without affecting the flow of the chapter.

where κ is the yield strength parameter associated with the resistance to plastic flow. This is a condition that ensures that the material is rate independent. A trivial example of such a function is $\zeta = \kappa \|\mathbf{D}_p\|$.

Then we invoke the maximum rate of dissipation criterion in the following form

---

THE MAXIMUM RATE OF DISSIPATION HYPOTHESIS

For given values of the state variables ϱ and $\mathbf{B}_e$ (and hence of ζ, and the stress $\mathbf{S}$, since they are functions of these variables), among all the possible values of $\mathbf{D}_p$ which satisfy (12.39), the actual value of $\mathbf{D}_p$ is the one that maximizes the rate of dissipation $\zeta(\mathbf{D}_p)$.

The above criterion will then deliver the constitutive equation for $\mathbf{D}_p$ although this is not obvious. We will show you how this occurs by means of an example and then we can consider the general result.

⊙ **12.10 (Exercise:). Normality rule in ξ space**
By following the standard procedure for constrained maximization using Lagrange multipliers (see Chapters 2 and 9 if you have forgotten), show that the maximum rate of dissipation criterion implies that

$$\mathbf{S} = \lambda \frac{\partial \zeta}{\partial \mathbf{D}_p} \quad \text{if } \mathbf{D}_p \neq 0. \tag{12.41}$$

where the value of λ is obtained from satisfying (12.39). Show that, if $\zeta = \kappa \|\mathbf{D}_p\|$ then we get

$$\mathbf{S} = \frac{\partial \zeta}{\partial \mathbf{D}_p} = \kappa \frac{\mathbf{D}_p}{\|\mathbf{D}_p\|} \quad \text{if } \mathbf{D}_p \neq 0. \tag{12.42}$$

Observe that the equation above can also be written as $\mathbf{D}_p = \phi \mathbf{S}$ which is a mathematical statement to the effect that "$\mathbf{D}_p$ and $\mathbf{S}$ are parallel".

In the above equation, the form of the rate of dissipation ζ is simple enough that we could obtain $\mathbf{D}_p$ as a function of $\mathbf{S}$. In general, inverting the expression $\mathbf{S} = \partial \zeta / \partial \mathbf{D}_p$ to obtain $\mathbf{D}_p$ in terms of $\mathbf{S}$ is not a simple task (since ζ could be any positively homogeneous function of $\mathbf{D}_p$) and requires a fair bit of convex analysis. The method for the construction of the yield function and the normality rule are stated briefly below:

THE YIELD FUNCTION

Following the procedure used in chapter 9 (see equation (9.80) on p.306) we first introduce the yield function by defining

$$f(\mathbf{S}, \kappa) = \max_{\|\mathbf{D}_p\|=1} \frac{\mathbf{S} \cdot \mathbf{D}_p}{\zeta(\mathbf{D}_p, \kappa)}. \tag{12.43}$$

Under rather mild conditions on ζ, it can be shown using convex analysis that the inequality $f(\mathbf{S}, \kappa) - 1 \leq 0$ defines a convex set in the six dimensional $\mathbf{S}$ space. Further if ζ is an isotropic function of $\mathbf{D}_p$ then f is an isotropic function of $\mathbf{S}$. The condition (12.42) implies that

$$\mathbf{D}_p = \phi \frac{\partial f}{\partial \mathbf{S}} \Rightarrow \mathbf{A}_p = 2\phi \frac{\partial f}{\partial \mathbf{S}} \mathbf{B}_e \tag{12.44}$$

Note that there is a factor of two difference between the flow rule in terms of $\mathbf{D}_p$ and that in terms of $\mathbf{A}_p$. So the Kuhn-Tucker parameters for the two cases are NOT the same. Please take care when interpreting results as to which form of the flow rule is used.

Thus, we note that if we are able to obtain the yield function through using (12.43) then the normality rule follows. The function $f(\mathbf{S})$ is called the "polar" of the function $\zeta(\mathbf{D}_p)$ in the convex analysis literature and the polars of several common positively homogeneous functions are listed in [Rockafellar (1970)]). Note that *if the dissipation function ζ is an isotropic function of $\mathbf{D}_p$, then the yield function f will also be an isotropic function of $\mathbf{S}$. This in turn will imply that $\mathbf{D}_p$—which is along the gradient of f— will be coaxial with $\mathbf{S}$.* This fact is important because, being coaxial, they will commute and this will enable us to simplify the constitutive equations.

⊙ **12.11 (Exercise:).** Explicit construction of the yield function from ζ

Assume that $\zeta = \kappa \|\mathbf{D}_p\|$. Then show that (12.43), delivers

$$f(\mathbf{S}) = \frac{\|\mathbf{S}\|}{\kappa}. \tag{12.45}$$

Also show that

$$\mathbf{D}_p = \phi \mathbf{S}. \tag{12.46}$$

Hint: Show that the maximum on the right-hand side of (12.43) occurs when $\mathbf{D}_p$ is along $\mathbf{S}$.

Thus, for $\zeta = \kappa \|\mathbf{D}_p\|$ the maximum rate of dissipation criterion implies that the yield surface is $f(\mathbf{S}) = \|\mathbf{S}\|/\kappa = 1$ and the normality flow rule becomes the statement that $\mathbf{D}_p$ is proportional to $\partial f/\partial \mathbf{S} = \phi \mathbf{S}$.

Given this painful way of obtaining f what are you supposed to do? Nothing! Simply ignore $\zeta(\mathbf{D}_p)$ and assume a positively homogeneous convex function $f(\mathbf{S}, \kappa)$ and then use the normality condition (12.44). This will guarantee the satisfaction of the maximum dissipation criterion AND satisfy the second law of thermodynamics!

12.5.1 Isotropic hardening in finite plasticity

HARDENING AND PLASTIC ARC LENGTH

The finite deformation generalization of the plastic arc length parameter defined by $\dot{s} = \|\dot{\boldsymbol{\varepsilon}}_p\|$ is

$$\dot{s} = \|\mathbf{D}_p\| = \frac{1}{2}\|\mathbf{A}_p \mathbf{B}_e^{-1}\| = \frac{\phi}{2}\|\mathbf{P}\|, \qquad (12.47)$$

where we have used the normality rule in the form $\mathbf{A}_p = \phi \mathbf{P} \mathbf{B}_e$. We can then define the isotropic hardening parameter as usual as

$$\dot{\kappa} = \Theta(\kappa, s)\dot{s} = \frac{\phi}{2}\Theta\|\mathbf{P}\|. \qquad (12.48)$$

This implies that $\tilde{\Theta}$ defined in Eq. (12.18e) p. 415 is given by $\tilde{\Theta} = \Theta\|\mathbf{P}\|$.

Summarizing the developments so far, we have the following set of equations:

> **Thermodynamically consistent model for isotropic inelastic materials**
>
> List of state variables: $\mathbb{S} = \{\mathbf{B}_e, \boldsymbol{\sigma}\}$,
>
> The strain energy function: $\psi = \psi(\mathbf{B}_e)$,
>
> The stress response function: $\mathbf{S} = \dfrac{\boldsymbol{\sigma}}{\varrho} = 2\dfrac{\partial \psi}{\partial \mathbf{B}_e}\mathbf{B}_e$,
>
> The yield function: $f(\mathbf{S}, \kappa) - 1 \leq 0$
> $$\Rightarrow g(\mathbf{B}_e, \kappa) = f(\mathbf{S}(\mathbf{B}_e), \kappa) \leq 1,$$
>
> The plastic potential: $H(\mathbf{S}, \kappa) = f(\mathbf{S}, \kappa)$, $\mathbf{P} = \mathbf{N} = \dfrac{\partial f}{\partial \mathbf{S}}$, (12.49)
>
> The flow normality rule: $\mathbf{A}_p := \mathbf{L}\mathbf{B}_e + \mathbf{B}_e\mathbf{L}^t - \dot{\mathbf{B}}_e = \phi \mathbf{P}\mathbf{B}_e$,
>
> The plastic arc length: $\dot{s} := \dfrac{1}{2}\|\mathbf{A}_p\mathbf{B}_e^{-1}\|$
>
> The hardening rule: $\dot{\kappa} = \Theta(\kappa, s)\dot{s} = \dfrac{\phi}{2}\Theta \|\mathbf{P}\|$,
>
> The plastic flow conditions: $\phi \geq 0$, $f \leq 0$, $\phi f = 0$.

Comparison with the corresponding small strain equations equation (12.3) on p.408 demonstrates that the following analogies have been made:
(1) The elastic strain $\boldsymbol{\varepsilon}_e$ becomes $\mathbf{B}_e$,
(2) The driving force $\boldsymbol{\xi}$ now becomes $\mathbf{S} = \boldsymbol{\sigma}/\varrho$,
(3) The plastic strain rate $\dot{\boldsymbol{\varepsilon}}_p = \dot{\boldsymbol{\varepsilon}} - \dot{\boldsymbol{\varepsilon}}_e$ now becomes $\mathbf{D}_p = 1/2\mathbf{V}_e^{-1}(\mathbf{A}_p)\mathbf{V}_e^{-1}$.
The rest of the constitutive equation is the same.

⊙ **12.12 (Exercise:). Example constitutive equations**
By using (12.49c,d,e), derive explicit forms for $\mathbf{S}$, $\mathbf{P}$ in terms of $\mathbf{B}_e$ and κ, given the free energy function ψ and yield function f of the form

$$\psi = f(\bar{J}) + \mu/2(\bar{J}^{-1/3}(tr\mathbf{B}_e) - 3), \quad f(\mathbf{S}) = \boldsymbol{\tau}\cdot\boldsymbol{\tau} - \kappa^2, \qquad (12.50)$$

where $\boldsymbol{\tau}$ is the deviatoric part of $\mathbf{S}$ and is equal to $\mathbf{S} - (tr\mathbf{S}/3)\mathbf{I}$. This is a solid with a compressible neo-Hookean elastic behavior and a von Mises yield condition. Assume that $\Theta = 0$, i.e. no hardening.
Ans: $\mathbf{S} = 2I_3 f'\mathbf{I} + \mu I_3^{-2/3}\bar{\mathbf{B}}_e$ and $\mathbf{P} = 2\boldsymbol{\tau} = 2\mu I_3^{-1/3}\bar{\mathbf{B}}_e$, where $\bar{\mathbf{B}}_e$ is the deviatoric part of $\mathbf{B}_e$ and is defined by

$$\bar{\mathbf{B}}_e := \mathbf{B}_e - tr(\mathbf{B}_e)\mathbf{I}. \qquad (12.51)$$

12.6 Loading/unloading criteria and the value of ϕ; The advantages of strain space yield functions

The procedure for finding ϕ is the SAME as for small deformation: Differentiate the yield function, set it to zero and solve for ϕ. However, it is much simpler to rewrite the yield function in terms of $\mathbf{B}_e$ (rather than in terms of $\mathbf{S}$ and to work in the eigenspace of $\mathbf{B}_e$).

The final piece of the puzzle, before we can embark on the solution of initial and boundary value problems is the explicit characterization of ϕ. From (12.49), we note that ϕ is non-negative and is intimately connected with f. As with small strains, the trick to finding an acceptable value of ϕ is to note that as long as ϕ is nonzero, f must vanish. Thus, during a process, if plastic flow takes place, then $\dot{f}$ must be zero, since f is a function of $\boldsymbol{\xi}$ and κ, we differentiate it with respect to time and solve for ϕ. Since by now, you should have done this dozens of times, we will carry out this process in a series of exercises (with hints) so that you can test yourself on your understanding.

Here is where one of the major differences between small strain and finite strain inelasticity appears. In small strain inelasticity, it is convenient to do everything in terms of the stress (since the strains are small anyway and it is easy to write the strains in terms of the stress. This is NOT the case when the elastic strains are not small. In this case, it is better to write the yield function in terms of $\mathbf{B}_e$ and then do the differentiation. Why is this? The answer lies in the fact that the equation for the stress is nonlinear and is quite hard to manipulate. Substituting for $\mathbf{S}$ in terms of $\mathbf{B}_e$ and exploiting the fact that the material is isotropic will avoid tedious calculations involving the derivative of the stress with respect to $\mathbf{B}_e$ (which is never a fun thing to calculate) and leads to a rather direct way to find ϕ.

Let us see how this works with reference to the constitutive equation (12.49). We (by which we mean *you* gentle reader:-))) will do this in many steps so that the whole structure is revealed.

Step 1: Writing f in terms of $\mathbf{B}_e$ and obtaining a strain space yield function:

The idea here is to substitute for $\mathbf{S}$ in (12.49d) using the expression (12.49c) and thus obtaining

$$g(\mathbf{B}_e, \kappa) = f(\mathbf{S}(\mathbf{B}_e), \kappa). \tag{12.52}$$

⊙ **12.13 (Exercise:). Form for g**
Let $f = \boldsymbol{\tau} \cdot \boldsymbol{\tau} - \kappa^2$ where $\boldsymbol{\tau}$ is the deviatoric part of $\mathbf{S}$. Using the constitutive equations (12.18), show that

$$g(\mathbf{B}_e, \kappa) = \boldsymbol{\tau}(\mathbf{B}_e) \cdot \boldsymbol{\tau}(\mathbf{B}_e) - \kappa^2 = \mu^2 I_{3_e}^{-2/3} \bar{\mathbf{B}}_e \cdot \bar{\mathbf{B}}_e - \kappa^2, \qquad (12.53)$$

where $\bar{\mathbf{B}}_e$ is the deviatoric part of $\mathbf{B}_e$. What would be the form for g if we used the form (12.17) for the stress?

To simplify notation, and to be consistent with small strain plasticity, we will list and use the following variables

$$\text{Plastic flow direction: } \mathbf{A}_p = \phi \mathbf{P} \mathbf{B}_e \Leftrightarrow \mathbf{D}_p = \frac{1}{2} \phi \mathbf{P},$$

$$\text{Normal to yield surface in stress space: } \mathbf{N} = \frac{\partial f}{\partial \mathbf{S}},$$

$$\text{Tangent Modulus: } \mathbb{C} = \frac{\partial \mathbf{S}}{\partial \mathbf{e}_e} = \frac{\partial^2 \psi}{\partial \mathbf{e}_e^2},$$

$$\text{Normal to yield surafece in elastic strain space: } \mathbf{M} = \frac{\partial g}{\partial \mathbf{e}_e} = 2 \frac{\partial g}{\partial \mathbf{B}_e} \mathbf{B}_e = \mathbb{C} \mathbf{N}.$$

$$(12.54)$$

Step 2: Differentiating g:

We know that if $g = 0$ and $\phi > 0$ (i.e. if loading occurs, then from the Kuhn-Tucker conditions, we must have $\dot{g} = 0$. This simple condition allows us to compute ϕ. To see how this is done, first differentiate g with respect to time to get

$$\dot{g} = \frac{\partial g}{\partial \mathbf{B}_e} \cdot \dot{\mathbf{B}}_e - \frac{\partial g}{\partial \kappa} \dot{\kappa} = 0. \qquad (12.55)$$

We note that $\dot{s} = \|\mathbf{D}_p\| = 1/2 \phi \Theta \|\mathbf{P}\|$. Now upon using the evolution equations (12.49)f,h, in the above equation we will get[16]

$$\dot{g} = \mathbf{M} \cdot \mathbf{D} - 1/2 \phi (\mathbf{M} \cdot \mathbf{P} - 1/2 \|\mathbf{P}\| \Theta \partial g / \partial \kappa) = 0$$

$$\Rightarrow \phi = \frac{2 \mathbf{M} \cdot \mathbf{D}}{\mathbf{M} \cdot \mathbf{P} - \|\mathbf{P}\| \Theta \partial g / \partial \kappa} \qquad (12.56)$$

$$\Rightarrow \phi = \frac{2 \mathbf{M} \cdot \mathbf{D}}{\mathbf{M} \cdot \mathbf{N} - \|\mathbf{N}\| \Theta \partial g / \partial \kappa}.$$

The form listed above is very close to that for small strain plasticity as you can see by comparing (12.56) and the last of (12.3). Even though this equation appears simple, it is actually quite involved to calculate it for a specific model (compared to small strain plasticity) and will generally tax your skills in tensor algebra and

---
[16] After some manipulation, using the fact that g is an isotropic function of $\mathbf{B}_e$.

calculus. The main difficulty is the explicit calculation of two terms $\mathbf{M} = \partial g/\partial \mathbf{B}_e$ and $\mathbf{P} = \partial f/\partial \mathbf{S}$ and the dot products in the numerator and denominator. You can see this for yourself in the following exercise

⊙ **12.14 (Exercise:). Explicit form for ϕ**
Assume the free energy function is given by the compressible neo-Hookean model (12.18) and g is given by (12.53). Then show that

1. $\mathbf{M} = 4\mu^2 I_3^{-2/3} \bar{\mathbf{B}}_e \mathbf{B}_e - \dfrac{4}{3}\kappa^2 \mathbf{I}$.

2. $\mathbf{P} = \mathbf{N} = 2\boldsymbol{\tau} = 2\mu I_3^{-1/3} \bar{\mathbf{B}}_e$.

3. $2\mathbf{M}\cdot\mathbf{D} = 8\mu^2 I_3^{-2/3} \bar{\mathbf{B}}_e \mathbf{B}_e \cdot \mathbf{D} - \dfrac{8}{3}\kappa^2 tr\mathbf{D}$. (12.57)

4. $\mathbf{P}\cdot\mathbf{M} = 8\mu^2 I_3^{-2/3} \boldsymbol{\tau} \cdot \bar{\mathbf{B}}_e \mathbf{B}_e$
   $= 8\mu^3 I_3^{-1} \bar{\mathbf{B}}_e \mathbf{B}_e \cdot \bar{\mathbf{B}}_e$.

(Hint: You need to use the fact that $\mu^2 I_3^{-2/3} \bar{\mathbf{B}}_e \cdot \bar{\mathbf{B}}_e = \kappa^2$ and the fact that both $\bar{\mathbf{B}}_e$ (which is the deviatoric part of $\mathbf{B}_e$) and $\boldsymbol{\tau}$ are traceless. You will also need to use the fact that $\boldsymbol{\tau}, \mathbf{B}_e, \bar{\mathbf{B}}_e$ are coaxial and hence commute with each other. This is a hard calculation, and will test your manipulation skills quite a bit.)

By using the above results we can show that, for the special set of constitutive equations (12.50) and with g given by (12.53), we obtain the following explicit form for ϕ:

$$\phi = \frac{2\mu^2 I_3^{-2/3} \bar{\mathbf{B}}_e \mathbf{B}_e \cdot \mathbf{D} - \tfrac{2}{3}\kappa^2 tr\mathbf{D}}{2\mu^2 I_3^{-1} \bar{\mathbf{B}}_e \mathbf{B}_e \cdot \bar{\mathbf{B}}_e + \kappa^2 \Theta}. \qquad (12.58)$$

Step 3: Identifying loading criteria:

The solution (12.56) for ϕ is valid only so long as it is positive. Thus, we need to find out what happens to ϕ for other cases as well. The situation when $g < 0$ is simple: we simply have the condition $\phi = 0$. What happens when $g = 0$? There are two possibilities for ϕ namely it could be zero or positive. Thus, our solution (12.56) is clearly valid only as long as the ratio on the right-hand side is positive. So we must have[17]

$g < 0 \Rightarrow \phi = 0$ (purely elastic response)

$g = 0, \phi = 0 \Rightarrow \dot{g} < 0$, (unloading: going back into the elastic domain) (12.59)

$g = 0, \phi > 0 \Rightarrow \dot{g} = 0$, (plastic flow occurs)

We are now going to make a stronger statement:

---

[17] These are equivalent to the strain space yield conditions of Naghdi and Trapp [Naghdi and Trapp (1975c)], see also [Casey (2002)].

> **LOADING CRITERIA**
>
> It is possible to obtain a unique value for ϕ (irrespective of the value of $\mathbf{D}$) only if the denominator on the right-hand side of (12.56) is positive. If the denominator is negative, we will either get multiple solutions for ϕ or none at all.

Why is this so? We first observe that the numerator can be written as

$$\dot{g}|_{\text{no plastic flow}} = \frac{\partial g}{\partial \mathbf{B}_e} \cdot \dot{\mathbf{B}}_e|_{\text{no plastic flow}} \qquad (12.60)$$
$$= 2\mathbf{M} \cdot \mathbf{D}.$$

In other words *the numerator on the right-hand side of (12.56b) is nothing but the rate of change of g when no plastic flow occurs*[18]. *This term is usually referred to as $\hat{g}$ and is a very important quantity.*

Let us assume that the denominator on the RHS of (12.56b) is negative. Let us see what this will lead to. First note that the denominator is purely a function of the state and so if we know $\mathbf{B}_e$ and κ we can find its value. On the other hand, the value of the numerator depends linearly upon $\mathbf{D}$. So if we don't like the value of $\hat{g}$ we can change it by changing $\mathbf{D}$. Now let us assume that $\hat{g}$ is positive. Then this means that ϕ will be negative, which is not allowed. So the solution (12.56) is not valid and we will have to choose $\phi = 0$. But if you do that, then $\dot{g} = \hat{g} > 0$ which means that we will go outside the yield surface. This leads to a contradiction. There is no value of ϕ that will work here. Thus we have a condition that the denominator of (12.56b) must be positive; then the sign of the numerator $\hat{g}$ will decide whether plastic flow occurs or not.

Thus the various conditions for loading etc., can be put in a compact format as follows:

$$\phi = \frac{2h(g)\langle \mathbf{M} \cdot \mathbf{D} \rangle}{(\mathbf{P} \cdot \mathbf{M}) - (\partial g/\partial k)\Theta \|\mathbf{P}\|}, \quad \mathbf{M} = 2\frac{\partial g}{\partial \mathbf{B}_e}\mathbf{B}_e, \quad \mathbf{P} = \frac{\partial f}{\partial \mathbf{S}}, \qquad (12.61)$$

where, as usual, $h(g)$ is the Heaviside step function which is zero if $g < 0$ and is equal to one otherwise and the notation $\langle x \rangle$ stands for the ramp function and is a short form for $0.5 * (x + \|x\|)$.

We are now ready to carry out a complete example calculation to show how this works. Let us consider a model defined by the Hencky strain energy function and the von Mises yield condition, i.e. let

$$\psi = \frac{1}{2}\{c_1(\text{tr}\mathbf{e}_e)^2 + c_2\mathbf{e}_e\}, \quad \boldsymbol{\tau} = (\mathbf{S})_{dev},$$
$$f = \boldsymbol{\tau} \cdot \boldsymbol{\tau} - \kappa^2, \qquad (12.62)$$

---

[18] Moreover, by now you should be able to see that when no plastic flow occurs, $\dot{\mathbf{B}}_e = \mathbf{L}\mathbf{B}_e + \mathbf{B}_e\mathbf{L}^t$ without prompting from us.

where e_e is the logarithmic strain, $1/2 \ln \mathbf{B}_e$. As far as we know, this model is the closest that we can come to a Prandtl-Reuss theory. For this reason we could call it the neo-Prandtl-Hencky theory (similar to the use of the term neo-Hookean for the finite deformation extension of the Hookean elasticity model). For the case of the neo-Prandtl-Hencky model, it is possible to obtain an expression for ϕ using tensor notation as we did in the previous section, because the stress response as well as the yield function was given in tensor notation.

⊙ **12.15 (Exercise:). S and g for the neo-Prandtl-Hencky inelasticity model**
Show that, given the strain energy and yield functions as in (12.62), the Kirchhoff stress $\mathbf{S}$, its deviatoric part τ and the yield function g are given by, respectively,

$$\mathbf{S} = c_1(tr\mathbf{e}_e)\mathbf{I} + c_2 \mathbf{e}_e,$$
$$\tau = \mathbf{S}_{dev} = c_2(\mathbf{e}_e - \frac{1}{3} tr(\mathbf{e}_e)\mathbf{I}) = c_2 \mathbf{\Gamma}_e, \quad (12.63)$$
$$g = c_2^2 \|\mathbf{\Gamma}_e\|^2 - \kappa^2.$$

Compare the above form with that for the small strain Prandtl-Reuss equations (12.4). In what ways is it identical? What are the significant differences if any?

Now that we have a form for the yield function etc., we can calculate the Kuhn-Tucker parameter ϕ as follows:

⊙ **12.16 (Exercise:). The value of ϕ for the neo-Prandtl-Hencky inelasticity model**
Using the results of the last exercise, obtain an explicit expression for the value of ϕ in (12.49f). Compare this to the form for the small strain Prandtl-Reuss equations (see equation (6.61) on p.206).
(Hint: Find $\mathbf{M}$ and $\mathbf{N}$ by using their original definitions in terms of the logarithmic strain $\mathbf{e}_e = 1/2 \ln \mathbf{B}_e$. Also, try and use stresses instead of strains as much as possible.)
Ans:

$$\phi = \frac{h(g) 2 c_2 \langle \tau \cdot \mathbf{D} \rangle}{\kappa^2 (c_2 + \Theta)} \quad (12.64)$$

Why is there a factor of two difference between this equation and (6.61)? (Hint: Are the definitions of ϕ in the two cases the same when we assume small deformations?)

12.6.1 Results in terms of eigenvalues

Unlike in small strain plasticity, the expression (12.61) cannot be generally used easily to find ϕ since, in many cases, the function g may not be easily differentiable in terms of $\mathbf{B}_e$ or $\mathbf{e}_e$. Since g is obtained from f by substituting for $\mathbf{S}$, whether or not g can be differentiated depends upon
(1) The complexity of f, and
(2) The relationship between $\mathbf{S}$ and $\mathbf{B}_e$.

For the neo-Prandtl Hencky model, it was possible to obtain explicit tensorial expressions for ϕ. This is rather unusual for finite deformation models, however, since in these cases, we generally prefer to write both the strain energy function as well as the yield function in terms of eigenvalues. This approach allows us to consider a much wider class of models since we can combine network models for elasticity with advanced yield surface models to match experimental data.

The stored energy function and the yield function assumed to be of the form

$$\psi = \psi(E_e^i), S_i = \frac{\partial \psi}{\partial E_e^i}, f(\mathbf{S}) = f(S_i, \kappa), \qquad (12.65)$$

respectively, where E_e^i are the eigenvalues of the logarithmic stretch $\ln \mathbf{B}_e$ and S_i are eigenvalues of the Kirchhoff stress $\mathbf{S}$, their eigenvectors being the same.

To implement such models we will use the eigenvalue decomposition $\mathbf{B}_e = \sum_i^3 b_e^i \mathbf{b}_i \otimes \mathbf{b}_i$ and use the result (12.30). Since all the major tensors $\mathbf{B}_e$, $\mathbf{S}$, $\mathbf{D}_p$ etc., are coaxial, we can simply use their eigenvalue forms for the computation of ϕ.

Note, however, that $\dot{\mathbf{B}}_e$ is NOT coaxial with $\mathbf{B}_e$ so it cannot be reduced to just the diagonal terms. Nevertheless the diagonal terms of $\dot{\mathbf{e}}_e$ satisfy the following rather simple equation (cf. equation (12.30) on p.425)

$$\dot{e}_e^i = \mathbf{b}_i \bullet (\mathbf{D} - \mathbf{D}_p) \mathbf{b}_i, \qquad (12.66)$$

where $\mathbf{b}^i$ is the ith eigenvector of $\mathbf{B}_e$ and $e_e^i = \mathbf{b}_i \bullet (\mathbf{e}_e) \mathbf{b}_i$ is the ith eigenvalue of the logarithmic strain $\mathbf{e}_e$. Now, from the normality condition, we know that $\mathbf{D}_p$ is equal to $\phi \partial f / \partial \mathbf{S}$. Note that here *the value of ϕ is different from that in (12.49f) since there is a factor of 1/2 that has been swallowed up into it*. Thus so the rate of change of e_e^i simply becomes

$$\dot{e}_e^i = \mathbf{b}_i \bullet (\mathbf{D}) \mathbf{b}_i - \phi \frac{\partial f}{\partial S^i} = D_i - \phi \frac{\partial f}{\partial S^i}, \qquad (12.67)$$

where D_i is a short form for the normal component $\mathbf{b}_i \bullet (\mathbf{D}) \mathbf{b}_i$ of the stretching tensor along $\mathbf{b}_i$ and S^i is the ith eigenvalue of $\mathbf{S}$.

We are now in a position to state the general procedure for finding $\dot{\mathbf{B}}_e$ given $\mathbf{L}$ and $\mathbf{B}_e$ and κ at any instant of time. In order to keep the notation clean we will use the Einstein convention unless explicitly forbidden.

Procedure for finding $\dot{\mathbf{B}}_e$ in terms of eigenvalues and eigenvectors of $\mathbf{B}_e$.

The first part is just setting up the problem in a format that is convenient:
Preparatory calculations:

(1) Write the isothermal work function, the yield function and the hardening respectively as $\psi(e_e^1, e_e^2, e_e^3)$ and $f(S_1, S_2, S_3, \kappa)$.

(2) Write down the expression for the Kirchhoff stress response as

$$S_i = \left(2\left(\frac{\partial \psi}{\partial b_e^i}\right) b_e^i\right) = 2\frac{\partial \psi}{\partial e_e^i}. \tag{12.68}$$

You have done this many times, so this is a trivial exercise.

(3) Write down the expression for $g = g(E_e^i)$ and for the yield function.

(4) Calculate expressions for the eigenvalues of $\mathbf{P}$ and $\mathbf{M}$ respectively as

$$P_i = \frac{\partial f}{\partial S^i}, \quad M_i = \frac{\partial g}{\partial e_e^i}. \tag{12.69}$$

Now at each instant of time, given $\mathbf{B}_e(t)$ and $\mathbf{L}(t)$

(a) Compute the eigenvectors $\mathbf{b}_i$ and eigenvalues b_e^i of $\mathbf{B}_e$. Let $e_e^i = 1/2 \ln b_e^i$.

(b) Compute g, S_i, M_i and N_i using the formulae from steps 1-3 above.

(c) Compute $D_i = \mathbf{b}_i \cdot \mathbf{L} \mathbf{b}_i$, $(i = 1, 2, 3)$.

(d) Then compute ϕ from

$$D_i^p = \phi \frac{\partial f}{\partial \mathbf{S}} \Rightarrow \phi = \frac{H(g)\langle M_i D_i \rangle}{M_i P_i - \Theta \partial g/\partial \kappa \sqrt{P_i P_i}} = \frac{H(g)\langle [M]^t [D] \rangle}{[M]^t [N] - \Theta \partial g/\partial \kappa \sqrt{[P]^t [P]}}. \tag{12.70}$$

The last version, writing M_i and P_i as column vectors, is to help us write it in Matlab code.

(e) Then find $\dot{\mathbf{B}}_e = \mathbf{L}\mathbf{B}_e + \mathbf{B}_e \mathbf{L}^t - 2\phi \sum_i (P_i b_e^i) \mathbf{b}_i \otimes \mathbf{b}_i$.

(f) Update $\mathbf{B}_e$ and $\mathbf{L}$ and repeat for next instant.

Note 1: $\dot{\mathbf{B}}_e$ CANNOT be computed in eigenvalue form since it is NOT coaxial with $\mathbf{B}_e$. For this reason we have to revert to the global Cartesian axes (see step (e) above).

Note 2: The above procedure works only if ψ, and f are isotropic functions of their arguments.

⊙ **12.17 (Exercise:). Form for ϕ for the Hencky Model**
This is a multi-part exercise:
1. Show that for von-Mises yield condition for the material with a Hencky strain energy function, g can be written as

$$g = c_2^2 \sum_i \Gamma_e^i \Gamma_e^i - \kappa^2, \qquad (12.71)$$

where $\Gamma_e^i = e_e^i - (\sum_i e_e^i)/3$. (Hint: Simply substitute the eigenvalue decomposition $\mathbf{\Gamma}_e = \sum_i \Gamma_e^i \mathbf{b}_i \otimes \mathbf{b}_i$ into the yield function.)

⊙ **12.18 (Exercise:).**
2. Show that if $\mathbf{D}_p = \phi \partial f/\partial \mathbf{S}$ then,

$$\dot{\Gamma}_e^i = D_i - 1/3 \sum_i^3 (D_i) - 2\phi(S_i - 1/3 \sum_i^3 (S_i)). \qquad (12.72)$$

(Hint: Differentiate $\Gamma_e^i = E_e^i - (\sum_i E_e^i)/3$ and use (12.67).)

⊙ **12.19 (Exercise:).**
3. Using 1 and 2 show that,

$$\phi = \frac{2c_2^2 \sum_i (\Gamma_e^i \bar{D}_i)}{4c_2^3 \sum_i \Gamma_e^i \Gamma_e^i + 2\kappa\Theta} = \frac{c_2^2 \sum_i (\Gamma_e^i \bar{D}_i)}{\kappa^2(c_2 + \Theta)}, \qquad (12.73)$$

where $\bar{D}_i = D_i - 1/3 \sum_i D_i$ is the normal component of the deviatoric part of $\mathbf{D}$ in the direction of $\mathbf{b}_i$.
(Hint: Start from the eigenvalue form developed in 1 for g, differentiate it with time and then use the result 2 above to solve for ϕ.)
Compare the result with the form (12.4) for the small strain Prandtl Reuss equations. What features are the same? What features are different?

The implementation of this procedure is shown below.

12.7 Numerical implementation of the plastic flow equations using the Convex Cutting Plane Algorithm

Generally speaking, compared to exact solutions for small deformation plasticity (see e.g., [Prager and G (1951)]) there are very few attempts at exact solutions for finite deformation (see [Bonn and Haupt (1995)] for an example). As with the case of small deformations, we can use the convex cutting plane algorithm to find the elastic stretch $\mathbf{B}_e(t)$ given the deformation gradient $\mathbf{F}(t)$. The steps are similar to that for the small strain except for one major difference. This difference comes from the requirement of Objectivity. Generally it is dangerous to write a finite difference scheme for objective evolution equations since the discretization may violate the criteria of objectivity and give crazy answers when large rotations are involved. The book by Simo and Hughes ([Simo and Hughes (1998)]) includes an entire chapter on objective update algorithms that can be used. Rather than reproduce these results, we will refer to that book, focussing only on one item: implementing the flow rule using the Convex cutting plane algorithm.

Convex cutting plane algorithm for isotropic finite plasticity:

Given: $\mathbf{F}(i)$ at discrete time intervals $i = 0, 1, 2, 3, \ldots$ and $\mathbf{B}_e(0)$, we need to find $\mathbf{B}_e(i)$ by integrating the evolution equation

$$\dot{\mathbf{B}}_e = \mathbf{L}\mathbf{B}_e + \mathbf{B}_e\mathbf{L}^t - \phi\mathbf{P}\mathbf{B}_e, \phi \geq 0, f(\mathbf{S}, \kappa) \leq 0, \phi f = 0, \mathbf{P} = \partial f/\partial \mathbf{S}. \quad (12.74)$$

Before we embark on the procedure, we recall that if the material were elastic, i.e. $\phi = 0$ then it is possible to integrate the above equation exactly. In this case, we will get (see [Simo and Hughes (1998)] p. 315)

$$\phi = 0 \Rightarrow \mathbf{B}_e(i+1) = \mathbf{f}\mathbf{B}_e(i)\mathbf{f}^t, \text{ where } \mathbf{f} := \mathbf{F}(i+1)\mathbf{F}^{-1}(i). \quad (12.75)$$

In the following iterative procedure, the iteration number is shown as a superscript and the time instance is shown in brackets. Thus $x(n)$ means the value of x at time n while x^i means the value of x at the ith iteration. The procedure calculates $\mathbf{B}_e(n+1), \kappa(n+1)$ given $\mathbf{B}_e(n), \kappa(n), \mathbf{F}(n)$ and $\mathbf{F}(n+1)$.

(1) Set iteration number $i = 0, \phi^0 = 0$;
(2) Calculate $\mathbf{f} := \mathbf{F}(n+1)\mathbf{F}^{-1}(n)$.
(3) Calculate $\mathbf{B}_e^i := \mathbf{f}\mathbf{B}_e(n)\mathbf{f}^t$, $\kappa^i = \kappa(n)$. This is called the "elastic predictor".
(4) Find $g^i = (\mathbf{B}_e^i, \kappa^i)$;
(5) If $g^0 < Tol$ then set $(\cdot)(n+1) = (\cdot)^i$ and exit.
    ELSE
(6) Find (i) $\mathbf{S}^i = \partial\psi/\partial\mathbf{e}_e|^i$, (ii) $\mathbf{P}^i = \partial f/\partial\mathbf{S}|^i$, (iii) $\mathbf{M}^i = \partial g/\partial\mathbf{e}_e|^i$ (iv) $\partial g/\partial\kappa|^i$, where $1/2\mathbf{e}_e = \ln\mathbf{B}_e$, and (v) the hardening function $\Theta^i = \Theta(\mathbf{B}_e^i, \kappa^i)$. These calculations can be simplified for certain kinds of yield functions (such as the von Mises yield function). In other cases, you may need to carry out an eigenvalue decomposition of $\mathbf{B}_e$ for the calculation, This is the main step.
(7) Calculate

$$\triangle\phi = \frac{g^i}{\mathbf{M}^i \cdot \mathbf{P}^i - (\partial g/\partial\kappa|^i)\Theta^i \|\mathbf{P}^i\|}. \quad (12.76)$$

(8) Update:

$$\phi^{(i+1)} = \phi^i + \triangle\phi,$$
$$\mathbf{B}_e^{(i+1)} = \mathbf{B}_e^i(\mathbf{I} - \triangle\phi\mathbf{P}^i),$$
$$\kappa^{(i+1)} = \kappa^i + \triangle\phi\Theta^i, \qquad (12.77)$$
$$i \leftarrow i + 1.$$

GO TO 4.

The above iterative scheme for the neo-Prandtl-Hencky plasticity model together with the von Mises yield function implemented in a MATLAB program available online (see authors' websites in the preface). We have run this program from simple shear to illustrate the response. This is not the only algorithm that works for these models and is not even the most widely recognized one[19], and the literature is replete with different algorithms that work for different kinds of models (see e.g., [Nemat-Nasser (1991)]–large deformation J2 type yield functions with isotropic hardening, [Luhrs et al. (1997)]–finite strain viscoplasticity with isotropic and kinematic hardening with applications to metal forming, [Weber et al. (1990)]–objective time integration for hypoelastic type constitutive theories, [Papadopoulos and Lu (1998)]–framework for numerical solutions).

The result of running the program can be seen in Fig. 12.1

If you observe the result closely, you will note that the shear yield stress is NOT constant even though we have simulated a non-hardening material. This is not a mistake of the program, nor is it due to numerical errors. This is because of the fact that when a material undergoes finite simple shear with a *nonlinear* elastic response, it develops normal stresses. This is a singular feature of nonlinear materials and is referred to as the "Poynting effect" in the literature[20]. Due to this, we see that while the strain is a simple shear, the state of stress is both normal and shear stresses. Due to this combined loading, the shear yield stress will change with deformation in such a way that the von Mises yield condition is met.

12.8 Summary

- There are many different finite inelasticity models depending on the level of microstructural detail that you want to capture. Choose the one that is most suitable for your purpose.
- For process modeling where you are interested in gross shape change, forces generated, and estimation of springback, an isotropic inelasticity model can be adopted.

---

[19]See e.g., [Simo and Ortiz (1985); Simo (1992)] for return mapping based algorithms that are much more widely known.
[20]see [Srinivasa (2001)] for an illustration of this effect in plasticity.

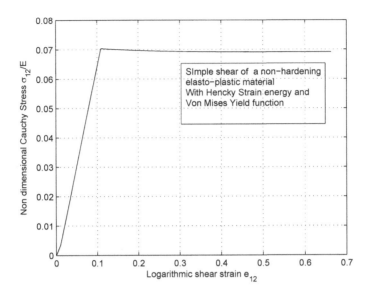

Fig. 12.1 Response of an isotropic elasto-plastic material under finite simple shear. Note that even though the yield stress is constant with no hardening, the shear stress decreases slightly with increasing deformation.

- It is easy to write down the constitutive equations for an isotropic inelastic model by following the form (12.17); (i) replace (12.17a,b) with your favorite strain energy function and its corresponding stress response; (ii) replace (12.17)a,b with your favorite yield function; (iii) substitute for **S** in (12.17c) with the expression for **S** from (12.17a); (iv) choose your form for the hardening function.

  If you want to find an explicit expression for ϕ then write **S** f and g in terms of eigenvalues and then use the expression (12.70).

- The key steps in arriving at this simple result are (i) extending elasticity by replacing **B** with $\mathbf{B}_e$ (ii) using the form for the rate of dissipation and the maximum rate of dissipation criterion to find a suitable measure for the plastic flow rate and the normality flow rule.

- Numerical methods for integrating elastoplastic equations for finite strains is still an ongoing effort and there are plenty of different approaches in the literature (see e.g., [Balendran and Nemat-Nasser (1995); Miehe (1996)]).

12.9 Moving natural states, aka multiplicative decompositions

This section can be skipped on first reading.

Well, we have developed a generalization of elastic materials to inelastic materials by replacing $\mathbf{B}$ with $\mathbf{B}_e$. But what is $\mathbf{B}_e$? In other words, what does $\mathbf{B}_e$ represent? We know that $\mathbf{B} = \mathbf{FF}^t$ and the related change is shape of a circle drawn in the reference configuration. Can we make a similar statement for $\mathbf{B}_e$? The answer is yes.

We will present this in a way that we think is relatively straightforward, freely borrowing from ideas introduced by [Eckart (1948); Lee (1969); Rajagopal and Srinivasa (1998); Rubin (2001)] and others[21]. As with any really good idea, a number of people have discovered it in their own way with certain subtle differences in philosophy. However the general structure and the final constitutive equations are, by now well accepted enough that we can present them in a straightforward way.

We begin by noting an obvious fact: a soft material such as copper, when bent or otherwise heavily deformed, appears to completely "forget" its original shape. This is a fundamental aspect of materials which is routinely exploited in all metal forming operations. In this aspect, such materials resemble fluids (which take on the shape of the container). It stands to reason that the original starting configuration, which was used to label the points and which —for elasticity— served as the basis for the measure of response of the material, loses its latter role in inelasticity. For an ordinary fluid such a water, deformation measures such as strain have NO role to play in the response functions. On the other hand, when a material such as aluminum is bent by a moment and then released, it does not retain its exact bent shape, neither does it go back to its original straight shape but it assumes an "intermediate shape" (neither completely straight nor completely bent) which depends upon the amount of external stimulus needed to bend it. This too is an industrially recognized fact (called springback) and forming dies are designed to compensate for this. It is this intermediate shape that serves as a basis for subsequent deformation, i.e. for subsequent deformations it is this intermediate shape that serves as a basis for measuring kinematical parameters. Again, further stimulus makes the material "forget" this intermediate shape and take on a new intermediate state. *In other words, an inelastic material is "forgetful" of its shape but not so much as a fluid.* However its forgetfulness is NOT a function of time, in contrast to that of a viscoelastic material—its forgetfulness is triggered only by external stimuli such as forces or temperatures[22]. It is for this reason that the configuration it retains is not one of its "past" configurations but one where there is no external stimulus inducing it to forget.

It thus stands to reason that we ought to

---

[21] In view of the rather extensive debate currently continuing regarding the nature of the interpretations, we hasten to add that we do not claim that this is the only or even the best interpretation, only that we found this to be appealing to us. We accept full responsibility for any misinterpretation: If we are indeed guilty of such, we urge the reader to write a clarifying paper on the subject because some of the points are perhaps not clear to us even after working in the area for 15 years or more. So imagine the plight of someone who just wants to find out how to use finite deformation inelasticity for his/her engineering needs.

[22] At an extreme, you can always melt it and thus "wipe out its memory completely and start again by casting a new shape".

- introduce a *changing or moving configuration*, deviations from which are a measure of the response of the material to the external stimulus,
- stipulate that this configuration is the one that is "remembered" or "retained" when there is no stimulus such as forces or heating or cooling of the body,
- suggest how this configuration changes when there is a stimulus on the body, and
- demand that when this configuration is fixed then the material is indistinguishable from an elastic material.

For motivational purposes, consider a soft material such as pizza dough. It starts out in a stress-free state. We extend it with our hand by applying forces. The deformation gradient $\mathbf{F}$ is a measure of this extension. Now if we let go, the dough shrinks back—but not *all the way back*! This observation is the key to our discussion: we introduce a measure of how much of the original deformation is recoverable upon unloading.

The trick to doing this, is to use $\mathbf{V}_e = \sqrt{\mathbf{B}_e}$ and, corresponding to any configuration $k(t)$ of the material, we define a *moving or changing configuration k_p* shown in Fig. 12.2, the deformation gradient from k_p to $k(t)$ being $\mathbf{V}_e$. in other words, if we were able to unload the material from its current configuration *in such a way that the orientation of the line elements coinciding with the eigenvectors of $\mathbf{V}_e$ in the current configuration are preserved*, the configuration that we get is k_p. Thus the deformation from k_p to $k(t)$ is the "elastic stretch". In other words, for plain elasticity, *we measure strains from a fixed reference or natural configuration* whereas for inelasticity *we measure strain from a moving reference or natural configuration*. This rather "obvious" observation has taken researchers in inelasticity many years to formalize properly.

Eckart [Eckart (1948)] appears to be the first one to recognize the role of a moving natural state. He described the geometry associated with such a moving state using convected coordinates and non-Riemannian geometry[23].

Q: If the approach is so good, why did you not start the chapter with this idea?

The pedagogical problem is that if we try to *motivate* inelasticity by using this idea, then we have to answer many tricky questions such as

- Can we really unload a material to zero stress without reverse plastic flow?
- Should the unloading be isothermal or adiabatic?
- How should the unloaded configuration be oriented?

---

[23] See also e.g., [Kroner (1960); Backman (1964); Fox (1968); Nematnasser (1979, 1982); Bertram (1999); Rajagopal and Srinivasa (1998)] who constitute a small sample of the reinterpreters of these notions in various ways. Readers are also referred to a very thorough survey of the use of the idea by [Lubarda (2004)], which also contains an extensive bibliography. See also [Rubin (2001)] for a well-argued reason NOT to introduce this notion.

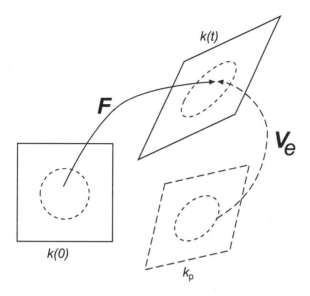

Fig. 12.2 A schematic figure illustrating the interpretation of $\mathbf{V}_e$ as the deformation from a moving "natural state". The body starts from a reference configuration $k(0)$ and is homogeneously deformed by $\mathbf{F}$ to the current configuration $k(t)$. If the external stimuli were removed, the body would "spring back" to a configuration k_p (shown by dotted lines). The orientation of k_p is chosen in such a way that the deformation gradient from k_p to $k(t)$ is the pure stretch $\mathbf{V}_e$. In other words, the orientation is prescribed by the requirement that the eigenvectors of $\mathbf{V}_e$ are fixed.

Also, while it provides a compelling rationale for choosing certain variables, it seems to raise as many questions as it answers, especially with thermal effects.

Thus, there are many subtle questions to be answered before we use this as the starting point. We were able to side-step these issues, because we are using this notion of natural state for "explaining" $\mathbf{B}_e$ and not for motivating it. We chose to give you a fast introduction to how finite inelasticity models are developed without a lot of "esoteric" and philosophical discussions about the model. We motivated $\mathbf{B}_e$ simply from small strain inelasticity and finite strain elasticity. Our experience in trying different ways of teaching this to students suggests that the approach in this book is easier to follow.

The idea of measuring from a moving natural state is one that is shared in common with finite viscoelasticity where, researchers in both small and finite deformation theories have *always* used the idea of measuring the strains from a moving configuration—this moving configuration being some past configuration of the body. An extensive body of literature exists dealing with the kinematics associated with such a notion. A similar form was suggested by [Pipkin and Rivlin (1965)] for inelasticity also. The main point to note here is that *for viscoelasticity, typically the moving configuration is one that was occupied by the body at some specified past time whereas for inelasticity, the moving configuration may not be occupied by the body at all but is specified by its own evolution law (see (12.3f))*.

What does plastic "flow" mean? We want it to describe a process by which the moving configuration k_p changes its "shape"—not just its orientation. Changes in orientation of k_p will not have any effect since the material is isotropic and because of the fact that the definition of k_p includes the requirement that the line elements along the eigenvectors of $\mathbf{V}_e$ are unchanged by the elastic deformation. So if you rotate k_p, k is also rotated by the same amount. What we want is a measure of "distortion" of k_p and $\mathbf{D}_p$ supplies just such a measure. To see this, let us define a tensor $\mathbf{F}_p$ by

$$\mathbf{F}_p = \mathbf{V}_e^{-1}\mathbf{F} \qquad (12.78)$$

and note that $\mathbf{F}$ is just the sequential action of $\mathbf{F}_p$ followed by $\mathbf{V}_e$ in other words, $\mathbf{F}_p$ is the deformation gradient (recall that we are only interested in homogeneous deformations) from k_0 to $k_p(t)$. In other words, in order to get to current configuration from the reference state, we carry out the following thought process: First deform and rotate the configuration by $\mathbf{F}_p$. Then apply a *pure stretch* by elongating or contracting it along the eigenvectors of $\mathbf{V}_e$. For example, if $\mathbf{F}_p = \mathbf{I} + 3\mathbf{e}_2 \otimes \mathbf{e}_1$ and $\mathbf{V}_e = 2\mathbf{e}_1 \otimes \mathbf{e}_1 + 1/2\mathbf{e}_2 \otimes \mathbf{e}_2 + \mathbf{e}_3 \otimes \mathbf{e}_3$, then $\mathbf{F} = \mathbf{V}_e\mathbf{F}_p$. This corresponds to (1) shear the material along the x_1 direction by 3 units (this is the "plastic flow") (2) then stretch along the x_1 direction by a factor of 2 and compress it along the x_2 direction by a factor of $1/2$ (this is the elastic deformation).

⊙ **12.20 (Exercise:). Describing elastic and plastic deformations**
Given that

$$\mathbf{F} = \mathbf{I} \quad \text{and} \quad \mathbf{V}_e = 2\mathbf{e}_1 \otimes \mathbf{e}_1 + \frac{1}{\sqrt{2}}\mathbf{e}_2 \otimes \mathbf{e}_2 + \frac{1}{\sqrt{2}}\mathbf{e}_3 \otimes \mathbf{e}_3, \qquad (12.79)$$

find $\mathbf{F}_p$. Describe the motion in terms of elastic deformation and plastic flow.

We hasten to add that, in reality, this *sequential split into an elastic and plastic parts is NOT the way inelasticity actually occurs*; rather, these two occur *simultaneously*. The split into elastic and plastic part is for visualization purposes only. The simultaneous action of elastic and plastic deformation is evident in the equation for the plastic flow rate in terms of the velocity gradient and the elastic stretch $\mathbf{B}_e$. To avoid this possible misinterpretation, we developed the expression (12.29) for the plastic flow rate directly, with no reference to any plastic strain.

Q: How can you define the plastic flow rate without defining the plastic strain?

This is an interesting question and can be thought of in two ways. In fluid mechanics, we introduce the velocity gradient $\mathbf{L}$ without reference to the deformation gradient

F *although the time derivative of* **F** *is related to* **L** *through* $\dot{\mathbf{F}} = \mathbf{LF}$. So why can't we introduce the plastic flow rate $\mathbf{D}_p$ without reference to any plastic strain?

The other point is that in one sense, $\mathbf{D}_p$ is a convenient shorthand for $1/2\mathbf{V}_e^{-1}(\mathbf{LB}_e + \mathbf{B}_e\mathbf{L}^t - \dot{\mathbf{B}}_e)\mathbf{V}_e^{-1}$ so we could have called it whatever we want and it is properly defined. So there is no problem regarding this.

Q: Can we actually deduce $\mathbf{V}_e$ from measurements?

In principle, if you measure the Cauchy stress and have the constitutive equation for an isotropic elastic material, you can deduce $\mathbf{V}_e$.

The above statement is oversimplified and there are a number of philosophical issues with what is meant by "measuring the stress".

In the previous examples, we have always assumed that we know **F** (how much we have deformed the material) and $\mathbf{V}_e$ is the elastic stretch, i.e. they are the independent (or input) variables in the problem whereas the stress is the dependent (or output) variable. The deformation gradient **F** can be measured, given the starting configuration by drawing grid lines on the body and measuring their distortions or by placing markers on the body and tracking their motion (this is now a routine procedure in the metal forming industry). However, measuring $\mathbf{V}_e$ is not so straightforward since we cannot really unload the body.

We can't measure $\mathbf{V}_e$ directly unless we unload the body (with all its attendant problems and questions). But, we can measure it indirectly. What we need here is the fact that, for an isotropic material, if we measure the value of the stress $\boldsymbol{\sigma}$ and we know the constitutive equation for $\boldsymbol{\sigma}$ then we can, in principle find $\mathbf{V}_e$.

It can be reduced to a problem of three equations for three unknowns: the eigenvectors of the stress are the same as that of $\mathbf{V}_e$, and the eigenvalues of $\boldsymbol{\sigma}$ and $\mathbf{V}_e$ are related by

$$\sigma^i = \varrho \frac{\partial \psi}{\partial V_e^i} V_e^i = \varrho \frac{\partial \psi}{\partial \ln V_e^i}, \qquad (12.80)$$

where V_e^i are the eigenvalues of $\mathbf{V}_e$. The above expression furnishes 3 equations for the three unknowns V_e^i. So given the stress, we can find V_e^i and hence $\mathbf{V}_e$. Actually, we can do a so-called Legendre transformation, and show that there exists a scalar complementary energy function $\Phi(\sigma^i)$ which is the negative of the Gibbs potential such that

$$\ln V_e^i = \frac{\partial \Phi}{\partial \sigma^i}. \qquad (12.81)$$

For example let as assume that

$$\psi = \frac{C_1}{2} \ln(V_e^1 V_e^2 V_e^3)^2 + C_2 \sum_i^3 (\ln V_e^i)^2. \qquad (12.82)$$

This is nothing but the Hencky strain energy.

⊙ **12.21 (Exercise:). Finding V_e from T**
Given the form for ψ as in (12.82), show that

$$\sigma_i = C_1(\ln(V_e^1 V_e^2 V_e^3)) + 2C_2(\ln V_e^i) \tag{12.83}$$

and hence obtain $\ln V_e^i$ in terms of σ_i.
(Hint: This is the same as linear elasticity with the strains replaced by $\ln V_e^i$, and C_1 and C_2 being the Lamé constants. First solve for $\ln V_e^i$. The inversion will involve the Young's modulus and the Poisson's ratio.
In general we will assume invertibility between the eigenvalues of the stress and that of $\mathbf{V}_e$ (and so equivalently $\mathbf{B}_e$).)

12.10 Geometrical significance of $\mathbf{D}_p$

Now that you are comfortable (we hope) with the idea that (1) the elastic stretch $\mathbf{V}_e$ can be obtained from $\boldsymbol{\sigma}$ and (2) we can use $\mathbf{V}_e^{-1}$ to define an intermediate configuration, we now hope to convince you that $\mathbf{D}_p$ defined in (12.14) is a measure of how the configuration k_p changes its shape.

Here is where we use the special advantage provided by the association of configurations to kinematical quantities. We can happily use the tools of continuum mechanics to describe changes of shape and orientation etc. Specifically, we are sure that you will agree with us that $\mathbf{C}_p = \mathbf{F}_p^t \mathbf{F}_p$ is a measure of the shape of k_p relative to k_0. To be precise, recall that the referential squared stretch tensor $\mathbf{C}$ gives you the dimensions and angles of the parallelepiped obtained by homogeneously deforming a unit cube in the reference configuration. Thus, the rate of change of $\mathbf{C}_p$ measures the changes in the shape and size of the intermediate configuration.

⊙ **12.22 (Exercise:). $\mathbf{D}_p$ measures plastic shape change**
Using the definition (12.78) for $\mathbf{F}_p$, show that

$$\mathbf{B}_e = \mathbf{F}\mathbf{C}_p^{-1}\mathbf{F}^t. \tag{12.84}$$

By taking the time derivative of $\mathbf{C}_p^{-1}$ show that

$$\mathbf{A}_p = \mathbf{L}\mathbf{B}_e + \mathbf{B}_e\mathbf{L}^t - \dot{\mathbf{B}}_e = \mathbf{F}\overline{\dot{\mathbf{C}_p^{-1}}}\mathbf{F}^t. \tag{12.85}$$

(Hint: You will need to repeatedly use $\dot{\mathbf{A}^{-1}} = -\mathbf{A}^{-1}\dot{\mathbf{A}}\mathbf{A}^{-1}$ and the fact that $\dot{\mathbf{F}} = \mathbf{L}\mathbf{F}$).

The above equation clearly demonstrates that $\mathbf{A}_p$ is zero if there is no shape change (i.e. if $\dot{\mathbf{C}}_p$ is zero) and is nonzero if there is shape change. On the other hand, it is NOT a direct measure, it is quite indirect.

We will show that $\mathbf{D}_p$ defined by (12.29) is a much more direct and understandable measure of plastic flow and this is why $\mathbf{D}_p$ is preferred over $\mathbf{A}_p$ for purposes of understanding.

⊙ **12.23 (Exercise:). $\mathbf{D}_p$ is the stretching tensor**
Using (12.29), (12.78) and the fact that $\mathbf{B}_e = \mathbf{V}_e^2$ show that

$$\mathbf{D}_p = 1/2 \mathbf{V}_e^{-1} \mathbf{A}_p \mathbf{V}_e^{-1} = (\mathbf{V}_e^{-1}(\mathbf{L}\mathbf{V}_e - \dot{\mathbf{V}}_e))_{symm} \tag{12.86}$$

where the notation $(\cdot)_{symm}$ stands for the symmetric part of the tensor defined by $1/2((\cdot) + (\cdot)^t)$.
Start with the definition (12.78) and show that

$$(\dot{\mathbf{F}}_p \mathbf{F}_p^{-1})_{symm} = \mathbf{D}_p. \tag{12.87}$$

Using your knowledge of kinematics of continua, interpret this result and justify the comment that "$\mathbf{D}_p$ is an understandable measure of plastic flow. (Hint: what is the physical meaning of $\dot{\mathbf{F}}_p \mathbf{F}_p^{-1}$? Compare with the meaning of $\mathbf{L}$.)

12.11 The model can be extended to viscoplasticity and other dissipative responses

We can generate a whole host of different fluid models by generalizing the normality rule (12.38).

The starting point for modeling other inelastic materials (other than rate independent plasticity) is the equation for the dissipation of mechanical power, namely

$$\frac{\partial \psi}{\partial \mathbf{B}_e} \cdot \mathbf{A}_p = \mathbf{S} \cdot \mathbf{D}_p = \zeta, \tag{12.88}$$

with all the symbols retaining the same meanings as for inelastic materials. Recall that, for, elasto-plasticity, we assumed that ζ was *positively homogeneous of degree one in* $\mathbf{D}_p$. This condition is necessary and sufficient to get rate independent behavior. On the other hand, if we remove this condition, we can get a much richer class of models.

> ⊙ 12.24 (Exercise:). Viscoplastic fluids
> *Assume that $\zeta = \mu \|\mathbf{D}_p\|^2$. By maximizing the dissipation ζ subject to the constraint (12.88), show that*
>
> $$\mathbf{S} = \mu \mathbf{D}_p. \qquad (12.89)$$
>
> *Show that the above constitutive equation can be written as*
>
> $$\dot{\mathbf{B}}_e - \mathbf{L}\mathbf{B}_e - \mathbf{B}_e \mathbf{L}^t = 2\frac{\mathbf{S}\mathbf{B}_e}{\mu}. \qquad (12.90)$$
>
> *(Hint: Substitute for $\mathbf{D}_p$ in terms of $\dot{\mathbf{B}}_e$ in (12.89) and use the fact that $\mathbf{S}$ is an isotropic tensor function of $\mathbf{B}_e$.)*

The exercise above shows an interesting feature namely that, by using (12.89), it is possible to rewrite ζ in terms of $\mathbf{S}$ as $\zeta = f(\mathbf{S}) = (\|\mathbf{S}\|^2)/\mu$. Then, the evolution equation for $\mathbf{D}_p$ can be written as

$$\mathbf{A}_p = \dot{\mathbf{B}}_e - \mathbf{L}\mathbf{B}_e - \mathbf{B}_e \mathbf{L}^t = \phi \frac{\partial f}{\partial \mathbf{S}} \mathbf{B}_e, \qquad (12.91)$$

where ϕ is obtained from the requirement that

$$\frac{\partial \psi}{\partial \mathbf{B}_e} \bullet \mathbf{A}_p = \zeta = f(\mathbf{S}). \qquad (12.92)$$

This is trick that works NOT ONLY for this case, but for all the models that satisfy the maximum rate of dissipation assumption.

In other words,

GENERAL RATE DEPENDENT MODEL

We can generate a wide class of nonlinear rate dependent dissipative models by assuming that

$$\psi = \psi(\mathbf{B}_e), \quad \mathbf{S} = 2\frac{\partial \psi}{\partial \mathbf{B}_e}\mathbf{B}_e,$$

$$\mathbf{A}_p := \mathbf{L}\mathbf{B}_e + \mathbf{B}_e\mathbf{L}^t - \dot{\mathbf{B}}_e = \phi\frac{\partial f}{\partial \mathbf{S}}\mathbf{B}_e, \quad \zeta = \frac{\partial \psi}{\partial \mathbf{B}_e}\cdot\mathbf{A}_p = f(\mathbf{S}) \geq 0. \tag{12.93}$$

Here $f(\mathbf{S})$ can be any positive function of $\mathbf{S}$, and need not be positively homogeneous in $\mathbf{S}$ and ϕ is calculated from (12.93c,d) as

$$\phi = \frac{f}{\mathbf{S}\cdot\partial f/\partial \mathbf{S}}. \tag{12.94}$$

Note that this gives an EXPLICIT expression for ϕ directly without the need for any complex manipulations of the constitutive equations. This always gives you time dependent behavior.

⊙ **12.25 (Exercise:). A Maxwell-like fluid**
Assume that ψ is the Hencky Potential (with $\mathbf{B}_e$ instead of $\mathbf{B}$ of course) And that $f(\mathbf{S})$ is given by $\boldsymbol{\tau}\cdot\boldsymbol{\tau}$. Obtain the constitutive equation for $\mathbf{A}_p$.
Will this model show yielding behavior? (Hint: Consider a "vanishingly small value" of $\boldsymbol{\tau}$, does $\mathbf{A}_p$ vanish? Or is it nonzero? Why does a nonzero value of $\mathbf{A}_p$ mean that the material is flowing?)

Now, how does this relate to the maximum rate of dissipation? It turns out the constitutive equations (12.93c,d) also satisfy the maximum rate of dissipation criteria, except that ζ is not a homogeneous function of $\mathbf{D}_p$ as before.

We are going to leave the proof of this as an exercise. If you can prove this, you can consider yourself a "master" of modeling dissipative behavior and having truly understood the material.

If you want to model a material that is not a viscoelastic fluid but is a viscoplastic solid, i.e. one with a yield surface and an elastic domain but which is rate dependent, then consider the following exercise.

⊙ **12.26 (Exercise:). A finite strain viscoplastic model**
Assume that ψ is the Hencky Potential and that $f(\mathbf{S})$ is given by

$$f(\mathbf{S}) = \mu_0 \langle \|\boldsymbol{\tau}\| - \kappa \rangle^2, \boldsymbol{\tau} = \mathbf{S}_{dev}, \tag{12.95}$$

where we have used the Macauley bracket notation, i.e.

$$\langle x \rangle^n = \begin{cases} x^n, & \text{if } x > 0., \\ 0, & \text{otherwise.} \end{cases} \tag{12.96}$$

Obtain the evolution equation for $\dot{\mathbf{B}}_e$ using (12.93).
Ans:

$$\mathbf{A}_p = \phi \langle \|\boldsymbol{\tau}\| - \kappa \rangle \frac{\boldsymbol{\tau}}{\|\boldsymbol{\tau}\|} \mathbf{B}_e. \tag{12.97}$$

You might be wondering, "If this is a book on inelasticity, why did we spend the bulk of our time of plasticity and only a short time on viscoplasticity, viscoelasticity etc.?" The answer, is that compared to the constitutive equations for rate independent plasticity, (primarily finding the value of ϕ), the constitutive relations for viscoplasticity are simple. Solving boundary value problems are a different matter altogether. The reason for the complexity of rate independent plasticity is the rather unique rate of dissipation function for this model.

If you can cast your mind back to your statics and dynamics courses, you might remember that you spent several chapters on friction (and perhaps left the class with a mixture of trepidation and relief—trepidation that you didn't quite "get" friction and relief that you did not have to see it again). Plasticity is the continuum equivalent of dry friction and so it is much harder to "get," than viscosity.

If you want to learn about viscoelastic and other rate dependent models, here are a few references— each dealing with a different aspect of inelasticity—that you could use as a starting point for further investigations ([Amirkhizi et al. (2006); Anand (1982); Anand and Su (2005); Baig et al. (2006); Cheng et al. (2001); Henann and Anand (2008); Holzapfel and Simo (1996); Masud and Chudnovsky (1999)]).

Q: What do we do about anisotropic materials?

It is not so simple to extend the considerations of this chapter to anisotropic materials. There appear to be three major ways to deal with anisotropic materials:

- An extension of the approach developed here where we generalize the notion of

F, the deformation gradient and introduce an elastic deformation gradient $\mathbf{F}_e$ or equivalently a triad of vectors. This approach is explained in Chapter 13.

- A closely related approach is by using the notion of elastic unloading and defining a plastic deformation gradient $\mathbf{F}_p$. This is a very popular approach and is being investigated by a number of researchers. (See also [Papadopoulos and Lu (2001); Lu and Papadopoulos (2004)] in this regard.)
- A final approach is to go to a different approach utilizing the Gibbs potential as a function of a suitable measure of stress. This is a relatively straightforward approach and we will illustrate it with a case study in Chapter 14 section 2. There have been a number of researchers who are investigating this approach (see e.g., [Hackl et al. (2001); Iwakuma and Nematnasser (1984)]). Interestingly, the original idea proposed by [Green and Naghdi (1965)] can be considered as an example of this approach where the Gibbs potential is assumed to be a function of the symmetric Piola-Kirchhoff stress tensor. This is a comparatively new idea and there is, as yet, only limited results available. In fact we can generate a wide range of models by using different measures of stress and their *duals* (see [Hill (1987)] for a description of the duals of different strain and stress measures that are common in plasticity).

Most of these extensions are still under active research considerations and we invite you to join the research groups and add your ideas to the mix.

12.12 Homework projects and exercises

(1) Modify the program for finite plasticity by introducing a suitable isotropic hardening response.
(2) Consider the following deformation:

$$\mathbf{F} = \begin{pmatrix} \cos(wt) & \sin(wt) & a(1 - \cos(wt)) \\ -\sin(wt) & \cos(wt) & -a\sin(wt) \\ 0 & 0 & 1 \end{pmatrix}, \quad (12.98)$$

where w and a are constants. The motion is a combination of rotation and shear. Simulate the response of the neo-Prandtl-Hencky material (see [Pradeep et al. (2005)] for a discussion of this motion).
(3) Write a Matlab program to find $\boldsymbol{\sigma}(t)$ and $\mathbf{F}(t)$ given $\mathbf{L}$ and n for a "fluid model" of the form

$$\boldsymbol{\sigma} = \mu \|\mathbf{D}\|^n \mathbf{D} \quad (12.99)$$

where μ and n are constants. Use this to find $\boldsymbol{\sigma}$ and $\mathbf{F}$ as functions of time for a flow problem of a "fluid" with $n = 1/2$ and $\mu = 1$, with

$$\mathbf{L} = \begin{pmatrix} 0 & 1 & 0 \\ 2 & 0 & 3 \\ 0 & 1 & 0 \end{pmatrix}. \quad (12.100)$$

(4) Start with the Blatz-Ko model for a compressible foam, and extend it to a finite inelasticity model using the yield function for a compressible foam (with $\mathbf{S}$ instead of $\boldsymbol{\sigma}$).

(5) Find an explicit expression for ϕ for the model above by following the procedure listed in this chapter.

(6) Team Project: Observe that we have not included kinematic hardening in the developments. Derive the Constitutive equations for linear kinematic hardening. This is not a simple project and you need to consider objective rates etc. when writing an evolution equation for the back stress $\boldsymbol{\alpha}$. Also many of the results that used the coaxiality of $\mathbf{S}$ and $\mathbf{B}_e$ will not carry over. So, additional care will be needed (see e.g., [Haupt and Tsakmakis (1986); Kamlah and Haupt (1997); Mollica (2001)] for ideas on how to include kinematic hardening in finite strain plasticity.

(7) Assume that we have an isotropic material whose Helmholtz potential is specified by $\psi = \psi(\mathbf{B}_e)$ and whose rate of dissipation is specified by $\zeta = \zeta(\mathbf{D}_p)$ which is NOT homogeneous in $\mathbf{D}_p$. By maximizing ζ subject to the constraint (12.39) with ξ and $\mathbf{D}_p$ given by (12.29), obtain an explicit expression for $\mathbf{S}$. Assuming it is invertible show that we will obtain (12.93). Why does this not work if ζ is a homogeneous function of $\mathbf{D}_p$? This is a hard project and could be used as a team endeavor. It requires some rather clever mathematics though it is not particularly complicated. The trick is guess a kind of "Legendre transformation".

(8) Survey the literature on finite strain anisotropic plasticity models and write a short report on the different approaches, highlighting similarities and differences.

Chapter 13

Inelasticity of Single Crystals

Learning Objectives

By learning the material in this chapter and doing the exercises you should be able to

(1) Derive the expression for the rate of plastic flow in terms of the lattice vectors and the reciprocal basis vectors.

(2) List the different slip systems for Face Centered Cubic (FCC) and Body Centered Cubic (BCC) materials and their representation in terms of the P matrices.

(3) Starting with a Helmholtz potential as a function of the Lattice vectors, systematically derive the equations for the stress as well as an expression for the resolved shear stress on the slip system.

(4) By considering each slip system as independent, choose meaningful rates of dissipation for each slip system and hence obtain the plastic flow rule.

(5) Postulate plausible hardening rules and evaluate different hardening rules in the literature.

(6) Modify the MATLAB program in this chapter to implement your own flow rule and/or hardening rule.

(7) Write your own MATLAB program to implement the response of a different FCC crystal and a BCC crystal.

> ## CHAPTER SYNOPSIS
>
> The central result of this chapter is that if we introduce the velocity gradient $\mathbf{L}_p$ associated with the plastic flow, then the evolution equations governing the motion of the lattice vectors are of the form
>
> Crystallographic slip rule: $\mathbf{L}_p = \sum_\alpha^N \dot{\gamma}(\alpha) \mathbf{s}(\alpha) \otimes \mathbf{n}(\alpha),$
>
> Plastic flow rule: $\dot{\gamma}(\alpha) = \mu \left(\dfrac{\|\tau(\alpha)\|}{\kappa(\alpha)} \right)^r \tau(\alpha),$
>
> Hardening Rule: $\dot{\kappa}(\alpha) = f[\kappa(\alpha)] \|\dot{\gamma}(\alpha)\| + g[\kappa(\alpha)] \sum_\beta \|\dot{\gamma}(\beta)\|,$
>
> (13.1)
>
> where N is the number of slip planes, $\mathbf{n}(\alpha)$ is the current normal to the slip plane associated with the αth slip system, while $\mathbf{s}(\alpha)$ is the direction of the slip, and $\tau(\alpha) = \mathbf{s}(\alpha) \cdot \boldsymbol{\sigma} \mathbf{n}(\alpha)$ is the "resolved shear stress" on the slip system. Thus we see that the response is akin to N dashpots which have power law viscous behavior, with r being the power law exponent, which is typically very large. The hardening parameters $\kappa(\alpha)$, which in the crystal plasticity literature, are called the "critical resolved shear stresses (CRSS)" evolve in a complex way depending upon how many slip systems are active.

Chapter roadmap

The key idea of the chapter is that *the plastic flow of most single crystals is due to the collective motion of layers of atoms along specific planes (called slip planes) in specific directions (called slip directions). The rate of this sliding motion depends upon the component of the shear stress on the plane in the direction of motion. We can imagine this motion to be like the sliding of a deck of cards. The combination of slip plane and slip direction is called a slip system. The net plastic flow is the result of the* **simultaneous sliding on all slip systems.**

A crystalline material is one in which the atoms are arranged in a periodic pattern (like patterned wall paper) called a crystalline lattice. The patterns can be different and also oriented differently. Most readily available metallic materials are made up of a large number of small crystals each of them being differently oriented.

Polycrystalline aggregates can be approximately simulated as a collection of single crystals (all with the same velocity gradient (the so-called Taylor assumption) or all with the same stress (the so-called Sach's assumption)). There have been attempts to improve on these two assumptions for various special conditions.

13.1 Why is the study of the plasticity of single crystals important?

Imagine that you are working for a Materials processing company and are making a high tech alloy product for a very special application (such as waveguides for electromagnetic waves or special shape memory alloys). Your task is to get the maximum performance out of it by processing the material properly. This is because, many of the mechanical, thermal and electrical properties are actually quite sensitive to the arrangement of the crystals in the component. In other words *you can improve the performance of your component by controlling the processing operation at the crystal level*. Note that the task here is quite different from that considered in the previous chapter. There we were trying to optimize the parameters associated with the Processing equipment (loads, shapes etc.) and we couldn't care less if the resulting component has the right crystal orientation. This is why we could get away with ignoring the crystals but we have to pay attention to the die geometry etc.

On the other hand here, while we are interested in approximating the geometrical changes but are really interested in the crystallites. The reason is the following. The crystalline nature of the materials result in strong orientation dependence of the properties of the material (such as modulus, hardness, conductivity etc.). However, due to the fact that there are distributed orientations in the polycrystalline material, the orientation dependence of mechanical properties is attenuated or reduced. If the polycrystal were truly random, then the mechanical properties would be orientation independent at the macrolevel although the properties at the crystal level are strongly anisotropic. For many practical applications, we may want to enhance the properties in certain directions. This is achieved by introducing preferred orientations to the crystallites in the material. The particular kind of orientation dependence is called "texturing" of the material. For many physical properties, such as modulus or conductivity etc., which cannot be easily altered without altering the crystal structure itself (or making different alloys), texturing is just about the only way to get better properties.

Specifically "texture" is defined as the distribution of the orientations of various crystals in a polycrystalline aggregate. A material having a strong texture has a majority of the crystals having similar orientation. This gives rise to anisotropy within the material. On the other hand, a material having a weak texture has the crystals oriented in a much more random manner. We need to be able to predict the texture that a material will attain after certain forming processes are performed on it. This will enable us to create processes and materials that take advantage of the texture of the material. Some examples where the ability to predict the crystal response would be an asset are

(1) **Increasing buckling strength:** In the case of a simple pinned-pinned col-

umn, for a homogeneous material with fixed geometry, the critical buckling load is seen to be a linear function of the Youngs modulus of the material. For copper polycrystals, this modulus can vary from 66.7 GPa to 156.4 GPa [Mason and Maudlin (1999)]. For designing a simple pinned-pinned copper column, it can be seen that a "wire" texture, with a Youngs modulus of 156.4 GPa, would be preferred as it would be able to withstand a higher load.

(2) **Stabilizing nanocystalline materials:** In nanocrystalline materials, it has been shown that non-equilibrium grain boundaries in polycrystalline materials can form equilibrium grain boundaries by the process of diffusive flow of atoms towards the grain boundaries ([Bachurin et al. (2003)]). The flow of atoms is governed by the difference in energy between two neighboring crystals, which in turn depends upon the stress in the crystal. Texturing the crystals in a certain way can reduce this tendency for the grain boundaries to grow and thus obtain a stable nanocrystalline material.

(3) **Reducing grain growth:** In polycrystalline materials subjected to arbitrary strains and temperatures, a phenomenon known as grain growth takes place. During grain growth, grains with a lower energy grow at the expense of grains having a higher energy. The process takes place in such a way that the total energy of the polycrystalline material is lowered ([Ono et al. (1999)]). By successfully simulating the response of polycrystalline aggregates, it is possible to predict the grain growth within a polycrystalline material and thus minimize this effect by finding textures that are good.

(4) **Improving smart materials:** Shape Memory Alloys (SMAs) display a marked difference in the shape-memory effect in different directions ([Yuan and Wang (2002)]). By knowing the texture of an SMA after it has been processed (i.e. deformed plastically), we would be able to use the SMA optimally.

(5) **Minimizing fatigue:** Fatigue failure in titanium alloyed turbine blades is a major problem. Historically, the approach has been to minimize the effects of texture by making the materials have a highly random texture. However, researchers are now looking into creating directionally processed titanium alloys, in which the best property directions are matched to the most critical loading conditions ([Bachurin et al. (2003)]).

(6) **Reducing earing:** If metal sheets with preferred orientation (texture) are used in stamping circular cups, the sides of the cup are uneven, a phenomenon known as earing ([Inal et al. (2000)]). This effect becomes particularly problematic during deep-drawing applications, and extra processing steps need to be designed to accommodate for the change in height of the material due to earing. In this case, the metal sheet needs to be processed in such a way that it has a texture which will minimize the effects of earing.

(7) **Reducing transmission losses:** Researchers have determined that a "Goss"-type texture is desired in materials for transformer cores to reduce power losses during magnetization ([Rollett et al. (2001)]).

In order to accurately predict the textures of polycrystalline materials, it is necessary to know

- The elastic and plastic response of a single crystal.
- A hardening law that describes the hardening of slip systems in the single crystal.
- A polycrystalline material model that relates the deformation of single crystals to polycrystalline aggregates.

Thus, the study of the plastic flow of single crystals and crystalline aggregates is vital to optimizing material properties. Not surprisingly there are a large number of books that deal with this phenomenon in great detail. We can only introduce you to the topic with just enough information that you can read these books and articles and make intelligent decisions about how to use these models. We are going to present only the barest essence of the models—giving you enough information to learn the material in the books and papers devoted to the simulation of textures and their evolution.

13.2 Crystals and lattice vectors

G. I. Taylor may be regarded as the founder of the modeling of single crystalline metals when he, along with C. F. Elam reported on the response of aluminum crystals (see e.g., [Taylor (1938); Taylor and Elam (1938)]). Interestingly, the study of the finite deformation of single crystals superseded that of macroscopic polycrystalline metals by several decades, and, in some ways may be considered to be simpler.

13.2.1 *A word regarding our approach*

We present the entire constitutive theory of the crystals by making use of the *current* lattice vectors and temperature of the crystal as variables following an approach advocated by [Rubin (1994)] who was primarily motivated by the work of [Besseling (1968); Mandel (1973)]. This is equivalent to the usual approach that uses the *reference* lattice of the crystal and an assumption regarding the decomposition of the deformation gradient (see e.g., [Clayton and McDowell (2003)]). While much of one of the authors' PhD work was based on this latter idea, we find that, in our opinion, the approach based on using only the current lattice vectors has significant advantages since it does not require any notion of being able to unload to a stress-free configuration; neither does one have to bring in issues of the orientation of these configurations etc.[1]. The book by [Havner (1992)] (especially Chapters 3,

---

[1] Although our approach is philosophically similar to [Rubin (1996)] who presents a well-reasoned case for adopting this framework, we have followed a more applied and thermodynamic flavor in our developments.

4 and 5) contains an excellent and very clear account of the fundamentals of the elastoplasticity of crystals including details and analysis of experimental investigations.

This comes at a price however. We need to make extensive use of the lattice vectors as a sort of "moving local coordinate system"[2]. In the development of our approach using this moving coordinate system, we will have use for components of tensors with respect to various coordinate systems. Instead of using different symbols for these different components, we will adopt a notation from differential geometry and use superscript and subscript notations. Thus, we might have two possible components for vectors and four possible components for a second order tensor. Just keep in mind that for any tensor $\mathbf{S}$

$$S_{ij} \neq S_i^j \neq S_j^i \neq S^{ij} \tag{13.2}$$

although they will be related (being the components of the same tensor $\mathbf{S}$ with respect to different basis vectors). This is the only thing that you should watch out for very carefully. For the purposes of this chapter you don't need to worry about why some components are written as superscripts and others are written as subscripts, just remember that they are different components of the SAME tensor with respect to different basis vectors. If you are interested, perhaps this might encourage you to learn the beautiful concepts of differential geometry. In this chapter we will only touch upon this notion, there is a lot more to it than what we present here.

13.2.2 Introduction to crystal lattices

We will begin our study with a brief description of crystalline materials. As mentioned earlier, a crystal is an ordered arrangement of atoms into a lattice. This arrangement is represented mathematically by a "Bravais" Lattice, which is an infinite set of points in a three-dimensional space generated by discrete translation operations. The resulting set of points will repeat periodically and the lattice will look the same from any lattice point. Such Bravais Lattices are defined by three vectors $\mathbf{a}_i, (i = 1, 2, 3)$ and the position of any point in the lattice can be represented as $\mathbf{r} = l\mathbf{a}_1 + m\mathbf{a}_2 + n\mathbf{a}_3$ where l, m, n are integers. Such vectors (with integer coordinates with respect to $\mathbf{a}_i$) will be called lattice vectors. If these vectors are orthonormal then we will get a "cubic" lattice—by far the most common variety in metals. If they are not, then we will get other types of lattices. You can find detailed discussions of these issues in Buergers book on crystallography ([Buerger (1956)]).

The core concept is this:

---

[2] In the early development of continuum mechanics this approach was very common and has some significant advantages. This has been extensively developed by [Oldroyd (1950)].

> **LATTICE STRUCTURE**
>
> The lattice structure of the crystal is defined by these three lattice vectors $\mathbf{a}_i$. Everything about the lattice is given in terms of these lattice vectors. They form a set of basis vectors for all geometrical operations on lattices. We will use them to describe the phenomenon of plastic flow in crystals. You can think of the parallelepiped formed by the $\mathbf{a}_i$ as a sort of brick, the **smallest repeatable unit** of the lattice. By stacking the bricks together in rows, columns and layers, you can build up the entire lattice.

Associated with the lattice vectors, there is a "reciprocal" lattice defined by three other vectors $\mathbf{a}^i$, $(i = 1, 2, 3)$ which are given by

$$\mathbf{a}^1 = \frac{\mathbf{a}_2 \times \mathbf{a}_3}{\mathbf{a}_1 \cdot (\mathbf{a}_2 \times \mathbf{a}_3)} \tag{13.3}$$

with $\mathbf{a}^2$ and $\mathbf{a}^3$ defined by cyclically permuting $1, 2$ and 3. Note that the two sets of vectors are related by

$$\mathbf{a}^i \cdot \mathbf{a}_j = \delta^i_j. \tag{13.4}$$

In other words, each reciprocal lattice vector $\mathbf{a}^i$ is perpendicular to two of the three lattice vectors $\mathbf{a}_i$. These reciprocal lattice vectors can be easily calculated and are extremely useful in describing the crystallography of materials.

Before we plunge into more details, it is vital for you to become familiar with the tensor product between two vectors and to be able to manipulate them at will. So we will summarize the results here, leaving it to you to verify that they are true. In what follows, $\mathbf{a}, \mathbf{b}, \mathbf{c}$ and $\mathbf{d}$ are vectors and $\mathbf{A}, \mathbf{B}$, are tensors

$$\begin{aligned} (\mathbf{a} \otimes \mathbf{b})_{ij} &= a_i b_j \\ (\mathbf{a} \otimes \mathbf{b}) \cdot (\mathbf{c} \otimes \mathbf{d}) &= (\mathbf{a} \cdot \mathbf{c})(\mathbf{b} \cdot \mathbf{d}) = a_i c_i b_j d_j. \\ \mathbf{A}(\mathbf{a} \otimes \mathbf{b})\mathbf{B} &= (\mathbf{A}\mathbf{a}) \otimes (\mathbf{B}^t \mathbf{b}) \\ \mathbf{A} \cdot (\mathbf{a} \otimes \mathbf{b}) &= \mathbf{a} \cdot (\mathbf{A}\mathbf{b}). \end{aligned} \tag{13.5}$$

Please refer to this set of results in the future calculations so that we can do this in a simple way.

Common Lattice Structures

The two most common lattice types for metals are *Face Centered Cubic(FCC) lattices and Body Centered Cubic (BCC) lattices*. These are shown in Fig. 13.1.

These are not exactly the Bravais lattices (i.e. these are not the smallest repeatable unit) for these crystals but they are simpler to use than the actual Bravais lattice. For these cubic lattices, we will use the lattice vectors $\mathbf{a}_i$ which are shown in the figure.

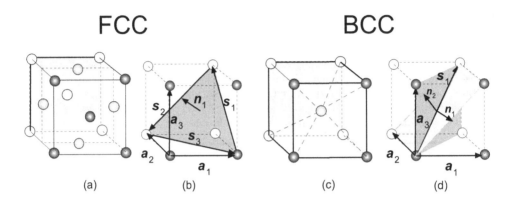

Fig. 13.1 Two common crystal structures and some example slip systems. (a) The FCC structure. It is a cube with an atom at each corner and middle of each face. (b) A plane along which the crystal will slide is shown in grey. The allowed directions of sliding are shown as s_i and the normal to the plane are shown as n_i. (c) A BCC structure: It is a cube with an atom at each corner and one in the center. (d) Two planes along which the crystal can slide are shown. The slip direction is the same for both planes.

We note that with these vectors, the atoms that are not at the corners of the cube will have components that are not integers, indicating that they are not the "smallest repeatable unit" of the structure. In other words we have joined the "bricks" together into a cubic super brick that is easier to handle (the original smallest brick was not cubic so stacking it to form a space filling thick structure is not easy to visualize).

Thus, lattice vectors and their reciprocals are the fundamental entities that allow us to describe behavior of single crystals.

13.3 Lattice deformation and crystallographic slip

Now that we have got all these preliminaries out of the way, it is time to plunge into the details of the modeling of deformation of these lattices.

This is a tricky thing to explain and it was not until the early 1900s that this phenomenon was satisfactorily explained by Ewing and Rosenhain [Ewing and Rosenhain (1000)]. We will paraphrase their picturesque description: If you stack the "FCC bricks" together, you will get layer upon layer of atoms. The atoms can be viewed as sheets lying on top of each other. As you can see in Fig. 13.2(a), we can view them as sheets in many different orientations. Imagine that the sheets are not

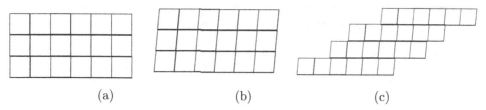

Fig. 13.2 Schematic representation of slip in a crystal. (a) Original undistorted crystal lattice. (b) Distortion of the lattice due to pure elastic response. (c) Distortion of the lattice accompanied by slip.

properly cemented together; if you now apply a force on the wall, the sheets don't deform much, they simply slide past one another like a deck of cards. This is the origin of plastic flow. Which sheets will slide? Will it slide in the direction a or b in the figure? One simple and reasonable answer is that whichever sheets are *most widely separated* will slide relative to each other.

However, this idea has an important caveat: The layers do not slide past each other like rigid planes—this requires too much energy: ALL the bonds have to be *simultaneously* broken. Based on this erroneous idea, if you make an approximate calculation of the yield strength, it will be of order of a tenth of the shear modulus. If you actually test the yield strength, it will turn out to be about a thousandth of the shear modulus! We are clearly way off. It was later realized that the sheets do not move like rigid bodies but they actually "creep along" like a caterpillar. You create a ripple and propagate the ripple. The ripples move along rapidly and the layer creeps easily. Later, you will see that as far as the flow of crystals go, these ripples (called dislocations) are a mixed blessing. They will "interfere with each other" and are the origins of hardening. At a macroscopic level, this does not change the result though, the crystal slips slowly on each slip plane and in the slip direction.

Thus, the "sliding tendencies" of a given crystal are defined by two vectors: the sliding direction $\mathbf{s}$ and the sliding plane whose normal is $\mathbf{n}$. The combination of a sliding direction and a sliding plane makes a "slip system". For the FCC crystal, there are 12 such slip systems (combination of slip direction and slip plane). Figure 13.1(b) shows three slip systems. We will number the possible slip systems from 1 to 12 and represent the slip systems by slip tensors $\mathbf{P}(\alpha) := \mathbf{s}(\alpha) \otimes \mathbf{n}(\alpha)$, where $\alpha = 1, \cdots, 12$ is the slip system number.

> ## SLIP DIRECTIONS AND PLANES
>
> For a given crystal, the slip directions and the normal vectors to the slip planes can be specified once and for all time in terms of the lattice vectors. Specifically, any slip direction $\mathbf{s}(\alpha)$ for the αth slip direction is represented as
>
> $$\mathbf{s}(\alpha) := s^i(\alpha)\mathbf{a}_i, \qquad (13.6)$$
>
> where $s^i(\alpha)$ are constants and we have used the summation convention. Similarly, the normals $\mathbf{n}(\alpha)$ to the slip planes are denoted using the *reciprocal vectors* $\mathbf{a}^i$ as follows,
>
> $$\mathbf{n}(\alpha) := n_i(\alpha)\mathbf{a}^i. \qquad (13.7)$$
>
> Please observe that *neither the slip vectors $\mathbf{s}$ nor the normal vectors $\mathbf{n}$ are unit vectors, they are always reported in terms of the smallest integer lattice units possible.*

For example, the slip direction $\mathbf{s}(1) = \mathbf{a}_1 + \mathbf{a}_2$ will be called the (110) direction, the numbers corresponding to the coordinates $s^i(1)$.

⊙ 13.1 (Exercise:). Slip directions
Find the vector notation for the other slip directions marked in Fig. 13.1(b),(d). Ans: The slip direction in (d) is (111).

For the slip plane shown in Fig. 13.1(b), the normal $\mathbf{n}(1)$ to the slip plane is given by

$$\mathbf{n}(1) = \mathbf{a}^1 + \mathbf{a}^2 - \mathbf{a}^3. \qquad (13.8)$$

This will be written as $[\mathbf{n}(1)] = [11\bar{1}]$ in crystallographic notation[3]. Notice that the crystallographic notation for lattice directions uses "round brackets" while that for the normals uses "square brackets". Also, in this notation the overbar is used to signify negative numbers[4].

It is actually quite easy to find the normal to a crystallographic plane. Find the locations (in lattice units) where it meets the x, y and z axes. Let them be a, b and c. Then the normal will be $\mathbf{n} = (1/a)\mathbf{a}^1 + (1/b)\mathbf{a}^2 + (1/c)\mathbf{a}^3$. You have to be careful about planes going through the origin: simply shift the plane away from the origin.

---

[3] You might have noticed that the components of the slip vectors are written with a superscript s^p while those of the normal vector are written as a subscript n_p. This is immaterial for our purposes, but those who are used to differential geometry will recognize contravariant and covariant components. There is a very beautiful connection between ideas of non-Riemannian geometry and crystallography. The interested reader can look up the work of e.g., [Eckart (1948); Bilby et al. (1955); Eshelby (1956); Naghdi and Srinivasa (1993)].

[4] Incidentally, it appears that the use of $\bar{a}$ to signify a negative number was found in ancient Indian works of mathematics and predates the use of $-a$. It is of significant help in doing mental mathematics.

Also if a plane is parallel to an axis, its intercept is ∞.

> ⊙ **13.2 (Exercise:). Normals directions**
> Find the vector notation for the normals marked in Fig. 13.1(d).
> Ans: $[1\bar{1}0], [10\bar{1}]$

A key result is that a normal vector $[a, b, c]$ will be perpendicular to a slip vector (l, m, n) if and only if $la + mb + nc = 0$ irrespective of the basis vectors $\mathbf{a}_i$.

> ⊙ **13.3 (Exercise:). normal vectors are ⊥ slip vectors**
> Verify that the normal vectors that you found in the previous exercise are indeed perpendicular to the slip vector (111).

The slip planes and slip directions in an FCC crystal are shown in Fig. 13.3.

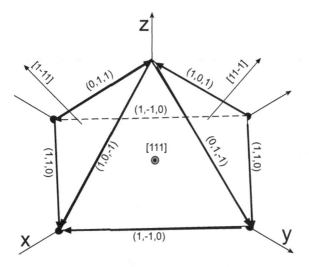

Fig. 13.3 The four slip planes and the slip directions. Note that the slip planes are in the form of a pyramid with a square base. The pyramid is oriented such that normal [111] is pointing straight out of the page. Each slip plane has three slip directions (forming the edges of the triangles). Note that each slip direction is shared by two slip planes. The angle between any two slip directions is either 60° or 90°. The angle between the slip plane normals is 70.53°.

13.4 Deformation and slip of single crystals

Now we have all that we need for discussing crystallographic slip. When a crystal is subjected to a force, it deforms; this deformation is composed of two parts, the distortion of the lattice and the slip of the atoms on the allowed slip planes. The net deformation is a combination of these two effects. It should be clear to you that the deformation of the lattice should be related to the stretching of the bonds and so should give rise to elastic behavior, whereas the sliding of the atoms should result in plastic flow and dissipative behavior. Moreover, most metallic materials are very "stiff", i.e. their lattice distorts only very slightly and the bulk of the deformation is due to plastic flow.

To understand this further, consider a FCC crystal initially in its stress free state. Its lattice vectors form a cubic structure. Now we gradually apply forces to it. When the forces are low, it should be evident to you that the lattice will become distorted[5] so that the base vectors are not of the same length anymore and the lattice has become a parallelepiped. If the atoms don't slide, then the distortion of the lattice is equal to the distortion of the whole body and we will have the following relation between the lattice vectors $\mathbf{a}_i(t)$ at time t to their initial values $\mathbf{a}_i(0)$:

$$\text{Purely elastic response:} \quad \Rightarrow \mathbf{a}_i(t) = \mathbf{F}(t)\mathbf{a}_i(0) \quad \Rightarrow \quad \dot{\mathbf{a}}_i = \mathbf{L}\mathbf{a}_i, \tag{13.9}$$

where $\mathbf{F}$ is the deformation gradient and $\mathbf{L}$ is the velocity gradient. If you compare this to the results in equation (11.13) on p.362, (established for line elements), you will see that

---

ELASTIC RESPONSE OF CRYSTALS

For a purely elastic response, the lattice vectors $\mathbf{a}_i$ behave exactly like line elements so that their rate of change is related to the velocity gradient $\mathbf{L}$ through (13.9).

Now what happens when the atoms start sliding past each other like a deck of cards? In this case, the overall distortion is much more than just what happens to the lattice. In other words, while the lattice still distorts slightly, the total distortion is different from that of the lattice. How do we model this? There are two equivalent ways of doing this. One is based on the velocity gradient and is founded on the assumption that as the body distorts, the sliding and the lattice distortion happens simultaneously[6]. In this case, as discussed in Chapter 11 equation (11.23) on p.366, we must add the velocity gradients due to distortion and due to sliding.

---

[5]Imagine a rickety scaffolding subject to lateral forces, you can see in your minds eye how the original scaffolding will lose its rectangular shape and become like a parallelogram, right?

[6]We discussed the notions of sequential and simultaneous action in Chapter 11. This is a good time to review that chapter.

We can do this by first introducing the "the lattice or elastic velocity gradient" $\mathbf{L}_e$:

LATTICE VELOCITY GRADIENT

The motion of the lattice vectors are defined by introducing a velocity gradient-like quantity $\mathbf{L}_e$ through

$$\mathbf{L}_e = \dot{\mathbf{a}}_i \otimes \mathbf{a}^i \ \Rightarrow \ \dot{\mathbf{a}}_j = \mathbf{L}_e \mathbf{a}_j, \qquad (13.10)$$

where we have used the reciprocal basis vectors $\mathbf{a}^i$. We will then define the plastic flow rate by

$$\mathbf{L}_p := \mathbf{L} - \mathbf{L}_e. \qquad (13.11)$$

That is it. That is all! As soon as we relate the plastic flow rate to crystallographic slip, we are done with the kinematics. This is again a straightforward process since we know that plastic flow in a single crystal occurs by simultaneous slipping on the allowed slip systems.

Note that we have the notion of "simultaneous action". If you recall from our discussion of continuum mechanics, this can be represented by "adding velocity gradients" as described in equation (11.23) on p.366. We now have to develop the expression for flow due to simultaneous sliding on different slip systems; this is something standard in continuum mechanics:

CRYSTALLOGRAPHIC SLIP

If a body deforms by shearing on a plane whose normal is $\mathbf{m}$ in a direction $\mathbf{n}$, then its velocity gradient is given by $\mathbf{L} = \dot{\gamma} \mathbf{n} \otimes \mathbf{m}$. The parameter $\dot{\gamma}$ is called the rate of shear.

Thus, since $\mathbf{L}_p$ is composed of simultaneous shearing on all allowable slip systems, then it stands to reason that (using equation (11.23) on p.366)

$$\mathbf{L}_p = \sum_{\alpha}^{N} \dot{\gamma}(\alpha) \mathbf{s}(\alpha) \otimes \mathbf{n}(\alpha), \qquad (13.12)$$

where $\mathbf{s}(\alpha)$ and $\mathbf{n}(\alpha)$ are defined in terms of $\mathbf{a}_i$ by (13.6) and (13.7) respectively.

Recalling the definitions (13.6) for the slip direction and (13.7) for the normal to the slip plane, we will define the "slip matrix" $[P](\alpha)$ corresponding to the αth

slip system (in terms of the components of the slip directions and slip normals) as

$$P(\alpha)^p_q = s^p(\alpha) n_q(\alpha). \tag{13.13}$$

where $s^p(\alpha)$ and $n_q(\alpha)$ are the components of the slip direction and slip plane normal for the αth slip system. Note that since these are constants, *the slip matrices $[P](\alpha)$ are constant matrices.*

> ⊙ **13.4 (Exercise:). Slip Matrices and Slip Systems**
> Show that
> $$\mathbf{s}(\alpha) \otimes \mathbf{n}(\alpha) = P(\alpha)^m_n \mathbf{a}_m \otimes \mathbf{a}^n. \tag{13.14}$$
> (Hint: Substitute $\mathbf{s} = s^m \mathbf{a}_m$ and $\mathbf{n} = n_p \mathbf{a}^p$.)
> In other words, the slip systems can be written in terms of the slip matrices and the slip vectors.

For the case of an FCC crystal, there are 12 slip matrices. They are listed in Fig. 13.4.

$$P(1) = \begin{pmatrix} 1 \\ -1 \\ 0 \end{pmatrix} \otimes \begin{bmatrix} 1 & 1 & 1 \end{bmatrix} = \begin{pmatrix} 1 & 1 & 1 \\ -1 & -1 & -1 \\ 0 & 0 & 0 \end{pmatrix},\ P(2) = \begin{pmatrix} 1 \\ 0 \\ -1 \end{pmatrix} \otimes \begin{bmatrix} 1 & 1 & 1 \end{bmatrix} = \begin{pmatrix} 1 & 1 & 1 \\ 0 & 0 & 0 \\ -1 & -1 & -1 \end{pmatrix},\ P(3) = \begin{pmatrix} 0 \\ 1 \\ -1 \end{pmatrix} \otimes \begin{bmatrix} 1 & 1 & 1 \end{bmatrix} = \begin{pmatrix} 0 & 0 & 0 \\ 1 & 1 & 1 \\ -1 & -1 & -1 \end{pmatrix},$$

$$P(4) = \begin{pmatrix} 1 \\ -1 \\ 0 \end{pmatrix} \otimes \begin{bmatrix} 1 & 1 & -1 \end{bmatrix} = \begin{pmatrix} 1 & 1 & -1 \\ -1 & -1 & 1 \\ 0 & 0 & 0 \end{pmatrix},\ P(5) = \begin{pmatrix} 1 \\ 0 \\ 1 \end{pmatrix} \otimes \begin{bmatrix} 1 & 1 & -1 \end{bmatrix} = \begin{pmatrix} 1 & 1 & -1 \\ 0 & 0 & 0 \\ 1 & 1 & -1 \end{pmatrix},\ P(6) = \begin{pmatrix} 0 \\ 1 \\ 1 \end{pmatrix} \otimes \begin{bmatrix} 1 & 1 & -1 \end{bmatrix} = \begin{pmatrix} 0 & 0 & 0 \\ 1 & 1 & -1 \\ 1 & 1 & -1 \end{pmatrix},$$

$$P(7) = \begin{pmatrix} 1 \\ 1 \\ 0 \end{pmatrix} \otimes \begin{bmatrix} 1 & -1 & 1 \end{bmatrix} = \begin{pmatrix} 1 & -1 & 1 \\ 1 & -1 & 1 \\ 0 & 0 & 0 \end{pmatrix},\ P(8) = \begin{pmatrix} 1 \\ 0 \\ -1 \end{pmatrix} \otimes \begin{bmatrix} 1 & -1 & 1 \end{bmatrix} = \begin{pmatrix} 1 & -1 & 1 \\ 0 & 0 & 0 \\ -1 & 1 & -1 \end{pmatrix},\ P(9) = \begin{pmatrix} 0 \\ 1 \\ -1 \end{pmatrix} \otimes \begin{bmatrix} 1 & -1 & 1 \end{bmatrix} = \begin{pmatrix} 0 & 0 & 0 \\ 1 & -1 & 1 \\ -1 & 1 & -1 \end{pmatrix},$$

$$P(10) = \begin{pmatrix} 1 \\ 1 \\ 0 \end{pmatrix} \otimes \begin{bmatrix} -1 & 1 & 1 \end{bmatrix} = \begin{pmatrix} -1 & 1 & 1 \\ -1 & 1 & 1 \\ 0 & 0 & 0 \end{pmatrix},\ P(11) = \begin{pmatrix} 1 \\ 0 \\ 1 \end{pmatrix} \otimes \begin{bmatrix} -1 & 1 & 1 \end{bmatrix} = \begin{pmatrix} -1 & 1 & 1 \\ 0 & 0 & 0 \\ -1 & 1 & 1 \end{pmatrix},\ P(12) = \begin{pmatrix} 0 \\ 1 \\ -1 \end{pmatrix} \otimes \begin{bmatrix} -1 & 1 & 1 \end{bmatrix} = \begin{pmatrix} 0 & 0 & 0 \\ -1 & 1 & 1 \\ 1 & -1 & -1 \end{pmatrix}$$

Fig. 13.4 The complete list of slip system matrices defined in equation (13.13) for a FCC crystal. The slip direction is written to the right and the slip plane are written above the matrix. Note that the trace of each matrix is zero, indicating the each slip mechanism is a volume preserving flow.

How did we get the fact that there are 12 slip matrices? Refer back to Fig 13.1b and observe that there the slip plane normal was along a body diagonal of the cube. It should be easy for you to see that there are four such body diagonals and hence 4 slip planes. Associated with each slip plane there are three slip systems corresponding to three of the six face diagonals. So we have 4 slip planes with three slip directions on each giving rise to a total of 12 slip systems. Each of the body diagonals are represented by vectors of the form [111] or [1$\bar{1}$1] etc, while the face diagonals are represented by (1$\bar{1}$0) etc. (see Fig. 13.3 p. 465).

We obtain the slip matrices by taking the outer product of the appropriate slip and normal vectors. It is an easy matter to obtain the slip directions associated with a slip plane for the FCC material. For example, if the slip plane normal is the body diagonal [1$\bar{1}$1], then its associated slip directions are (110), (10$\bar{1}$) and (011). How did we get this? We want vectors that have two ones and a zero (representing face diagonals). Also they must be perpendicular to the slip normal. The rest is simple counting.

> ⊙ **13.5 (Exercise:). Slip systems for FCC crystals**
> *List the normals to the four slip planes for a FCC material. for each slip plane, list the three slip directions.*
> *Ans: There are several places on the web where these are listed.*

For implementation purposes, it is more convenient to list these 12 slip matrices as a single 12×9 matrix with each row containing the nine components of each slip matrix written one column at a time. For the FCC crystal, this takes the form

$$\mathbb{P} = \begin{pmatrix} 1 & -1 & 0 & 1 & -1 & 0 & 1 & -1 & 0 \\ 1 & 0 & -1 & 1 & 0 & -1 & 1 & 0 & -1 \\ 0 & 1 & -1 & 0 & 1 & -1 & 0 & 1 & -1 \\ 1 & -1 & 0 & 1 & -1 & 0 & -1 & 1 & 0 \\ 1 & 0 & 1 & 1 & 0 & 1 & -1 & 0 & -1 \\ 0 & 1 & 1 & 0 & 1 & 1 & 0 & -1 & -1 \\ 1 & 1 & 0 & -1 & -1 & 0 & 1 & 1 & 0 \\ 1 & 0 & -1 & -1 & 0 & 1 & 1 & 0 & -1 \\ 0 & 1 & 1 & 0 & -1 & -1 & 0 & 1 & 1 \\ -1 & -1 & 0 & 1 & 1 & 0 & 1 & 1 & 0 \\ -1 & 0 & -1 & 1 & 0 & 1 & 1 & 0 & 1 \\ 0 & -1 & 1 & 0 & 1 & -1 & 0 & 1 & -1 \end{pmatrix}. \quad (13.15)$$

This may be confusing to describe but by comparing the components of the rows of the matrix $\mathbb{P}$ with that of each of the slip matrices, we hope you realize how this is listed. Any particular row of the $\mathbb{P}$ matrix is written as[7]

$$[n_1 \mathbf{s}^t, n_2 \mathbf{s}^t, n_3 \mathbf{s}^t], \quad (13.16)$$

where $(\cdot)^t$ stands for the transpose.

We are now done with the kinematics! Our central result is the following:

---

[7] For inquiring minds who want to know this is called the Kroenecker product between two matrices. MATLAB implements it as "$kron(n, s^t)$".

> ### DECOMPOSITION OF THE VELOCITY GRADIENT
>
> The velocity gradient of a single crystal is composed of two parts, that due to the distortion of the lattice and that due to the slip of the atoms. Since these act simultaneously, we have the following form for the velocity gradient $\mathbf{L} = \nabla \mathbf{v}$.
>
> $$\nabla \mathbf{v} = \mathbf{L}_e + \mathbf{L}_p = \mathbf{L}_e + \sum_\alpha \dot{\gamma}(\alpha) \mathbf{s}(\alpha) \otimes \mathbf{n}(\alpha) = \mathbf{L}_e + \sum_i \dot{\gamma}(\alpha) P_n^m(\alpha) \mathbf{a}_m \otimes \mathbf{a}^n, \qquad (13.17)$$
>
> where $\mathbf{L}_e$ is related to $\dot{\mathbf{a}}_i$ through (13.10) and $\mathbf{s}$ and $\mathbf{n}$ are defined in terms of $\mathbf{a}_i$ by (13.6) and (13.7) respectively. We thus have to find constitutive equations for $\mathbf{a}_i$ and $\dot{\gamma}_i$ to get a completely determined system of equations.

Thus the total velocity gradient has been split into a part $\mathbf{L}_e$ that is due to lattice distortions and a part $\mathbf{L}_p$ due to sliding of atoms. Since lattice distortions are supposed to be elastic, we would expect them to be determined by an equation of state. On the other hand, atomic sliding is supposed to be dissipative, so we would expect it to be determined by a kinetic equation. We will see in the next few sections that such considerations are indeed true.

13.5 The Helmholtz potential and equation of state for a crystal

When the lattice of a crystal is distorted, the bonds are stretched and we would expect the strain energy of the crystal to increase. Based on the considerations of continuum mechanics, the distortion of the lattice can be completely characterized by the six scalar invariants c_{ij}^e defined by

$$c_{ij}^e := \mathbf{a}_i \cdot \mathbf{a}_j \qquad (13.18)$$

The terms c_{11}^e, c_{22}^e and c_{33}^e are the squares of the lengths of the lattice vectors while the terms c_{12}^e etc. are related to the cosine of the angles between the lattice vectors. Students who are familiar with continuum mechanics will probably recognize that c_{ij}^e represent the components of a referential squared stretch matrix associated with the lattice distortion. We can hence define a "squared lattice strain components" E_{ij}^e through

$$E_{ij}^e = \frac{1}{2}(c_{ij}^e - \delta_{ij}) = \frac{1}{2}(\mathbf{a}_i \cdot \mathbf{a}_j - \delta_{ij}). \qquad (13.19)$$

The diagonal components of E_{ij}^e are the changes in the squares of the lengths of the lattice vectors while the off diagonal terms are related to the cosines of the angles between the lattice vectors. These definitions are useful primarily for cubic crystals since the lattice vectors in the reference state can be chosen to be orthonormal. For other lattices, the definitions must be modified.

> ⊙ **13.6 (Exercise:). Rate of change of E_{ij}^e**
> By differentiating (13.18) that
> $$\dot{E}_{ij}^e = \mathbf{D}_e \bullet (\mathbf{a}_i \otimes \mathbf{a}_j)_{symm}, \quad \dot{C}_{ij}^e = 2\dot{E}_{ij}^e, \quad (13.20)$$
> where $\mathbf{D}_e$ is the symmetric part of $\mathbf{L}_e$.
> (hint: use (13.9))

In general, the Helmholtz potential depends upon the current lattice vectors and is of the form $\psi = \psi(c_{ij}^e, T)$. For most materials (other than those that undergo phase changes) it suffices to consider c_{ij}^e values "close" to where ψ is a minimum and carry out a Taylor expansion keeping only quadratic terms; we will address this issue shortly after discussing cubic crystals.

For a cubic crystal, ψ is minimum for $c_{ij}^e = \delta_{ij}$ corresponding to a unit cubic lattice. Then for small distortions, the strain energy can be considered to be due to three distinct effects: Change in the volume of the lattice, differential changes in the lengths of the individual lattice vectors and change in the angle between the lattice vectors. Corresponding to these three distinct modes of lattice deformation, a cubic crystal has only three distinct elastic moduli (under small deformations) and referred to the axes of the crystal: the bulk modulus, a Young's modulus along the crystal axes and a shear modulus.

Based on these rather elementary considerations, and assuming that the lattice vectors at time $t = 0$ are orthonormal and hence of unit length, we will stipulate that the Helmholtz potential of the crystal is of the form (for a material with cubic symmetry)

$$\psi(E_{ij}^e) = 1/2 \left(k_1 (E_{11}^e + E_{22}^e + E_{33}^e)^2 + k_2 \{(E_{11}^e)^2 + (E_{22}^e)^2 + (E_{33}^e)^2\} \right. \\ \left. + k_3 ((E_{12}^e + E_{21}^e)^2 + (E_{23}^e + E_{32}^e)^2 + (E_{23}^e + E_{32}^e)^2) \right), \quad (13.21)$$

where k_i are constants and are the three moduli of the crystal.

Please remember that the above assumption is NOT meant to be realistic for anything but small lattice distortions. For small deformations, these constants can be related to the usual elastic constants. The table below shows some of the values for the constants.

Table 13.1: Elastic constants of some common metals

Metal	k_1 GPa	k_2 GPa	k_3 Gpa
Aluminum(FCC)	6.13	4.69	2.85
Copper(FCC)	12.14	4.70	7.54
Iron(BCC)	14.10	9.60	11.60
Nickel	14.73	9.92	12.47
Tungsten(BCC)	19.8	30.3	15.14

13.5.1 Thermomechanical equation of state

The above idea can easily be extended to other crystal types and also to account for thermal expansion in the following way: Let us assume that we are given $\psi(C_{ij}^e, T)$. Let $C_{ij}^0(T)$ be that value of C_{ij}^e where ψ is a minimum at any temperature. We can then write

$$\psi = \psi^0(T) + \mathbb{C}^{ijkl}(C_{ij}^e - C_{ij}^0(T))(C_{kl}^e - C_{kl}^0(T)) \tag{13.22}$$

where $\mathbb{C}$ is the elastic stiffness matrix. This is a straightforward generalization of what we get for linear elasticity except that C_{ij}^e need not be small strains. The only assumption required is that the lattice distortion be "small" from the minimum energy state, which is generally true other than for twinning or other more complex behavior that we are not considering here. The idea presented here is different (and quite a bit more natural and simple) from some of the recent literature where a "thermal deformation gradient" $\mathbf{F}^\theta$ is introduced. We think that this is completely unnecessary, since we have a relatively straightforward way of introducing this from the minima of the Helmholtz potential itself. This avoids a number of controversies and complications on this issue in a rather simple way.

For future reference, we will define a stress like quantity S^{ij} by

$$S^{ij} = 2\frac{\partial \psi}{\partial C_{ij}^e} \tag{13.23}$$

In anticipation of future results, these are the components of the Kirchhoff stress with respect to the lattice vectors, i.e. we will see shortly that the Kirchhoff stress can be written as,

$$\mathbf{S} = S^{ij}\mathbf{a}_i \otimes \mathbf{a}_j. \tag{13.24}$$

We are now in a position to develop the constitutive theory further by considering the dissipative response.

13.6 The dissipation function and constitutive equations for slip

As we have done in the previous chapters, our starting point is the mechanical dissipation equation, (equation (11.46) on p.382), which is of the form

$$\mathbf{S} \cdot \mathbf{L} - \dot{\psi} = \zeta, \tag{13.25}$$

where $\mathbf{S} = \boldsymbol{\sigma}/\varrho$ is the Kirchhoff stress and $\mathbf{L}$ is the velocity gradient and ζ is the rate of dissipation. Recall that, in words the above equation states that "the total mechanical power supplied less the rate of increase of the recoverable work is equal to the irrecoverable power loss".

Now we use the assumption (13.17) to obtain[8]

$$\mathbf{S} \cdot \mathbf{D}_e + \sum_\alpha \tau(\alpha)\dot{\gamma}(\alpha) - \dot{\psi} = \zeta, \tag{13.26}$$

---

[8]Don't take our word for it, actually substitute and obtain the result for yourself.

where $\tau(\alpha)$ is called the "generalized resolved shear stress" on the αth slip system and is defined as

$$\tau(\alpha) := \mathbf{S} \bullet \mathbf{s}(\alpha) \otimes \mathbf{n}(\alpha) = \mathbf{s}(\alpha) \bullet \mathbf{S}\mathbf{n}(\alpha) \qquad (13.27)$$

Why is this called the "resolved shear stress"? Recall that $\mathbf{n}$ is perpendicular to the slip plane, so $\mathbf{S}\mathbf{n}$ is the traction on the slip plane[9]. Thus $\mathbf{s} \bullet (\mathbf{S}\mathbf{n})$ is the component in the direction of $\mathbf{s}$ of the force per unit area on a surface whose normal is $\mathbf{n}$. This is called "generalized resolved shear stress conjugate to $\dot{\gamma}$" since the product of this with $\dot{\gamma}$ will give you the stress power. Since this is a mouthful we will simply refer to it as "resolved shear stress".

We now have to compute $\dot{\psi}$ to obtain a specific form for the left-hand side of the above equation. Since $\psi = \psi(C^e_{ij}, T)$, we see that

$$\dot{\psi} = \frac{\partial \psi}{\partial C^e_{ij}} \dot{C}^e_{ij} = \frac{1}{2} S^{ij} \dot{C}^e_{ij}. \qquad (13.28)$$

Now by using (13.20), we can rewrite it as

$$\dot{\psi} = S^{ij}(D_e \bullet (\mathbf{a}_i \otimes \mathbf{a}_j)). \qquad (13.29)$$

⊙ **13.7 (Exercise:). Rate of change of the Helmholtz potential**
Show by using (13.20) and (13.21), that, for a cubic crystal $\dot{\psi}$ can be written in the form

$$\dot{\psi} = \mathbf{D}_e \bullet ((k_1 E^e_{ii} + k_2 E^e_{11})\mathbf{a}_1 \otimes \mathbf{a}_1 + (k_1 E^e_{ii} + k_2 E^e_{22})\mathbf{a}_2 \otimes \mathbf{a}_2$$
$$+ (k_1 E^e_{ii} + k_2 E^e_{33})\mathbf{a}_3 \otimes \mathbf{a}_3 + k_3 E^e_{12}(\mathbf{a}_1 \otimes \mathbf{a}_2 + \mathbf{a}_2 \otimes \mathbf{a}_1) \qquad (13.30)$$
$$+ k_3 E^e_{23}(\mathbf{a}_2 \otimes \mathbf{a}_3 + \mathbf{a}_3 \otimes \mathbf{a}_2) + k_3 E^e_{31}(\mathbf{a}_3 \otimes \mathbf{a}_1 + \mathbf{a}_1 \otimes \mathbf{a}_3)).$$

Hint: This is mostly just differentiation.

We are now in a position to postulate a constitutive assumption for the rate of dissipation. Here we will assume that each slip system operates as a degree of freedom and that the rate of dissipation for the overall problem is additive, i.e. we will assume that

$$\zeta = \sum_{\alpha} \kappa(\alpha) \dot{\gamma}_0 \left\| \frac{\dot{\gamma}(\alpha)}{\dot{\gamma}_0} \right\|^{(1+r)}, r > 0. \qquad (13.31)$$

The parameters $\kappa(\alpha)$ are referred to as the "critical resolved shear stresses" (CRSS) and represent the resolved stress needed for substantial plastic flow to occur as we shall see shortly. Also, $\dot{\gamma}_0$ is called the "nominal strain rate" and is the strain rate at which the resolved shear stress will be roughly equal to the CRSS. The parameters $\kappa(\alpha)$ have the dimensions of stress and $\dot{\gamma}_0$ has the dimensions of strain rate. This way

---
[9] This is not precisely correct because $\mathbf{n}$ is not a UNIT vector, but it will differ only by a scaling constant that we don't worry about.

of writing the constants will ensure that the final constitutive equations have easily describable parameters irrespective of the value of r. Otherwise, the dimensions will depend upon r which could take a positive value, resulting in complicated dimensions.

> ⊙ **13.8 (Exercise:). MRDH and its consequences**
> Now invoke the maximum rate of dissipation hypothesis (MRDH) and maximize ζ over all possible values of $\mathbf{D}_e$ and $\dot{\gamma}(\alpha)$ subject to the constraint (13.26) and show that we get
>
> $$\mathbf{S} = (k_1 E^e_{ii} + k_2 E^e_{11})\mathbf{a}_1 \otimes \mathbf{a}_1 + (k_1 E^e_{ii} + k_2 E^e_{22})\mathbf{a}_2 \otimes \mathbf{a}_2$$
> $$+ (k_1 E^e_{ii} + k_2 E^e_{33})\mathbf{a}_3 \otimes \mathbf{a}_3 + k_3 E^e_{12}(\mathbf{a}_1 \otimes \mathbf{a}_2 + \mathbf{a}_2 \otimes \mathbf{a}_1) \quad (13.32)$$
> $$+ k_3 E^e_{23}(\mathbf{a}_2 \otimes \mathbf{a}_3 + \mathbf{a}_3 \otimes \mathbf{a}_2) + k_3 E^e_{31}(\mathbf{a}_3 \otimes \mathbf{a}_1 + \mathbf{a}_1 \otimes \mathbf{a}_3)$$
>
> and further that
>
> $$\tau(\alpha) = \kappa(\alpha) \left\| \frac{\dot{\gamma}(\alpha)}{\dot{\gamma}_0} \right\|^r \operatorname{sgn} \dot{\gamma}(\alpha). \quad (13.33)$$
>
> Invert the above expression to get $\dot{\gamma}(\alpha)$ as a function of $\tau(\alpha)$:
>
> $$\dot{\gamma}(\alpha) = \dot{\gamma}_0 \left(\frac{\|\tau(\alpha)\|}{\kappa(\alpha)} \right)^{1/r} \operatorname{sgn}(\tau(\alpha)). \quad (13.34)$$

We can now see why $\kappa(\alpha)$ is called a critical resolved shear stress. To begin with, we will assume that $r > 1$. Now from (13.34) we see that if the resolved shear stress $\|\tau(\alpha)\|$ is less than $\kappa(\alpha)$ then $\dot{\gamma}(\alpha)$ will be zero [10]. So not much flow will occur unless $\tau(\alpha)$ becomes comparable to or greater than $\kappa(\alpha)$. Thus $\kappa(\alpha)$ is the "critical" stress that must be exceeded. At this critical value, the shear rate will be $\dot{\gamma}_0$, thus justifying the name characteristic shear rate for this constant.

An important difference between the thermodynamical approach here and the approach found in Khan and Huang [Khan and Huang (1995)] section 10.3 as well as [Hill (1965, 1966); Hill and Rice (1972); Asaro (1983)] and a sizeable portion of the literature on the subject is that, in these papers instead of obtaining the stress as the derivative of the Helmholtz potential with respect to a suitable measure of lattice distortion, they simply postulate a constitutive equation for the Jaumann rate of the stress. This is not a good idea from the point of view of thermodynamics, since, in general, these constitutive equations are inconsistent with the existence of a Helmholtz potential.

---

[10] raising a number that is less than zero to a power that is greater than one will make it much smaller.

13.6.1 Independence of slip systems and other equations for the rate of slip

It is worth observing from (13.34) that, *the slip rate $\dot\gamma(\alpha)$ on any particular slip system depends only upon the resolved shear stress $\tau(\alpha)$ on that slip system.* For the particular form of the rate of dissipation function considered here, this was a natural consequence of the use of the MRDH. On the other hand, we can impose this condition as a requirement and then we can get other types of models in a relatively simple way. We can do this by requiring that *for each α*,

$$\tau(\alpha)\dot\gamma(\alpha) = \xi_\alpha(\dot\gamma(\alpha)) \tag{13.35}$$

where ξ_α is the rate of dissipation of the αth slip system. Note that the total rate of slip is $\zeta = \sum_\alpha \xi_\alpha$. How is the requirement (13.35) different from the requirement (13.26)? Clearly, if (13.35) is satisfied, then so is (13.26) provided the stress **S** is given by (13.32). However, the requirement (13.35) is much more stringent than (13.26) since *each slip system has to satisfy its own dissipation equation.* This stringent requirement actually makes it easier to find the equation for $\dot\gamma(\alpha)$ since each of them has to satisfy its only scalar equation, i.e. every slip rate is already specified. Note that $\dot\gamma(\alpha) = 0$ automatically satisfies (13.35).

Thus, the condition (13.35) further implies that

$$\dot\gamma(\alpha) \neq 0 \Rightarrow \tau(\alpha) = \frac{\xi_\alpha}{\dot\gamma\alpha}. \tag{13.36}$$

Which of these two possibilities (i.e. $\dot\gamma =$ or $\dot\gamma \neq 0$) should be chosen? The MRDH criterion merely requires that if there is a nonzero solutions to (13.35) then they must be chosen over the zero solution.

To make this idea very clear, consider the example where

$$\xi_\alpha = \kappa(\alpha)\left(\left\|\frac{\dot\gamma(\alpha)}{\dot\gamma_0}\right\| + \left\|\frac{\dot\gamma(\alpha)}{\dot\gamma_0}\right\|^{1+r}\right), \tag{13.37}$$

where $\kappa(\alpha)$ and $\dot\gamma_0$ are constants.

Using (13.36) we see that

$$\dot\gamma_\alpha \neq 0 \Rightarrow \tau = \kappa\left\{1 + \left\|\frac{\dot\gamma(\alpha)}{\dot\gamma_0}\right\|^r\right\} \text{sgn}(\dot\gamma_\alpha). \tag{13.38}$$

Since the terms inside the curly brackets are always greater than 1, we see that the above equation cannot be solved for $\dot\gamma(\alpha)$ unless $\tau(\alpha) > \kappa(\alpha)$. We thus get

$$\dot\gamma(\alpha) = \begin{cases} -\left\{\left\|\frac{\tau}{\kappa(\alpha)}\right\| - 1\right\}^{1/r}, & \text{if } \tau < -\kappa(\alpha); \\ 0, & \text{if } -\kappa(\alpha) < \tau < \kappa(\alpha); \\ \left\{\left\|\frac{\tau}{\kappa(\alpha)}\right\| - 1\right\}^{1/r}, & \text{if } \tau > \kappa(\alpha). \end{cases} \tag{13.39}$$

Thus the threshold or yield stress is $\kappa(\alpha)$ for each slip system. The above function can be written in more compact form as

$$\dot\gamma(\alpha) = f(\tau(\alpha)) = \left\langle\left\|\frac{\tau}{\kappa(\alpha)}\right\| - 1\right\rangle^{1/r} \text{sgn}\left(\frac{\tau(\alpha)}{\kappa(\alpha)}\right), \tag{13.40}$$

where the notation $\langle x \rangle$ stands for the Macaulay brackets.

Other forms can be obtained depending on the forms chosen for $\xi(\alpha)$. One of the more common ones is to have different threshold stresses for different slip systems and also for forward and reverse slips. We will thus assume that

$$\dot{\gamma}(\alpha) = \dot{\gamma}(\tau(\alpha)) := \begin{cases} -\left\{\left\|\frac{\tau}{\kappa_-(\alpha)}\right\| - 1\right\}^{1/r}, & \text{if } \tau < -\kappa_-(\alpha); \\ 0, & \text{if } -\kappa_-(\alpha) < \tau < \kappa_+(\alpha); \\ \left\{\left\|\frac{\tau}{\kappa_+(\alpha)}\right\| - 1\right\}^{1/r}, & \text{if } \tau > \kappa_+(\alpha), \end{cases} \quad (13.41)$$

where $\kappa_\pm(\alpha)$ are the forward and backward thresholds for each slip system. We can easily generate many other types of kinetic equations by choosing other kinds of dissipation functions. Note that in these cases, we have directly specified the full constitutive equations with very little role for the maximum rate of dissipation condition, which is used to simply distinguish between slip and no-slip.

13.6.2 Explicit expressions for the resolved shear stress in terms of the lattice vectors

Now that we have the form (13.32) for the stress, we can obtain an explicit expression for the resolved shear stress $\tau(\alpha)$ in terms of the lattice strains E^e_{ij}. We note that the resolved shear stress is given in terms of the slip systems which in turn are given in terms of the lattice vectors.

Preparatory to obtaining an explicit solution for the resolved shear stress, we will need the following expression:

⊙ **13.9 (Exercise:). Stress components in terms of lattice vectors**
Show that the constitutive assumption (13.32) implies that the components S^j_i defined by $S^j_i := \mathbf{a}_i \cdot \mathbf{S} \mathbf{a}^j$ is

$$S^j_i = C^e_{im} S^{mj} = 2C^e_{im} \frac{\partial \psi}{\partial C^e_{ij}} = C^e_{im}(k_1 E^e_{pp} \delta_{mj} + k_3 E^e_{mj} + \\ (k_2 - k_3)(\delta_{i1}\delta_{j1} E^e_{11} + \delta_{i2}\delta_{j2} E^e_{22} + \delta_{i3}\delta_{j3} E^e_{33}), \quad (13.42)$$

where we have used the summation convention. This is an important relation since it will come in handy when we compute resolved shear stresses as you can see from the next exercise.

For future convenience, we will write S^j_i for a cubic crystal in matrix form (so that it becomes easy to implement in MATLAB) as

$$[S^j_i] = \begin{pmatrix} c^e_{11} & c^e_{12} & c^e_{13} \\ c^e_{12} & c^e_{22} & c^e_{23} \\ c^e_{13} & c^e_{23} & c^e_{33} \end{pmatrix} \begin{pmatrix} k_2 E^e_{11} + k_1 E^e_{kk} & k_3 E^e_{12} & k_3 E^e_{13} \\ k_3 E^e_{12} & k_2 E^e_{22} + k_1 E^e_{kk} & k_3 E^e_{23} \\ k_3 E^e_{13} & k_3 E^e_{23} & k_2 E^e_{33} + k_1 E^e_{kk} \end{pmatrix}. \quad (13.43)$$

If we want to consider a general crystal with thermomechanical effects, then we obtain

$$S_i^j = C_{im}^e \frac{\partial \psi}{\partial C_{mj}^e} = C_{im}^e \mathbb{C}^{mjpq}(C_{pq}^e - C_{pq}^0), \qquad (13.44)$$

where $\mathbb{C}^{mjpq}$ are the elastic constants and C_{pq}^0 represents the Squared stretch matrix associated with the equilibrium lattice configuration of the crystal at that temperature.

> ⊙ **13.10 (Exercise:). Resolved shear stresses**
> Using the definition (13.13) for the "slip matrix $P(\alpha)$", the definition (13.27) for the resolved shear stress and the result (13.42), show that
>
> $$\tau(\alpha) = P(\alpha)_j^i S_i^j = A(\alpha)_i^j \left(C_{jm}^e \frac{\partial \psi}{\partial C_{im}^e} \right). \qquad (13.45)$$
>
> Of course, for the cubic crystals, S_i^j is given by the matrix (13.43).
> We have thus succeeded in obtaining an expression for the resolved shear stresses in terms of the lattice vectors $\mathbf{a}_i$. Show that this can be further simplified as
>
> $$[\tau] = [\mathbb{P}](S) \qquad (13.46)$$
>
> where $[\tau]$ is the set of resolved shear stresses written as a column vector of length 12 (one for each slip system), $[\mathbb{P}]$ is the 12×9 slip matrix defined in (13.15) and (S) are the stress components (13.43) written as a column vector of length 9.

13.6.3 The evolution equation for $\mathbf{a}_i$

Although it is not obvious yet, we now have all the ingredients for the governing equations for the behavior of single crystals. To see this explicitly, we will focus on the lattice vectors and figure out how much they will change: in other words we will use the results developed to write an equation in the form $\dot{\mathbf{a}}_i = f(\mathbf{a}_i, t)$.

To obtain this, we first observe (from (13.10) and (13.17)) that

$$\dot{\mathbf{a}}_i = \mathbf{L}_e \mathbf{a}_i = \mathbf{L}\mathbf{a}_i - \mathbf{L}_p \mathbf{a}_i. \qquad (13.47)$$

Now substitute (13.12) for $\mathbf{L}_p$ and (13.34) for $\dot{\gamma}(\alpha)$ and we will find that $\mathbf{L}_p \mathbf{a}_i$ is given by the expression

$$\mathbf{L}_p \mathbf{a}_i = \dot{\gamma}_0 \left[\sum_\alpha^N f[\tau(\alpha)](\mathbf{s}(\alpha) \otimes \mathbf{n}(\alpha)) \right] \mathbf{a}_i, \qquad (13.48)$$

where $f(\tau(\alpha))$ is given by the expressions (13.34) or (13.40) or (13.41), depending upon the evolution equation that you choose to implement. Note that all the terms—including $\tau(\alpha)$ through the constitutive relation (13.45)— on the right-hand side of the equation (13.48) are functions only of $\mathbf{a}_i$.

Thus, if **L** is prescribed, equations (13.47) and (13.48), when combined, will give us equations of the form $\dot{\mathbf{a}}_i = f(\mathbf{a}_i, t)$. In other words, we can now solve the above equation using MATLAB for any prescribed velocity gradient. This is not to say that this is an "easy" implementation. There are many complex calculations that have to be made in order to solve for $\mathbf{a}_i$. For one thing, we need to be able to "package" the three lattice vectors into one giant 9-dimensional vector for MATLAB (or any solver for that matter).

Thus, in the interests of speedy implementation, we need to make some further manipulations as follows:

(1) We first note that

$$(\mathbf{s}(\alpha) \otimes \mathbf{n}(\alpha))\mathbf{a}_i = (A_n^m \mathbf{a}_m \otimes \mathbf{a}^n)\mathbf{a}_i = \mathbf{a}_m P_i^m(\alpha), \quad (13.49)$$

where we have used $\mathbf{a}_i \cdot \mathbf{a}^j = \delta_i^j$ between the lattice vector and its reciprocal.

(2) The next step is to write the three lattice vectors as columns of a single 3×3 matrix. For this, we will need to use the three base vectors $\mathbf{i}_i$ of the laboratory reference frame $\mathbb{F}_r$. Obtaining the form of this matrix is the next exercise:

> ⊙ **13.11 (Exercise:). Slip in tensor form**
> Show that if we introduce tensors $\mathbf{A} = \mathbf{a}_i \otimes \mathbf{i}_i$ and $\mathbf{P}(\alpha) = P_i^j(\alpha)\mathbf{i}_i \otimes \mathbf{i}_j$, then the equation (13.49) can be rewritten as
>
> $$(\mathbf{s}(\alpha) \otimes \mathbf{n}(\alpha))\mathbf{A} = \mathbf{A}\mathbf{P}(\alpha). \quad (13.50)$$
>
> (Hint: Note that $\mathbf{A}\mathbf{i}_j = \mathbf{a}_j$. In matrix notation, the columns of $\mathbf{A}$ are simply the three lattice vectors.)

This is the point where we make contact with the usual approach to the subject of the plasticity of crystals (for example, as presented in the book by Khan and Huang [Khan and Huang (1995)]). These approaches START with the definition **A** and call it the elastic deformation $\mathbf{F}^e$. In our opinion, this has some philosophical difficulties since we haven't been told how to carry out this elastic deformation: is it isothermal, isentropic etc. In the approach here, we have neatly overcome this issue by NOT introducing any notion of elastic deformation but working with the lattice vectors directly, only introducing **A** as a matter of "computational convenience".

(3) Finally, the next result is the Grand finale and the one that we have been working steadily towards all this time:

By using (13.50), we can rewrite the differential equation (13.47) can be written in compact form which is highlighted below

> ### EVOLUTION EQUATION FOR THE LATTICE VECTORS
>
> If we define a tensor $\mathbf{A}$ through $\mathbf{A} = \sum_i \mathbf{a}_i \otimes \mathbf{i}_i$ then
>
> $$\dot{\mathbf{A}} = \mathbf{LA} - \mathbf{A}\sum_\alpha^N \dot{\gamma}(\alpha)\mathbf{P}(\alpha) = \mathbf{LA} - \mathbf{A}\sum_\alpha^N f[\tau(\alpha)]\mathbf{P}(\alpha) \qquad (13.51)$$
>
> where $f[\tau(\alpha)]$ is of the form (13.34) or (13.40) or (13.41).

⊙ **13.12 (Exercise:). Simple expression for $\dot{\mathbf{A}}$**
Prove equation (13.51).
Hint: first substitute $\mathbf{A} = \mathbf{a}_m \otimes \mathbf{i}_m$ into the left-hand side of (13.51). Next substitute $\mathbf{P} = P_n^m \mathbf{i}_m \otimes \mathbf{i}_n$ into the RHS. Finally take the inner product of this expression with $\mathbf{i}_i$ and use (13.49).

The main advantage of the above expression are that it gives all three lattice vectors in one simple compact form. Next it involves only a minimum of operations. Note especially that $\mathbf{P}(\alpha)$ are *fixed tensors*! They are defined entirely in terms of the slip systems. The point is that once $\dot{\gamma}(\alpha)$ is found, the rest is simple.

It is time to summarize the entire structure of the constitutive equations:

Structure of the constitutive equations for the elasto-plastic deformation of single crystals:

(1) **Input data:** We are given (1) The initial lattice vectors (at $t = 0$) written in matrix form with each column being one of the three lattice vectors: $\mathbf{A}_0 = \mathbf{A}(t=0)$, (2) the Helmholtz potential $\psi(c_{ij}^e)$, (3) the 12×9 slip matrix $\mathbb{P}$, (4) the critical resolved shear stresses $\kappa(\alpha)$ of each slip system and (5) the rate sensitivity index r.

(2) **Stress Components:** At each time instant, find the components of the stress with respect to the current lattice basis $\mathbf{a}_i \otimes \mathbf{a}^j$ as

$$S_i^j = C_{im}^e 1/2 \frac{\partial \psi}{\partial c_{mj}^e} = \frac{1}{2} C_{im}^e C^{mjpq}(C_{pq}^e - C_{pq}^0) \qquad (13.52)$$

where $C_{pq}^e = \mathbf{a}_p \cdot \mathbf{a}_q$ and C_{pq}^0 is the (temperature dependent) value of C_{pq}^e at which the free energy is a minimum. This would be the "stable configuration of the lattice at this temperature".

(3) **Resolved shear stress:** Find the resolved shear stresses as

$$\tau(\alpha) = \mathbf{P}(\alpha)_j^i S_i^j, \quad (\tau) = [\mathbb{P}](S) \qquad (13.53)$$

This step can be done in MATLAB if we write S_i^j as a 9×1 vector.

(4) **Finding the plastic flow terms:** Find the rate of slip on each system using
$$\dot{\gamma}(\alpha) = f[\tau(\alpha)] \tag{13.54}$$
where $f[\tau(\alpha)]$ is of the form (13.34) or (13.40) or (13.41). Then we have
$$\dot{\mathbf{A}} = \mathbf{L}\mathbf{A} - \mathbf{A}\sum_\alpha \dot{\gamma}(\alpha)\mathbf{P}(\alpha) \tag{13.55}$$

Before we show you the MATLAB file that implements this, it is important to know how lattice orientations are displayed. Since this is a three-dimensional rotation, it is impossible to do this in any simple way on the surface of a sheet of paper. However, there is a method called "stereographic projection" that will project any surface drawn on a sphere onto a two-dimensional surface by projecting lines [11].

In the days before computers, there were extensive geometrical constructions that needed to be made (using Wulff Nets etc.) in order to extract information. However, now there is even a free MATLAB toolbox that will do it for you (and a lot more!)

Our simple introduction to this is shown in Fig. 13.5 on p. 481.

If you have a crystallographic direction or normal to a plane and you want to plot it—

- Orient the crystal with the X,Y, and Z crystallographic axes marked as shown in the figure. Draw a sphere with radius 1 with center at the origin.
- Draw a line along any direction from the origin and find the point P when it intersects the unit sphere.
- Now from the North Pole of the sphere, draw a line that goes through P.
- Extend the line until it meets the plane parallel to the south pole at a point P'. This is the projection.

You can do this construction for any direction or for the normal to any plane (since every direction is the normal to some plane). Mathematically, a vector (x, y, z) is projected onto a point (y', z') where
$$(y', z') = \lambda(y, z), \ \lambda = \frac{2}{r+x}, \ r = \sqrt{x^2 + y^2 + z^2}. \tag{13.56}$$

⊙ **13.13 (Exercise:). Stereographic projections**
Consider the following directions : $[100], [110], [101], [011]$. Find and plot their corresponding stereographic projections. Compare your answers with Fig. 13.5.

If you plot the orientation of the normals of two planes of a crystal this way, you can figure out the orientation of the crystal. The MATLAB program available

---
[11] If a picture is equivalent to a thousand words, a video is the equivalent of a thousand pictures. Go to Youtube and search for "stereographic projections". Look for the one on the Riemann Sphere and you will get an idea of how this is done. See also the Mathematica webpage on stereographic projections by doing a Google search.

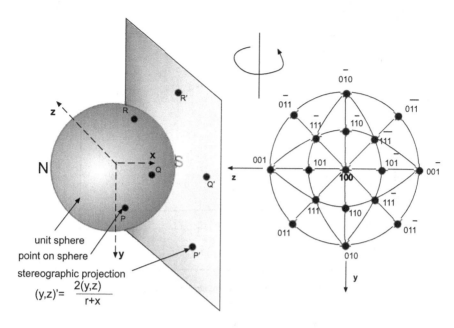

Fig. 13.5 Schematic of how a stereographic project works. We have the origin of the coordinate system at the center of the sphere. TO project any vector which ends at a point such as P. Draw a ray from the top (north pole) of the sphere through the tip P of the vector and extend it all the way until it hits the plane tangent to the south pole at P'. This is the projection. The projection of a variety of directions corresponding to locations on a cubic lattice are also shown

online (see authors' websites listed in the preface) can be used to see an animation of the deformation of the crystal and the rotation of lattice vectors and plots the stress response as well as the stereographic projection of the three crystal axes. Study the program to see how it works and then you can modify it to suit your purposes.

Now that we have all the paraphernalia, we will simulate the response of a single crystal under isochoric extension along the x-axis (which is aligned with the [1,0,0] direction of the crystal). The simulation is carried out using a MATLAB program available online (see authors' websites listed in the preface) with the CRSS $\kappa(\alpha)$ set to a constant.

The resulting stress strain graph is shown Fig 13.6:

The response of a real crystal is significantly different however, as shown in the schematic diagram shown in Fig. 13.7 p. 483. Clearly, we are missing most of the response characteristics of the crystal. The reason for this is that we haven't accounted for *hardening*, i.e. changes in the CRSS due to deformation and temperature.

The response shows three recognizable regions: (1) an "easy-glide" regime (usually appealing only if slip occurs on a single slip system) followed by a linear hardening regime (which is usually temperature independent). This appears with multislip

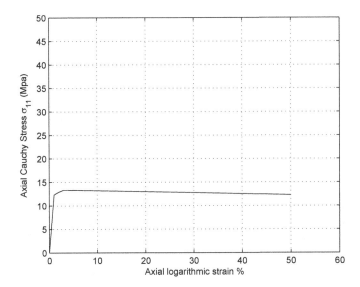

Fig. 13.6 An example of the response of a non-hardening crystal under isochoric simple extension. Note that the normal stress σ_{11} actually *decreases*! This is because of the fact that initially the crystal was not in a favorable orientation so it is hard to cause plastic flow. However, with increasing plastic flow, the lattice rotates to a more favorable orientation and it becomes easier to shear it (a phenomenon called geometrical softening). As soon as the orientation stabilizes, the normal stress becomes constant.

and is called Stage II hardening. After a substantial amount of linear hardening, we get a gradual "leveling off" which is called Stage III hardening and is sometimes referred to as the "dynamic recovery range". This is typically strongly temperature dependent.

In order to model this in a phenomenological way, we need to assume that κ is not constant but changes with deformation. This will be our next major development. Before we plunge in, we wish to emphasize that there is such a vast literature on this subject (see for example [Weng (1987); Cuitino and Ortiz (1993); Kocks and Mecking (2003)]) that we cannot hope to do justice to this at all. We will simply suggest a simple phenomenological model that can "simulate" the effect, making no claims to any deep physical significance. For Stage III hardening, Kocks and coworkers ([Mecking *et al.* (1986); Kocks and Mecking (2003)]) have developed universal scaling laws that show the remarkable similarity in the scaled stage III response of a vast number of different metals. Readers are urged to at least glance at the figures in these papers to see the power of non dimensionalization which, (sadly) unlike in fluid mechanics, still remains at a rudimentary level in solids.

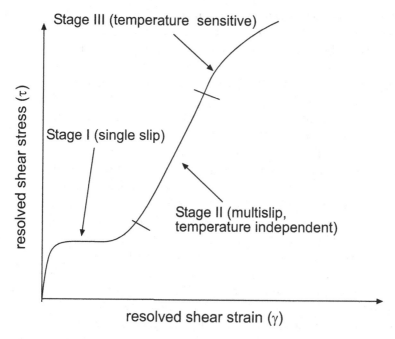

Fig. 13.7 A schematic figure of the response of a single crystal showing three distinct hardening stages: (1) Stage I: an easy glide region that appears for single slip (2) Stage II: a temperature independent linear hardening region which appears along with multiple slip (iii) Stage III:a temperature and rate dependent decrease in hardening.

13.7 Hardening

The developments so far have been focussed on just the constitutive equations for finding the rate of change of the lattice vectors and assume that the critical stresses $\kappa(\alpha)$ are constants. In reality, these also change with time. Why? We are not absolutely sure about all the details yet, but the general story is based on the existence and changes in lattice defects.

What do we mean by that? A real crystal is rarely "perfect", the arrangement of the lattice is full of "errors". These errors, in a single crystal, are of four kinds:

- Atoms missing from lattice sites (called vacancies), (think of a vacant lot in a new housing development).
- Atoms "freelancing" and wandering outside the lattice site (these are the "nomads" also-called "interstitials").
- Atoms that are too big or too small compared to the surrounding atoms (think of a mansion in the middle of a community of 2 or 3 bedroom houses) these are called "substitutional defects".
- Finally defects in the arrangements of rows of atoms itself (like errors in a grid of roads). These errors are called dislocations.

To understand this phenomenon, get a map of San Francisco near Market Street (Google "aerial map of san Francisco"). You will see that the roads form a rectangular lattice (sort of) but there are numerous errors. Market street itself forms the boundary between two differently oriented rectangles: This is called a twin boundary. Look at the streets near Market, there are a number of errors in the grid you will see many dead ended streets. No wonder you get "dislocated" if you travel these streets. You might take three right turns but end up somewhere else than you started. This is what a real crystal looks like.

On the other hand, the streets of Chennai (the city the two authors grew up in) are a great example of an amorphous solid. There may be some small regions of grid, but the roads are randomly oriented: as anyone who has traveled in Chennai knows, navigation by map is meaningless—you just have to know where to go.

Now plastically deforming the crystal is akin to relocating the streets; the streets are realigned by first creating a lattice defect and then propagating this to the required part. We will not describe this too much, since there are excellent materials science books (e.g., [Hertzberg (1976); Dieter (1986)]) that describe these features. There is also a very nice website http://www.doitpoms.ac.uk/tlplib/dislocations/index.php where you can find animations and experiments on dislocations. This is an excellent resource to see how these defects move.

Just as broken down cars cause traffic jams, dislocations that are unable to move will cause other dislocations to "pile up" and make it harder for slip to occur. Generally these pileups occur at places where there are substitutional atoms or because of "entanglements". Thus a dislocation marching happily on a slip plane creating slip will actually cause entanglements and problems for not only other dislocations on the slip plane (so-called self hardening) but also for other dislocations in other planes as well (so-called cross or latent hardening).

It is almost impossible for us to figure out exactly how these mechanisms operate since they all occur simultaneously although recent simulations of individual dislocations have thrown quite a bit of light on this phenomenon (see eg. [Moldovan et al. (2003); Benzerga et al. (2005); Wang et al. (2007)]). Also other ideas such as continuous distributions of dislocations ([Naghdi and Srinivasa (1994)]), statistical approaches ([Ortiz and Popov (1982)]) as well as phase field theories ([Koslowski et al. (2002)]) are being developed to simulate these phenomena in detail. In this introductory textbook, we cannot describe all of this. We will just have to model the phenomenon approximately in order to give you a flavor for the response.

Thus we now assume that the critical resolved shear stresses $\kappa_{\pm}(\alpha)$ are not constant, but change with time, i.e. the material hardens.

Qualitative observations with regard to hardening are as follows:

- Due to the interaction between dislocations, the critical shear stress required for slip $\kappa(\alpha)$ increases with increasing slipping on the slip system α. This is called "self hardening".

- What is not so obvious is that even if slipping occurs on the slip system α, the critical shear stress on the other slip systems also increase! This is called "cross" or "latent" hardening. This effect is not small by any means.
- Finally beyond a certain value, the effect of latent hardening levels off for most materials.

The reason for this cross hardening phenomenon is as follows: Let us focus on a single slip plane: There will be dislocations that lie on this plane and will spread on the application of stress causing plastic flow (imagine an ever growing ripple in a pond). But the motion of this "ripple" will be disturbed by dislocations that are perpendicular to this plane (imagine a wave trying to propagate through a "forest" of immobile trees). The main point is that A dislocation that is the cause of plastic flow on one slip system is a "tree" in the forest that retards plastic flow on other slip planes! This is what causes cross hardening. But beyond a certain level, the energy of the ripples are so large that they overwhelm the trees (this is called cross slip) and thus the effect of the forest levels off.

13.7.1 Phenomenological model of hardening: Single slip

To begin our analysis of hardening, we will consider each individual slip system in turn and pretend that no other slip system is active. We see that then it is like a one dimensional problem where the forward yield strength is $\kappa_+(\alpha)$ and the reverse yield strength is $\kappa_-(\alpha)$. Let us assume that $\dot{\gamma}(\alpha) > 0$; then we would clearly expect the forward yield stress to increase.

What if $\dot{\gamma}(\alpha) < 0$? If we wanted purely "isotropic" hardening, then the forward yield stress will continue to increase. A simple form of a hardening law that will account for this is

$$\dot{\kappa}_+(\alpha) = \dot{\kappa}_-(\alpha) = a \, \|\dot{\gamma}(\alpha)\|. \tag{13.57}$$

Said differently, we will stipulate something similar to one dimensional isotropic hardening where $\kappa_+(\alpha)$ is equivalent of the hardening parameter κ. In this case, the integral of $\|\dot{\gamma}(\alpha)\|$ represents the "amount of slip that has taken place" on the αth slip system and is a generalization of the notion of plastic arc length.

From equation (13.57), the yield stress increases at the rate a. In a typical material science paper or book this is written (with some abuse of notation) as

$$\Theta(\alpha) := \frac{d\kappa_+(\alpha)}{d\gamma(\alpha)} := \frac{\dot{\kappa}_+(\alpha)}{\|\dot{\gamma}(\alpha)\|} \tag{13.58}$$

where $\Theta(\alpha)$ is called a hardening modulus. Note that even though it is *written* as $d\kappa_+(\alpha)/d\gamma(\alpha)$ it is not meant to be a function of $\gamma(\alpha)$ alone, it is just a notation used in the literature. Moreover, most of these works do not make a difference between the applied stress σ and the hardening parameter κ since, during plastic flow, they happen to be equal. Please be aware of this when reading the literature and translating the information back and forth between the notation in the book and in the materials science papers. We have consistently used the notation κ for hardening parameters throughout this book so that there is continuity between macroscale continuum plasticity and crystal plasticity.

Returning to the discussion of hardening, the simplest possibility is to put $\Theta = Const$. Leading to a linear hardening of the material. This is not a bad assumption for the initial stages of hardening under single slip (the so-called Stage 1 hardening) for which $a \approx \mu/10000$, where μ is the shear modulus of the material. Typically however, when multiple slip occurs, the rate of hardening, i.e. Θ increases sharply to a value that is about $\mu/300$, before beginning to plateau out. Thus a typical resolved shear stress versus amount of slip plot would look like a combination of these two values of hardening with the latter kicking in when multiple slip occurs. During the later stages of multiple slip, the value of Θ should begin to drop linearly from $\mu/300$ to a much lower value.

We will now consider how to model this.

13.7.2 Hardening with multiple slip

An "isotropic" hardening model

The simplest extension of (13.57) to multiple slip is to assume that

$$\kappa_+(\alpha) = \kappa_-(\alpha) = \kappa(\alpha);\ ,\ \dot{\kappa}(\alpha) = a \sum_\beta \|\dot{\gamma}(\beta)\| \qquad (13.59)$$

In other words, we assume that *the CRSS for both forward and reverse slips are equal and further that all the slip systems contribute the same amount*. This is called the "isotropic hardening". Let us see what we can get out of this:

To understand the response, let us assume that there are only two slip systems, then the hardening modulus of the first slip system $\Theta(1)$ is given by

$$\Theta(1) := \frac{\dot{\kappa}(1)}{\|\dot{\gamma}(1)\|} = a\left(1 + \frac{\|\dot{\gamma}(2)\|}{\|\dot{\gamma}(1)\|}\right) \qquad (13.60)$$

Note that now if $\dot{\gamma}(2)$ is nonzero, i.e. if double slip occurs, the hardening modulus increases by a factor of $1 + \|\dot{\gamma}(2)/\dot{\gamma}(1)\|$, in a manner that is qualitatively similar to what occurs in stage II. However even if $\dot{\gamma}(1) = \dot{\gamma}(2)$, the hardening rate only doubles! It does not increase by orders of magnitude. This does not agree all that well with experiments.

Do not pooh-pooh the model (13.59) even though it does not do a good job with transitions from single to multiple slip; remember, models have to be chosen based on need. The advantage of this model is its utter simplicity! If you are developing a crystal plasticity model as a part of simulation of a polycrystal, then single slip is not important at all! Even if we assume that a is just a constant (say $\mu/300$) then you will be able to model responses up to strains of 20% with reasonable accuracy. That is pretty good for a very simple hardening rule!

An improved model

If you want to reflect what happens to a single crystal more accurately, we need more control over the individual terms in the summation that appears in (13.59).

Thus, a significantly better result can be obtained if we assume that for each slip system, the "self hardening modulus" is different from the "cross or latent hardening modulus", i.e. if we assume that

$$\dot{\kappa}(\alpha) = a \sum_{\beta} \|\dot{\gamma}(\beta)\| + (b - a) \|\dot{\gamma}(\alpha)\| \tag{13.61}$$

To see how this model behaves, again consider only two slip systems: A quick calculation yields

$$\Theta(1) := \frac{\dot{\kappa}(\alpha)}{\dot{\gamma}(\alpha)} = b + a \frac{\|\dot{\gamma}(2)\|}{\|\dot{\gamma}(1)\|} \tag{13.62}$$

Now, the hardening modulus for single slip (when $\dot{\gamma}(2) = 0$ will be b while that for equal double slip when $(\dot{\gamma}(2) = \dot{\gamma}(1))$ will be $a + b$. Since we can now choose a and b independently, if we choose $b = \mu/10000$ and $a = \mu/300$ we will approximately model stage I and stage II hardening. Note that if multiple slip occurs, then the slope keeps getting steeper and steeper as more slip systems become active. This phenomenon has indeed been observed (see e.g., [Kocks *et al.* (1998)]).

Accounting for stage III hardening

Beyond a certain stress, the dislocations are able to "cross slip" and slide around the obstacles, so that, beyond a certain stress, the hardening tapers off. This is a thermally activated phenomenon and hence this behavior (called dynamic recovery or stage III hardening) is quite temperature sensitive. We need to model this phenomenon too. Thus, to account for these stages, we can consider the following generalization:

$$\dot{\kappa}(\alpha) = \sum_{\beta \neq \alpha} \{a[\kappa(\beta), T] \|\dot{\gamma}(\beta)\|\} + b[\kappa(\alpha)] \|\dot{\gamma}(\alpha)\| \tag{13.63}$$

where a is a *cross hardening function* and indicates how much influence slip on the β system has on the hardening rate of the α system and b is the direct hardening of the α system. Just this simple generalization will allow us to explore a vast number of possibilities. We will consider some simple choices for the functions a and b and demonstrate that they are sufficient to simulate a wide a range of behavior [12].

---

[12] Please do not assume that the hardening function (13.63) is in any way "correct". We just made it up on the flimsiest of evidence just to be able to reproduce some aspects of single crystal behavior. Please feel free to explore the literature; there are a plethora of possibilities that are extensively discussed in the literature with varying degrees of fundamental support. Feel free to consider them and try your own simulations.

Please note that, if you are simulating polycrystalline materials, there is no point in sweating over the precise form of the hardening law for a single crystal. You will be forced to make such crude approximations about the interactions between the individual crystallites that there is no advantage to be gained in modeling the single crystal hardening very precisely.

With these caveats and observations, we will assume that the cross hardening rate, a is given by :

$$a[\kappa, T] = \frac{a_0}{1 + (a_1[T](\kappa - a_2))^n}, \quad a_1[T] = Ae^{\frac{-\Delta H}{kT}}, \tag{13.64}$$

where a_0 is related to the slope of the hardening curve in stage II hardening, a_1 is related to the stress needed for dynamic recovery (i.e. Stage III to begin) to occur at a given temperature and n is related to the rate of dynamic recovery. The shape of the function a as a function of κ is shown in Fig. 13.8.

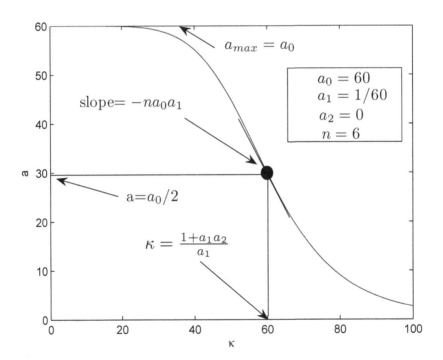

Fig. 13.8 A plot of the hardening function $a(\kappa, T)$ introduced in (13.64). Note that the parameters a_i and n together control (1) the maximum hardening rate, the slope of the stage III hardening and the location where the hardening decreases to half its maximum value. We have not shown the temperature dependence here.

Compared to the cross hardening rate, the self hardening rate is assumed to be small and constant i.e. $b = b_0 = const.$.

Some properties of the hardening function

When you are modeling the behavior of many materials, you have to frequently construct functions of various given characteristics. In the case of the hardening response, we need to construct a function that (1) has a high adjustable plateau for low values of κ indicating the almost constant Stage II hardening, and then drops in a linear way down to a much lower value (corresponding to Stage III) over a specified range. In other words the function is like an *inverse sigmoidal function*. We have to find a function in such a way that (1) the function has just a few adjustable parameters and (2) the parameters should be closely related to some physically observable characteristic of the function. This last characteristic is VITALLY important since it allows us to make qualitative judgements about the function.

This exercise is to show you how the parameters change the characteristics of the function defined by (13.64).

⊙ **13.14 (Exercise:). Exploring the Hardening Function**
Write a MATLAB program to plot the function defined in (13.64) as a function of $\kappa, a_i, n, \triangle H$ and T.
Keep the temperature at 300K, and investigate how the function changes for different values of $\triangle H$ (assume $A = 60$) and use the values of a_0, a_1 and a_2 as given in Fig. 13.8. By exploring the shape of the function for various values of the parameters, describe in words, the effect of the parameters.
Hint: The program for the figure was exactly 3 lines long. So it is not a major challenge. But describing the effect of the parameters is tricky.

The hardening law is then given by (13.63) together with (13.64). In order to implement this along with the flow equations (13.55), we need to not only solve a differential equations for the nine components of the lattice tensor **A** but also the twelve hardening parameters. So, we need to solve a set of 21 simultaneous nonlinear differential equations!

> ### GOVERNING EQUATIONS FOR SINGLE CRYSTALS
>
> Given $\mathbf{F}(t)$ and the initial values of the lattice vectors $\mathbf{a}_i$, the slip matrices $\mathbf{P}(\alpha)$, the Helmholtz potential ψ and the hardening parameters $\kappa(\alpha)$, the governing equations are
>
> The stress components: $S^i_j = C^e_{im} \dfrac{\partial \psi}{\partial C^e_{mj}}$,
>
> The resolved shear stresses: $\tau(\alpha) = \mathbf{P}(\alpha)^i{}_j S^j{}_i$
>
> The slip rate : $\dot{\gamma}(\alpha) = f[\tau(\alpha)] = \left\langle \left\| \dfrac{\tau}{\kappa(\alpha)} \right\| - 1 \right\rangle^{1/r} \operatorname{sgn}\left(\dfrac{\tau(\alpha)}{\kappa(\alpha)}\right)$, (13.65)
>
> Evolution equation for lattice vectors: $\dot{A} = L\mathbf{A} - \mathbf{A} \sum_\alpha \dot{\gamma}(\alpha) \mathbf{P}(\alpha)$
>
> Hardening law: $\dot{\kappa}(\alpha) = \sum_{\beta \neq \alpha} \{a[\kappa(\beta), T] \, \|\dot{\gamma}(\beta)\|\} + b[\kappa(\alpha)] \, \|\dot{\gamma}(\alpha)\|$.
>
> As usual, the notation $\langle x \rangle$ stands for the Macaulay brackets.
> We can get different kinds of responses by choosing different functions for ψ, for the slip rate and for the hardening laws. The other equations are fundamental to the development of crystal plasticity presented here.

The resulting response curve for a simulated isochoric extension, simulated using the MATLAB program available online (see authors' websites listed in the preface.) for three different orientations of the crystal are shown in Fig. 13.9; The program simulates the stress response, shows an animation of the deformation of the crystal as well as the lattice vectors and the stereographic projection of the lattice orientation.

Note the response anisotropy: The [111] crystal[13] has the highest hardening (due to the activation of many slip planes) while the [100] crystal has the lowest. Also note that there is no Stage I hardening since there is hardly any single slip. Note that dynamic recovery (or stage III) occurs at around 60-70 MPa. The numbers have been chosen to mimic the response of copper single crystals and the dots on the figure are based on the data reported in [Kocks and Mecking (2003)].

The program is provided online (see authors' websites in the preface) and can be modified to deal with other problems also. We show the results of the program for an Aluminum single crystal under plane strain compression matched with the experimental data of [Hosford Jr (1966)] in Fig. 13.10. The various parameters were chosen to fit the data for the $(1\bar{1}0), [110]$ oriented crystal (note that for FCC

---

[13]The reciprocal vector [111] is the direction of the tensile axis at the beginning of loading. Since we are dealing with isochoric extension, the orientation of the crystal around the tension axis is immaterial.

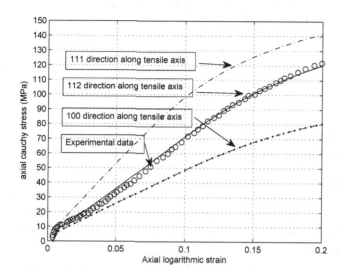

Fig. 13.9 Simulated hardening response of a copper single crystal.

crystals, the normals to planes coincide with crystallographic planes). We then test the predictions for other directions of compression. The predictions for the other orientations are quite good. We leave it as a project for you, dear reader, to modify the programs and the parameters to reproduce (or better still, improve upon) these results. This will be an interesting exercise since you will have to find the elastic constants, orient the crystal properly, etc., and so we urge you to try this task seriously.

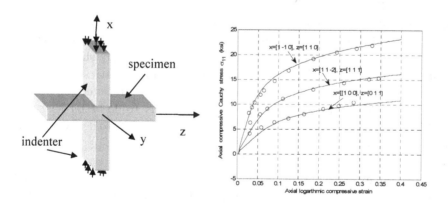

Fig. 13.10 Plane strain compression of a single crystal and the simulated response. The dots are data based on the paper by Hosford but modified to get tension data.

For further reading on different models for hardening and how well they compare with experiments, see [Liang and Khan (1999); Cuitino and Ortiz (1993)].

13.8 Concluding remarks

At this stage there might be a number of questions in your mind:

Q: Why did we have a rate dependent theory for single crystals and not a rate independent theory?

The reason is simple: rate dependent theory of a crystal is a LOT easier to simulate than a purely rate independent theory. There are numerous knotty issues that all go away when you consider a rate dependent theory (however, see [Havner (1992)] where there is an excellent description of the model for rate independent materials, and [Schurig and Bertram (2003)] who develop a rate independent model that deals with some of the issues) . Also compared to structural problems, heating and temperature dependence (and hence rate dependence is a lot more important for those who model single crystals (see e.g., [Canova *et al.* (1988)] to see the importance of rate dependence in single crystals). Because of this, most practical models of crystals are rate dependent. However, a word of caution, while the theory presented here was quite reasonable, our solution procedure is too dependent on MATLAB's ode solver. If you want to write your own solver then you should consider the paper by [Sarma and Zacharia (1999)] they show how to do these problems by using a Newton Raphson scheme that is very rapidly convergent. Other algorithms that are available are, for example ([Maniatty *et al.* (1992)]; NematNasser and Okinaka (1996); McGinty and McDowell (2006)].

Q: How does one simulate polycrystal response?

Here again, there is a vast literature: the simplest and most widely used idea is to consider a polycrystal to be overlapping single crystals with different orientations all undergoing the same deformation. In other words, you simply simulate 80 or 100 crystals with different starting orientations and average the stress. There are more sophisticated techniques broadly referred to as self consistent techniques (see e.g. [Hutchinson (1976); Jiang and Weng (2004)]) that try to do better than this simple averaging procedure and invoke other semi empirical ideas to deal with polycrystalline materials. Again this is beyond the scope of this book.

You can try an example of showing what happens if you choose 10 crystals that are initially oriented randomly and then deformed by extension. The idea is to run the program 10 times with different initial orientations but with the same deformation and average the stresses. You can also track the final orientations. This is also a good project for you to try.

Do you expect the final crystal orientation to be random? The non-random orientation of the crystals in a material is referred to as crystallographic "texture". Different "textures" are characteristic of different forming processes and crystal types. [Hertzberg (1976)] has a very approachable introduction to this topic. We are not going to spend much more time on this subject since this is an area of specialization in itself since it is an important way to obtain anisotropic response. You will notice that we have been saying that more and more things are beyond the scope of this book. This is because we have come to the end of what we can do in this book. It is time for you to explore the literature and find the details that fit your need.

There are several major trends in research that is currently underway: There is the metal forming community primarily focussed on sheet metals (see e.g., [Hosford (1993); Kocks et al. (1998); Banabic et al. (2000)]), there is the polycrystal simulation community (see e.g., [Anand and Kothari (1996); Anand (1999, 2004); Balasubramanian and Anand (1998); Bronkhorst et al. (1992)])

13.9 Exercises and projects

(1) A MATLAB program that displays a stereographic projection of the rotation of the lattice vectors with respect to the global xyz axes can be written (available online – see authors' websites listed in the preface)). It is customary in crystallography literature to display the rotation of the xyz axes with respect to the lattice vectors. Write a MATLAB program to do this.

(2) Polycrystal response: Find the place in the program where the initial orientation $R0$ is set. Run the program with the value of $R0$ set to ten different randomly chosen orientations, (in each case ensure that $R0$ is an orthogonal tensor) average the stresses from all the runs and plot the response. See if it matches the polycrystal copper response at room temperature. Does it come close? Does it over predict the response? What do you think is missing? Are your assumptions reasonable?

This way of simulating polycrystal response is called the Taylor assumption and is widely used in the crystal plasticity community.

(3) The hardening law used here does not account for kinematic hardening. Study the paper by [Weng (1987)] and implement the hardening rule that is described there. Note that you have to track 24 hardening parameters.

(4) In this chapter we described the response of copper under isochoric tension. Study the paper by [Hosford Jr (1966)] on the plane strain compression of Aluminum crystals. Write a program to simulate the response either by modifying the program included here or by writing your own program.

(5) Develop constitutive equations for BCC materials following the same lines as done for FCC materials, here. (You need to find the elastic moduli, the slip

systems, the **P** matrices, the plastic potential $f[\tau(\alpha)]$ and the hardening rules. (see e.g., [Nemat-Nasser et al. (1998); Stainier et al. (2002)] for single crystals and [Liang and Khan (1999); Khan and Liang (2000); Guo et al. (2003)] for polycrystalline materials).

Chapter 14

Advanced Case Studies

Learning Objectives

After going through this chapter, you should be able to

(1) get an idea of how different applications that involve inelasticity are, in general, idealized to extract information necessary to arrive at appropriate engineering process or design parameters.

(2) identify what characteristics to pay attention to while modeling so that meaningful results can be obtained to answer the relevant design questions that arise in the application.

(3) apply the principles learnt such as identifying the lumped parameter equivalent and extend the principles of continuum finite deformation inelasticity modeling learnt in the book.

(4) do preprocessing before conducting an appropriate analysis, use a standard commercial tool available and extract important numbers necessary for making the engineering design decisions.

CHAPTER SYNOPSIS

In this chapter, we select some advanced real life applications involving mainly continuum finite deformation inelasticity and illustrate how one goes about modeling, analyzing and solving the problem from a design perspective.

Three applications are illustrated here: (1) shot peening to induce compressive residual stresses on the surface of a metallic component to enhance the fatigue strength, (2) Equal channel angular extrusion process for effecting extreme shears that render a material with high strength and ductility, and (3) aging of a facial tissue to understand how sagging of the facial tissue occurs over age.

Each of these applications involve different situations necessitating different simplifications and decisions on the models to use and extract appropriate information from the results of a finite deformation analysis using a standard package such as ABAQUS.

Chapter roadmap

This chapter describes three basically different real life problems that deal with different decisions to be made in terms of modeling and analysis based on the nature of the problem and answers to the design questions sought.

The first application discussed in section 14.1 is the shot peening application in which compressive residual stresses are introduced by impinging shots on to the surface of the component. This is a typical industrial application where the interest is in finding out parameters that affect the final product and their effect on the introduction of residual stress. Decisions such as how to approximate the problem into a meaningful boundary value problem by assuming a representative volume for analysis and the application of suitable boundary conditions are paramount considerations here. Since this involves a rate-dependent simple hardening without any plastic cycling, a model available in a standard commercial finite element package such as ABAQUS is sufficient to simulate the process. The aim here is to be able conduct a *parametric study*—numerically finding the optimal set of values for the influencing parameters in the process for the introduction of the required residual stresses.

Section 14.2 discusses an application in a research project setting that concerns a metal forming application in which extreme shearing is carried out using a equal channel angular extrusion (ECAE) process. Unlike the shot peening application, a canned model in ABAQUS is not suitable, necessitating the development of a constitutive model that involved modeling anisotropy of the inelastic deformations. It is shown how Gibbs potential approach to finite deformation plasticity can be formulated and used in analyzing the inhomogeneity in the deformation. Here the emphasis is on the ability of the model to simulate the process so that several "what if?" questions can be answered.

Then, we move on to an application involving a time dependent deformation in a purely research setting. The long term degradation in terms of loss of elasticity and gradual sagging in a facial tissue is discussed in section 14.3. It describes how the model is tuned to avoid any sharp yield condition. As has been described in the earlier chapters, an explicit exponential form of expression is provided for the "plastic multiplier" in terms of the total strain. The user subroutine option in ABAQUS is utilized.

We have provided only a limited number of examples and that too to the point of interest to the material covered in this book. More real life examples will be added in the website.

14.1 Case Study I: Shot peening - A Process for Creating Wear Resistant Surfaces

14.1.1 Background and problem statement

This is an application of inelasticity in which residual stresses are introduced at the surface of a component to improve its fatigue life [Bozdana (2005)]. It is a common knowledge that the fatigue strength increases with mean compressive stresses but decreases with mean tensile stresses. Therefore, introducing compressive stresses on the surface layers is an established way of improving the fatigue strength. This also helps annulling the tensile stresses developed due to machining operations on the surface of the component. One of the popular ways of introducing the compressive stresses on the surface is the *shot peening* in which shots, typically made of steel, glass or ceramics, of size 0.3 to 1.5mm in diameter, are impinged on the surface with a high velocity causing local plastic deformation on the metallic surface leading to an elastic rebound leaving a high compressive residual stress in a thin layer of the beneath the surface [Kirk (1999); Baskaran and Srinivasan (2008)]. These residual stresses increase the fatigue strength by 10% to 100% (see Fig. 14.1).

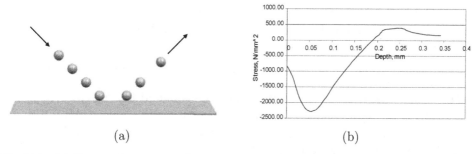

Fig. 14.1 (a) Shot peening process: shots of dia. 0.3 to 1.5mm hit the surface to be peened with a velocity of 25 to 100m/sec causing plastic deformations on the outer layer of the surface. This plastic deformation and the elastic rebound inner layers and surrounding material causes high residual compressive stresses at the surface locally typically over a of depth around 0.1mm in a plate as shown in the stress-distribution in (b). The self-equilibrating tensile stresses however are low but distributed over thicker inner layer below the outer layer.

The design engineer prefers to include the residual stresses in the design of parts to extract the benefit of residual compressive stresses so that the weight and the cost can be optimized. However, there are several input parameters that control the shot peening response. Typically the size, velocity, angle of impact, friction, material characteristics of the shot and the target are the key input parameters.

The surface residual stress, in terms of *the maximum of residual stress and the depth of maximum residual stress*, affects the fatigue behavior of materials. The peening intensity which is a measure of residual stress is influenced by these that are, in effect, determined by the amount of cold work. Typically, in the industry, strips of metal called the *Almen strips* are peened along with the parts peened. The cold work causes these thin strips to curve upwards and the magnitude of such a permanent deformation (called the *Almen intensity*) calibrated appropriately is used as a measure of peening intensity.

The number of input variables and the variation of each of them have made determination of the residual stresses through experimental studies very expensive and time consuming since it is difficult to cover the entire gamut of variation of all input variables. These reasons make it necessary to focus our efforts to explore the peening process numerically. Besides, such numerical studies help us to understand the physics of shot peening better and engineer to optimize the process appropriate to the requirements.

Given this background, the task is to *simulate the shot peening process for different conditions* of the input variables and parameters, understand and predict their effects on a component to eventually optimize the process for a given design condition.

14.1.2 Simplifications and assumptions on modeling the system

Based on the problem statement above, we need to make several decisions on simplifications and modeling of the system that we will be solving or simulating in order to answer the design questions.

The material model:

Unlike the other cases dealt with in this chapter, the standard material model options available in the finite element software package ABAQUS are used [ABAQUS (2008)]. The following are some of the salient features used in relation to the material model with suitable justifications.

- Heat produced due to the plastic deformation is usually small conducted away quickly by the metallic material. Hence, isothermal conditions are assumed.
- Since the process involves mainly a monotonic plastic response of the material followed by mainly elastic unloading and then reloading-unloading cycles due to the impact of the subsequent shots, a simple hardening model that is capable of simulating a single monotonic plastic response would suffice. In this sense, a purely isotropic hardening model can be used without paying much attention to the details of Bauschinger effect (kinematic hardening).
- The material used is strain-rate dependent. Hence, a strain-rate dependent analysis is option is used to obtain a realistic picture of residual stress generation. The stress-strain curves for different strain rates that are available are input in a tabular form in ABAQUS as material model properties for the analysis. ABAQUS uses a rate dependent power law type overstress material model

given by

$$\dot{\varepsilon}^p = D\left(q/\sigma_0 - 1\right)^n \qquad (14.1)$$

where D, n are found out internally in ABAQUS from the input mentioned above. σ_0 is the static yield stress (see ABAQUS theory manual in [ABAQUS (2008)]).

- the plastic deformations at the surface during the shot peening process are of the order of 20-30% which is in the finite deformation range.

ABAQUS uses in its metal plasticity (isotropic rate-dependent hardening) option, a standard additive decomposition of total strain into elastic strain and plastic strain with an assumption that the elastic strains are very small compared to the plastic strains. A multiplicative decomposition of deformation gradient of the form, $F = F^e F^p$ is used in their formulation (see ABAQUS theory manual [ABAQUS (2008)]). A von Mises yield function, an associated flow rule and a Hencky type elasticity model are used.

Preparation for numerical simulation:
Features such as impact (including multi-ball impact) and contact mechanics options available in ABAQUS are used. The following are some of the points related to the simulations:

- Since we are looking at the shot peening process as a random distribution of shots over the surface that introduces the residual compressive stresses at the surface, *a unit cell approach* which has dimensions reasonable to accommodate the distribution involved in this random process would suffice to understand the process of shot peening and its effect on introducing compressive stresses on the surface. Typically, in unit cell approach, a discrete volume of the problem is taken and the boundary conditions are assumed that simulate the actual conditions reasonably accurately. This helps in reducing the size of the problem.
- An in-built rate-dependent elastoplastic isotropic hardening finite deformation plasticity model is used in the elasto-plastic analysis. Details of this material model are described separately.
- though, in reality, the shot also undergoes both elastic and plastic deformations, considering the numerical efficiency, the shot is assumed to be rigid. It is expected that the results are not unduly affected by this assumption.
- 'general contact' option in ABAQUS is used for the contact analysis.
- Explicit dynamics method that is widely used for impact of shots on a target surface is used in this simulation. ABAQUS/Explicit uses this method.
- shot peening on a typical plate is modeled using a square unit cell.
- unless mentioned explicitly, the shot is assumed to impact normal to the square unit cell of the plate.
- coulomb friction is used to model the contact friction.

The finite element model
To start with, it is important to understand the effect of single shot impact and

then move to multi-ball impact to understand the effect of juxtaposing.

A unit cell of dimension $1mm \times 1mm \times 2mm$ is taken up for shot peening analysis (for a shot size of 0.3mm dia). Due to the symmetry of the unit cell and the shot, only a quarter of the cell need be modeled for the single shot analysis. Solid continuum elements (C3D8R elements in ABAQUS/Explicit [ABAQUS (2008)]) have been chosen for the analysis. The typical element size in the target plate is made to be very fine of approximately 0.0125 mm on each side. Such a fine mesh is arrived at after a mesh convergence study to capture the variation of plastic strains in the surface layers while keeping the computational costs at optimum. The bottom surface is supported in the vertical direction. The non-symmetry planes are assumed to be free surfaces.

The single shot analysis is useful in obtaining the essential numbers related to the residual stress, amount of plastic strain, the penetration of the residual stress, coefficient of restitution etc. Using these numbers, it will be possible to carry out a more rigorous multi-ball analysis for estimation of possible residual stresses in the target material.

14.1.3 Sample results of the analysis

As discussed earlier, we wish to find out the distribution of the residual stresses, the plastic strain distribution that indicates indirectly the capability of storing the residual stresses, the coefficient of restitutions that indicates to a certain extent the amount of energy used up in plastic deformation process, influence of a single shot - the vicinity of the short range effect of residual stress build up, the effect of rate dependency of the material, etc.

Equivalent plastic strain distribution (PEEQ) and residual compressive stress (RCS) variation across the depth of the target material (see Fig. 14.2): The amount of 'cold work' or plastic strain distribution indicates that there is a small layer of high compressive residual stresses near the surface that dies down very rapidly away from the surface. Only the centerline section is shown since with complete coverage of the shots. This is the typical distribution one would see.

The plastic strain reaches a maximum value and drops to zero magnitude steeply. The RCS and PEEQ contours are shown in Fig. 14.3.

Theoretically, the area under compressive RCS curve with respect to depth can be used as a measure and the peening parameters can be optimized for the maximum area under this curve. In general, higher the depth and/or higher the RCS magnitude, the higher the fatigue strength. To increase the area under the RCS curve, shot size or velocity, for example, can be increased. This, however, can result in surface cracks and they can cause early fatigue failure beyond optimum levels. To simulate proper coverage, multiple shots can be used on the unit cell to obtain a random impacts on the surface. Optimization needs further analysis and a parametric study for different values of the affecting parameters.

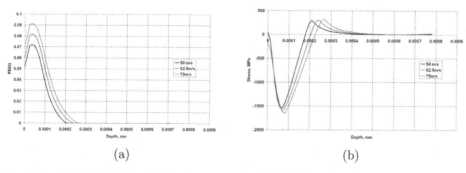

Fig. 14.2 (a) Plastic strain distribution across the depth. These directly correlate to the compressive residual stress build up across the depth as shown in (b). The residual stress builds up to high compressive value rapidly beyond the surface while the tensile residual stresses distribute evenly to compensate for self-equilibrium.

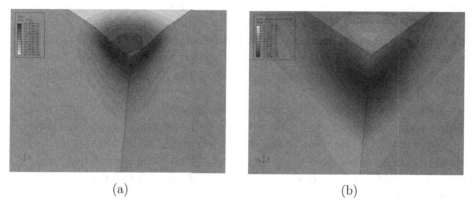

Fig. 14.3 Contour plots of (a) plastic strain distribution (b) residual stress distribution over the width and depth near the shot impact area.

14.2 Case Study II: Equal Channel Angular Extrusion - A Materials Processing Route

14.2.1 Background and problem statement

This case study grew out of the work of Dr. Anshul Kaushik, who was a graduate student with one of the authors at Texas A&M University. The results cited here are partly based on his Ph.D Thesis ([Kaushik (2007)]).

In recent years, there has been a considerable interest in the use of *very large strain (of the order of 100% or more per cycle), cyclic plastic deformation* as a way to generate ultra fine grained materials. This method of processing a material is called "severe plastic deformation" or SPD. How does this occur? A simplistic explanation is that due to the SPD, the microstructure is severely damaged, to the extent that the individual grains breakup into smaller subgrains through the formation of dislocation cells. Note that if material is not under large confining pressures, the damage can be severe enough that the material forms cracks and other defects. So one of the challenges of SPD is to deform the material enough to form subgrains but not cracks and other undesirable phenomena.

There are several methods of doing SPD, the most popular ones being High Pressure Torsion (HPT) and equal channel angular extrusion (ECAE) or equal channel angular pressing (ECAP). Of these, ECAE was invented at Texas A&M University by Segal ([Segal (1995); Ferrasse et al. (1997)]. There is an extensive literature on this method of processing (see e.g., [Valiev and Langdon (2006)] for a recent review). The process essentially consists of using a plunger to force a prismatic bar of material through a sharp corner formed by two intersecting channels with equal cross-sections as seen in Fig 14.4. As the billet passes through, it is severely deformed within a small area at the intersection of the two channels.

The main advantage of ECAE over conventional extrusion is that, unlike conventional extrusion where the cross section of the body is changed, the cross section of the material undergoing ECAE is unchanged since the two channels are of equal cross sectional area. Thus the process can be repeated to carry out cyclic loading by passing it through the same channel in different orientations. In addition, the deformation is quite uniform as compared to that from the conventional deformation processes.

Due to the attractiveness of the process, the question arises: why start with solid billets and refine the grains? why not start with nanopowders and use it for "green compaction", i.e. we want to use the pressures and shear forces generated by the ECAE process to compact the grains but not sinter them (i.e. we don't want to "cook" the particles in high heat because then grain growth will take place and destroy our intent of making fine grained materials.). This approach has been tried and has been very successful (see the growing body of literature in this area, e.g., [Karaman et al. (2003); Robertson et al. (2003); Yoon and Kim (2006)]).

Of course, since the channel is open in the bottom, you can't just fill it with powder and then just use the plunger—the powder, not being confined, will simply dribble out the other end. So researchers, primarily Drs. Hartwig and Karaman at Texas A&M University ([Karaman et al. (2003); Robertson et al. (2003)]) came up with the idea of using special "cans" to contain the powder. The cans are extruded and the compacted fine-grained material is extracted from it. There are a number of processing issues that have to dealt with and a brute force experimentation is NOT the answer. What is needed is a simulation to ask a bunch of "what if" questions.

Given this background the task was **to simulate the powder compaction process in ECAE**, with a view to answering a number of questions:

- How do the loads used in ECAE compare with that for other compaction methods? We can use this information to determine the press capacity required to operate the plunger.
- What is the effect of can material, wall thickness etc., on the product? This

information is needed to reduce the cost of the can which is a consumable.
- What is the reason for the presence of high porosity in some regions of the compacted materials? Porosity is the bane of any powder compaction operation. We want to get very low porosity in the sample and that too in a uniform way. This requires knowledge of the parameters that affect porosity.

14.2.2 Need analysis and specification of the system

Based on the problem statement we have to make a number of decisions:

(1) **Simplification of the external stimuli and geometry:** We decided to study square dies with circular cylindrical cavities where the powder was filled. A study of the ECAE process indicated that friction between the can and the powder as well as the can and the walls were important to the process. So the simulation had to be done in a way that allowed different friction models to be implemented. The plunger was much stiffer than the billet so that it could be assumed to be rigid and since it was hydraulically operated, a displacement control on the top of the die was assumed. During the process, due to the right angled turn, there would be changes in contact conditions. This had to be accounted for. Finally, there was some interest in applying pressure to the bottom end of the can (so-called back pressure). This had to be considered. Finally, there were numerical instability problems at the sharp reentrant corner; we had to choose a fillet radius to prevent inter-penetration of material at the corner.

Finally, there was much discussion about the fact that initial porosity of the powder could be as high as 50-60% since this was just filling a can with powder. Also the powder, being ultrafine particles had a tendency to clump together. After much discussion, we decided to focus only on the process parameters and, for the initial runs, ignore the variations in porosity of the powder.

(2) **The material model:** Material models were needed for the can and the powder. Since the can was either of copper or nickel, we were able to use the existing plasticity models that were built into ABAQUS. For the copper can, we used a plasticity model Young's modulus of 124GPa and initial yield stress of 80MPa increasing to 300MPa. One disadvantage of using a "canned" model for the can is that we do not know precisely what constitutive equation was actually implemented and how it was done (Caveat Emptor!).

The challenge was much more severe for the powder. Much of the literature on sintering and compaction was focussed only on pure compression and not much was available on the effect of shear on such powders[1]. The most commonly used model was the "Gurson model" ([Gurson (1977)]) which was widely implemented in a variety of applications. However it was generally assumed to be valid for porosities of the order of 10% and not 50%. So a new model had

---
[1] See the recent comparative study by [Biswas (2005)].

to be created; This was done by studying the works of [Duva and Crow (1992); Kim and Carroll (1987); Gu et al. (2001)] and choosing aspects of these works that were most useful for the task at hand.

Most of the analysis of ECAE processes had assumed plane strain conditions; but the dimensions of the channel and more importantly the fact that there is a can within which there is powder invalidated any 2-D simplification and a full 3-D simulation needed to be carried out. This was a major complication since complexity grows as the Power of the dimension; we were limited as the fineness of the element grid and the run time. This was a huge challenge for accurate prediction. Clearly, writing our own software code was out of the question due to the complex friction conditions etc.

Finally, with a view to future developments such as (1) dealing with anisotropy and inhomogeneity of the powder and billet (2) possible grain growth mechanics etc. it was decided to use a Gibbs potential approach to developing the constitutive theory (since it is easiest to formulate grain growth kinetics in terms of Gibbs potential).

So the starting point was the generalization of the Gibbs potential approach to small deformation plasticity (see Section 9.4, Section 9.4.2 and the exercise on p. 281 of Chapter 9).

Compared to the Helmholtz formulation, the Gibbs formulation has some significant advantages: They are

(a) All the independent variables are defined in terms of measurable variables, such as stress, temperature etc. *without the use of the notion of elastic strains.*
(b) Because of this, it provides a means for sidestepping some raging controversies in the field regarding proper formulations for anisotropic materials.
(c) It is a very convenient way to proceed if one were interested in such areas as diffusion and chemoplasticity (i.e. if you want to incorporate diffusion phenomena or chemical reactions).
(d) As we shall see soon, it becomes easy to extend anisotropic constitutive equations from small strain plasticity to large strains.

However, it has major difficulties too. For starters, *unless you are very careful to use special objective tensors, it is NOT possible to recover conventional finite elasticity as a special case!* (see e.g., [Xiao et al. (2006); Hackl et al. (2001); Miehe et al. (2002)]) as well as the remarks in [Simo and Hughes (1998)] Section 7.3 regarding some of the problems that are encountered with some (though not all) of these models. The resulting algorithm is actually quite difficult to integrate and special "incrementally objective algorithms" are needed.

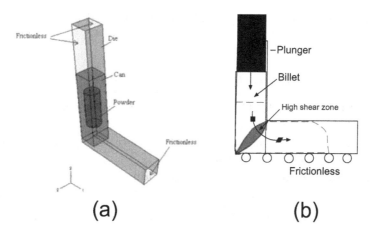

Fig. 14.4 (a) Schematic of the powder compation process using ECAE. The can is pushed into the channel by a plunger that moves in from the top. (b) schematic of the deformation of a portion of the material as it travels through the billet. Note the large amount of shear and the rotation of the element as it passes through the billet. Reprinted from A. Kaushik, Ph.D thesis, 2007, Texas A&M University.

14.2.3 Embodiment of the model

Since we are introducing the notion of a Gibbs potential, we write it in terms of the Kirchhoff stress $\mathbf{S}$ as described in Sections 9.4 and 9.4.2. However, since we are dealing with finite deformations, we recall from our discussion of objective tensors in Chapter 11, section 11.4 beginning on p.389 and especially the boxed result regarding objective constitutive theories on page 395, we will replace the Kirchhoff stress (which is an objective quantity) with the *rotated kirchhoff stress tensor* $\mathbf{S}^*$ through

$$\mathbf{S}^* := \mathbf{R}^t \mathbf{S} \mathbf{R}, \ \Rightarrow \ \dot{\mathbf{S}}^* = \mathbf{R}^t (\dot{\mathbf{S}} - \Omega \mathbf{S} + \mathbf{S}\Omega) \mathbf{R} = \mathbf{R}^t \overset{\square}{\mathbf{S}} \mathbf{R}. \tag{14.2}$$

where the notation $\overset{\square}{(\cdot)}$ is the Green-Naghdi-Mcinnis (GNM) rate of any objective second order tensor. Why did we not use other rates such as the Logarithmic rotation rate [Xiao et al. (2006, 2007b,a)]) or Jaumann rate? First we know that the GNM rate is easy to visualize (we are measuring everything from the corotational frame) and it gives reasonable results for simple shear (unlike the Jaumann rate see [Khan and Huang (1995)]). Also—and this is an engineering and not a scientific consideration—all the models in ABAQUS are implemented using the GNM rate and so implementation issues are simpler.

We are now ready to assume forms for the constitutive equations

(1) **Equations of state:** Thus, following the principles of Chapter 9 and Chapter 12, and recalling that the Gibbs potential is the negative of the complementary energy function, the constitutive equations for the state variables are given by

(2) **The Gibbs potential G and the Helmholtz potential ψ:**

$$G = -\frac{1}{2}\mathcal{K}(\mathbf{S}^*)\bullet\mathbf{S}^*$$
$$\Rightarrow \psi = G - \mathbf{S}^*\bullet\frac{\partial G}{\partial \mathbf{S}^*} = \frac{1}{2}\mathcal{K}(\mathbf{S}^*)\bullet\mathbf{S}^*, \qquad (14.3)$$

where $\mathcal{K}$ is the fourth order Compliance tensor (which is the inverse of the elastic modulii tensor) Note that *the Helmholtz potential is now a function of the rotated Kirchhoff stress and so it is not necessary to introduce any notion of elastic strain at all in this formulation.*

(3) **The mechanical energy dissipation equation:** By standard arguments that should be familiar to you now, we require that

$$\boldsymbol{\sigma}\bullet\mathbf{D} - \varrho\dot{\psi} = \varrho\zeta \geq 0. \qquad (14.4)$$

By dividing through by ϱ and using the definition of the rotated Kirchhoff stress, we can rewrite the above equation as

$$\mathbf{S}^*\bullet\mathbf{D}^* - \dot{\psi} = \zeta \geq 0 \qquad (14.5)$$

where $\mathbf{D}^* = \mathbf{R}^t\mathbf{D}\mathbf{R}$ is the "rotated rate of deformation". Now, all the tensors that appear in the constitutive formulation are invariant so that they can be differentiated without worry.

Now substituting for ψ and grouping terms, we obtain

$$\mathbf{S}^*\bullet\left(\mathbf{D}^* - \mathcal{K}\dot{\mathbf{S}}^*\right) = \zeta. \qquad (14.6)$$

It is now a trivial matter to introduce the plastic flow rate as

$$\mathbf{D}_p^* := \mathbf{D}^* - \mathcal{K}\dot{\mathbf{S}}^*. \qquad (14.7)$$

If you compare this with the corresponding equation for small deformation (equation (9.56) on p.294), we see that $\dot{\boldsymbol{\varepsilon}}$ becomes $\mathbf{D}^*$ and $\dot{\boldsymbol{\varepsilon}}_p$ becomes $\mathbf{D}_p^*$ and $\mathbf{S}$ becomes $\mathbf{S}^*$—a rather simple and direct analogy. The difficulties in the theory stems from the fact that while the various small deformation quantities are related to the derivatives of state variables, the same is NOT true for this version of finite deformation theory.

(4) **Constitutive equation for $\mathbf{D}_p^*$:** The conditions for $\mathbf{D}_p^*$ are stipulated as follows:

Since there was no clear and reliable model for such a wide change in porosity (starting from powder and ending up with a completely densified solid), it was decided to explore the Duva and Crowe Model ([Duva and Crow (1992)]) for porous materials. Duva and Crow originally suggested a model to study the densification of powders undergoing hot isostatic pressing. However the original model was developed as a rigid plasticity model along the lines described in Section 12.2. The model includes the effect of permanent volume change and dilatation. We extended it to elastoplastic materials in a simple way as shown below:

(a) The material is assumed to be a rate dependent material. whose rate of dissipation is given by

$$\zeta = \frac{d_0 s_0}{n+1} \left\{ \frac{d}{d_0} \right\}^{n+1}, \qquad (14.8)$$

where d_0, s_0 and n are constants and s is a scalar function of $\mathbf{D}_p^*$ which is given by

$$d^2 = \frac{(\mathbf{D}_p^*)_{dev} \cdot (\mathbf{D}^*)_{dev}}{a} + \frac{(\mathrm{tr}(\mathbf{D}_p^*))^2}{b}, \qquad (14.9)$$

where a and b are functions of the relative density (RD) of the material, which is the volume fraction of voids present in the material. This form of the rate of dissipation function is similar to the power law models discussed in Chapter 13 (see equation (13.31) on p.473).

(b) We assume the following condition when there is no dissipation:

$$\text{no dissipation implies } \mathbf{D}^* = \mathcal{K}\dot{\mathbf{S}}^* \qquad (14.10)$$

Clearly, this is NOT finite deformation elasticity but is actually a model for Hypoelasticity, *although it has been derived from an acceptable Gibbs potential and hence stress cycles are non-dissipative.* This means that you can have response models that are completely non-dissipative but do not coincide with traditional elasticity (see [Rajagopal and Srinivasa (2009)] for further discussion on such models). Whether we choose to accept it as being valid is still a matter of debate at the time of the writing of this book. But we can wear the engineers' hat and say, "for now we proceed with caution since we want to find answers to pressing questions". (pun intended!:-))

(c) When there is dissipation, we use the maximum rate of dissipation hypothesis to find the constitutive equation for $\mathbf{D}_p$; we maximize ζ over all values of $\mathbf{D}_p^*$ that satisfy (14.6). A routine calculation by using the method of Lagrange multipliers and after some extensive algebra, we get

$$\mathbf{S}^* = \lambda \frac{\partial \zeta}{\partial \mathbf{D}_p^*}$$
$$\Rightarrow \mathbf{D}_p^* = A s^{1/n-1} \left(a(\mathbf{S}^*)_{dev} + \frac{\mathrm{tr}\mathbf{S}^*}{3} \mathbf{I} \right), \qquad (14.11)$$

where $A = d_0(n+1/s_0)^{1/n}$ is a constant and λ is found by satisfying (14.6).

(5) Finally we specify the evolution of the void density and the parameters a and b through

$$\overline{\dot{RD}} = -RD\,\mathrm{tr}\mathbf{D}_p^*$$
$$a = \frac{3 + 2(1 - RD)}{RD^{2n/(n+1)}}, \quad b = \frac{3}{4n^2} \left[\frac{n(1 - RD)}{[1 - (1 - RD)^{1/n}]^n} \right]^{2/(n+1)}. \qquad (14.12)$$

The forms for a and b might seem odd, but they are based on the idea that when we consider a power law viscous material with a single spherical cavity under uniform compression, the value of the average stress calculated must be identical if we replace $\mathbf{D}_p^*$ with $\mathbf{D}$.

The powder material model is characterized by the elastic compliance $\mathcal{K}$ which, for an isotropic material, is characterized by two constants, and just three other constants—the power law exponent n, and $\dot{\epsilon}_0$ and s_0 which are related to the threshold stress for significant plastic flow and the corresponding strain rate. So, in a sense, it is a minimalistic model.

(6) **Boundary conditions:** This is where there are a number of issues that have to be dealt with. First, we note from Fig. 14.4a that there are several faces which are frictionless. On the surfaces with friction, it was assumed that it was coulomb friction with a friction coefficient that varies with the material surfaces in contact. The contact conditions between the powder and can were unknown. So we made different assumptions and compared with experiments were that were run with copper cans and copper powder in Nickel cans. Based on the results and in view of the simplicity, perfect bonding between the can and the powder was assumed.

(7) **Implementation:** The various material constants were determined and the results were implemented by using a user subroutine in ABAQUS Explicit with a VUMAT user subroutine (available online – see authors' websites listed in the preface.). Experiments were carried out using 10 to 40 micron copper powders. The can was square in cross section of side length 2.5 cm and of length 10 cm and the die was a cylindrical hole of 2 cm diameter and 5 cm length.

We show some of the results on the density profile in Fig 14.5. Note, in these figures, the can is not shown and only the powder part is shown.

The main result is that the bottom part of the powder does not consolidate well because the ECAE process does not build up sufficient compressive forces on the bottom part of the can (bout 550 MPa at the top of the can compared to only about 350MPa towards the bottom). This compressive force difference caused a variety of cracks and cracklike features on the bottom part of the can. Another interesting feature is that *the ECAE process requires only about 1/2 the pressure of a conventional compaction process.* This is because, in the ECAE process, the compaction is enhanced by both the compressive and shear forces, hence facilitating better compaction compared to just pure compression.

14.2.4 *Concluding remarks*

In this case study, some of the tradeoffs on the modeling side were illustrated. We had to make many compromises in order to complete the project on time; all the compromises were generally associated with our prejudices and are not fundamental requirements of thermodynamics; We did not make compromises on fundamen-

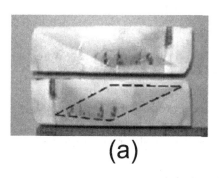 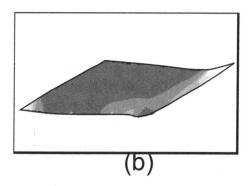

Fig. 14.5 (a) Extruded can with the compacted region marked. (b) densification profile obtained from ABAQUS. Note the fact that the simulations agree well with the observed densification patterns. Reprinted from A. Kaushik PhD. Thesis, Texas A&M University.

tal physical principles—these invariably lead to completely unacceptable behavior. This case study also illustrates the Gibbs potential approach to finite plasticity. We will readily admit that we are not completely comfortable with the approach but it has some powerful advantages: not the least of which is the ease of extending small strain plasticity. There is a considerable interest in this area and new models for powder processing are constantly being developed.

14.3 Case Study III: Modeling of an Aging Face - An Application in Biomechanics

14.3.1 *Background and problem statement*

The case study described here is from the works of [Mazza *et al.* (2005, 2007)] who carried out simulations (using a inelasticity model) of how our face ages with time. [LaTrenta (2004)] states in his book that "the most commonly held theory is that facial aging is the result of progressive gravimetric soft tissue descent. Over time, the soft tissues of the face simply sag off the bones of the face, forming the distinctive wrinkles, furrows, folds, and eventual tissue redundancy of the aged face. Gravimetric soft tissue descent is complex, however, and encompasses several distinct processes. One of the most important processes is actinic damage or solar elastosis.... Wrinkles become apparent in a womans skin in her mid-30s as estrogen levels begin to decline from their peak. The dermis begins to lose collagen, elastin, etc. Fat, unlike muscle, is supported solely by facial ligaments. After years of being pulled and stretched, these facial ligaments never regain their "tautness."

Given this background the task was **to simulate the aging of the face of a real person over time.**

14.3.2 Need analysis and specification of the system

Based on the problem statement we have to make a number of decisions:

(1) **Simplification of the external stimuli:** Based on the study of literature on aging, it was seen that aging is the result of loss of elasticity of the facial tissue due to exposure to the sun and other damage, resulting in the sagging of the face due to gravity. Thus gravitational forces are the primary external force on the face. Since the exact process of damage by the sun was not known and could not be modeled, it was decided to ignore the detailed mechanism and simulate the effect of the sun's damage.

(2) **Time of simulation:** The simulation was done for 30 years of aging assuming that the face is vertical about 14 hrs a day.

(3) **Geometry of the tissue:** While the surface of the face can be accurately mapped, it is hard to find the thickness. Also the face is made up of many layers as shown in Fig 14.6b.

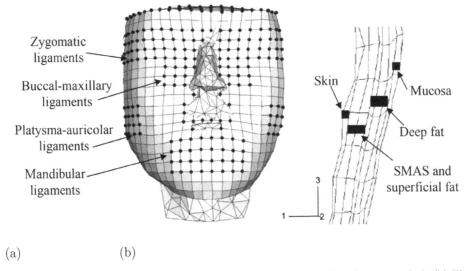

(a)   (b)

Fig. 14.6 (a) A digitized representation of a face with the points of anchorage marked. (b) The four layers of facial tissue. SMAS stands for "superficial musculoaponeurotic system". Figure reprinted from the work of Mazza et al., (Journal of Biomechanical Engineering Vol 129, 691–623, 2007) with permission from ASME International.

(4) **The material model:** Since much of the literature on tissue is focused on short-term effects, soft tissue is generally considered elastic and only elastic moduli are available. This is not useful here. Since we are interested in permanent, time dependent shape change, a large deformation viscoplastic model (shown schematically as a spring and dashpot model in Fig. 14.7). Further, to simulate the effects of the sun, the elastic moduli of the one of the elements was assumed to change slowly with time.

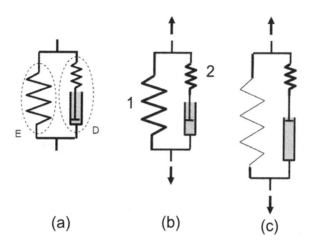

Fig. 14.7 Spring-dashpot analogy for the tissue response. Note that the short-term sagging (b) is primarily due to the viscoelastic properties of the material, while the long term "sagging" of the tissue (c) is influenced by the degradation of the modulus of the spring 1. Figure reprinted from the work of Mazza et al. [Mazza et al. (2007)] with permission from ASME International.

14.3.3 Embodiment of the model

Based on the spring dashpot analogy, we introduce the current squared stretch tensor $\mathbf{B} = \mathbf{F}\mathbf{F}^t$ as a measure of the deformation of spring 1 (which coincides with the deformation of the whole body) and the *elastic squared stretch tensor* $\mathbf{B}_e$) as *the measure of the deformation of the second spring*. As discussed in chapter 12, $\mathbf{B}_e$ needs an evolution equation that accounts for the short term dissipative behavior of the system.

(1) **Equations of state:** Thus, following the principles of chapter 12, the constitutive equations for the state variables are given by
(2) The Helmholtz potential:

$$\psi = \frac{\mu_0}{2q}[e^{qf(\mathbf{B},\mathbf{B}_e)} - 1]$$

$$f(\mathbf{B}, \mathbf{B}_e) := 2m_1(J - 1 - \ln J) + m_2(\beta - 3) + \gamma - 3,$$

$$J = \sqrt{\det \mathbf{B}}, \quad \beta = J^{-2/3}\mathrm{tr}(\mathbf{B}) = \mathrm{tr}(\bar{\mathbf{B}}_e), \quad \bar{\mathbf{B}}_e = J^{-2/3}\mathbf{B}, \quad \gamma = \sqrt{\frac{2}{3}\mathbf{B}_e \bullet \mathbf{B}_e}.$$

(14.13)

In order to account for the degradation of the modulus, we will assume that m_2 is a function of time.

(3) **The mechanical energy dissipation equation:** By standard arguments that should be familiar to you now, we require that

$$\boldsymbol{\sigma} \bullet \mathbf{D} - \varrho \dot{\psi} = \varrho \zeta \geq 0.$$

(14.14)

Substituting for ψ and grouping terms—recalling that ψ depends upon both $\mathbf{B}$

and $\mathbf{B}_e$—we obtain

$$\left(\mathbf{S} - 2\frac{\partial \psi}{\partial \mathbf{B}}\mathbf{B} - 2\frac{\partial \psi}{\partial \mathbf{B}_e}\mathbf{B}_e\right)\cdot\mathbf{D} - \frac{\partial \psi}{\partial \mathbf{B}_e}\cdot\mathbf{A}_p - \frac{\partial \psi}{\partial m_2}\dot{m}_2 = \zeta, \qquad (14.15)$$

where $\mathbf{A}_p$ is the negative of the Oldroyd upper convected derivative of $\mathbf{B}_e$ defined by

$$\mathbf{A}_p = \mathbf{L}\mathbf{B}_e + \mathbf{B}_e\mathbf{L} - \dot{\mathbf{B}}_e. \qquad (14.16)$$

As usual we will set the first terms in the bracket of (14.15) to zero and, after some manipulation, we get

$$\boldsymbol{\sigma} = -\mu(m_1[1/J - 1]\mathbf{I} + m_2 J^{-1}[(\bar{\mathbf{B}}_e)_{dev}] + J^{-1}[(\bar{\mathbf{B}}_e)_{dev}]), \qquad (14.17)$$

where $(\cdot)_{dev}$ stands for the deviatoric part of the tensor and $\mu = \mu_0 e^{qf}$ is a nonlinear shear modulus.

(4) **Evolution equation for $\mathbf{B}_e$:** The conditions for $\mathbf{B}_e$ are stipulated as follows:
- $\mathbf{B}_e$ must be unimodular, i.e. $\det \mathbf{B}_e = 1 \Rightarrow \mathbf{B}_e \cdot \mathbf{B}_e^{-1} = 0$
- When there is no dissipation, $\mathbf{B}_e$ must behave exactly like the $\bar{\mathbf{B}}_e$, i.e. it should satisfy[2]

$$\text{no dissipation implies } \dot{\mathbf{B}}_e = \mathbf{L}\mathbf{B}_e + \mathbf{B}_e\mathbf{L}^t - \frac{2}{3}(\text{tr}\mathbf{D})\mathbf{B}_e \qquad (14.18)$$

- When there is dissipation, we have the following evolution equation for $\mathbf{B}_e$:

$$\mathbf{L}\mathbf{B}_e + \mathbf{B}_e\mathbf{L}^t - \frac{2}{3}(\text{tr}\mathbf{D})\mathbf{B}_e - \dot{\mathbf{B}}_e = \phi\tilde{\mathbf{P}}$$

$$\tilde{\mathbf{P}} = \mathbf{B}_e - \left\{\frac{3}{\text{tr}\mathbf{B}_e^{-1}}\right\}\mathbf{I}, \quad \phi = [\Gamma_1 + \Gamma_2\dot{\epsilon}]e^{\left[-\frac{1}{2}\{\frac{\kappa}{\gamma}\}^{2n}\right]} \qquad (14.19)$$

$$\dot{\epsilon} = \sqrt{\frac{2}{3}\mathbf{D}_{dev}\cdot\mathbf{D}_{dev}}, \quad \dot{\kappa} = \phi\left[\frac{a_1 a_2 + a_2\dot{\epsilon}}{a_3 + \dot{\epsilon}}\right]\gamma - a_4\kappa^{a_5}.$$

where Γ_1, Γ_2 and $a_i (1 = 1, \cdots, 5)$ are constants.

We will make some general observations about the form of the equation, leaving you to explore the papers cited at the beginning of the section to understand each term and look up the specific constants.

The first two of the above equations are exactly in the tradition of what we discussed in Chapter 12. However, since we are constructing a rate dependent theory, as mentioned in Chapter 12 Section 12.11, we will propose an explicit form for ϕ. The material has no sharp yield condition, but note that because of the exponential dependence of ϕ on $1/\gamma$, the value of ϕ will go from practically zero below when γ, which is the norm of the elastic stretch $\mathbf{B}_e$ becomes less than κ. On the other hand ϕ rises to a value of about $(\Gamma_1 + \Gamma_2\dot{\epsilon})$ when γ becomes lager than κ. So it is somewhat similar to the constitutive equation equation (12.97) on p.452. Note that ϵ is a measure of the total "amount" of

---
[2] The extra term involving $\text{tr}\mathbf{D}$ is to ensure that $\mathbf{B}_e$ remains unimodular. You can check for yourself that $\dot{\mathbf{B}}_e \cdot \mathbf{B}_e^{-1} = 0$ as required.

distortion that takes place. As usual κ is the hardening parameter and has its own evolution equation.

With the assumed for the evolution equation for $\dot{\mathbf{B}}_e$, and substituting the specific form for the Helmholtz potential, the mechanical dissipation equation becomes,

$$\frac{\mu}{2J}[\phi \operatorname{tr}\tilde{P} - \dot{m}_2(\beta - 3)] = \zeta \geq 0 \qquad (14.20)$$

It is a straightforward substitution to see that $\phi > 0$ and $\operatorname{tr}\tilde{P} > 0$ so that the first term in the above equation is positive[3]. The second term is due to the long term degradation and requires additional considerations.

(5) **Long term degradation:** In the model, the loss of elasticity of the tissue is modeled by allowing m_2 to decrease slowly over time to account for the gradual loss of elasticity of the tissues. Thus, since $\dot{m}_2$ is negative, the mechanical dissipation equation (14.20) will again be met (since $\beta > 3$).

Based on experimental observations, it is stipulated that

$$m_2 = k_1 \left(\frac{k_2^p}{\delta^p + k_2^p}\right), \quad \dot{\delta} = \Gamma_3 + \Gamma_4 \epsilon \qquad (14.21)$$

which provides an S shaped curve similar to Fig.13.8 on p.488. Note that m_2 goes from value of k_1 when δ is smaller than k_2 to a value of 0 as δ increases, the crossover happening around $\delta = k_2$. To rate equation for the variable δ is a "creep" type equation where δ gradually increases with age irrespective of any deformation, the rate of increase being controlled by the parameter Γ_3. The degradation is accelerated by the amount of prior deformation that has taken place (controlled by the parameter Γ_4).

(6) **Boundary conditions:** The facial skin is anchored at only a few places to the facial bones as shown in Fig. 14.6a and these points are assumed to be immobile. The primary cause of deformation is the gravitational force on the face, which is represented as a body force field.

(7) **Implementation:** The various material constants were determined and the results were implemented by using a user subroutine in ABAQUS with about 3840 nodes and 11520 DOFs. We show how the face changes shape before and after in Fig. 14.8. You can see the sag in the unsupported areas.

14.3.4 Concluding remarks

Note that the approach used for the modeling of the face is quite similar to that discussed in Chapter 12 but does not make use of the maximum rate of dissipation hypothesis. Nevertheless the authors were careful to ensure that the second law of thermodynamics was met by the constitutive equation chosen. You can verify

---

[3] this is good exercise for you to check yourself on. It is easiest to prove using the eigenvalues of $\mathbf{B}_e$.

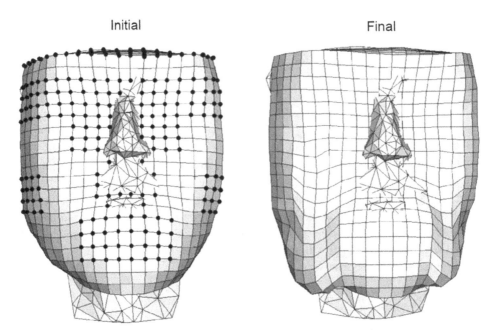

Fig. 14.8 The face simulation before and after aging. You can see the noticeable sag in the unsupported regions. Figure reprinted from the work of Mazza et al. [Mazza *et al.* (2007)] with permission from ASME International.

this by yourself. Note that the simple spring dashpot model can be used in an approximate way to get an idea of the deformations to be expected, since, we can do a simple calculation with an estimate of the force to see if the degradation is approximately correct. However, it ignores the rejuvenation of the cells of the tissue completely (which would not be a good thing if you were in the cosmetics business). Whether this effect is significant or not can only be determined by further research. One of the most difficult tasks in this work is the calculation of the material moduli; this requires extensive experimentation. The same is true of the finite element discretization and implementation in ABAQUS; we have skipped over these issues since we have neither the expertise nor the time to discuss them. Interested readers can refer to the original papers and learn more.

Bibliography

ABAQUS (2008). Abaqus online analysis users manual ver.6.6.
Amirkhizi, A. V., Isaacs, J., McGee, J. and Nemat-Nasser, S. (2006). An experimentally-based viscoelastic constitutive model for polyurea, including pressure and temperature effects, *Philosophical Magazine* **86**, 36, pp. 5847–5866.
Anand, L. (1979). Hencky,h. approximate strain-energy function for moderate deformations, *Journal of Applied Mechanics-Transactions of the Asme* **46**, 1, pp. 78–82.
Anand, L. (1982). Constitutive-equations for the rate-dependent deformation of metals at elevated-temperatures, *Journal of Engineering Materials and Technology-Transactions of the Asme* **104**, 1, pp. 12–17.
Anand, L. (1999). Polycrystal plasticity, *Mechanics and Materials* , pp. 231–269.
Anand, L. (2004). Single-crystal elasto-viscoplasticity: application to texture evolution in polycrystalline metals at large strains, *Computer Methods in Applied Mechanics and Engineering* **193**, 48-51, pp. 5359–5383.
Anand, L. and Kothari, M. (1996). A computational procedure for rate-independent crystal plasticity, *Journal of the Mechanics and Physics of Solids* **44**, 4, pp. 525–558.
Anand, L. and Su, C. (2005). A theory for amorphous viscoplastic materials undergoing finite deformations, with application to metallic glasses, *Journal of the Mechanics and Physics of Solids* **53**, 6, pp. 1362–1396.
Armstrong, P. J. and Frederick, C. O. (1966). C.e.g.b. report rd/b/n731: A mathematical representation of the multiaxial bauschinger effect, Tech. rep., Berkeley Nuclear Laboratories.
Asaro, R. J. (1983). Micromechanics of crystals and polycrystals, in *Advances in Applied Mechanics*, Vol. 23, pp. 1–115.
Auricchio, D., F; Fugazza and Desroches, R. (2006). Earthquake performance of steel frames with nitinol braces, *J. EARTHQU. ENG.* **10**, Sp. iss. 1, pp. 45–66.
Bache, M., Evans, W., Suddell, B. and Herrouin, F. (2001). The effects of texture in titanium alloys for engineering components under fatigue. *International Journal of Fatigue* **23**, pp. S153–S159.
Bachurin, D., Nazarov, A., Shenderova, O. and Brenner, D. (2003). Diffusion-accomodated rigid-body translations along grain boundaries in nanostructured materials. *Materials Science and Engineering A-Structural Materials Properties Microstructure and Processing* **A359**, pp. 247–252.
Backman, M. E. (1964). The relation between stress and finite elastic and plastic strains under impulsive loading, *Journal of Applied Physics* **35**, pp. 2524–2433.
Baig, C., Jiang, B., Edwards, B. J., Keffer, D. J. and Cochran, H. D. (2006). A comparison of simple rheological models and simulation data of n-hexadecane under shear and

elongational flows, *Journal of Rheology* **50**, 5, pp. 625–640.

Balasubramanian, S. and Anand, L. (1998). Polycrystalline plasticity: Application to earing in cup drawing of al2008-t4 sheet, *Journal of Applied Mechanics-Transactions of the Asme* **65**, 1, pp. 268–271.

Balendran, B. and Nematnasser, S. (1995). Integration of inelastic constitutive-equations for constant velocity-gradient with large rotation, *Applied Mathematics and Computation* **67**, 1-3, pp. 161–195.

Banabic, D., J., B. H., Pohlandt, K. and Tekkaya, A. E. (2000). *Formability of metallic materials : plastic anisotropy, formability testing, forming limits* (Springer, Berlin).

Baratta, A. and Corbi, O. (2002). On the dynamic behaviour of elastic-plastic structures equipped with pseudoelastic sma reinforcements, *Comput. Mater. Sci.* **25**, 1-2, pp. 1 – 13, doi:DOI:10.1016/S0927-0256(02)00245-8.

Barlat, F. (1987). Crystallographic texture, anisotropic yield surfaces and forming limits of sheet metals. *Materials Science and Engineering A-Structural Materials Properties Microstructure and Processing* **91**, pp. 55–72.

Barlat, F., Maeda, Y., Chung, K., Yanagawa, M., Brem, J., Hayashida, Y., Lege, D., , Matsui, K., Murtha, S., Hattori, S., Becker, R. and Makosey, S. (1997). Yield function development for aluminum alloy sheets, *Journal of the Mechanics and Physics of Solids* **45**, pp. 1727–1763.

Bartera, F. and Giacchetti, R. (2004). Steel dissipating braces for upgrading existing building frames, *J. Constr. Steel Res.* **60**, 3-5, pp. 751–769.

Baskaran, B. and Srinivasan, S. M. (2008). Overview of the effects of shot peening on plastic strain, work hardening, and residual stresses, in *Computational Materials, W.U. Oster* (NOVA Science Publishers, NY).

Bate, P. and Wilson, D. V. (1986). Analsis of the bauschinger effect, *Acta Metallurgica* **36**, 6, pp. 1097–1105.

Bauschinger, J. (1881). Ueber die veraenderung der elasticitactgrenze und des elasticitaetmodulus verschiedener metalle, *Civilingenieur* **27**, pp. 289–348.

Belytschko, T., Krongauz, Y., Organ, D., Fleming, M. and Krysl, P. (1996). Meshless methods: An over view and recent developments, *Comput. Methods Appl. Mech. Engg.* **139**, 1, pp. 3 – 47.

Belytschko, T., Lu, Y. and Gu, L. (1994). Element free galerkin method, *International Journal for numerical methods in engineering* **37**, 1, pp. 229 – 256.

Benzerga, A. A., Brechet, Y., Needleman, A. and Van der Giessen, E. (2005). The stored energy of cold work: Predictions from discrete dislocation plasticity, *Acta Materialia* **53**, 18, pp. 4765–4779.

Benzerga, A. A. and Shaver, N. F. (2006). Scale dependence of mechanical properties of single crystals under uniform deformation, *Scripta Materialia* **54**, 11, pp. 1937–1941.

Bertram, A. (1999). An alternative approach to finite plasticity based on material isomorphisms, *International Journal of Plasticity* **15**, 3, pp. 353–374.

Bertram, A. (2003). Finite thermoplasticity based on isomorphisms, *International Journal of Plasticity* **19**, 11, pp. 2027–2050.

Besseling, J. F. (1968). A thermodynamic approach to rheology, in H. Parkus and L. I. Sedov (eds.), *IUTAM symposium on Irreversible Aspects of Continuum Mechanics* (Springer, Vienna).

Bever, M., Holt, D. L. and Tichener, A. L. (1973). The stored energy of cold work, in B. Chalmers, J. W. Christian and T. B. Massalski (eds.), *Progress in Materials Science* (Pergamon, Oxford).

Bilby, B. A., Bullough, R. and Smith, E. (1955). Continuous distributions of dislocations - a new application of the methods of non-riemannian geometry, *Proceedings of the*

Royal Society of London Series a-Mathematical and Physical Sciences **231**, 1185, pp. 263–273.

Biswas, K. (2005). Comparison of various plasticity models for metal powder compaction processes, *Journal of Materials Processing Technology* **166**, 1, pp. 107–115.

Bohlke, T., Risy, G. and Bertram, A. (2006). Finite element simulation of metal forming operations with texture based material models, *Modelling and Simulation in Materials Science and Engineering* **14**, 3, pp. 365–387.

Bohlke, T., Risy, G. and Bertram, A. (2008). A micro-mechanically based quadratic yield condition for textured polycrystals, *Zamm-Zeitschrift Fur Angewandte Mathematik Und Mechanik* **88**, 5, pp. 379–387.

Bonn, R. and Haupt, P. (1995). Exact-solutions for large elastoplastic deformations of a thick-walled tube under internal-pressure, *International Journal of Plasticity* **11**, 1, pp. 99–118.

Bozdana, A. (2005). On the mechanical surface enhancement techniques in aerospace industry - a review of technology, *Aircraft Engineering and Aerospace Technology* **77**, 4, pp. 279–292.

Bridgman, P. W. (1964). *Studies in large plastic flow and fracture, with special emphasis on the effects of hydrostatic pressure.*, 2nd edn. (Harvard University Press, Cambridge, Mass).

Bronkhorst, C. A., Kalidindi, S. R. and Anand, L. (1992). Polycrystalline plasticity and the evolution of crystallographic texture in fcc metals, *Philosophical Transactions of the Royal Society of London Series a-Mathematical Physical and Engineering Sciences* **341**, 1662, pp. 443–477.

Buerger, M. (1956). *Elemenetary Crystallography* (John Wiley and Sons, New York).

Callen, H. B. (1985). *Thermodynamics and an introduction to thermostatics*, 2nd edn. (John Wiley and Sons).

Canova, G. R., Fressengeas, C., Molinari, A. and Kocks, U. (1988). Effect of rate sensitivity on slip system activity and lattice rotation, *Acta Metallurgica* **8**, pp. 1961–1970.

Casey, J. (2002). On loading criteria in plasticity, *Comptes Rendus Mecanique* **330**, 4, pp. 285–290.

Casey, J. and Naghdi, P. M. (1981). On the characterization of strain-hardening in plasticity, *Journal of Applied Mechanics-Transactions of the Asme* **48**, 2, pp. 285–296.

Casey, J. and Naghdi, P. M. (1983). A remark on the definition of hardening, softening and perfectly plastic behavior, *Acta Mechanica* **48**, 1-2, pp. 91–94.

Caulk, D. A. and Naghdi, P. M. (1978). Hardening response in small deformation of metals, *Journal of Applied Mechanics-Transactions of the Asme* **45**, 4, pp. 755–764.

Chaboche, J. L. (1989a). Constitutive-equations for cyclic plasticity and cyclic viscoplasticity, *International Journal of Plasticity* **5**, 3, pp. 247–302.

Chaboche, J. L. (1989b). A new kinematic hardening rule with discrete memory surfaces, *Recherche Aerospatiale* , 4, pp. 49–69.

Chaboche, J. L. (1991). On some modifications of kinematic hardening to improve the description of ratchetting effects, *International Journal of Plasticity* **7**, 7, pp. 661–678.

Chaboche, J. L. (1993a). Cyclic viscoplastic constitutive-equations, .1. a thermodynamically consistent formulation, *Journal of Applied Mechanics-Transactions of the Asme* **60**, 4, pp. 813–821.

Chaboche, J. L. (1993b). Cyclic viscoplastic constitutive-equations .2. stored energy comparison between models and experiments, *Journal of Applied Mechanics-Transactions of the Asme* **60**, 4, pp. 822–828.

Chadwick, P. (1976). *Continuum mechanics : concise theory and problems* (Allan & Unwin,

London).

Chakrabarthy, J. (2000). *Applied Plasticity*, Mechanical Engineering Series (Springer-Verlag, New York).

Chapra, S. (2006). *Applied numerical methods with MATLAB for engineers and scientists* (McGraw-Hill College).

Cheng, J. Y., Nemat-Nasser, S. and Guo, W. G. (2001). A unified constitutive model for strain-rate and temperature dependent behavior of molybdenum, *Mechanics of Materials* **33**, 11, pp. 603–616.

Choi, S., Cho, J., Barlat, F., Chung, K., Kwon, J. and Oh, K. (1999). Prediction of yield surfaces of textured sheet metals, *Metallurgical and Materials Transactions A-Physical Metallurgy and Materials Science* **30**, pp. 377–386.

Clayton, J. D. and McDowell, D. L. (2003). A multiscale multiplicative decomposition for elastoplasticity of polycrystals, *International Journal of Plasticity* **19**, 9, pp. 1401–1444.

Coleman, B. D. and Noll, W. (1963). The thermodynamics of elastic materials with heat conduction and viscosity, *Archive for Rational Mechanics and Analysis* **13**, pp. 167–178.

Collins, I. F. and Houlsby, G. T. (1997). Application of thermomechanical principles to the modelling of geotechnical materials, *Proceedings of the Royal Society A-Mathematical Physical and Engineering Sciences* **453**, pp. 1975–2001.

Criscione, J. C., Humphrey, J. D., Douglas, A. S. and Hunter, W. (2000). An invariant basis for natural strain which yields orthogonal stress response terms in isotropic hyperelasticity, *Journal of the Mechanics and Physics of Solids* **48**, pp. 2445–2465.

Cuitino, A. M. and Ortiz, M. (1993). Computational modeling of single-crystals, *Modelling and Simulation in Materials Science and Engineering* **1**, 3, pp. 225–263.

de Andres, A., Perez, J. L. and Ortiz, M. (1999). Elastoplastic finite element analysis of three-dimensional fatigue crack growth in aluminum shafts subjected to axial loading, *International Journal of Solids and Structures* **36**, 15, pp. 2231–2258.

Desroches, R. and Smith, B. (2004). Shape memory alloys in seismic resistant design and retrofit: A critical review of their potential and limitations, *J. EARTHQU. ENG.* **8**, 3, pp. 415–429.

Dieter, G. E. (1986). *Mechanical Metallurgy* (MeGraw-Hill, New York).

Ding, H. Z., Mughrabi, H. and Hoppel, H. W. (2002). A low-cycle fatigue life prediction model of ultrafine-grained metals, *Fatigue & Fracture of Engineering Materials & Structures* **25**, 10, pp. 975–984.

Dixit, P. M. and Dixit, U. S. (2008). *Modeling of Metal Forming and Machining Processes: by Finite Element and Soft Computing Methods (Engineering Materials and Processes)*.

Dolce, M. and Cardone, D. (2001). Mechanical behaviour of shape memory alloys for seismic applications 1. martensite and austenite niti bars subjected to torsion 2. austenite niti wires subjected to tension, *International Journal of Mechanical Sciences* **43**, 11, pp. 2631–2677.

Dorris, J. F. and Nematnasser, S. (1982). A plasticity model for flow of granular-materials under triaxial stress states, *International Journal of Solids and Structures* **18**, 6, pp. 497–531.

Drucker, D. C. and Prager, W. (1952). Soil mechanics and plastic analysis or limit design, *Quarterly of Applied Mathematics* **10**, 2, pp. 157–165.

Drucker, D. C. and Tachau, H. (1945). A new design criterion for wire rope, *Journal of Applied Mechanics-Transactions of the Asme* **12**, 1, pp. A31–A38.

Du, Q., Faber, V. and Gunzburger, M. (1999). Centroidal voronoi testellations :applica-

tions and algorithms, *SIAM Rev* **41**, pp. 637–676.
Duva, J. M. and Crow, P. D. (1992). The densification of powders by power-law creep during hot isostatic pressing, *Acta Materialia* **40**, 1, pp. 31–35.
Eckart, C. (1948). The thermodynamics of irreversible processes iv. the theory of elasticity and anelasticity, *Physical Review* **73**, pp. 373–382.
Eshelby, J. D. (1956). The continuum theory of lattice defects, in F. Seitz and D. Turnbull (eds.), *Progress in Solid State Physics*, Vol. 3 (Academic Press), pp. 79–156.
Eve, R. A., Reddy, B. D. and Rockafellar, R. T. (1990). An internal variable theory of elastoplasticity based on the maximum plastic work inequality, *Quarterly of Applied Mathematics* **48**, 1, pp. 59–83.
Ewing, J. A. and Rosenhain, W. (1900). Bakerian lecture. the crystalline structure of metals, *Philosophical Transactions of the Royal Society of London Series a-Containing Papers of a Mathematical or Physical Character* **193**, pp. 353–401.
Ferrasse, S., Segal, V. M., Hartwig, K. T. and Goforth, R. E. (1997). Microstructure and properties of copper and aluminum alloy 3003 heavily worked by equal channel angular extrusion, *Metallurgical and Materials Transactions A-Physical Metallurgy and Materials Science* **28**, 4, pp. 1047–1057.
Fotiu, P. A. and NematNasser, S. (1996). A universal integration algorithm for rate-dependent elastoplasticity, *Computers & Structures* **59**, 6, pp. 1173–1184.
Fowler, H. W. (1926). *A Dictionary of Modern English Usage* (Oxford University Press).
Fox, N. (1968). On the continuum theories of dislocations and plasticity, *Quarterly Journal of Mechanics and Applied Mathematics* **21**, 1, pp. 319–338.
Fox, R. W., McDonald, A. T. and Pritchard, P. J. (2006). *Introduction to Fluid Mechanics*, 6th edn. (John Wiley, Hoboken, New Jersey).
Freed, A. D., Chaboche, J. L. and Walker, K. P. (1991). A viscoplastic theory with thermodynamic considerations, *Acta Mechanica* **90**, 1-4, pp. 155–174.
Gauss, C. F. (1829). Uber ein neues allgemeines grundgesetz der mechanik, *Crelles Journal* **4**, pp. 232–235.
Gijsbertus, d. W. (2006). *Structure, Deformation, and Integrity of Materials: Volume I: Fundamentals and Elasticity / Volume II: Plasticity, Visco-elasticity, and Fracture* (Wiley-VCH).
Gilat, A. (2007). *Matlab: An introduction with applications* (John Wiley & Sons Inc.).
Gotoh, M. (1977). A theory of plastic anisotropy based on a yield function of fourth order (plane stress state). parts i and ii. *International Journal of Mechanical Sciences* **19**, pp. 505–520.
Green, A. E. and McInnis, B. C. (1965). Generalized hypo-elasticity, *Proceedings of the Royal Society of Edinburg Section a-Mathematical and Physical Sciences* **67**, p. 220.
Green, A. E. and Naghdi, P. M. (1965). A general theory of an elasticplastic continuum, *Archive for Rational Mechanics and Analysis* **18**, pp. 251–281.
Green, A. E. and Naghdi, P. M. (1973). Rate-type constitutive equations and elastic-plastic materials, *International Journal of Engineering Science* **11**, 7, pp. 725–734.
Grmela, M. and Carreau, P. J. (1987). Conformation tensor rheological models, *Journal of Non-Newtonian Fluid Mechanics* **23**, pp. 271–294.
Groot, d. S. R. and Mazur, P. (1962). *Non-equilibrium thermodynamics* (Interscience Publishers).
Gu, C., Kim, M. and Anand, L. (2001). Constitutive equations for metal powders: Application to powder forming processes, *International Journal of Plasticity* **17**, 2, pp. 147–209.
Guo, W. G., Li, Y. L. and Nemat-Nasser, S. (2003). The plastic flow behavior and model of bcc polycrystalline metals with application to pure ta, mo, nb and v, *Progress in*

Experimental and Computational Mechanics in Engineering **243-2**, pp. 445–450.

Gurson, A. L. (1977). Continuum theory of ductile rupture by void nucleation and growth: Part iyield criteria and flow rules for porous ductile materials, *Journal of Engineering Materials and Technology-Transactions of the Asme* **99**, pp. 2–15.

Guruprasad, P. J. and Benzerga, A. A. (2008). Size effects under homogeneous deformation of single crystals: A discrete dislocation analysis, *Journal of the Mechanics and Physics of Solids* **56**, 1, pp. 132–156.

Hackl, K., Miehe, C. and Celigoj, C. (2001). Theory and numerics of anisotropic materials at finite strains (euromech colloquium 394), *International Journal of Solids and Structures* **38**, 52, pp. 9421–9421.

Hardesty, S., Drucker, D. C. and Tachau, H. (1946). A new design criterion for wire rope, *Journal of Applied Mechanics-Transactions of the ASME* **13**, 1, pp. A75–A76.

Hartmann, S., Luhrs, G. and Haupt, P. (1997). An efficient stress algorithm with applications in viscoplasticity and plasticity, *International Journal for Numerical Methods in Engineering* **40**, 6, pp. 991–1013.

Haupt, P. and Kamlah, M. (1995). Representation of cyclic hardening and softening properties using continuous-variables, *International Journal of Plasticity* **11**, 3, pp. 267–291.

Haupt, P. and Tsakmakis, C. (1986). On kinematic hardening and large plastic-deformations, *International Journal of Plasticity* **2**, 3, pp. 279–293.

Havner, K. S. (1992). *Finite Plastic Deformations of Crystalline Solids*, Cambridge Monographs on Mechanics and Applied Mathematics (Cambridge University Press, Cambridge, U.K.).

Henann, D. and Anand, L. (2008). A constitutive theory for the mechanical response of amorphous metals at high temperatures spanning the glass transition temperature: Application to microscale thermoplastic forming, *Acta Materialia* **56**, 13, pp. 3290–3305.

Hertzberg, R. W. (1976). *Deformation and fracture mechanics of engineering materials* (Wiley, New York).

Hill, R. (1965). Continuum micro-mechanics of elastoplastic polycrystals, *Journal of the Mechanics and Physics of Solids* **13**, 2, p. 89.

Hill, R. (1966). Generalized constitutive relations for incremental deformation of metal crystals by multislip, *Journal of the Mechanics and Physics of Solids* **14**, pp. 95–102.

Hill, R. (1987). Constitutive dual potentials in classical plasticity, *Journal of the Mechanics and Physics of Solids* **35**, 1, pp. 23–33.

Hill, R. (1990). Constitutive modeling of orthotropic plasticity in sheet metals, *Journal of the Mechanics and Physics of Solids* **38**, 3, pp. 405–417.

Hill, R. (1993). A user-friendly theory of orthotropic plasticity in sheet metals, *International Journal of Mechanical Sciences* **35**, 1, pp. 19–25.

Hill, R. and Rice, J. (1972). Constitutive analysis of elastic-plastic crystals at arbitrary strain, *Journal of the Mechanics and Physics of Solids* **20**, pp. 401–413.

Hill, R. A. (1948). Theory of yielding and plastic flow of anisotropic metals. *Proceedings of the Royal Society A-Mathematical Physical and Engineering Sciences* **193**, pp. 281–297.

Hofstetter, G., Simo, J. C. and Taylor, R. L. (1993). A modified cap model - closest point solution algorithms, *Computers & Structures* **46**, 2, pp. 203–214.

Holzapfel, G. A. and Simo, J. C. (1996). A new viscoelastic constitutive model for continuous media at finite thermomechanical changes, *International Journal of Solids and Structures* **33**, 20-22, pp. 3019–3034.

Hosford, W. (1993). *The mechanics of crystals and textured polycrystals*, Oxford Engineering Science Series (Oxford University Press, New York).

Hosford Jr, W. F. (1966). Plane-strain compression of aluminum crystals, *Acta Metallurgica* **14**, 9, pp. 1085–1094.

Hosford Jr, W. F. (1972). A generalized isotropic yield criterion, *Journal of Applied Mechanics-Transactions of the Asme* **39**, pp. 607–623.

Houlsby, G. T. and Puzrin, A. M. (2000). A thermomechanical framework for constitutive models for rate-independent dissipative materials, *International Journal of Plasticity* **16**, 9, pp. 1017–1047.

Hutchinson, J. (1976). Bounds and self-consistent estimates for creep of polycrystalline materials. *Proceedings of the Royal Society A-Mathematical Physical and Engineering Sciences* **A348**, pp. 101–127.

Il'iushin, A. A. (1961). On a postulate of plasticity, *Journal of Applied Mathematics and Mechanics (Translation of PMM)* **25**, pp. 746–750.

Inal, K., Wu, P. and Neale, K. (2000). Simulation of earing in textured aluminum sheets, *International Jorunal of Plasticity* **16**, pp. 635–648.

International, A. (2002). *Atlas of stress-strain curves* (ASM International, Materials Park, OH.).

Iwakuma, T. and Nematnasser, S. (1984). Finite elastic plastic-deformation of polycrystalline metals, *Proceedings of the Royal Society of London Series a-Mathematical Physical and Engineering Sciences* **394**, 1806, pp. 87–119.

Jaunzemis, W. (1967). *Continuum mechanics* (Macmillan, New York).

Jiang, B. and Weng, G. J. (2004). A generalized self-consistent polycrystal model for the yield strength of nanocrystalline materials, *Journal of the Mechanics and Physics of Solids* **52**, 5, pp. 1125–1149.

Joshi, R. B., Bayoumi, A. E. and Zbib, H. M. (1992). The use of digital processing in studying stretch-forming sheet-metal, *Experimental Mechanics* **32**, 2, pp. 117–123.

Juleff, G. (1996). An ancient wind powered iron smelting technology in sri lanka, *Nature* **379**, 3, pp. 60–63.

Kachanov, L. M. (2004). *Fundamentals of the theory of plasticity* (Dover publications).

Kamlah, M. and Haupt, P. (1997). On the macroscopic description of stored energy and self heating during plastic deformation, *International Journal of Plasticity* **13**, 10, pp. 893–911.

Karafillis, A. and Boyce, M. (1993). A general anisotropic yield criterion using bounds and a transformation weighting tensor, *Journal of the Mechanics and Physics of Solids* **41**, pp. 1859–1886.

Karaman, I., Haouaoui, M. and Maier, H. J. (2003). Nanoparticle consolidation using equal channel angular extrusion at room temperature, *Journal of Materials Science* **42**, 5, pp. 1561–1576.

Kaushik, A. (2007). *SImulation of ECAE*, Ph.D. thesis, Texas A & M University.

Khan, A. S., Chen, X. and Abdel-Karim, M. (2007). Cyclic multiaxial and shear finite deformation response of ofhc: Part i, experimental results, *International Journal of Plasticity* **23**, 8, pp. 1285–1306.

Khan, A. S. and Cheng, P. (1996). Study of three elastic-plastic constitutive models by nonproportional finite deformations of ofhc copper, *International Journal of Plasticity* **12**, 6, pp. 737–759.

Khan, A. S. and Huang, S. (1995). *Continuum Theory of plasiticity* (John Wiley and Sons, New York).

Khan, A. S. and Liang, R. Q. (2000). Behaviors of three bcc metals during non-proportional multi-axial loadings: experiments and modeling, *International Journal of Plasticity*

16, 12, pp. 1443–1458.

Khan, A. S. and Parikh, Y. (1986). Large deformation in polycrystalline copper under combined tension-torsion, loading, unloading and reloading or reverse loading - a study of 2 incremental theories of plasticity, *International Journal of Plasticity* **2**, 4, pp. 379–392.

Kim, K. T. and Carroll, M. M. (1987). Compaction equations for strain-hardening porous materials, *International Journal of Plasticity* **3**, 1, pp. 63–73.

Kirk, D. (1999). Shot peening, *Aircraft Engineering and Aerospace Technology* **71**, 4, pp. 349–361.

Klisinski, M. and Mroz, Z. (1988). Description of inelastic deformation and degradation of concrete, *International Journal of Solids and Structures* **24**, 4, pp. 391–416.

Kocks, U. and Mecking, H. (2003). Physics and phenomenology of strain hardening: the fcc case, *Progress in Materials Science* **48**, 171–273, p. 171.

Kocks, U., Tome, C. and Wenk, H. (1998). *Texture and anisotropy: preferred orientations in polycrystals and their effect on material properties.* (Cambridge University Press, Cambridge, UK.).

Koslowski, M., Cuitino, A. M. and Ortiz, M. (2002). A phase-field theory of dislocation dynamics, strain hardening and hysteresis in ductile single crystals, *Journal of the Mechanics and Physics of Solids* **50**, 12, pp. 2597–2635.

Kratochvil, J. and Dillon, O. W. (1969). Thermodynamics of elastic-plastic materials as a theory with internal state variables, *Journal of Applied Physics* **40**, 8, p. 3207.

Kroner, E. (1960). Allegemeine kontinuumstheorie der versetzungen und eigenspannungen, *Archive for Rational Mechanics and Analysis* **4**, pp. 273–334.

Lambrecht, M. and Miehe, C. (1999). Two non-associated isotropic elastoplastic hardening models for frictional materials, *Acta Mechanica* **135**, 1-2, pp. 73–90.

Lancaster, P. and Salkauskas, K. (1981). Surfaces generated by moving least square methods, *Mathematical Computations* **37**, 1, pp. 141 – 158.

LaTrenta, G. (2004). *Aesthetic Face and Neck Surgery* (Saunders, Philadelphia, USA.).

Lee, E. (1969). Elastic-plastic deformation at finite strains, *Journal of Applied Mechanics-Transactions of the Asme* **36**, pp. 1–6.

Leonov, A. I. (1987). On a class of constitutive equations for viscoelastic liquids, *Journal of Non-Newtonian Fluid Mechanics* **25**, 1, pp. 1–59.

Leonov, A. I. (1992). Analysis of simple constitutive equations for viscoelastic liquids, *Journal of Non-Newtonian Fluid Mechanics* **42**, 3, pp. 323–350.

Leonov, A. I. and Padovan, J. (1996). On a kinetic formulation of elasto-viscoplasticity, *International Journal of Engineering Science* **34**, 9, pp. 1033–1046.

Liang, R. Q. and Khan, A. S. (1999). A critical review of experimental results and constitutive models for bcc and fcc metals over a wide range of strain rates and temperatures, *International Journal of Plasticity* **15**, 9, pp. 963–980.

Lim, T. J. and McDowell, D. L. (2002). Cyclic thermomechanical behavior of a polycrystalline pseudoelastic shape memory alloy, *Journal of the Mechanics and Physics of Solids* **50**, 3, pp. 651–676.

Lin, H. C. and Naghdi, P. M. (1989). Necessary and sufficient conditions for the validity of a work inequality in finite plasticity, *Quarterly Journal of Mechanics and Applied Mathematics* **42**, pp. 13–21.

Lu, J. and Papadopoulos, P. (2004). A covariant formulation of anisotropic finite plasticity: theoretical developments, *Computer Methods in Applied Mechanics and Engineering* **193**, 48-51, pp. 5339–5358.

Lu, M. W. and Sayir, M. (1982). An analytical asymptotic approximation to the deformation of an elastic-plastic plate, *Zeitschrift Fur Angewandte Mathematik Und Physik*

33, 4, pp. 443–460.

Lubarda, V. A. (2004). Constitutive theories based on the multiplicative decomposition of deformation gradient: Thermoelasticity, elastoplasticity, and biomechanics, *Applied Mechanics Reviews* **57**, 2, pp. 95–108.

Luhrs, G., Hartmann, S. and Haupt, P. (1997). On the numerical treatment of finite deformations in elastoviscoplasticity, *Computer Methods in Applied Mechanics and Engineering* **144**, 1-2, pp. 1–21.

Makosey, S. J. and Rajagopal, K. R. (2001). The application of ideas associated with materials with memory to modeling the inelastic behavior of solid bodies, *International Journal of Plasticity* **17**, pp. 1087–1116.

Mandel, J. (1973). Equations constitutives et directeurs dans les milieux plastiques et viscoplastiques. *International Journal of Solids and Structures* **9**, pp. 725–740.

Maniatty, A., Dawson, P. and Lee, Y. (1992). A time integration algorithm for elastoviscoplastic cubic crystals applied to modeling polycrystalline deformation, *International Jorunal of Numerical Methods in Engineering* **35**, pp. 1565–1588.

Manoach, E. and Karagiozova, D. (1993). Dynamic response of thick elastic-plastic beams, *International Journal of Mechanical Sciences* **35**, 11, pp. 909–919.

Marciniak, Z. and Kuczinsky, K. (1967). Limit strains in the processes of stretch-forming sheet metal. *International Journal of Mechanical Sciences* **9**, pp. 609–621.

Martinelli, L., Mulas, M. G. and Perotti, F. (1996). The seismic response of concentrically braced moment-resisting steel frames, *Earthquake Engineering & Structural Dynamics* **25**, 11, pp. 1275–1299.

Mason, T. and Maudlin, P. a. (1999). Effects of higher-order anisotropic elasticity using textured polycrystals in three-dimensional wave propagation problems, *Mechanics and Materials* **31**, pp. 861–882.

Masud, A. and Chudnovsky, A. (1999). A constitutive model of cold drawing in polycarbonates, *International Journal of Plasticity* **15**, 11, pp. 1139–1157.

Mazza, E., Papes, O., Rubin, M. B., Bodner, S. and Binur, N. S. (2007). Simulation of the aging face, *Journal of Biomechanical Engineering-Transactions of the Asme* **129**, pp. 619–623.

Mazza, E., Papes, O., Rubin, M. B., Bodner, S. R. and Binur, N. S. (2005). Nonlinear elastic-viscoplastic constitutive equations for aging facial tissues, *Biomechanics and Modeling in Mechanobiology* **4**, 2-3, pp. 178–189.

McGinty, R. D. and McDowell, D. L. (2006). A semi-implicit integration scheme for rate independent finite crystal plasticity, *International Journal of Plasticity* **22**, 6, pp. 996–1025.

Mecking, H., Nicklas, B., Zarubova, N. and Kocks, U. (1986). A "universal" temperature scale for plastic flow, *Acta Metallurgica* **34**, pp. 527–535.

Mendelson, A. (1983). *Plasticity Theory and Application* (Krieger Publishing Company).

Miehe, C. (1993). Computation of isotropic tensor functions, *Communications in Numerical Methods in Engineering* **9**, 11, pp. 889–896.

Miehe, C. (1996). Numerical computation of algorithmic (consistent) tangent moduli in large-strain computational inelasticity, *Computer Methods in Applied Mechanics and Engineering* **134**, 3-4, pp. 223–240.

Miehe, C., Apel, N. and Lambrecht, M. (2002). Anisotropic additive plasticity in the logarithmic strain space: modular kinematic formulation and implementation based on incremental minimization principles for standard materials, *Computer Methods in Applied Mechanics and Engineering* **191**, 47-48, pp. 5383–5425.

Mises, R., von (1913). Mechanik der festen lorper in plastisch deformbllem zust, *Nacr. Gess. Wiss.*, p. 582.

Moldovan, D., Wolf, D. and Phillpot, S. R. (2003). Linking atomistic and mesoscale simulations of nanocrystalline materials: quantitative validation for the case of grain growth, *Philosophical Magazine* **83**, 31-34, pp. 3643–3659.

Mollica, F. (2001). *Evolving anisotropy during metal forming*, Ph.D. thesis, Texas A&M University.

Mollica, F. and Srinivasa, A. R. (2002). A general framework for generating convex yield surfaces for anisotropic metals, *Acta Mechanica* **154**, 1-4, pp. 61–84.

Mughrabi, H. (1977). What causes metal fatigue, *Umschau in Wissenschaft Und Technik* **77**, 3, pp. 80–81.

Mughrabi, H. (1993). Cyclic plasticity and fatigue of metals, *Journal De Physique Iv* **3**, C7, pp. 659–668.

Mughrabi, H. and Christ, H. J. (1997). Cyclic deformation and fatigue of selected ferritic and austenitic steels: Specific aspects, *Isij International* **37**, 12, pp. 1154–1169.

Naghdi, P. M. and Nikkel, D. J. (1984). Calculations for uniaxial-stress and strain cycling in plasticity, *Journal of Applied Mechanics-Transactions of the Asme* **51**, 3, pp. 487–493.

Naghdi, P. M. and Nikkel, D. J. (1986). Two-dimensional strain cycling in plasticity, *Journal of Applied Mechanics-Transactions of the Asme* **53**, 4, pp. 821–830.

Naghdi, P. M. and Srinivasa, A. R. (1993). A dynamical theory of structured solids .1. basic developments, *Philosophical Transactions of the Royal Society of London Series a-Mathematical Physical and Engineering Sciences* **345**, 1677, pp. 425–458.

Naghdi, P. M. and Srinivasa, A. R. (1994). Characterization of dislocations and their influence on plastic-deformation in single-crystals, *International Journal of Engineering Science* **32**, 7, pp. 1157–1182.

Naghdi, P. M. and Trapp, J. A. (1975a). Nature of normality of plastic strain rate and convexity of yield surfaces in plasticity, *Journal of Applied Mechanics-Transactions of the Asme* **42**, 1, pp. 61–66.

Naghdi, P. M. and Trapp, J. A. (1975b). Restrictions on constitutive equations of finitely deformed elastic-plastic materials, *Quarterly Journal of Mechanics and Applied Mathematics* **28**, FEB, pp. 25–46.

Naghdi, P. M. and Trapp, J. A. (1975c). Significance of formulating plasticity theory with reference to loading surfaces in strain space, *International Journal of Engineering Science* **13**, 9-10, pp. 785–797.

Nematnasser, S. (1979). Decomposition of strain measures and their rates in finite deformation elasto-plasticity, *International Journal of Solids and Structures* **15**, 2, pp. 155–166.

Nematnasser, S. (1982). On finite deformation elasto plasticity, *International Journal of Solids and Structures* **18**, 10, pp. 857–872.

Nematnasser, S. (1991). Rate-independent finite-deformation elastoplasticity - a new explicit constitutive algorithm, *Mechanics of Materials* **11**, 3, pp. 235–249.

NematNasser, S. and Okinaka, T. (1996). A new computational approach to crystal plasticity: Fcc single crystal, *Mechanics of Materials* **24**, 1, pp. 43–57.

Ogden, R. W. (1984). *Non-linear elastic deformations* (Halsted Press).

Oldroyd, J. G. (1950). On the formulation of rheological equations of state, *Proceedings of the Royal Society A-Mathematical Physical and Engineering Sciences* **200**, pp. 523–591.

Onat, E. T. (1968). The notation of state and its implications in thermodynamics of inelastic solids, in H. Parkus and L. I. Sedov (eds.), *IITAM Symposium on Irreversible Aspects of Continuum Mechanics* (Vienna).

Onat, E. T. and Prager, W. (1953). Limit analysis of arches, *Journal of the Mechanics and Physics of Solids* **1**, 2, pp. 77–89.

Onat, E. T. and Shield, R. T. (1953). Remarks on combined bending and twisting of thin tubes in the plastic range, *Journal of Applied Mechanics-Transactions of the Asme* **20**, 3, pp. 345–348.

Ono, N., Kimura, K. and Watanabe, T. (1999). Monte carlo simulation of grain growth with the full spectra of grain orientation and grain boundary energy, *Acta Materialia* **47**, pp. 1007–1017.

Onsager, L. (1931). Reciprocal relations in irreversible thermodynamics-i, *Physical Review* **37**, pp. 405–426.

O'Reilly, O. M. and Srinivasa, A. R. (2001). On a decomposition of generalized constraint forces, *Proceedings of the Royal Society of London Series a-Mathematical Physical and Engineering Sciences* **457**, 2016, pp. 3052–3052.

Ortiz, M., Pinsky, P. M. and Taylor, R. L. (1983). Operator split methods for the numerical-solution of the elastoplastic dynamic problem, *Computer Methods in Applied Mechanics and Engineering* **39**, 2, pp. 137–157.

Ortiz, M. and Popov, E. P. (1982). A statistical-theory of polycrystalline plasticity, *Proceedings of the Royal Society of London Series a-Mathematical Physical and Engineering Sciences* **379**, 1777, pp. 439–458.

Ortiz, M. and Simo, J. C. (1986). An analysis of a new class of integration algorithms for elastoplastic constitutive relations, *International Journal for Numerical Methods in Engineering* **23**, 3, pp. 353–366.

Papadopoulos, P. and Lu, J. (1998). A general framework for the numerical solution of problems in finite elasto-plasticity, *Computer Methods in Applied Mechanics and Engineering* **159**, 1-2, pp. 1–18.

Papadopoulos, P. and Lu, J. (2001). On the formulation and numerical solution of problems in anisotropic finite plasticity, *Computer Methods in Applied Mechanics and Engineering* **190**, 37-38, pp. 4889–4910.

Pastor, M., Zienkiewicz, O. C. and Chan, A. H. C. (1990). Generalized plasticity and the modeling of soil behavior, *International Journal for Numerical and Analytical Methods in Geomechanics* **14**, 3, pp. 151–190.

Pearson, G. (1795). Experiments and observations to investigate the nature of a kind of steel, manu-factured at bombay, and there called wootz: with remarks on the properties and composition of the different states of iron, *Philosophical Transactions of the Royal Society of London* **85**, pp. 322–346.

Pietruszczak, S. and Mroz, Z. (1981). Finite-element analysis of deformation of strain-softening materials, *International Journal for Numerical Methods in Engineering* **17**, 3, pp. 327–334.

Pietruszczak, S. and Mroz, Z. (1983). On hardening anisotropy of k0-consolidated clays, *International Journal for Numerical and Analytical Methods in Geomechanics* **7**, 1, pp. 19–38.

Pipkin, A. C. and Rivlin, R. S. (1965). Mechanics of rate-independent materials, *Zeitschrift Fur Angewandte Mathematik Und Physik* **16**, 3, p. 313.

Pradeep, H., Rajagopal, K. R. and Srinivasa, A. R. (2005). On the deformation of an elastoplastic body when placed between flat plates rotating about non-coincident axes, *International Journal of Plasticity* **21**, 12, pp. 2255–2276.

Prager, W. (1975). Bauschinger adaptation of plastic, kinematically hardening solid, *Comptes Rendus Hebdomadaires Des Seances De L Academie Des Sciences Serie B* **280**, 19, pp. 585–587.

Prager, W. (1976). Geometric discussion of optimal design of a simple truss, *Journal of Structural Mechanics* **4**, 1, pp. 57–63.

Prager, W. and G, H. J. P. (1951). *Theory of perfectly plastic solids* (Wiley, New York).

Prager, W. and Rozvany, G. I. N. (1975). Plastic design of beams - optimal locations of supports and steps in yield moment, *International Journal of Mechanical Sciences* **17**, 10, pp. 627–631.

Prandtl, L. (1924). Spannungsverteilung in plastischen koerpern, in *1st International Congress in Applied Mechanics* (Delft), pp. 43–54.

Pratap, R. (2005). *Getting Started With Matlab 7: A Quick Introduction For Scientists And Engineers* (Oxford University Press).

Qiang, D., Gunzburger, M. and Ju, L. (2002). Meshfree, probabilistic determination of point sets and support regions for meshless computing, *Computer Methods in Applied Mechanics and ¿Enginnering* **191**, pp. 1349–1366.

Rajagopal, K. R. (1995). Report no. 6: Multiple configurations in continuum mechanics, Tech. rep., Institute for Computational and Applied Mechanics, University of Pittsburgh.

Rajagopal, K. R. and Srinivasa, A. R. (1998). Mechanics of the inelastic behavior of materials - part 1, theoretical underpinnings, *International Journal of Plasticity* **14**, 10-11, pp. 945–967.

Rajagopal, K. R. and Srinivasa, A. R. (2004a). On the thermomechanics of materials that have multiple natural configurations - part i: Viscoelasticity and classical plasticity, *Zeitschrift Fur Angewandte Mathematik Und Physik* **55**, 5, pp. 861–893.

Rajagopal, K. R. and Srinivasa, A. R. (2004b). On the thermomechanics of materials that have multiple natural configurations - part ii: Twinning and solid to solid phase transformation, *Zeitschrift Fur Angewandte Mathematik Und Physik* **55**, 6, pp. 1074–1093.

Rajagopal, K. R. and Srinivasa, A. R. (2004c). On thermomechanical restrictions of continua, *Proceedings of the Royal Society of London Series a-Mathematical Physical and Engineering Sciences* **460**, 2042, pp. 631–651.

Rajagopal, K. R. and Srinivasa, A. R. (2009). On a class of non-dissipative materials that are not hyperelastic, *Proceedings of the Royal Society A-Mathematical Physical and Engineering Sciences* **465**, 2102, pp. 493–500.

Reibold, M., Paufler, P., Levin, A. A., Kochmann, W., Patzke, N. and Meyer, D. C. (2006). Materials: Carbon nanotubes in an ancient damascus sabre, *Nature* **444**, p. 286.

Reuss, E. (1939). Beriicksichtigung der elastischen formanderung in der plastizita tstheorie, *Zeitschrift Fur Angewandte Mathematik Und Mechanik* **10**, pp. 266–274.

Rice, J. R. (1971). Inelastic constitutive relations for solids - an internal-variable theory and its application to metal plasticity, *Journal of the Mechanics and Physics of Solids* **19**, 6, p. 433.

Robertson, J., Im, J. T., Karaman, I., Hartwig, K. T. and Anderson, I. E. (2003). Consolidation of amorphous copper based powder by equal channel angular extrusion, *Journal of Non-Crystalline Solids* **317**, 1–2, pp. 144–151.

Rockafellar, R. T. (1970). *Convex Analysis*, Princeton Landmarks in Mathematics (Princeton University Press).

Rollett, A., Storch, M., Hilsinki, E. and Goodman (2001). Approach to saturation in textured soft magnetic materials, *Metallurgical and Materials Transactions A-Physical Metallurgy and Materials Science* **32**, pp. 2595–2603.

Rubin, M. B. (1994). Plasticity theory formulated in terms of physically based microstructural variables-part i. theory, *International Journal of Solids and Structures* **31**, 19, pp. 2615–2634.

Rubin, M. B. (1996). On the treatment of elastic deformation in finite elastic-viscoplastic theory, *International Journal of Plasticity* **12**, 7, pp. 951–965.

Rubin, M. B. (2001). Physical reasonse for abandoning plastic deformation measures in plasticity and viscoplasticity theory, *Archives of Mechanics* **53**, 4–5, pp. 519–539.

Runesson, K. and Mroz, Z. (1989). A note on nonassociated plastic-flow rules, *International Journal of Plasticity* **5**, 6, pp. 639–658.

Saadat, S., Salichs, J., Noori, M., Hou, Z., Davoodi, H., Suzuki, Y. and Masuda, A. (2002). An overview of vibration and seismic applications of niti shape memory alloy, *Smart Mater. Struct.* **11**, 2, pp. 218–229.

Sadd, M. H. (2005). *Elasticity, Theory, Applications and Numerics*, Amsterdam, Boston (Elsevier Butterworth Heinemann).

Sarma, G. and Zacharia, T. (1999). Integration algorithm for modeling the elasto-viscoplastic response of polycrystalline materials, *Journal of the Mechanics and Physics of Solids* **47**, pp. 1219–1238.

Schurig, M. and Bertram, A. (2003). A rate independent approach to crystal plasticity with a power law, *Computational Materials Science* **26**, pp. 154–158.

Segal, V. M. (1995). Materials processing by simple shear, *Materials Science and Engineering A-Structural Materials Properties Microstructure and Processing* **197**, pp. 157–164.

Segel, L. (2007). *Mathematics applied to continuum mechanics* (Society for Industrial and Applied Mathematics).

Seyman, O. and Esendemir, U. (2002). An elasticplastic stress analysis of simply supported metal-matrix composite beams under a transverse uniformly distributed load, *Composites Science and Technology* **62**, pp. 265–273.

Simo, J. C. (1988). A framework for finite strain elastoplasticity based on maximum plastic dissipation and the multiplicative decomposition 2. computational aspects, *Computer Methods in Applied Mechanics and Engineering* **68**, 1, pp. 1–31.

Simo, J. C. and Hughes, T. J. R. (1998). *Computational inelasticity*, Interdisciplinary applied mathematics ; v. 7 (Springer, New York).

Spencer, A. J. M. (1980). *Continuum Mechanics* (Longman, New York).

Srinivasa, A. R. (1997). On the nature of the response functions in rate-independent plasticity, *International Journal of Non-Linear Mechanics* **32**, 1, pp. 103–119.

Srinivasa, A. R. (2001). Large deformation plasticity and the poynting effect, *International Journal of Plasticity* **17**, 9, pp. 1189–1214.

Suresh, S. (1998). *Fatigue of Materials* (Cambridge University Press, New York).

Tamai, H. and Kitagawa, Y. (2002). Pseudoelastic behavior of shape memory alloy wire and its application to seismic resistance member for building, *Comput. Mater. Sci.* **25**, 1-2, pp. 218–227.

Taylor, G. and Elam, C. (1938). The plastic extension and fracture of aluminium crystals, *Proceedings of the Royal Society a-Mathematical Physical and Engineering Sciences* **A108**, pp. 28–51.

Taylor, G. I. (1938). Plastic strain in metals, *Journal of the Institute of Metals* **62**, pp. 307–324.

Taylor, G. I. and Quinney, M. A. (1934). The latent energy remaining in a metal after cold working, *Proceedings of the Royal Society A-Mathematical Physical and Engineering Sciences* **143**, p. 307.

Timoshenko, S. P. and Goodier, J. N. (1970). *Theory of elasticity*, Engineering societies monographs: engineering mechanics series (McGraw Hill, Singapore).

Tresca, H. (1864). Mmoire sur l'coulement des corps solides soumis de fortes pressions. c.r. acad. sci. paris, vol. 59, p. 754. *Comptes Rendus De L Academie Des Sciences* **59**, p. 754.

Truesdell, C. (1977). *A first course in rational continuum mechanics* (Academic Press,

New York).

Truesdell, C. and Noll, W. (2004). *The non-linear field theories of mechanics*, 3rd edn. (Springer, Berlin).

Valanis, K. C. (1980). Fundamental consequences of a new intrinsic time measure of plasticity as a limit of the endochronic theory, *Archives of Mechanics* **32**, p. 171.

Valiev, R. Z. and Langdon, T. G. (2006). Developments in the use of ecap processing for grain refinement, *Reviews on Advanced Materials Science* **13**, 1, pp. 15–26.

Verhoeven, J. (1987). Damascus steel. i. indian wootz steel, *Metallography* **20**, 2, pp. 145–151.

Verhoeven, J., Pendray, A. H. and Dauksch, W. E. (1998). The key role of impurities in ancient damascus steel blades, *Journal of Metals* **50**, 9, pp. 58–64.

Voyiadjis, G. Z. and Sivakumar, S. M. (1991). A robust kinematic hardening rule for cyclic plasticity with ratchetting effects. 1. theoretical formulation, *Acta Mechanica* **90**, 1-4, pp. 105–123.

Voyiadjis, G. Z. and Sivakumar, S. M. (1994). A robust kinematic hardening rule for cyclic plasticity with ratchetting effects. 2. application to nonproportional loading cases, *Acta Mechanica* **107**, 1-4, pp. 117–136.

Wang, A. J. and McDowell, D. L. (2005). Yield surfaces of various periodic metal honeycombs at intermediate relative density, *International Journal of Plasticity* **21**, 2, pp. 285–320.

Wang, Z. Q., Beyerlein, I. J. and Lesar, R. (2007). Dislocation motion in high strain-rate deformation, *Philosophical Magazine* **87**, 16-17, pp. 2263–2279.

Weber, G. G., Lush, A. M., Zavaliangos, A. and Anand, L. (1990). An objective time-integration procedure for isotropic rate-independent and rate-dependent elastic plastic constitutive-equations, *International Journal of Plasticity* **6**, 6, pp. 701–744.

Weng, G. J. (1987). Anisotropic hardening of single crystals and the plasticity of polycrystals, *International Journal of Plasticity* **3**, pp. 315–339.

Weng, G. J. and Phillips, A. (1977a). Investigation of yield surfaces based on dislocation mechanics .1. basic theory, *International Journal of Engineering Science* **15**, 1, pp. 45–59.

Weng, G. J. and Phillips, A. (1977b). Investigation of yield surfaces based on dislocation mechanics .2. application, *International Journal of Engineering Science* **15**, 1, pp. 61–70.

Winter, G., Drucker, D. C., Horne, M. R., Popov, E. P., Vallerga, B. A. and Hrennikoff, A. (1948). Theory of inelastic bending with reference to limit design - discussion, *Transactions of the American Society of Civil Engineers* **113**, pp. 248–268.

Xiao, H., Bruhns, O. T. and Meyers, A. (2006). Elastoplasticity beyond small deformations, *Acta Mechanica* **182**, pp. 31–111.

Xiao, H., Bruhns, O. T. and Meyers, A. (2007a). The integrability criterion in finite elastoplasticity and its constitutive implications, *Acta Mechanica* **188**, pp. 227–244.

Xiao, H., Bruhns, O. T. and Meyers, A. (2007b). Thermodynamic laws and consistent eulerian formulation of finite elastoplasticity with thermal effects, *Journal of the Mechanics and Physics of Solids* **55**, pp. 338–365.

Yoon, S. C. and Kim, H. S. (2006). Equal channel angular pressing of metallic powders for nanostructured materials, *Materials Science Forum* **503**, pp. 221–226.

Yuan, W. and Wang, J. (2002). Anisotropy of the phase-transformation plasticity in textured cuznal shape-memory sheets, *Journal of Materials Processing Technology* **123**, pp. 31–35.

Zenkour, A. M. (1999). Transverse shear and normal deformation theory for bending analysis of laminated and sandwich elastic beams, *Mechanics of Composite Materials* and Strcutures **6**, pp. 267–283.

Ziegler, H. (1963). Some extremism principles in irreversible thermodynamics, in I. N. Sneddon and R. Hill (eds.), *Progress in Solid Mechanics*, Vol. 4.

Ziegler, H. (1983). *An Introduction to Thermodynamics*, North-Holland Series in Applied Mathematics and Mechanics (North-Holland, Amsterdam).

Ziegler, H. and Wehrli, C. (1987). The derivation of constitutive equations from the free energy and the dissipation function, in T. Y. Wu and J. W. Hutchinson (eds.), *Advances in Applied Mechanics*, Vol. 25 (Academic Press, New York).

Index

Π plane, 244
t-distribution, 332
1-D continuum, 95

ABAQUS
    contact friction, 499
ABAQUS
    analysis software, 499
    explicit analysis, 499
accumulated plastic strain, 105
adjugate
    see area element, 361
Airy stress function
    biharmonic equation, 227
    example of boundary conditions for, 228
    general expression for boundary conditions, 229
    in plasticity, 227
    solving using finite differences, 231
Almen strips, 498
area element
    deformation of, 361
    rate of change of, 363
associative plasticity
    see normality rule, 241

back stress
    in small strain plasticity, 252
backward euler scheme, 341
balance laws for a continuum
    small deformation, 271
bar
    inelastic
        formulation of inelastic laws, 103
        strain hardening, 108, 110
            prescribed strain rate, 110
            prescribed stress rate, 108
bars
    inelastic, 93
        list of variables, 93
        simplifications, 93
Bauschinger effect
    1-dimensional, 112
    history dependent Helmholtz potentials, 310
    modeled with history dependent potentials, 312
    qualitative description of, 309
    see kinematic hardening, 257
beam
    Gibbs potential, 135
beams
    axial stress to bending moment relationship, 135
    balance laws, 133
    boundary conditions, 129, 133
    constitutive model, 129
    decomposition of total strain, 136
    determinate, 139
    elastoplastic, 125
        list of variables, 130
    energy balance, 133
    Euler-Bernoulli, 131
    example, 140
        no axial confinement, 140
    general beam problems, 145
        solution procedure, 147
    Governing differential equations, 139
    hard constraints, 132
    idealization of geometry, 128
        defeaturing, 129
    indeterminate, 145

solution procedure, 147
kinematics, 132
Kuhn-Tucker conditions, 137
mass balance, 133
momentum balance, 133
plastic hinge, 142
plastic strain distribution
    axial confinement, 143
    iterative procedure, 141, 143
plastic strain update, 148
simplifications for analysis, 131
thermal response, 138
with axial confinement, 143
Bernoulli-Euler beam, 131
biharmonic equation
    finite difference scheme for, 231
    see Airy stress function, 227
bulge test, 25

Castigliano's I theorem, 38
coaxial tensors, 374
coefficient of restitution, 500
Coffin-Manson rule, 119
cold working, 105
compatibility
    equation for plane problems, 227
    equations for small strain plasticity, 217
    see plane stress or plane strain, 227
compressible neo-Hookean material, 410
configuration
    definition of reference, 358
consistency condition, 14
constitutive equation
    definition of objective constitutive equations, 395
    for an isotropic elastic material, 386
    for thermoelasticity, 384
    see crystal, 384
continuum
    classical, 364
convergence criteria, 339
convex cutting plane algorithm
    in finite plasticity, 439
    tension-torsion example, 345
crane girder, 153
creep, 19, 32
crystal
    constitutive equations, 490
    constitutive equations for, 479
    equation of state, 472

hardening due to multiple slip, 486
hardening functions, 487
hardening laws, 483
hardening modulus, 485
hardening stages, 481
latent hardening, 485
latent hardening function, 488
rate independent theory, 492
self hardening, 484
stereographic projection, 480
cyclic loading phenomena, 257

damage
    aging of tissue, 509
damage parameter, 105
damping, 24
damping ratio, 173
dashpot
    3-D
        Kuhn-Tucker conditions, 185
    3-D linear, 184
decomposition
    of the velocity gradient, 372
        see crystal, 372
    polar, 368
deformation
    definition of deformation gradient, 361
    of area element, 361
    of line element, 361
    of volume element, 361
    time rate of change of deformation gradient, 360
displacement
    equations of equilibrium for plane stress, 223
    equations of equilibrium in terms of, 216
    gradient of, 359
    plain strain equations of equilibrium, 221
dissipation
    plastic
        maximization of, 297
    posutlates
        comparison of, 297
dissipation equation, 81
dissipation, rate of
    see rate of dissipation, 288
dissipative forces, 54
dissipative potential, 54

driving force
  definition of, 293
  in finite inelasticity, 420
Drucker's stability postulate, 298
dynamic response, 168

effective stress, 162
elastic strain
  derived from Gibbs potential, 287
  finite deformation, 411
  in small strain plasticity, 275
elasticity
  rate form for finite isotropic Green elasticity, 412
elastomeric spring, 74
endochronic theory, 314
energy
  and Gibbs potential, 287
  heating and working, definitions of, 279
  total energy per unit mass, 271
energy absorption, 158
energy release rate
  see rate of dissipation, 306
entropy
  and Helmholtz potential, 284
  and production of "heat", 287
  and second law, 277
  and temperature, 278
  and uncertainty regarding microstates, 276
  balance equation, 289
  balance law with history dependence, 312
  configurational, 276
  derived from Gibbs potential, 287
  derived from Helmholtz potential, 283
  entropic equation of state, 278
  exampsl of different kinds, 277
  of mixing, 276
  relationship to Helmholtz potential, 381
  some characteristics of, 278
  statistical motivation for, 276
  thermal, 276
equivalent damping ratio, 173
equivalent plastic strain, 105
  see plastic arc length, 255
equivalent viscous damping, 160
Euler-Bernoulli beam, 131

fail-safe, 5, 7

failure
  fail-safe, 5
  material modeling, 8
  observation, experimentation, 8
  safe-fail, 7
  types of, 182
failure analysis, 6
fatigue strength, 497
finite deformations
  multiplicative decomposition, 499
finite element method
  for solving plasticity problems, 233
finite plasticity
  anisotropic, 452
    see crystal, 452
  classification of models for, 405
  comparison with small strain, 416
  different models, 402
  driving forces in, 420
  example of model using stress rates, 505
  hardening, 414
  in metal forming technology, 403
  isotropic
    $F_e F_p$ decomposition, 445
    2nd law of thermodynamics, 419
    comparison with small strain, 424, 431
    constitutive equations in terms of eigenvalues, 437
    constitutive equations, summary of, 431
    elastic strain in, 411
    examples of constitutive equations, 414
    flow rule, 413
    hardening function, 430
    Helmholtz potential for, 411
    Kuhn-Tucker parameter, calculation of, 434
    Kuhn-Tucker parameters in terms of eigenvalues, 438
    loading criteria, 432, 434
    natural configuration, moving, 444
    normality rule, 426
    numerical method, 439
    physical description of $\mathbf{D}_p$, 425
    plastic are length, 430
    plastic flow rate, 413, 421
    relation between plastic flow rate and $F_e F_p$ decomposition, 448

yield condition in strain space, 432
isotropic materials
    model development procedure, 410
    maximum rate of dissipation
        hypothesis(MRDH), 427
    normality rule, 415
    rate of dissipation, 418
    using Gibbs potential, 505
flow rule, 14
Fourier's Law, 138
frame
    corotational, 371
    global, 357
    inertial, 357
    other rotating frames, 396
frame indifference
    see objectivity, 390
friction block analogy, 14
frictional material
    yield function for, 251

gauge function
    and its polar, 307
    used in plasticity, 305
Gauss' theory of least constraint, 29
general loading conditions
    one-dimensional, 119
Gibbs potential, 68
    and elastic strain, 287
    and energy, 287
    and entropy, 287
    and equations for plastic flow rate, 294
    and Legendre transforms, 285
    and the heat equation, 294
    and work of a dead load, 281
    application to finite plasticity models, 505
    derived from Helmholtz potential, 285
governing differential equations
    small deformations, 323
gradient
    referential versus spatial, 359
    relationship between referential and spatial, 360
ground motion, 170

hardening, 17
    crystal, 483
    definition, 203
    functions for single crystal, 487

    in finite plasticity, 414
    in small strain plasticity, 256
    kinematic
        see kinematic hardening, 257
    multiple slip, 486
    power law, 107
    stages in single crystal, 481
    work hardening, 105
hardening function
    crystal, 485
    in finite plasticity, 430
    in small strain plasticity, 256
hardening laws
    isotropic, 106
hardening modulus, 105
hardening rate, 105
heat
    and entropy production, 287
    and mechanical dissipation, 287
    and rate of change of energy, 280
    and second law, 280
    and thermal entropy, 276
    equation in terms of Gibbs potential, 294
    equation in terms of Helmholtz potential, 289
    equation with history dependence, 312
    heating and working, definition of, 288
heat equation, 79
    one-dimensional, 102
Helmholtz potential, 65
    and entropy, 283, 284, 381
    and Gibbs potential, 285
    and internal energy, 381
    and Legendre transformation, 282
    and maximum isothermal work, 281
    and the heat equation, 289
    evolution equation for, 311
    example for an isotropic material, 388
    extension of neo-Hookean, 388
    for a continuum, 381
    for isotropic finite plasticity, 411
    for thermoplasticity, 284
    history dependence of, 310
    history dependent
        integral form for, 316
    of a cubic crystal, 471
Hencky material, 410
Hencky potential
    in finite inelasticity, 414

hysteretic
    definition, 5

Il'iushin's's postulate
    see work inequality, 298
incompressibility
    plastic, 191
inelastic bars, 93
inelasticity
    definition, 5
integration
    backward Euler scheme, 341
    convex cutting plane algorithm, 342, 344
        examples, 345
integration of plastic flow equations, 339
internal energy
    relationship to Helmholtz potential, 381
    spring, 70
invariance
    see tensors, 390
isothermal work function
    see Helmholtz potential, 381
isothermal work potential, 65

J2 plasticity
    elasto-plasticity model, 193
        small deformations, 190
    flow rule, 192
    hardening, 201
    hardening law, 208
    hardening model, 206
        tangent modulus, 207
    Kuhn-Tucker conditions, 193
    nondimensionalization, 200
    plastic incompressibility, 191
    rigid plastic, 182
        examples, 188
        standard form, 187
    tangent modulus, 194
        finding, 199
    viscous fluid theory, 185
    von Mises criterion, 186
    yield function, 186
    yield strength vs. κ, 193
    yield surface, 186
J2 plasticity equations
    general, 323
jacobian
    gradient of, 362
    see volume element, 361

kinematic hardening
    Armstrong-Frederick rule, 261
    in small strain plasticity, 252
    one-dimensional, 112
Kuhn-Tucker
    beams, 137
Kuhn-Tucker conditions
    for small strain plasticity, 239
    relationship to Lagrange multipliers, 302
Kuhn-Tucker parameter
    general expression for
        in finite plasticity, 433
    in finite plasticity
        example calculation of, 434
        in terms of eigenvalues, 438
    in terms of the rate of deformation, 433

Lagrange multiplier
    determining the value of, 303
    technique for constrained maximization, 302
Lagrange multiplier technique, 29
latent energy of transformation, 160
lattice
    bravais, 460
    cubic, 461
    evolution equations for, 477, 479
    plastic flow rate of , 467
    reciprocal vector, 461
    relation to non-Riemannian geometry, 464
    vector, 460
    Velocity Gradient of, 467
Law of conservation of energy, 72
laws
    balance laws for a continuum, finite deformation, 379
    difference between conservation and balance laws, 272
    of thermodynamics for a continuum, 380
least squares, 327
Legendre transform
    and Gibbs potential, 285
Legendre transformation
    and generalized Massieu functions, 282
    and Helmholtz potential, 282

of the equation of state, 281
line element
   deformation of, 361
   rate of change of, 362
   see lattice, 361
linear viscous fluid, 184
loading
   simplifications, 322
loading condition, 14
   strain space
      finite plasticity, 433
loading criteria
   in finite plasticity, 432
logarithmic strain, 410
low cycle fatigue, 119
LPM, 31
   2 dimensions, 43
   2-dimensions
      example, 44
   Castigliano's I theorem, 38
   conservation of energy, 72
   constitutive relations, 36
   dashpot
      3-D dashpot, 184
   dashpots
      common dissipative materials, 39
      constitutive equations, 42
      creep functions, 40
      frictional, 40
      types of, 39
      viscous, 39
   degrees of freedom, 50
   dissipative potential, 54
   DOFs, 50
   elastomers, 74
   energy formulation, 47
   entropy, 75
   equations of motion, 34, 61
   FAQ
      thermomechanical systems, 83
   frequently asked questions see LPM
      FAQ, 52
   frictional dashpots, 40
   Gauss' principle of least constraint, 58
   Gibbs potential, 66, 68
      example, 87
      expression for, 76
   heat equation, 79
   Helmholtz
      examples, 78
   Helmholtz and Gibbs, 68
   Helmholtz potential, 65
      expression for, 76
   inertialess
      thermomechanical equations, 82
   inertialess system, 61
   inertialess systems, 43
   internal energy, 70
   isothermal work potential, 65
   Kelvin-Voigt model, 33
   Lagrange multiplier technique, 29
   maximum rate of dissipation, 57
   maxwell model, 33
   mechanical dissipation equation, 81
   mechanical power flow, 52
   mechanical power theorem, 49
   MRDH see LPM
      maximum rate of dissipation, 57
   nonlinear spring-dashpot models, 43
   ODE solver, 85
   plant models, 36
   power considerations, 49
   power theorem
      example, 49
      generalizability, 52
      generalization, 69
      Gibbs potential, 69
   solving evolution equations, 85
   spring element, 11
   springs
      types of, 37
   standard solid model, 33
   strain energy $\neq$ total stored energy, 70
   thermodynamics, 74
      second law, 80
   thermomechanical system
      example, 83
   variational
      2-D example, 62
   variational principle, 58
LPM for elastoplastic bar, 101
lumped parameter model, 11, 31
lumped parameter model: see LPM, 31

magnification factor, 153
materials processing
   application of plasticity to, 501
matrix functions of tensors, 374
maximum plastic work postulate, 297

maximum rate of dissipation hypothesis, 57
maximum rate of dissipation hypothesis(MRDH)
   and generalized normality rule, 303
   and non-associative flow rules, 308
   and plastic potentials, 308
   conclusions drawn from, 308
   for small deformation, 297
   in finite plasticity, 427
   pictorial view of, 304
   statement of, 301
   use of Lagrange multipliers in, 302
mean stress relaxation, 114
mechanical dissipation, 81
mechanistic approach, 32
mesh convergence, 500
meshfree methods, 326
meshless
   compact support, 331
   domain of influence, 332
   elastoplastic, 334
   enforcement of BCs, 332
   Kroenecker delta property, 333
   local support, 331
   scaling parameter, 332
   smoothness, 331
   strain-displacement, 335
   transformation method, 333
   weight funcion, 331
   weight function
      student's t-distribution, 332
meshless methods, 326
MLS, 327
motion
   homogeneous, 363
      governing equations for, 262
   infinitesimal, 365
   rigid body, 367
   sequential versus simultaneous, 365, 366
moving least squares, 327, 329

natural configuration
   definition of, 273
   in finite plasticity, 444
   moving, 274
natural frequency, 168
Navier-Stokes fluid, 184
neo-Hookean
   compressible
      see compressible neo-Hookean material, 410
neo-Hookean potential
   in finite inelasticity, 414
Newton's law of cooling, 138
Newton-Raphson
   convergence criteria, 339
Newton-Raphson iteration, 337
non-associative flow rule
   and maximum rate of dissipation hypothesis (MRDH), 308
non-associative plasticity
   examples of, 249
non-dimensional plasticity equations, 340
non-dimensionalization, 339
nonlinear spring mass system, 170
normality rule, 267
   derived from the maximum rate of dissipation hypothesis, 308, 427
   generalized, derived from maximization of the rate of dissipation, 303
   in finite plasticity, 415, 426
   in small strain plasticity, 241
   in terms of eigenvalues, 246
   relation to dissipation postulates, 297
numerical examples of IBVP, 345
numerical methods
   convex cutting plane algorithm for finite plasticity, 439
   forward difference for plastic flow, 232
   homogeneous deformation governing equations, 262
   small strain plasticity solutions, 216
   using finite difference method, 231
   using MATLABS FEM solver, 233
numerical solution
   1-D rod, 346
   constitutive model, 322
   convergence criteria, 339
   general
      steps involved, 325
   plate with a hole, 347
numerical techniques
   choice of, 324
   displacement approximation, 324
   finite element methods, 324

objectivity
   corotational rate, 395

definition of objective constitutive equations, 395
definition of objective quantities, 393
importance and use of objective rates, 396
rates of objective tensors, 394
spring mass systems, 390
ODE solver, 122
one-dimension
 general loading conditions, 119
one-dimensional
 heat equation, 102
one-dimensional continuum, 95
 balance laws, 97
 deformation gradient, 96
 energy conservation, 98
 energy equation, 99
 kinematics, 95
 mass conservation, 97
 momentum conservation, 98
 stretch, 96

passive damping, 158
PDE Tool
 solving plasticity problems with, 233
peening intensity, 497
perfectly plastic, 16
plane strain
 compatibility equations, 227
 equations of equilibrium, 221
plane stress
 compatibility equation, 227
 equations of equilibrium, 223
 plastic flow equations, 225
plant models, 36
plastic arc length
 in finite plasticity, 430
 in small strain plasticity, 255
plastic flow rate
 forward difference scheme for, 232
 in finite plasticity, 421
 in plane stress, 225
 in terms of Gibbs potential, 294
plastic hinge, 142
plastic potential
 and maximum rate of dissipation hypothesis(MRDH), 308
 examples of, 249
 in small strain plasticity, 241
plastic work
 maximization of, 297
 see rate of dissipation, 295
plastic working, 105
porous materials
 see also Strength Differential(S-D) effect, 251
 yield functions for, 251
power law hardening, 107
power law type overstress model, 499
power theorem
 mechanical, 49
Pratt truss, 152
prescribed general loading, 119
principle of virtual work, 334
projection
 stereographic, 480
projection operator, 341

quasi-static, 10

ratchetting, 113, 257
rate dependent power law, 499
rate of dissipation
 and plastic work, 295
 constitutive equation for, 300
 constitutive equations
  examples of, 300
 deriving yield functions from, 306
 equatin for, 288
 equation for a continuum, 382
 equation for, in small deformation, 289
 example of calculations of, 291
 for a single crystal, 473
 general expression for, 290
 in finite plasticity, 418
 maximization, 297
 maximization for a crystal, 474
 maximization using Lagrange multipliers, 302
 maximum rate of dissipation hypothesis, 301
 maximum rate of dissipation hypothesis (MRDH), 297
 positive homogeneity of, 305
 specific forms for a single crystal, 475
rate of the rotation tensor, 372
re-centering, 158
relaxation, 23
residual stress, 497
resonance, 167

retardation, 32
reverse transformation, 161
rubber band, 74

safe-life, 7
seismic isolation, 158
self-equilibrating tensile stresses, 497
shakedown, 114, 257
    elastic, 114
    plastic, 114
shape memory alloys, 158
shear stress
    CRSS, 456
    in terms of lattice vectors, 476
    resolved, 456
    resolved, in a single crystal, 473
shot peening, 497
simple shear, 188
simplifications in geometry
    for general problems, 322
single DOF system, 166
slip direction, 464
slip normal, 464
slip plane, 456
slip system, 456, 463
    FCC, 465
    in matrix form, 467
SMA braced frame, 164
SMA bracings, 158
small deformations
    governing differential equations, 323
small strain plasticity
    Airy stress function, 227
    and generalization of spring dashpot
        models, 291
    anisotropic
        constitutive equations,example of, 259
    anisotropic yield function
        generating, 248
        quadratic, 247
    Armstrong-Frederick rule for kinematic
        hardening, 261
    biharmonic equation, 227
    compatibility equations in 3-D, 217
    compatibility for plane problems, 227
    crushable foam, 251
    Drucker-Prager yield function, 251
    equations of equilibrium, 216

    example of boundary conditions for
        plane problems, 228
    general boundary condition for plane
        problems, 229
    general constitutive equations for, 267
    governing equations for homogeneous
        deformations, 262
    isotropic hardening
        examples of, 256
    isotropic yield function
        examples of, 243, 245
        general form for, 243
        normality rule, 246
        visualization of, 244
    Kuhn-Tucker conditions, 239
    models with additional variables, 254
    models with multiple plastic strain
        tensors, 291
    non-associative, 249
    normality rule, 241
    plane strain, reduced constitutive
        equations, 219
    plane stress
        plastic flow equations, 225
    plane stress problems, 221
    plastic potential, 241
        examples of, 249
    rate dependent, 295
    solving equations of equilibrium, 216
    solving suing finite elements, 233
    solving using a finite difference scheme, 231
    solving using forward difference, 232
    Strength Differential (S-D) effect,
        example of, 242
    thermoplastic Helmholtz potential, 284
    with history dependent Helmholtz
        potential, 315
    without yield function,example of, 296
    yield function
        deriving from the rate of
            dissipation, 306
        with back stress, 252
spin tensor, 372
spring dashpot models
  continuum generalizations of, 291
spring element, 11
squared stretch
    of a single crystal, 470
state

equation of
    and Legendre transformation, 281
    crystal, 472
    in terms of energy, 279
    in terms of entropy, 278
    in terms of temperature, 282
    with history dependence, example of, 312
  list of state variables, 276
  meaning of macrostate variables, 273
  state variables for thermoelasticity, 381
state variables, 31
statically determinate beams, 139
statically indeterminate beams, 145
stiffness matrix
  global stiffness, 338
strain
  common measures of, 375
  measures of, 373
Strength Differential (S-D) effect
  example of, 242
stress
  Cauchy, 377
  engineering, 377
  in finite plasticity, 412
stress relaxation, 32
stretch
  definition of pure stretch, 367
  definition of stretch tensors, 368
  eigenvalues and eigenvectors of, 369
  rate of, 373
  visualization of, 370
stretch ratio, 362
stretching
  isochoric, 189
  volume preserving, 189
stretching tensor, 372
superposed rigid body motions
  see objectivity, 390
symmetry, 500

tangent stiffness matrix, 337
temperature
  and entropy, 278
  definition of absolute temperature, 278
tensor
  corotational rate of, 395
  definition of objective, 393
  invariant, 392
  invariants of, 386

objective
  see objectivity, 389
  properties of objective tensors, 393
  rates of objective tensors, 394
testing protocols, 9
texture, 457
thermal response
  beams, 138
thermodynamics
  and absolute temperature, 278
  entropy, 75
  heating and working, definition of, 288
  review, 74
  second law, 80
    and entropy, 277
    and heat, 280
    in finite inelasticity, 419
    Kuhn-Tucker, 185
thermoplasticity
  Helmholtz potential for, 284
  without yield function, example of, 296
transformation strain, 160
transformation stress, 160
Tresca criterion, 185
truss
  apparent hardening, 105
  Bauschinger effect, 116
  collapse load, 153
  cyclic loading, 116
  determinate, 91
    onset of yield, 103
    strain control, 110
    strain hardening bars, 108
  excessive deflections, 153
  inelastic response, 89
  initial yield, 156
  low cycle fatigue, 92
  static strength failure, 92
  strain cycling, 117
  stress controlled, 100
  stress cycling, 116

uncertainty
  and relation to entropy, 276
unit cell approach, 499

vector
  basis vectors for reference frame, 357
velocity
  angluar, 367

    gradient in a continuum, 359
    velocity gradient for a single crystal,
        470
visco-elastic, 22
viscoplasticity
    application to powder compaction, 501
    at finite strain, 509
    example of application to biomaterials,
        509
    finite strain
        general formulation, 449
    for small deformations, 295
    rate dependent constitutive equations
        for, 451
    with yield condition, 452
viscous
    Kuhn-Tucker form, 185
viscous fluid theory, 185
volume element
    deformation of, 361
    rate of change of, 363
von Mises criterion, 186
vorticity, 372

weak damper, 168
weak form of equilibrium equations, 334
weight function
    student's t-distribution, 332
weighted least squares, 329
work hardening, 105
work inequality
    statement of, 298

yield, 32
yield condition, 14
yield function
    see small strain plasticity, 251

Series on Advances in Mathematics for Applied Sciences

Editorial Board

N. Bellomo
Editor-in-Charge
Department of Mathematics
Politecnico di Torino
Corso Duca degli Abruzzi 24
10129 Torino
Italy
E-mail: nicola.bellomo@polito.it

F. Brezzi
Editor-in-Charge
IMATI - CNR
Via Ferrata 5
27100 Pavia
Italy
E-mail: brezzi@imati.cnr.it

M. A. J. Chaplain
Department of Mathematics
University of Dundee
Dundee DD1 4HN
Scotland

C. M. Dafermos
Lefschetz Center for Dynamical Systems
Brown University
Providence, RI 02912
USA

J. Felcman
Department of Numerical Mathematics
Faculty of Mathematics and Physics
Charles University in Prague
Sokolovska 83
18675 Praha 8
The Czech Republic

M. A. Herrero
Departamento de Matematica Aplicada
Facultad de Matemáticas
Universidad Complutense
Ciudad Universitaria s/n
28040 Madrid
Spain

S. Kawashima
Department of Applied Sciences
Engineering Faculty
Kyushu University 36
Fukuoka 812
Japan

M. Lachowicz
Department of Mathematics
University of Warsaw
Ul. Banacha 2
PL-02097 Warsaw
Poland

S. Lenhart
Mathematics Department
University of Tennessee
Knoxville, TN 37996–1300
USA

P. L. Lions
University Paris XI-Dauphine
Place du Marechal de Lattre de Tassigny
Paris Cedex 16
France

B. Perthame
Laboratoire J.-L. Lions
Université P. et M. Curie (Paris 6)
BC 187
4, Place Jussieu
F-75252 Paris cedex 05, France

K. R. Rajagopal
Department of Mechanical Engrg.
Texas A&M University
College Station, TX 77843-3123
USA

R. Russo
Dipartimento di Matematica
II University Napoli
Via Vivaldi 43
81100 Caserta
Italy

Series on Advances in Mathematics for Applied Sciences

Aims and Scope

This Series reports on new developments in mathematical research relating to methods, qualitative and numerical analysis, mathematical modeling in the applied and the technological sciences. Contributions related to constitutive theories, fluid dynamics, kinetic and transport theories, solid mechanics, system theory and mathematical methods for the applications are welcomed.

This Series includes books, lecture notes, proceedings, collections of research papers. Monograph collections on specialized topics of current interest are particularly encouraged. Both the proceedings and monograph collections will generally be edited by a Guest editor.

High quality, novelty of the content and potential for the applications to modern problems in applied science will be the guidelines for the selection of the content of this series.

Instructions for Authors

Submission of proposals should be addressed to the editors-in-charge or to any member of the editorial board. In the latter, the authors should also notify the proposal to one of the editors-in-charge. Acceptance of books and lecture notes will generally be based on the description of the general content and scope of the book or lecture notes as well as on sample of the parts judged to be more significantly by the authors.

Acceptance of proceedings will be based on relevance of the topics and of the lecturers contributing to the volume.

Acceptance of monograph collections will be based on relevance of the subject and of the authors contributing to the volume.

Authors are urged, in order to avoid re-typing, not to begin the final preparation of the text until they received the publisher's guidelines. They will receive from World Scientific the instructions for preparing camera-ready manuscript.

SERIES ON ADVANCES IN MATHEMATICS FOR APPLIED SCIENCES

Published*:

Vol. 65 Mathematical Methods for the Natural and Engineering Sciences
    by R. E. Mickens

Vol. 66 Advanced Mathematical and Computational Tools in Metrology VI
    eds. P. Ciarlini et al.

Vol. 67 Computational Methods for PDE in Mechanics
    by B. D'Acunto

Vol. 68 Differential Equations, Bifurcations, and Chaos in Economics
    by W. B. Zhang

Vol. 69 Applied and Industrial Mathematics in Italy
    eds. M. Primicerio, R. Spigler and V. Valente

Vol. 70 Multigroup Equations for the Description of the Particle Transport in Semiconductors
    by M. Galler

Vol. 71 Dissipative Phase Transitions
    eds. P. Colli, N. Kenmochi and J. Sprekels

Vol. 72 Advanced Mathematical and Computational Tools in Metrology VII
    eds. P. Ciarlini et al.

Vol. 73 Introduction to Computational Neurobiology and Clustering
    by B. Tirozzi, D. Bianchi and E. Ferraro

Vol. 74 Wavelet and Wave Analysis as Applied to Materials with Micro or Nanostructure
    by C. Cattani and J. Rushchitsky

Vol. 75 Applied and Industrial Mathematics in Italy II
    eds. V. Cutello et al.

Vol. 76 Geometric Control and Nonsmooth Analysis
    eds. F. Ancona et al.

Vol. 77 Continuum Thermodynamics
    by K. Wilmanski

Vol. 78 Advanced Mathematical and Computational Tools in Metrology and Testing
    eds. F. Pavese et al.

Vol. 79 From Genetics to Mathematics
    eds. M. Lachowicz and J. Miękisz

Vol. 80 Inelasticity of Materials: An Engineering Approach and a Practical Guide
    by A. R. Srinivasa and S. M. Srinivasan

*To view the complete list of the published volumes in the series, please visit:
http://www.worldscibooks.com/series/samas_series.shtml